Mechatronics

Principles and Applications

Mechatronics

Principles and Applications

Godfrey C. Onwubolu
Professor of Engineering
The University of the South Pacific, Fiji

ELSEVIER
BUTTERWORTH
HEINEMANN

AMSTERDAM • BOSTON • HEIDELBERG • LONDON • NEW YORK • OXFORD
PARIS • SAN DIEGO • SAN FRANCISCO • SINGAPORE • SYDNEY • TOKYO

Elsevier Butterworth-Heinemann
Linacre House, Jordan Hill, Oxford OX2 8DP
30 Corporate Drive, Burlington, MA 01803

First published 2005

British Library Cataloguing in Publication Data
A catalogue record for this book is available from the British Library

Library of Congress Cataloguing in Publication Data
A catalogue record for this book is available from the Library of Congress

ISBN 0 7506 6379 0

For information on all Elsevier Butterworth-Heinemann publications visit our website at http://books.elsevier.com

Printed and bound in Great Britian by Biddles Ltd, King's Lynn, Norfolk

Contents

Preface

Introduction

With the advent of integrated circuits and computers, the borders of formal engineering disciplines of electronic and mechanical engineering have become fluid and fuzzy. Most products in the marketplace are made up of interdependent electronic and mechanical components, and electronic/electrical engineers find themselves working in organizations that are involved in both mechanical and electronic or electrical activities; the same is true of many mechanical engineers. The field of mechatronics offers engineers the expertise needed to face these new challenges.

Mechatronics is defined as the synergistic combination of precision mechanical, electronic, control, and systems engineering, in the design of products and manufacturing processes. It relates to the design of systems, devices and products aimed at achieving an optimal balance between basic mechanical structure and its overall control. Mechatronics responds to industry's increasing demand for engineers who are able to work across the boundaries of narrow engineering disciplines to identify and use the proper combination of technologies for optimum solutions to today's increasingly challenging engineering problems. Understanding the synergy between disciplines makes students of engineering better communicators who are able to work in cross-disciplines and lead design teams which may consist of specialist engineers as well as generalists. Mechatronics covers a wide range of application areas including consumer product design, instrumentation, manufacturing methods, motion control systems, computer integration, process and device control, integration of functionality with embedded microprocessor control, and the design of machines, devices and systems possessing a degree of computer-based intelligence. Robotic manipulators, aircraft simulators, electronic traction control systems, adaptive suspensions, landing gears, air-conditioners under fuzzy logic control, automated diagnostic systems, micro electromechanical systems (MEMS), consumer products such as VCRs, driver-less vehicles are all examples of mechatronic systems. These systems depend on the integration of mechanical, control, and computer systems in order to meet demanding specifications, introduce 'intelligence' in mechanical hardware, add versatility and maintainability, and reduce cost.

Competitiveness requires devices or processes that are increasingly reliable, versatile, accurate, feature-rich, and at the same time inexpensive. These objectives can be achieved by introducing electronic controls and computer technology as integrated parts of machines and their components. Mechatronic design results in improvements both to existing products, such as in microcontrolled drilling machines, as well as to new products and systems. A key prerequisite in building successful mechatronic systems is the fundamental understanding of the three basic elements of mechanics, control, and computers, and the synergistic application of these in designing innovative products and processes. Although all three building blocks are very important, mechatronics focuses explicitly on their interaction, integration, and synergy that can lead to improved and cost-effective systems.

Aims of this book

This book is designed to serve as a mechatronics course text. The text serves as instructional material for undergraduates who are embarking on a mechatronic course, but contains chapters suitable for senior undergraduates and beginning postgraduates. It is also valuable resource material for practicing electronic, electrical, mechanical, and electromechanical engineers.

Overview of contents

The elements covered include electronic circuits, computer and microcontroller interfacing to external devices, sensors, actuators, systems response, modeling, simulation, and electronic fabrication processes of product development of mechatronic systems. Reliability, an important area missed out in most mechatronic textbooks, is included.

Detailed contents – A route map

The book covers the following topics. Chapter 1 introduces mechatronics. Chapter 2 provides the reader with a review of electrical components and circuit elements and analysis. Chapter 3 presents semiconductor electronic devices. Chapter 4 covers digital electronics. Chapter 5 deals with analog electronics. Chapter 6 deals with important aspects of microcontroller architecture and programming in order to interface with external devices. Chapter 7 covers data acquisition systems. Chapter 8 presents various commonly used sensors in mechatronic systems. Chapters 9 and 10 present electrical and mechanical external devices, respectively, for actuating mechatronic systems. Chapter 11 deals with interfacing microcontrollers with external devices for actuating mechatronic systems; this chapter is the handbook for practical applications of most integrated circuits treated in this book. Chapter 12 deals with the modeling aspect of control theory, which is of considerable importance in mechatronic systems. Chapter 13 presents the analysis aspect of control theory, while Chapter 14 deals with graphical techniques in control theory. Chapter 15 presents robotic system fundamentals, which is an important area in mechatronics. Chapter 16 presents electronic fabrication process, which those working with mechatronic systems should be familiar with. Chapter 17 deals with reliability in mechatronic systems; a topic often neglected in mechatronics textbooks. Finally Chapter 18 presents some case studies.

The design process and the design of machine elements are important aspects of mechatronics. While a separate chapter is not devoted to these important areas, which are important in designing mechatronic systems, the appendices present substantial information on design principles and mechanical actuation systems design and analysis.

Additional features and supplements

Specific and practical information on mechatronic systems that the author has been involved in designing are given throughout the book, and a chapter has been devoted to hands-on practical guides to interfacing microcontrollers and external actuators, which is fundamental to a mechatronic system.

End-of chapter problems

All end-of-chapter problems have been tested as tutorials in the classroom at the University of the South Pacific. A fully worked Solutions Manual is available for adopting instructors.

Online supplements to the text

For the student:

- Many of the exercises can be solved using MATLAB® and designs simulated using Simulink® (both from MathWorks Inc.). Copies of MATLAB® code used to solve the chapter exercises can be downloaded from the companion website http://books.elsevier.com/companions.

For the instructor:

- An Instructor's Solutions Manual is available for adopting tutors. This provides complete worked solutions to the problems set at the end of each chapter. To access this material please go to http://textbooks.elsevier.com and follow the instructions on screen.

- Electronic versions of the figures presented are available for adopting lecturers to download for use as part of their lecture presentations. The material remains copyright of the author and may be used, with full reference to their source, only as part of lecture slides or handout notes. They may not be used in any other way without the permission of the publisher.

Acknowledgments

This textbook evolved out of a necessity for the Department of Engineering at the University of the South Pacific to propose and teach mechatronics as a postgraduate course. The draft of this book was therefore the first lecture note material of the course, 'Mechatronic Applications'. The nature of the Department of Engineering at the University of the South Pacific is remarkable because it is one that combines the four disciplines of mechanical, manufacturing, electrical and electronic engineering into one small department. Consequently, this structure, which initially seemed disadvantageous, turned out to be beneficial because it was easy to see the place of mechatronics in such a setup. Therefore, I am appreciative to the University, Faculty members, and students for making it possible and relatively easy for me to undertake teaching mechatronics and writing this textbook. In particular, my former graduate student, Shivendra Kumar, who was a student on the first 'Mechatronic Applications' course, is highly acknowledged. He solved most of the problems in chapters 2–7 as tutorials for the course and had significant input to the projects described in Chapter 18 as part of his undergraduate and postgraduate projects, which I supervised. He is now a faculty member of the same department. Alok Sharma, a colleague in the department, answered some of my queries on MATLAB®, while Hamendra Reddy answered some of my questions on electric motors. Ravinesh Singh, a colleague who teaches microprocessor applications, was useful in my endeavor to utilize microcontrollers for mechatronic applications. I also thank all my graduate and undergraduate students who worked on different aspects of the case studies under my supervision.

The University of the South Pacific funded the mechatronic projects described in Chapter 18 under different research grant titles. This book would have been incomplete but for the funds provided by the Research Committee for various mechatronic projects that I undertook.

I am appreciative of the rigor and standard of education which I received at the University of Benin, where I undertook my undergraduate program. Without such an exposure, it would not have been possible to write this book. My graduate studies at the University of Aston in Birmingham, UK, also prepared me to undertake this project.

I appreciate the efforts of Catherine Shaw at Elsevier and owe much to the enthusiasm and energy of my Editor, Jonathan Simpson, to whom I express much gratitude for taking this project through review process and publication. I would also like to thank the copy-editor, Alex Sharpe, and also Miranda Turner and Renata Corbani of Elsevier.

I acknowledge the contributions of the reviewers of the initial proposal of this book. Their suggestions greatly improved the book and gave me insight into inclusion of topics which have significantly improved it.

This development and writing of the book has taken much more of my time than my other books. The effect of this was that my family had to bear with my long times at work and little time to spend with them. Their patience and forbearance, which made it possible for me to commence, continue and conclude this book, is greatly appreciated. My sincere thanks to my wife, Ngozi, and our children: Chioma, Chineye, Chukujindu, Chinwe, and Chinedu.

I owe God much appreciation for His immense providence and I dedicate this book to Him.

Godfrey C. Onwubolu
May 2004

CHAPTER 1

Introduction to mechatronics

Chapter objectives

When you have finished this chapter you should be able to:

- trace the origin of mechatronics;
- understand the key elements of mechatronics systems;
- relate with everyday examples of mechatronics systems;
- appreciate how mechatronics integrates knowledge from different disciplines in order to realize engineering and consumer products that are useful in everyday life.

1.1 Historical perspective

Advances in microchip and computer technology have bridged the gap between traditional electronic, control and mechanical engineering. Mechatronics responds to industry's increasing demand for engineers who are able to work across the discipline boundaries of electronic, control and mechanical engineering to identify and use the proper combination of technologies for optimum solutions to today's increasingly challenging engineering problems. All around us, we can find mechatronic products. Mechatronics covers a wide range of application areas including consumer product design, instrumentation, manufacturing methods, motion control systems, computer integration, process and device control, integration of functionality with embedded microprocessor control, and the design of machines, devices and systems possessing a degree of computer-based intelligence. Robotic manipulators, aircraft simulators, electronic traction control systems, adaptive suspensions, landing gears, air conditioners under fuzzy logic control, automated diagnostic systems, micro electromechanical systems (MEMS), consumer products such as VCRs, and driver-less vehicles are all examples of mechatronic systems.

The genesis of mechatronics is the interdisciplinary area relating to mechanical engineering, electrical and electronic engineering, and computer science. This technology has produced many new products and provided powerful ways of improving the efficiency of the products we use in our daily life. Currently, there is no doubt about the importance of mechatronics as an area in science and technology. However, it seems that mechatronics is not clearly understood; it appears that some people think that mechatronics is an aspect of science and technology which deals with a system that includes mechanisms, electronics, computers, sensors, actuators and so on. It seems that most people define mechatronics by merely considering what components are included in the system and/or how the mechanical functions are realized by computer software. Such a definition gives the impression that it is just a collection of existing aspects of science and technology such as actuators, electronics, mechanisms, control engineering, computer technology, artificial intelligence, micro-machine and so on, and has no original content as a technology. There are currently several mechatronics textbooks, most of which merely summarize the subject picked up from existing technologies. This structure also gives people the impression that mechatronics has no unique technology. The definition that mechatronics is simply the combination of different technologies is no longer sufficient to explain mechatronics.

Mechatronics solves technological problems using interdisciplinary knowledge consisting of mechanical engineering, electronics, and computer technology. To solve these problems, traditional engineers used knowledge provided only in one of these areas (for example, a mechanical engineer uses some mechanical engineering methodologies to solve the problem at hand). Later, due to the increase in the difficulty of the problems and the advent of more advanced products, researchers and engineers were required to find novel solutions for them in their research and development. This motivated them to search for different knowledge areas and technologies to develop a new product (for example, mechanical engineers tried to introduce electronics to solve mechanical problems). The development of the microprocessor also contributed to encouraging the motivation. Consequently, they could consider the solution to the problems with wider views and more efficient tools; this resulted in obtaining new products based on the integration of interdisciplinary technologies.

Mechatronics gained legitimacy in academic circles with the publication of the first refereed journal: *IEEE/ASME Transactions on Mechatronics*. In it, the authors worked tenaciously to define mechatronics. Finally they coined the following:

> The synergistic combination of precision mechanical engineering, electronic control and systems thinking in the design of products and manufacturing processes.

This definition supports the fact that mechatronics relates to the design of systems, devices and products aimed at achieving an optimal balance between basic mechanical structure and its overall control.

1.2 Key elements of a mechatronic system

It can be seen from the history of mechatronics that the integration of the different technologies to obtain the best solution to a given technological problem is considered to be the essence of the discipline. There are at least two dozen definitions of mechatronics in the literature but most of them hinge around the 'integration of mechanical, electronic, and control engineering, and information technology to obtain the best solution to a given technological problem, which is the realization of a product'; we follow this definition. Figure 1.1 shows the main components of a mechatronic system. This book covers the principles and applications of mechatronic systems based on this framework. As can be seen, the key element of mechatronics are electronics, digital control, sensors and actuators, and information technology, all integrated in such a way as to produce a real product that is of practical use to people.

The following subsections outline, very briefly, some fundamentals of these key areas. For fuller discussions the reader is invited to explore the rich and established information sources available on mechanics, electrical and electronic theory, instrumentation and control theory, information and computing theory, and numerical techniques.

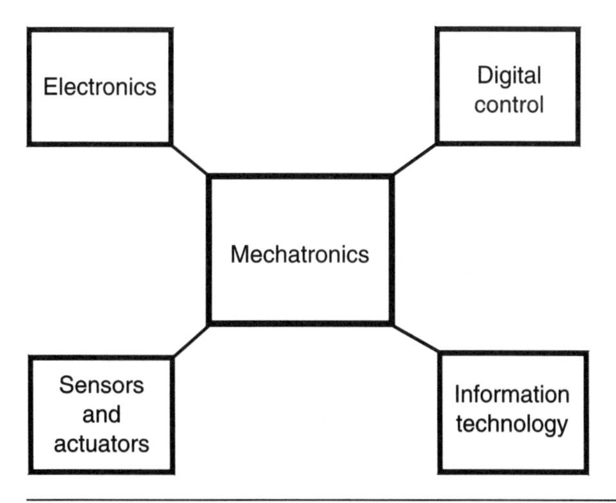

Figure 1.1 Main components of a mechatronic system.

1.2.1 Electronics

1.2.1.1 *Semiconductor devices*

Semiconductor devices, such as diodes and transistors, have changed our lives since the 1950s. In practice, the two most commonly used semiconductors are germanium and silicon (the latter being most abundant and cost-effective). However, a semiconductor device is not made from simply one type of atom and impurities are added to the germanium or silicon base. These impurities are highly purified tetravalent atoms (e.g. of boron, aluminum, gallium, or indium) and pentavalent atoms (e.g. of phosphorus, arsenic, or antimony) that are called the doping materials. The effects of doping the semiconductor base material are 'free' (or unbonded) electrons, in the case of pentavalent atom doping, and 'holes' (or vacant bonds), in the case of tetravalent atoms.

An n-type semiconductor is one that has an excess number of electrons. A block of highly purified silicon has four electrons available for covalent bonding. Arsenic, for, example, which is a similar element, has five electrons available for covalent bonding. Therefore, when a minute amount of arsenic is mixed with a sample of silicon (one arsenic atom in every 1 million or so silicon atoms), the arsenic atom moves into a place normally occupied by a silicon atom and one electron is left out in the covalent bonding. When external energy (electrical, heat, or light) is applied to the semiconductor material, the excess electron is made to 'wander' through the material. In practice, there would be several such extra negative electrons drifting through the semiconductor. Applying a potential energy source (battery) to the semiconductor material causes the negative terminal of the applied potential to repulse the free electrons and the positive terminal to attract the free electrons.

If the purified semiconductor material is doped with a tetravalent atom, then the reverse takes place, in that now there is a deficit of electrons (termed 'holes'). The material is called a p-type semiconductor. Applying an energy source results in a net flow of 'holes' that is in the opposite direction to the electron flow produced in n-type semiconductors.

A semiconductor diode is formed by 'joining' a p-type and n-type semiconductor together as a p–n junction (Figure 1.2).

Initially both semiconductors are totally neutral. The concentration of positive and negative carriers is quite different on opposite sides of the junction and a thermal energy-powered diffusion of positive carriers into the n-type material and negative carriers into the p-type material occurs. The n-type material acquires an excess of positive charge near the junction and the p-type material acquires an excess of negative charge. Eventually diffuse charges build up and an electric field is created which drives the minority charges and eventually equilibrium is reached. A region develops at the junction called the *depletion layer*. This region is essentially 'un-doped' or just intrinsic silicon. To complete the diode conductor, lead materials are placed at the ends of the p–n junction.

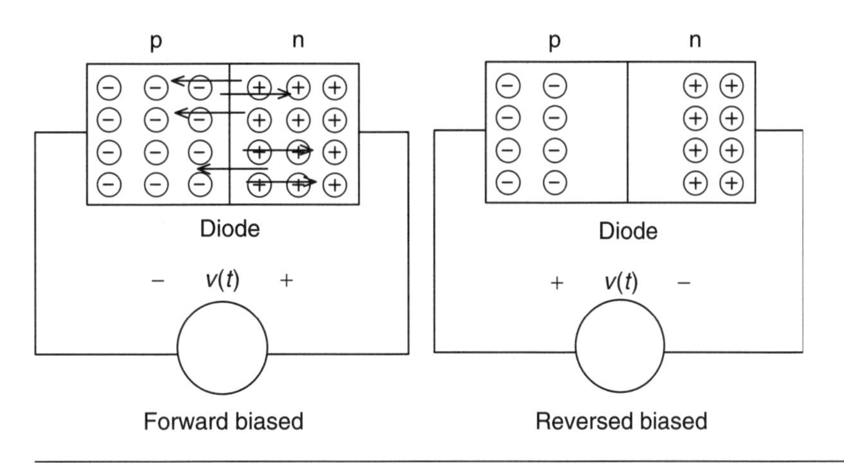

Figure 1.2 p–n junction diode.

Transistors are active circuit elements and are typically made from silicon or germanium and come in two types. The bipolar junction transistor (BJT) controls current by varying the number of charge carriers. The field-effect transistor (FET) varies the current by varying the shape of the conducting volume.

By placing two p–n junctions together we can create the bipolar transistor. In a pnp transistor the majority charge carriers are holes and germanium is favored for these devices. Silicon is best for npn transistors where the majority charge carriers are electrons.

The thin and lightly doped central region is known as the *base* (B) and has majority charge carriers of opposite polarity to those in the surrounding material. The two outer regions are known as the *emitter* (E) and the *collector* (C). Under the proper operating conditions the emitter will emit or inject majority charge carriers into the base region, and because the base is very thin, most will ultimately reach the collector. The emitter is highly doped to reduce resistance. The collector is lightly doped to reduce the junction capacitance of the collector–base junction.

The schematic circuit symbols for bipolar transistors are shown in Figure 1.3. The arrows on the emitter indicate the current direction, where $I_E = I_B + I_C$. The collector is usually at a higher voltage than the emitter. The emitter–base junction is forward biased while the collector–base junction is reversed biased.

1.2.2 Digital control

1.2.2.1 *Transfer function*

A transfer function defines the relationship between the inputs to a system and its outputs. The transfer function is typically written in the frequency (or *s*) domain,

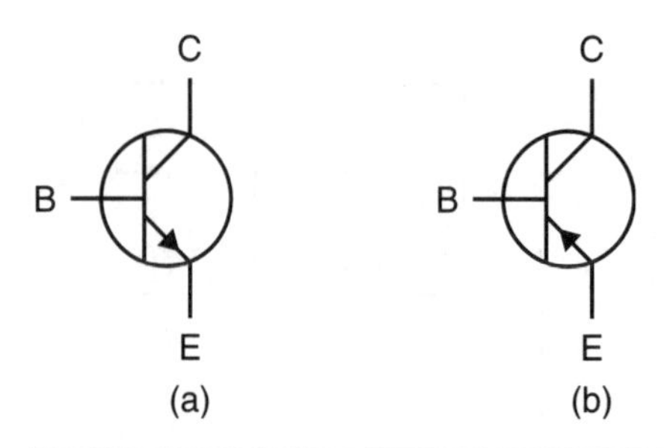

Figure 1.3 (a) npn bipolar transistor; (b) pnp bipolar transistor.

rather than the time domain. The Laplace transform is used to map the time domain representation into the frequency domain representation.

If $x(t)$ is the input to the system and $y(t)$ is the output from the system, and the Laplace transform of the input is $X(s)$ and the Laplace transform of the output is $Y(s)$, then the transfer function between the input and the output is

$$\frac{Y(s)}{X(s)}. \tag{1.1}$$

1.2.2.2 ***Closed-loop system***

A closed-loop system includes feedback. The output from the system is fed back through a controller into the input to the system. If $G_u(s)$ is the transfer function of the uncontrolled system, and $G_c(s)$ is the transfer function of the controller, and unity (negative) feedback is used, then the closed-loop system block diagram (Figure 1.4) is expressed as:

$$Y(s) = \frac{G_c(s)G_u(s)}{1 + G_c(s)G_u(s)} X(s). \tag{1.2}$$

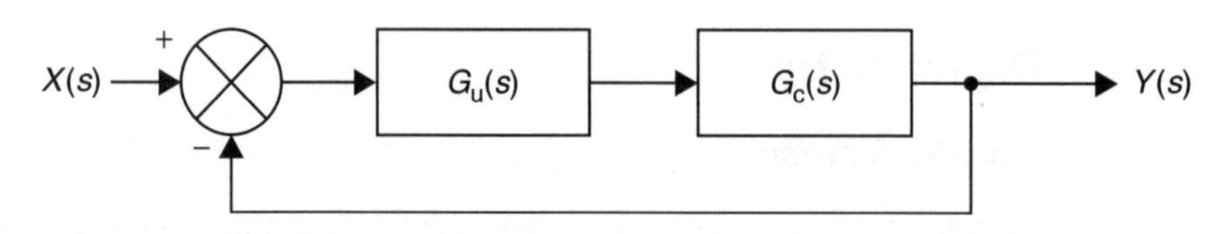

Figure 1.4 Block diagram of closed-loop system with unity gain.

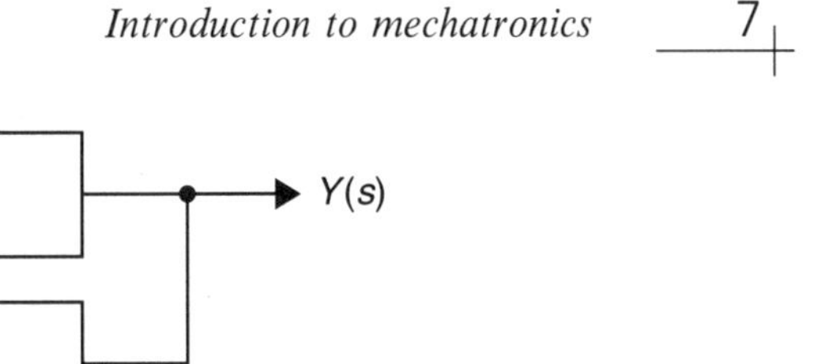

Figure 1.5 Block diagram of closed-loop system with transfer function in feedback loop.

Sometimes a transfer function, $H(s)$, is included in the feedback loop (Figure 1.5). For negative feedback this is expressed as:

$$Y(s) = \frac{G(s)}{1 + H(s)G(s)} X(s). \tag{1.3}$$

1.2.2.3 *Forward-loop system*

A forward-loop system (Figure 1.6) is a part of a controlled system. As the name suggests, it is the system in the 'forward' part of the block diagram shown in Figure 1.4. Typically, the forward-loop includes the uncontrolled system cascaded with the controller. Closing the loop around this controller and system using unity feedback gain yields the closed-loop system. For a system with controller $G_c(s)$ and system $G_u(s)$, the transfer function of the forward-loop is:

$$Y(s) = G_c(s)G_u(s)X(s). \tag{1.4}$$

1.2.2.4 *Open-loop system*

An open-loop system is a system with no feedback; it is an uncontrolled system. In an open-loop system, there is no 'control loop' connecting the output of the system to its input. The block diagram (Figure 1.7) can be represented as:

$$Y(s) = G(s)X(s). \tag{1.5}$$

Figure 1.6 Forward-loop part of Figure 1.4.

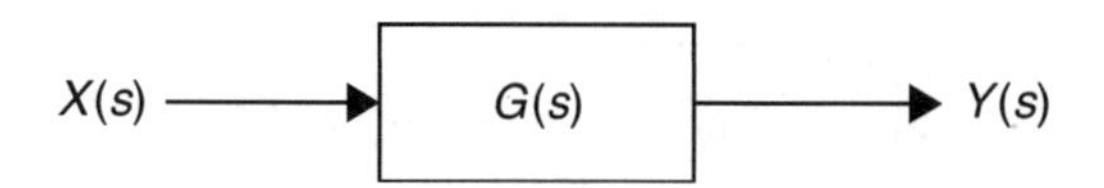

Figure 1.7 Block diagram of open-loop system.

1.2.3 Sensors and actuators

1.2.3.1 *Sensors*

Sensors are elements for monitoring the performance of machines and processes. The common classification of sensors is: distance, movement, proximity, stress/strain/force, and temperature. There are many commercially available sensors but we have picked the ones that are frequently used in mechatronic applications. Often, the conditioned signal output from a sensor is transformed into a digital form for display on a computer or other display units. The apparatus for manipulating the sensor output into a digital form for display is referred to as a measuring instrument (see Figure 1.8 for a typical computer-based measuring system).

1.2.3.2 *Electrical actuators*

While a sensor is a device that can convert mechanical energy to electrical energy, an electrical actuator, on the other hand, is a device that can convert electrical energy to mechanical energy. All actuators are transducers (as they convert one form of energy into another form). Some sensors are transducers (e.g. mechanical actuators), but not all. Actuators are used to produce motion or action, such as linear motion or angular motions. Some of the important electrical actuators used in mechatronic systems include solenoids, relays, electric motors (stepper, permanent magnet, etc.). These actuators are instrumental in moving physical objects in mechatronic systems.

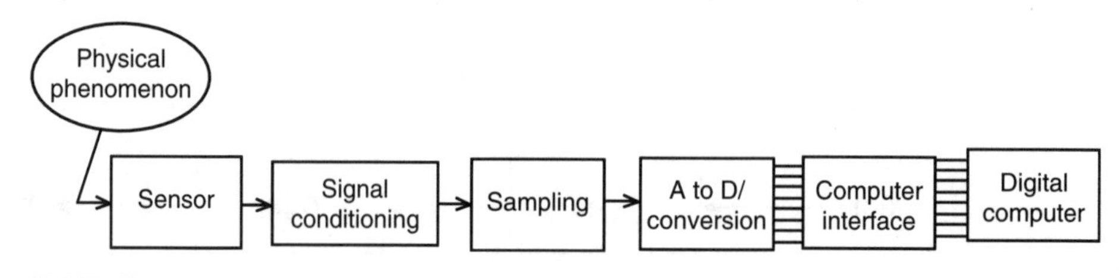

Figure 1.8 Measurement system.

1.2.3.3 *Mechanical actuators*

Mechanical actuators are transducers that convert mechanical energy into electrical energy. Some of the important mechanical actuators used in mechatronic systems include hydraulic cylinders and pneumatic cylinders.

1.2.4 Information technology

1.2.4.1 *Communication*

Signals to and from a computer and its peripheral devices are often communicated through the computer's serial and parallel ports. The parallel port is capable of sending (12 bits per clock cycle) and receiving data (up to 9 bits per clock cycle). The port consists of four control lines, five status lines, and eight data lines. Parallel port protocols were recently standardized under the IEEE 1284 standard. These new products define five modes of operation such as:

- Compatibility mode
- Nibble mode
- Byte mode
- EPP mode (enhanced parallel port)
- ECP mode (extended capabilities mode)

This is the concept on which the PC printer operates. Therefore, the code required to control this port is similar to that which makes a printer operate. The parallel port has two different modes of operation: The standard parallel port (SPP) mode and the enhanced parallel port (EPP) mode. The SPP mode is capable of sending and receiving data. However, it is limited to only eight data lines.

The EPP mode provides 16 lines with a typical transfer rate in the order of $500\,\mathrm{kB\,s^{-1}}$ to $2\,\mathrm{MB\,s^{-1}}$ (WARP). This is achieved by hardware handshaking and strobing of the data, whereas, in the SPP mode, this is software controlled.

In order to perform a valid exchange of data using EPP, the EPP handshake protocol must be followed. As the hardware does all the work required, the handshake only needs to work for the hardware. Standard data read and write cycles have to be followed while doing this.

Engineers designing new drivers and devices are able to use the standard parallel port. For instance, EPP has its first three software registers as Base + 0, Base + 1, Base + 2 as indicated in Table 1.1. EPP and ECP require additional hardware to handle the faster speeds, while Compatibility, Byte, and Nibble mode use the hardware available on SPP.

Compatibility modes send data in the forward direction at a rate of $50\text{–}150\,\mathrm{kb\,s^{-1}}$, i.e. only in data transmission. In order to receive the data the

Table 1.1 EPP address, port name, and mode of operation

Address	*Port name*	*Read/Write*
Base + 0	Data Port (SPP)	Write
Base + 1	Status Port (SPP)	Read
Base + 2	Control Port (SPP)	Write
Base + 3	Address Port (SPP)	Read/Write
Base + 4	Data Port (SPP)	Read/Write
Base + 5, 6, 7	16–32 bits	

mode must change to Nibble or Byte mode. Nibble mode can input 4 bits in the reverse direction and the Byte mode can input 8 bits in the reverse direction. EPP and ECP increase the speed of operation and can output at 1–2 MB s^{-1}. Moreover ECP has the advantage that data can be handled without using an input/output (I/O) instruction. The address, port name, and mode of operation of EPP are shown in Table 1.1.

1.3 Some examples of mechatronic systems

Today, mechatronic systems are commonly found in homes, offices, schools, shops, and of course, in industrial applications. Common mechatronic systems include:

- Domestic appliances, such as fridges and freezers, microwave ovens, washing machines, vacuum cleaners, dishwashers, cookers, timers, mixers, blenders, stereos, televisions, telephones, lawn mowers, digital cameras, videos and CD players, camcorders, and many other similar modern devices;
- Domestic systems, such as air conditioning units, security systems, automatic gate control systems;
- Office equipment, such as laser printers, hard drive positioning systems, liquid crystal displays, tape drives, scanners, photocopiers, fax machines, as well as other computer peripherals;
- Retail equipment, such as automatic labeling systems, bar-coding machines, and tills found in supermarkets;
- Banking systems, such as cash registers, and automatic teller machines;
- Manufacturing equipment, such as numerically controlled (NC) tools, pick-

and-place robots, welding robots, automated guided vehicles (AGVs), and other industrial robots;

- Aviation systems, such as cockpit controls and instrumentation, flight control actuators, landing gear systems, and other aircraft subsystems.

Problems

Q1.1 What do you understand by the term 'mechatronics'?

Q1.2 What are the key elements of mechatronics?

Q1.3 Is mechatronics the same as electronic engineering plus mechanical engineering?

Q1.4 Is mechatronics as established as electronic or mechanical engineering?

Q1.5 List some mechatronic systems that you see everyday.

Further reading

[1] Alciatore, D. and Histand, M. (1995) Mechatronics at Colorado State University, *Journal of Mechatronics*, Mechatronics Education in the United States issue, Pergamon Press.
[2] Jones, J.L. and Flynn, A.M. (1999) *Mobile Robots: Inspiration to Implementation*, 2nd Edition, Wesley, MA: A.K. Peters Ltd.
[3] Onwubolu, G.C. *et al.* (2002) Development of a PC-based computer numerical control drilling machine, *Journal of Engineering Manufacture, Short Communications in Manufacture and Design,* 1509–15.
[4] Shetty, D. and Kolk, R.A. (1997) *Mechatronics System Design*, PWS Publishing Company.
[5] Stiffler, A.K. (1992) *Design with Microprocessors for Mechanical Engineers*, McGraw-Hill.
[6] Bolton, W. (1995) *Mechatronics – Electronic Control Systems in Mechanical Engineering*, Longman.
[7] Bradley, D.A., Dawson, D., Burd, N.C. and Leader, A. J. (1993) *Mechatronics – Electronics in Products and Processes*, Chapman & Hall.
[8] Fraser, C. and Milne, J. (1994) *Integrated Electrical and Electronic Engineering for Mechanical Engineers*, McGraw-Hill.

[9] Rzevski, G. (Ed). (1995) *Perception, Cognition and Execution – Mechatronics: Designing Intelligent Machines, Vol. 1*, Butterworth-Heinemann.
[10] Johnson, J. and Picton, P. (Eds) (1995) *Concepts in Artificial Intelligence – Mechatronics: Designing Intelligent Machines, Vol. 2.*
[11] Miu, D. K. (1993) *Mechatronics: Electromechanics and Contromechanics*. Springer-Verlag.
[12] Auslander, D. M. and Kempf, C. J. (1996) *Mechatronics: Mechanical System Interfacing*, Prentice Hall.
[13] Bishop, R. H. (2002) *The Mechatronics Handbook (Electrical Engineering Handbook Series)*, CRC Press.
[14] Braga, N.C. (2001) *Robotics, Mechatronics and Artificial Intelligence: Experimental Circuit Blocks for Designers*, Butterworth-Heinemann.
[15] Popovic, D. and Vlacic, L. (1999) *Mechatronics in Engineering Design and Product Development*, Marcel Dekker, Inc.

CHAPTER 2

Electrical components and circuits

Chapter objectives

When you have finished this chapter you should be able to:

- understand the basic electrical components: resistor, capacitor, and inductor;
- deal with resistive elements using the node voltage method and the node voltage analysis method;
- deal with resistive elements using the mesh current method, principle of superposition, as well as Thévenin and Norton equivalent circuits;
- deal with sinusoidal sources and complex impedances.

2.1 Introduction

Most mechatronic systems contain electrical components and circuits, hence a knowledge of the concepts of electric charge (Q), electric field (E), and magnetic field (B), as well as, potential (V) is important. We will not be concerned with a detailed description of these quantities but will use approximation methods when dealing with them. Electronics can be considered as a more practical approach to these subjects.

The fundamental quantity in electronics is electric charge, which, at a basic level, is due to the charge properties of the fundamental particles of matter. For all intents and purposes it is the electrons (or lack of electrons) that matter. The role of the proton charge is negligible.

The aggregate motion of charge, the current (I), is given as

$$I(t) = \frac{\mathrm{d}Q}{\mathrm{d}t}, \tag{2.1}$$

where dQ is the amount of *positive* charge crossing a specified surface in a time dt. It is accepted that the charges in motion are actually negative electrons. Thus the electrons move in the opposite direction to the current flow. The SI unit for current is the ampere (A). For most electronic circuits the ampere is a rather large unit so the milliampere (mA), or even the microampere (μA), unit is more common.

Current flowing in a conductor is due to a potential difference between its ends. Electrons move from a point of less positive potential to more positive potential and the current flows in the opposite direction.

It is often more convenient to consider the electrostatic potential (V) rather than the electric field (E) as the motivating influence for the flow of electric charge. The generalized vector properties of E are usually not important. The change in potential dV across a distance dx in an electric field is

$$dV = -E \times dx. \tag{2.2}$$

A positive charge will move from a higher to a lower potential. The potential is also referred to as the potential difference, or (incorrectly) as just voltage:

$$V = V_{21} = V_2 - V_1 = \int_{V_1}^{V_2} dV. \tag{2.3}$$

The SI unit of potential difference is the volt (V). Direct current (d.c.) circuit analysis deals with constant currents and voltages, while alternating current (a.c.) circuit analysis deals with time-varying voltage and current signals whose time average values are zero.

Circuits with time-average values of non-zero are also important and will be mentioned briefly in the section on filters. The d.c. circuit components considered in this book are the constant voltage source, constant current source, and the resistor.

Figure 2.1 is a schematic diagram consisting of *idealized* circuit elements encountered in d.c. circuits, each of which represents some property of the *actual* circuit.

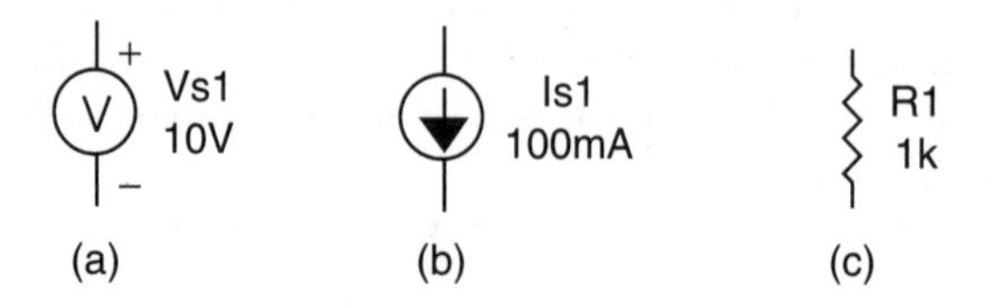

Figure 2.1 Common elements found in d.c. circuits: (a) ideal voltage source; (b) ideal current source; (c) resistor.

2.1.1 External energy sources

Charge can flow in a material under the influence of an external electric field. Eventually the internal field due to the repositioned charge cancels the external electric field resulting in zero current flow. To maintain a potential drop (and flow of charge) requires an electromagnetic force (EMF), that is, an external energy source (battery, power supply, signal generator, etc.).

There are basically two types of EMFs that are of interest:

- the *ideal voltage source*, which is able to maintain a constant voltage regardless of the current it must put out ($I \rightarrow \infty$ is possible);
- the *ideal current source*, which is able to maintain a constant current regardless of the voltage needed ($V \rightarrow \infty$ is possible).

Because a battery cannot produce an infinite amount of current, a suitable model for the behavior of a battery is an internal resistance in series with an ideal voltage source (zero resistance). Real-life EMFs can always be approximated with ideal EMFs and appropriate combinations of other circuit elements.

2.1.2 Ground

A voltage must always be measured relative to some reference point. We should always refer to a voltage (or potential difference) being 'across' something, and simply referring to voltage at a point assumes that the voltage point is stated with respect to ground. Similarly current flows through something, by convention, from a higher potential to a lower (do not refer to the current 'in' something). Under a strict definition, ground is the body of the Earth (it is sometimes referred to as earth). It is an infinite electrical sink. It can accept or supply any reasonable amount of charge without changing its electrical characteristics.

It is common, but not always ncccssary, to conncct somc part of thc circuit to earth or ground, which is taken, for convenience and by convention, to be at zero volts. Frequently, a common (or reference) connection from, and electrical current to, the metal chassis of a piece of equipment suffices. Sometimes there is a *common* reference voltage that is not at 0 V. Figure 2.2 show some common ways of depicting ground on a circuit diagram.

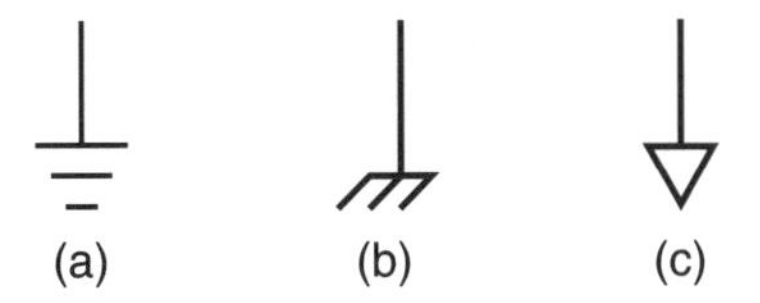

Figure 2.2 Some grounding circuit diagram symbols: (a) earth ground; (b) chassis ground; (c) common.

When neither a ground nor any other voltage reference is shown explicitly on a schematic diagram, it is useful for purposes of discussion to adopt the convention that the bottom line on a circuit is at zero potential.

2.2 Electrical components

The basic electrical components which are commonly used in mechatronic systems include resistors, capacitors, and inductors. The properties of these elements are now discussed.

2.2.1 Resistance

Resistance is a function of the material and shape of the object, and has SI units of ohms (Ω). It is more common to find units of kilohm (kΩ) and megohm (MΩ). The inverse of resistivity is conductivity.

Resistor tolerances can be as much as ±20 percent for general-purpose resistors to ±0.1 percent for ultra-precision resistors. Only wire-wound resistors are capable of ultra-precision accuracy.

For most materials:

$$V \propto I; \quad V = RI, \tag{2.4}$$

where $V = V_2 - V_1$ is the voltage *across* the object, I is the current *through* the object, and R is a proportionality constant called the resistance of the object. This is Ohm's law.

The resistance in a uniform section of material (for example, a wire) depends on its length L, cross-sectional area A, and the resistivity of the material ρ, so that

$$R = \rho \frac{L}{A}, \tag{2.5}$$

where the resistivity has units of ohm-m (Ω-m). Restivitiy is the basic property that defines a material's capability to resist current flow. Values of resistivity for selected materials are given in Table 2.1.

It is more convenient to consider a material as conducting electrical current rather than resisting its flow. The conductivity of a material, σ, is simply the reciprocal of resistivity:

$$\text{Electrical conductivity, } \sigma = \frac{1}{\rho}. \tag{2.6}$$

Table 2.1 Resistivity of selected materials

Material	*Resistivity (Ω-m)*
Conductors	10^{-8}
Aluminum	2.8
Aluminum alloys	4.0
Cast iron	65.0
Copper	1.7
Gold	2.4
Iron	9.5
Lead	20.6
Magnesium	4.5
Nickel	6.8
Silver	1.6
Steel, low C	17.0
Steel, stainless	70.0
Tin	11.5
Zinc	6.0
Carbon	5000
Semiconductors	10^1 to 10^5
Silicon	10×10^3
Insulators	10^{12} to 101^5
Natural rubber	1.0×10^{12}
Polyethylene	100×10^{12}

Conductivity has units of $(\Omega\text{-m})^{-1}$.

Table 2.2 shows the resistor color code. Using this table, it is easy to determine the resistance value and tolerance of a resistor that is color-coded (Figure 2.3).

Table 2.2 Resistor color code

Color	*Value*	*Color*	*Value*
Black	0	Gold	±5%
Brown	1	Silver	±10%
Red	2	nothing	±20%
Orange	3		
Yellow	4		
Green	5		
Blue	6		
Violet	7		
Gray	8		
White	9		

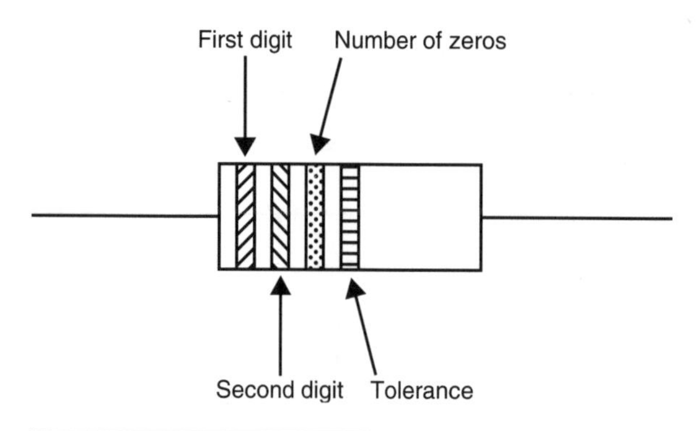

Figure 2.3 Resistor color code.

EXAMPLE 2.1

Resistance

Determine the resistance of a silver wire, which is 0.5 m long and 1.5 mm in diameter.

Solution

$$R = \rho \frac{L}{A} = 1.6 \times 10^{-8} \frac{0.500}{\pi \frac{(0.0015)^2}{4}} = 0.00453 = 4.5\ \text{m}\Omega \tag{2.6A}$$

EXAMPLE 2.2

Resistance color code

Determine the possible range of resistance values for the following color band: orange, gray, and yellow.

Solution

From Table 2.2, orange color has a value of 3, gray color has a value of 8, and yellow color has a value of 4. Hence, the resistance is 38×10^4 (380 kΩ), with tolerance of $\pm 20\% \times 380$, or (380 ± 76) kΩ, so that $304\ \text{k}\Omega \leq R \leq 456\ \text{k}\Omega$.

2.2.2 Capacitance

The fundamental property of a capacitor is that it can store charge and, hence, electric field energy. The capacitance C between two appropriate surfaces is found from

$$V = \frac{Q}{C}, \tag{2.7}$$

where V is the potential difference between the surfaces and Q is the magnitude of the charge distributed on either surface. In terms of current, $I = \mathrm{d}Q/\mathrm{d}t$ implies

$$\frac{\mathrm{d}V}{\mathrm{d}t} = \frac{1}{C}\frac{\mathrm{d}Q}{\mathrm{d}t} \tag{2.8}$$

In electronics, we take $I = I_{\mathrm{D}}$ (displacement current). In other words, the current flowing from or to the capacitor is taken to be equal to the displacement current through the capacitor. Consequently, capacitors add linearly when placed in parallel.

There are four principal functions of a capacitor in a circuit:

- since Q can be stored, a capacitor can be used as a (non-ideal) source of I;
- since E can be stored a capacitor can be used as a (non-ideal) source of V;
- since a capacitor passes alternating current (a.c.) but not direct current (d.c.) it can be used to connect parts of a circuit that must operate at different d.c. voltage levels;
- a capacitor and resistor in series will limit current and hence smooth sharp edges in voltage signals.

Charging or discharging a capacitor with a constant current results in the capacitor having a voltage signal with a constant slope, i.e.

$$\frac{\mathrm{d}V}{\mathrm{d}t} = \frac{I}{C} = \text{constant}, \tag{2.8A}$$

if I is a constant.

Some capacitors (electrolytic) are asymmetric devices with a polarity that must be taken into account when placed in a circuit. The SI unit for capacitance is the farad (F). The capacitance in a circuit is typically measured in microfarads (μF) or picofarads (pF). Non-ideal circuits will have stray capacitance, leakage currents and inductive coupling at high frequency. Although important in real circuit design, we will not go into greater detail of these aspects at this point.

Capacitors can be obtained in various tolerance ratings from ±20 percent to ±0.5 percent. Because of dimensional changes, capacitors are highly temperature dependent. A capacitor does not hold a charge indefinitely because the dielectric is never a perfect insulator. Capacitors are rated for leakage, the conduction through the dielectric, by the leakage resistance–capacitance product (MΩ–μF). High temperature increases leakage.

2.2.3 Inductance

Faraday's laws of electromagnetic induction applied to an inductor states that a changing current induces a back EMF that opposes the change. Putting this in another way,

$$V = V_A - V_B = L\frac{dI}{dt}, \qquad (2.9)$$

where V is the voltage across the inductor and L is the inductance measured in henries (H). The more common units encountered in circuits are the microhenry (μH) and the millihenry (mH). The inductance will tend to smoothen sudden changes in current just as the capacitance smoothens sudden changes in voltage. Of course, if the current is constant there will be no induced EMF. Hence, unlike the capacitor which behaves like an open-circuit in d.c. circuits, an inductor behaves like a short-circuit in d.c. circuits.

Applications using inductors are less common than those using capacitors, but inductors are very common in high frequency circuits. Inductors are never pure (ideal) inductances because they always have some resistance in and some capacitance between the coil windings. We will skip the effect these have on a circuit at this stage.

When choosing an inductor (occasionally called a choke) for a specific application, it is necessary to consider the value of the inductance, the d.c. resistance of the coil, the current-carrying capacity of the coil windings, the breakdown voltage between the coil and the frame, and the frequency range in which the coil is designed to operate. To obtain a very high inductance it is necessary to have a coil of many turns. Winding the coil on a closed-loop iron or ferrite core further increases the inductance. To obtain as pure an inductance as possible, the d.c. resistance of the windings should be reduced to a minimum. Increasing the wire size, which, of course, increases the size of the choke, is the means of achieving this. The size of the wire also determines the current-handling capacity of the choke since the work done in forcing a current through a resistance is converted to heat in the resistance. Magnetic losses in an iron core also account for some heating, and this heating restricts any choke to a certain safe operating current. The windings of the coil must be insulated from the frame as well as from each other. Heavier insulation, which necessarily makes the choke more bulky, is used in applications where there will be a high voltage between the frame and the winding. The losses sustained in the iron core increases as the frequency increases. Large inductors, rated in henries, are used principally in power applications. The frequency in these circuits is relatively low, generally 60 Hz or low multiples thereof. In high-frequency circuits, such as those found in FM radios and television sets, very small inductors (of the order of microhenries) are often used.

Now that we have briefly familiarized ourselves with these basic electrical elements, it is now necessary to consider the basic techniques for analyzing them.

2.3 Resistive circuits

The basic techniques for the analysis of resistive circuits are:

- node voltage and mesh current analysis;
- the principle of superposition;
- Thévenin and Norton equivalent circuits.

The principle of superposition is a conceptual aid that can be very useful in visualizing the behavior of a circuit containing multiple sources. Thévenin and Norton equivalent circuits are the reductions of an arbitrary circuit to an equivalent, simpler circuit. In this section it will be shown that it is generally possible to reduce all linear circuits to one of two equivalent forms, and that any linear circuit analysis problem can be reduced to a simple voltage or current divider problem.

2.3.1 Node voltage method

Node voltage analysis is the most general method for the analysis of electrical circuits. In this section its application to linear resistive circuits will be illustrated. The node voltage method is based on defining the voltage at each node as an independent variable. One of the nodes is selected as a reference node (usually, but not necessarily, ground), and each of the other node voltages is referenced to this node. Once each node voltage is defined, Ohm's law may be applied between any two adjacent nodes in order to determine the current flowing in each branch. In the node voltage method, each branch current is expressed in terms of one or more node voltages; thus, currents do not explicitly enter into the equations. Figure 2.4(a) illustrates how one defines branch currents in this method.

In the node voltage method, we define the voltages at nodes *a* and *b* as v_a and v_b, respectively; the branch current flowing from *a* to *b* is then expressed in terms of these node voltages.

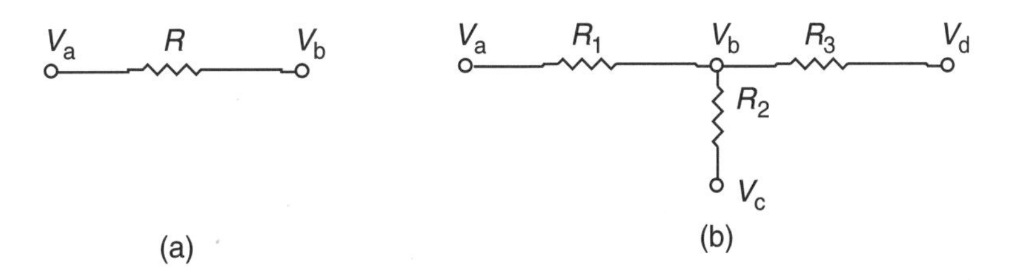

Figure 2.4 Use of Kirchhoff's current law in nodal analysis.

Once each branch current is defined in terms of the node voltages, Kirchhoff's current law (KCL) is applied at each node, so $\Sigma i = 0$.

Figure 2.4(b) illustrates this procedure for a more complex network. By KCL: $i_1 - i_2 - i_3 = 0$, where i_n is the current flowing through R_n. In the node voltage method, we express KCL by

$$\frac{v_a - v_b}{R_1} = \frac{v_b - v_c}{R_2} + \frac{v_b - v_d}{R_3} \tag{2.10}$$

Applying this method systematically to a circuit with n nodes would lead to obtaining n linear equations. However, one of the node voltages is the reference voltage and is therefore already known, since it is usually assumed to be zero. Thus, we can write $n - 1$ independent linear equations in the $n - 1$ independent variables (which, in this case, are the node voltages). Nodal analysis provides the minimum number of equations needed to solve the circuit, since any branch voltage or current may be determined from a knowledge of nodal voltages.

2.3.1.1 *Node voltage analysis method*

The steps involved in the node voltage analysis method are as follows:

1. Select a reference node (usually ground). Reference all other node voltages to this node.
2. Define the remaining $n - 1$ node voltages as the independent variables.
3. Apply KCL at each of the $n - 1$ nodes, expressing each current in terms of the adjacent node voltages.
4. Solve the linear system of $n - 1$ equations in $n - 1$ unknowns.

Let us now apply this method to a problem to illustrate the technique.

EXAMPLE 2.3

Node voltage analysis

In the circuit shown in Figure 2.5, $R_1 = 1\,\text{k}\Omega$, $R_2 = 2\,\text{k}\Omega$, $R_3 = 5\,\text{k}\Omega$, and $i_S = 50\,\text{mA}$. Determine the two node voltages.

Solution

The direction of current flow is selected arbitrarily (we assume that i_S is a positive current). We apply KCL at node *a*, to yield:

$$i_s - i_1 - i_2 = 0 \tag{2.11}$$

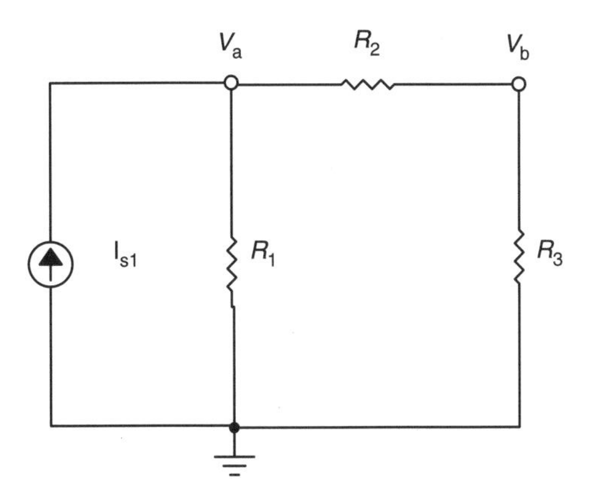

Figure 2.5 Example of nodal analysis.

Whereas, at node b,

$$i_2 - i_3 = 0 \tag{2.12}$$

There is no need to apply KCL at the reference node since the equation obtained at node c,

$$i_2 - i_3 = 0 \tag{2.13}$$

is not independent of Equations 2.11 and 2.12.

In a circuit containing n nodes, we can write at most $n-1$ independent equations.

When we apply the node voltage method, the currents i_1, i_2, and i_3 are expressed as functions of v_a, v_b, and v_c, the independent variables. Applying Ohm's law gives the following results:

$$i_1 = \frac{v_a - v_c}{R_1}, \tag{2.14}$$

since it is the potential difference, $v_a - v_c$, across R_1 that causes the current i_1 to flow from node a to node c. In the same manner,

$$\begin{aligned} i_2 &= \frac{v_a - v_b}{R_2} \\ i_3 &= \frac{v_b - v_c}{R_3}. \end{aligned} \tag{2.15}$$

Substituting the expression for the three currents in the nodal equations (equations 2.11 and 2.12, and noting that $v_c = 0$), leads to the following relationships:

$$i_s - \frac{v_a}{R_1} - \frac{v_a - v_b}{R_2} = 0 \tag{2.16}$$

and

$$\frac{v_a - v_b}{R_2} - \frac{v_b}{R_2} = 0. \tag{2.17}$$

We now solve these equations for v_a and v_b, for the given values of i, R_1, R_2, and R_3. The same equations are expressed as follows:

$$\begin{aligned} \left(\frac{1}{R_1} + \frac{1}{R_2}\right) v_a + \left(-\frac{1}{R_2}\right) v_b &= i_s \\ \left(-\frac{1}{R_2}\right) v_a + \left(\frac{1}{R_2} + \frac{1}{R_3}\right) v_b &= 0. \end{aligned} \tag{2.18}$$

On substituting the given values,

$$\begin{aligned} \left[\left(\frac{1}{1} + \frac{1}{2}\right) v_a + \left(-\frac{1}{2}\right) v_b\right] \times 10^{-3} &= 50 \times 10^{-3} \\ \left[\left(-\frac{1}{2}\right) v_a + \left(\frac{1}{1} + \frac{1}{2}\right) v_b\right] \times 10^{-3} &= 0, \end{aligned} \tag{2.18A}$$

yielding two simultaneous equations:

$$1.5 v_a - 0.5 v_b = 50$$

and

$$-0.5 v_a - 0.7 v_b = 0$$

Solving these two equations leads to the following node voltages: $v_a = 43.75$ V and $v_b = 31.25$ V.

2.3.2 Mesh current method

The second method of circuit analysis that we discuss employs the mesh currents as the independent variables; it is in many respects analogous to the method of node voltages. In this method, we write the appropriate number of independent equations, using mesh currents as the independent variables. Analysis by mesh currents consists of defining the currents around the individual meshes as the

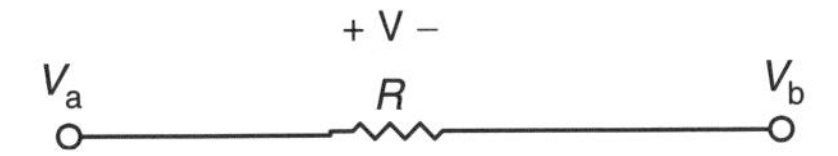

Figure 2.6 Basic principle of mesh analysis.

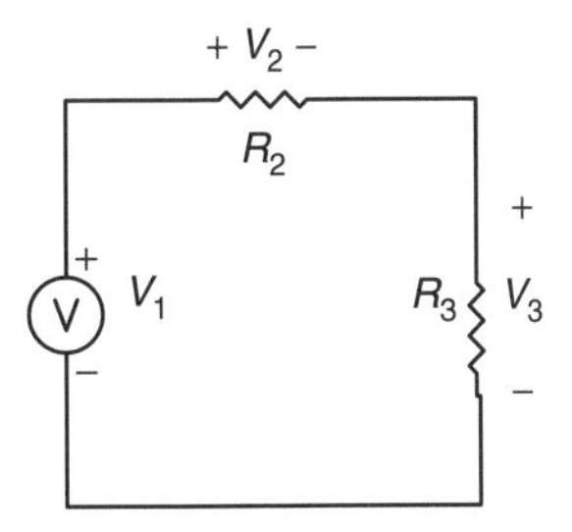

Figure 2.7 Use of Kirchoff's voltage law in mesh analysis.

independent variables. Then, the Kirchhoff's voltage law (KVL) is applied around each mesh to provide the desired system of equations.

In the mesh current method, we observe that a current flowing through a resistor in a specified direction defines the polarity of the voltage across the resistor, as illustrated in Figure 2.6, and that the sum of the voltages around a closed circuit must equal zero, by KVL. The current i, defined as flowing from left to right in Figure 2.6 establishes the polarity of the voltage across R. Once a convention is established regarding the direction of current flow around a mesh, simple application of KVL provides the desired equation. Figure 2.7 illustrates this point.

The number of equations obtained by this technique is equal to the number of meshes in the circuit. All branch currents and voltages may subsequently be obtained from the mesh currents. Since meshes are easily identified in a circuit, this method provides a very efficient and systematic procedure for the analysis of electrical circuits.

Once the direction of current flow has been selected, KVL requires that $v_1 - v_2 - v_3 = 0$.

2.3.2.1 *Mesh current analysis method*

The mesh current analysis method is described in the following steps:

1. Define each mesh current consistently. We shall always define mesh currents clockwise, for convenience.

2. Apply KVL around each mesh, expressing each voltage in terms of one or more mesh currents.
3. Solve the resulting linear system of equations with mesh currents as the independent variables.

In mesh analysis, it is important to be consistent in choosing the direction of current flow. To illustrate the mesh current method, consider the simple two-mesh circuit shown in Figure 2.8. This circuit will be used to generate two equations in the two unknowns, the mesh currents i_1 and i_2. It is instructive to first consider each mesh by itself.

Beginning with mesh 1, note that the voltages around the mesh have been assigned in Figure 2.8 according to the direction of the mesh current, i_1. Recall that as long as signs are assigned consistently, an arbitrary direction may be assumed for any current in a circuit; if the resulting numerical answer for the current is negative, then the chosen reference direction is opposite to the direction of actual current flow. Thus, one need not be concerned about the actual direction of current flow in mesh analysis, once the directions of the mesh currents have been assigned. The correct solution will result, eventually.

According to the sign convention, then, the voltages v_1 and v_2 are defined as shown. Now, it is important to observe that while mesh current i_1 is equal to the current flowing through resistor R_1 (and is therefore also the branch current through R_1), it is not equal to the current through R_2. The branch current through R_2 is the difference between the two mesh currents, $i_1 - i_2$. Thus, since the polarity of the voltage v_2 has already been assigned, according to the convention discussed in the previous paragraph, it follows that the voltage v_2 is given by:

$$v_2 = (i_1 - i_2)R_2 \tag{2.19}$$

Finally, the complete expression for mesh 1 is

$$v_s - i_1R_1 - (i_1 - i_2)R_2 = 0 \tag{2.20}$$

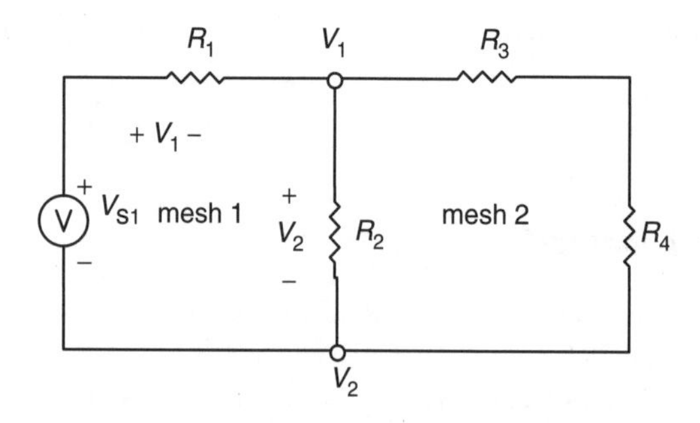

Figure 2.8 Assigning currents and voltages for mesh 1.

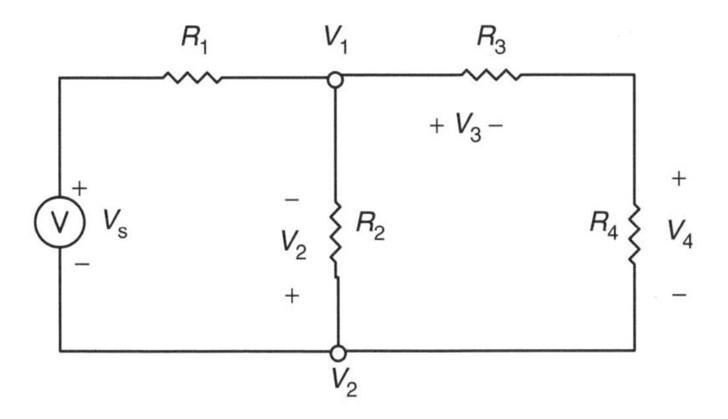

Figure 2.9 Assigning currents and voltages for mesh 2.

The same line of reasoning applies to the second mesh. Figure 2.9 depicts the voltage assignment around the second mesh, following the clockwise direction of mesh current i_2. The mesh current i_2 is also the branch current through resistors R_3 and R_4; however, the current through the resistor that is shared by the two meshes, R_2, is now equal to $(i_2 - i_1)$, and the voltage across this resistor is

$$v_2 = (i_2 - i_1)R_2 \tag{2.21}$$

and the complete expression for mesh 2 is

$$(i_2 - i_1)R_2 + i_2R_3 + i_2R_4 = 0 \tag{2.22}$$

Why is the expression for v_2 obtained in Equation 2.21 different from Equation 2.19? The reason for this apparent discrepancy is that the (clockwise) mesh current dictates the voltage assignment for each mesh. Thus, since the mesh currents flow through R_2 in opposing directions, the voltage assignments for v_2 in the two meshes will also be opposite. This is perhaps a potential source of confusion in applying the mesh current method; you should be very careful to carry out the assignment of the voltages around each mesh separately.

Combining the equations for the two meshes, we obtain the following system of equations:

$$\begin{aligned} (R_1 + R_2)i_1 - i_2R_2 &= v_s \\ -R_2i_1 + (R_2 + R_3 + R_4)i_2 &= 0 \end{aligned} \tag{2.23}$$

These equations may be solved simultaneously to obtain the desired solution, namely, the mesh currents, i_1 and i_2. You should verify that knowledge of the mesh currents permits determination of all the other voltages and currents in the circuit. The following example further illustrates some of the detail of this method.

EXAMPLE 2.4 Mesh current analysis

Figure 2.10 shows a circuit, in which node voltages are:

$$V_{s1} = V_{s2} = 120\,\text{V}$$
$$V_A = 100\,\text{V}$$
$$V_B = -115\,\text{V}$$

Determine the voltage across each resistor.

Solution

Assume a polarity for the voltages across R_1 and R_2 (e.g. from ground to node A, and from node B to ground). R_1 is connected between node A and ground; therefore, the voltage across R_1 is equal to this node voltage. R_2 is connected between node B and ground; therefore, the voltage across R_2 is equal to the negative of this voltage.

$$V_{R1} = V_A = 110\,\text{V}$$
$$V_{R2} = 0 - V_B = 115\,\text{V}$$

The two node voltages are with respect to the ground, which is given. Assume a polarity for the voltage across R_3 (e.g. from node B to node A). Then:

$$\text{By KVL:} \quad V_A + V_{R3} + V_B = 0\,\text{V}$$
$$V_{R3} = V_A + V_B = 110 - (-115) = 225\,\text{V}$$

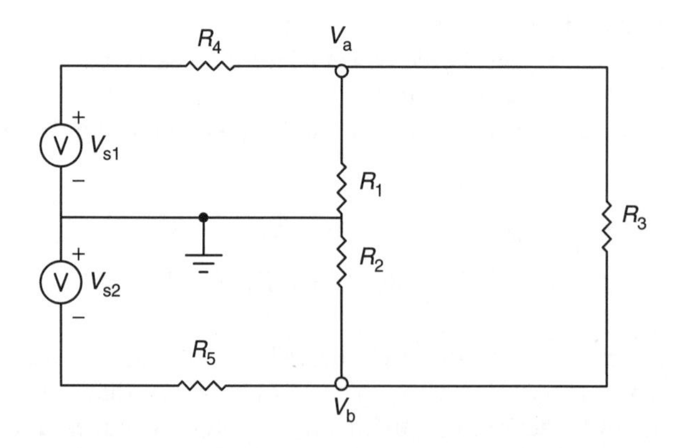

Figure 2.10 A circuit with three meshes.

Assume the polarities for the voltages across R_4 and R_5 (e.g. from node A to ground, and from ground to node B):

$$\text{By KVL}: \quad V_{s1} + V_{R4} + V_A = 0\,\text{V}$$

$$V_{R4} = V_{s1} - V_A = 120 - 110 = 10\,\text{V}$$

$$\text{Also by KVL}: \quad -V_{s2} - V_B - V_{R5} = 0\,\text{V}$$

$$V_{R5} = -V_{s2} - V_B = -120 - (-115) = -5\,\text{V}$$

2.3.3 The principle of superposition

This section briefly discusses a concept that is frequently called upon in the analysis of linear circuits. Rather than a precise analysis technique, such as the mesh current and node voltage methods, the principle of superposition is a conceptual aid that can be very useful in visualizing the behavior of a circuit containing multiple sources. The principle of superposition applies to any linear system and for a linear circuit may be stated as follows:

> In a linear circuit containing $N\psi$ sources, each branch voltage and current is the sum of $N\psi$ voltages and currents, each of which may be computed by setting all but one source equal to zero and solving the circuit containing that single source.

This principle can easily be applied to circuits containing multiple sources and is sometimes an effective solution technique. More often, however, other methods result in a more efficient solution. We consider an example.

EXAMPLE 2.5 Superposition

Figure 2.11 shows a circuit, in which

$$I_B = 10\,\text{A}; \quad V_G = 12\,\text{V};$$
$$R_B = 1.25\,\Omega; \quad R_G = 0.5\,\Omega; \quad R = 0.25\,\Omega$$

Determine the voltage across the resistor R.

Solution

Specify a ground node and the polarity of the voltage across R. Suppress the voltage source by replacing it with a short circuit. Redraw the circuit.

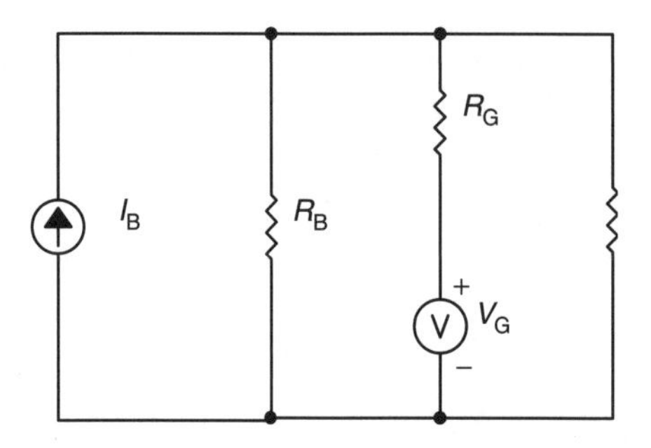

Figure 2.11 A circuit used to illustrate the superposition principle.

By KCL:

$$-I_B + \frac{V_{R-I}}{R_B} + \frac{V_{R-I}}{R_G} + \frac{V_{R-I}}{R} = 0 \tag{2.23A}$$

$$V_{R-I} = \frac{I_B}{\frac{1}{R_B} + \frac{1}{R_G} + \frac{1}{R}} = \frac{10}{\frac{1}{1.25} + \frac{1}{0.5} + \frac{1}{0.25}} = 1.47\ \text{V} \tag{2.23B}$$

Suppress the current source by replacing it with an open circuit.

By KCL:

$$\frac{V_{R-V}}{R_B} + \frac{V_{R-V} - V_G}{R_G} + \frac{V_{R-V}}{R} = 0 \tag{2.23C}$$

$$V_{R-V} = \frac{\frac{V_G}{R_G}}{\frac{1}{R_B} + \frac{1}{R_G} + \frac{1}{R}} = \frac{\frac{12}{0.5}}{\frac{1}{1.25} + \frac{1}{0.5} + \frac{1}{0.25}} = 3.53\ \text{V} \tag{2.23D}$$

$$V_R = V_{R-I} + V_{R-V} = 1.47 + 3.53 = 5\ \text{V} \tag{2.23E}$$

Note: Superposition essentially doubles the work required to solve this problem. The voltage across R can easily be determined using a single KCL.

2.3.4 Thévenin and Norton equivalent circuits

It is always possible to view even a very complicated circuit in terms of much simpler *equivalent* source and load circuits. The analysis of equivalent circuits is more easily managed than the original complex circuit. In studying node voltage and mesh current analysis, you may have observed that there is a certain correspondence (called *duality*) between current sources and voltage sources, on the one hand, and parallel and series circuits, on the other. This duality appears again very clearly in the analysis of equivalent circuits: it will shortly be shown that equivalent circuits fall into one of two classes, involving either a voltage or a current source and, respectively, either series or parallel resistors, reflecting this same principle of duality.

2.3.4.1 ***Thévenin's theorem***

As far as a load is concerned, an equivalent circuit consisting of an ideal voltage source, V_T, in series with an equivalent resistance R_T, may represent any network composed of ideal voltage and current sources, and of linear resistors.

2.3.4.2 ***Norton's theorem***

As far as a load is concerned, an equivalent circuit consisting of an ideal current source, I_N, in parallel with an equivalent resistance R_N, may represent any network composed of ideal voltage and current sources, and of linear resistors.

2.3.4.3 ***Determination of the Norton or Thévenin equivalent resistance***

The first step in computing a Thévenin or Norton equivalent circuit consists of finding the equivalent resistance presented by the circuit at its terminals. This is done by setting all sources in the circuit equal to zero and computing the effective resistance between terminals. The voltage and current sources present in the circuit are set to zero by the same technique used with the principle of superposition: voltage sources are replaced by short circuits, current sources by open circuits. The steps involved in the computation of equivalent resistance are as follows:

1. Remove the load.
2. Set all independent voltage and current sources to zero.
3. Compute the total resistance between load terminals, *with the load removed.* This resistance is equivalent to that which would be encountered by a current source connected to the circuit in place of the load.

To illustrate the procedure, consider the simple circuit of Figure 2.12; the objective is to compute the equivalent resistance the load R_L 'sees' at port *a–b*.

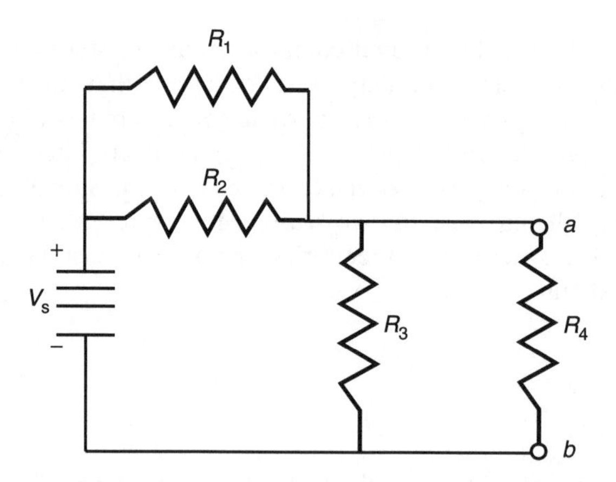

Figure 2.12 Network to illustrate the calculation of the Thévenin resistance.

In order to compute the equivalent resistance, we remove the load resistance from the circuit and replace the voltage source, V_S, by a short circuit. At this point, seen from the load terminals, the circuit appears as shown in Figure 2.13. You can see that R_1 and R_2 are in parallel, since they are connected between the same two nodes. If the total resistance between terminals a and b is denoted by R_T, its value can be determined as follows:

$$R_T = R_1 \| R_2 \| R_3 \qquad (2.23F)$$

The equivalent circuit is shown in Figure 2.14, with the source voltage in series with the equivalent Thévenin resistance, so that the voltage seen at a–b is obtained using

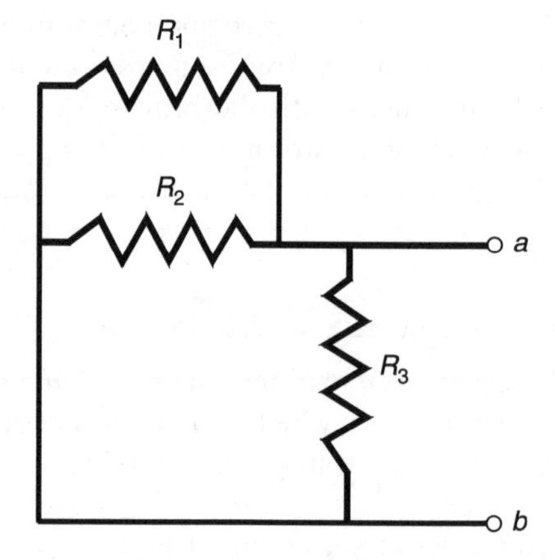

Figure 2.13 Equivalent resistance seen by the load.

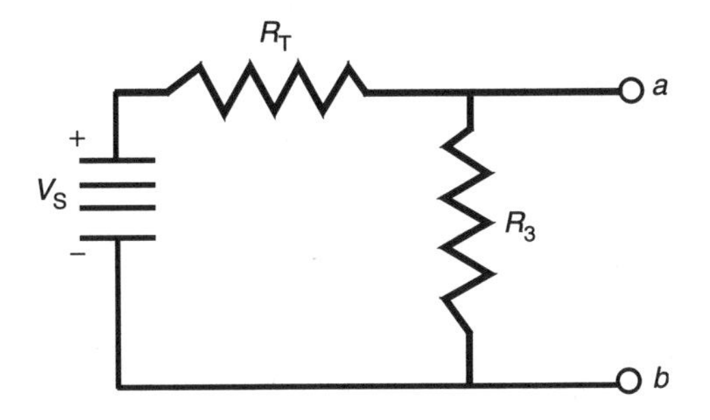

Figure 2.14 Thévenin equivalent circuit of Figure 2.12.

voltage divider equation as

$$V_{TH} = \frac{R_{TH} R_3}{R_{TH} + R_3} V_S \tag{2.23G}$$

Let us now apply this principle to solve a problem.

EXAMPLE 2.6 Thévenin equivalent circuit

For Figure 2.15 having:

$$V_B = 11\,\text{V}; \quad V_G = 12\,\text{V}$$
$$R_B = 0.7\,\text{V}; \quad R_G = 0.3\,\text{V}; \quad R_L = 7.2$$

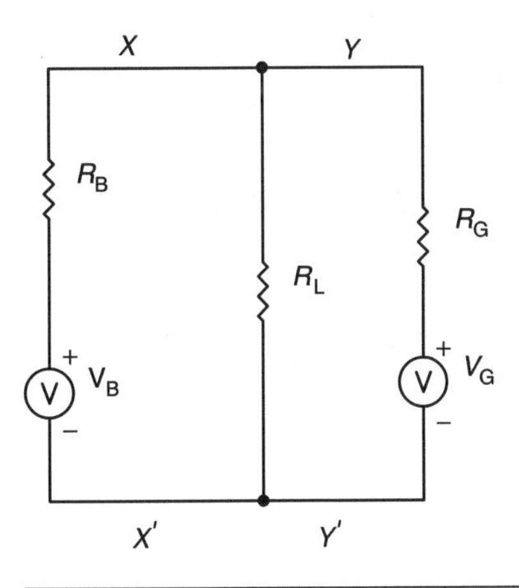

Figure 2.15 Two-mesh, two-source circuit.

Determine:

(a) the Thévenin equivalent of the circuit to the left of Y–Y';
(b) the voltage between Y–Y'.

Solution

(a) Specify the polarity of the Thévenin equivalent voltage:
Using the voltage divider expression:

$$V_{\mathrm{TH}} = \frac{V_{\mathrm{B}} R_{\mathrm{L}}}{R_{\mathrm{B}} + R_{\mathrm{L}}} = \frac{11 \times 7.2}{0.7 + 7.2} = 10.03\ \mathrm{V} \tag{2.23H}$$

Suppress the generator source:

$$R_{\mathrm{T}} = R_{\mathrm{L}} \| R_{\mathrm{B}} = \frac{7.2 \times 0.7}{0.7 + 7.2} = 638\ \mathrm{m\Omega} \tag{2.23I}$$

(b) Specify the polarity of the terminal voltage. Choose a ground.
Using KCL:

$$\frac{V_{\mathrm{T}} - V_{\mathrm{G}}}{R_{\mathrm{G}}} + \frac{V_{\mathrm{T}} - V_{\mathrm{T}}}{R_{\mathrm{T}}} = 0$$

$$V_{\mathrm{T}} = \frac{\dfrac{V_{\mathrm{G}}}{R_{\mathrm{G}}} + \dfrac{V_{\mathrm{T}}}{R_{\mathrm{T}}}}{\dfrac{1}{R_{\mathrm{G}}} + \dfrac{1}{R_{\mathrm{T}}}} = 11.37\ \mathrm{V} \tag{2.23J}$$

2.4 Sinusoidal sources and complex impedance

We now consider current and voltage sources with time average values of zero. We will use periodic signals but the observation time could well be less than one period. Periodic signals are also useful in the sense that arbitrary signals can usually be expanded in terms of a Fourier series of periodic signals. Let us start with the following:

$$\begin{aligned} v(t) &= V_{\mathrm{o}} \cos\ t(\omega t + \phi_{\mathrm{v}}) \\ i(t) &= I_{\mathrm{o}} \cos\ t(\omega t + \phi_{\mathrm{I}}) \end{aligned} \tag{2.24}$$

Notice that we have now switched to lowercase symbols. Lowercase is generally used for a.c. quantities while uppercase is reserved for d.c. values. Now is the time to get into complex notation, often used in electrical and electronic equations, since it will make our discussion easier. The above voltage and current signals can be written as

$$\begin{aligned} \bar{v}(t) &= V_o e^{j(\omega t + \phi_v)} \\ \bar{i}(t) &= I_o e^{j(\omega t + \phi_I)} \end{aligned} \tag{2.25}$$

In order to make things easier, we define one EMF in the circuit to have $\varphi = 0$. In other words, we will pick $t = 0$ to be at the peak of one signal. The vector notation is used to remind us that complex numbers can be considered as vectors in the complex plane. Although not so common in physics, in electronics we refer to these vectors as phasors. Hence the reader should now review complex notation. The presence of sinusoidal $\bar{v}(t)$ or $\bar{i}(t)$ in circuits will result in an inhomogeneous differential equation with a time-dependent source term. The solution will contain sinusoidal terms with the source frequency. The extension of Ohm's law to a.c. circuits can be written as

$$\bar{v}(\omega, t) = Z(\omega)\bar{i}(\omega, t), \tag{2.26}$$

where ω is the source frequency. The generalized resistance referred to as the impedance is represented by the letter Z. We can cancel out the common time-dependent factors to obtain

$$\bar{v}(\omega) = Z(\omega)\bar{i}(\omega) \tag{2.27}$$

and hence the power of the complex notation becomes obvious. For a physical quantity we take the amplitude of the real signal as follows

$$\left|\vec{v}(\omega)\right| = \left|Z(\omega)\right|\left|\vec{i}(\omega)\right| \tag{2.28}$$

We will now examine each circuit element in turn with a voltage source to deduce its impedance.

2.4.1 Resistive impedance

For a voltage source and resistor, the impedance is equal to the resistance, as expected, given as

$$Z = R \tag{2.29}$$

2.4.2 Capacitive impedance

For a voltage source and capacitor, the impedance is given as

$$Z = \frac{1}{j\omega C}. \tag{2.30}$$

For d.c. circuits $\omega = 0$ and hence $Z_c \to \infty$. The capacitor acts like an open circuit (infinite resistance) in a d.c. circuit.

2.4.3 Inductive impedance

For a voltage source and an inductor, the impedance is given as

$$Z = j\omega L \tag{2.31}$$

For d.c. circuits $\omega = 0$ and hence $Z_L \to \infty$. There is no voltage drop across an inductor in a d.c. circuit.

EXAMPLE 2.7 Alternating current circuit

For Figure 2.16 having:

$$R_1 = 100\,\Omega; \quad R_2 = 50\,\Omega; \quad L = 10\,\text{mH}; \quad C = 10\,\mu\text{F};$$

$$v_s(t) = 5\ \cos\left(10\,000t + \frac{\pi}{2}\right) \tag{2.31A}$$

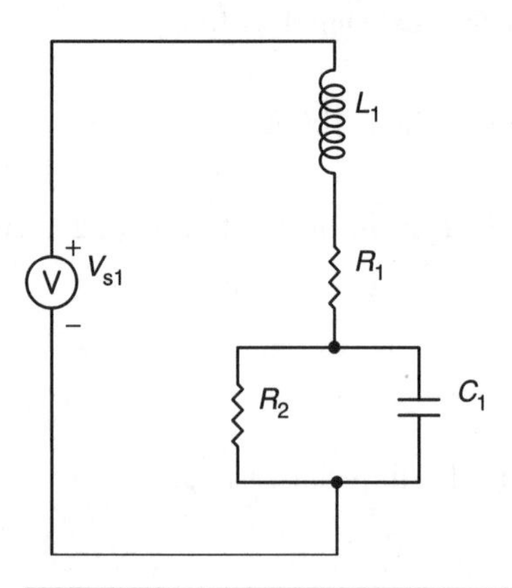

Figure 2.16 Alternating current example.

Determine:

(a) the equivalent impedance of the circuit;
(b) the source current.

Solution

Considering R_2 and C:

$$Z_C = \frac{1}{j\omega C} \tag{2.31B}$$

$$Z_{\|} = \frac{Z_R Z_C}{Z_R + Z_C} = \frac{R\frac{1}{j\omega C}}{R + j\omega C} = \frac{R}{1 + j\omega RC} \tag{2.31C}$$

$$Z_{\|} = \frac{50}{1 + j \times 10^4 \times 50 \times 10 \times 10^{-6}}$$

$$= \frac{50}{1 + j5} = \frac{50(1 - j5)}{(1 + j5)(1 - j5)} = 1.92 - j9.62$$

$$r = \sqrt{(1.92^2 + 9.62^2)} = 9.81;$$

(2.31D)

$$\theta = \tan^{-1}\left(\frac{-9.62}{1.92}\right) = -78.7^\circ = \frac{-78.7\pi}{180} = -1.3734^c$$

$$Z_{\|} = 9.81\angle -1.3734$$

$$Z = Z_{R1} + Z_L + Z_{\|}$$

$$= 100 + j10^4 \times 10 \times 10^{-3} + 1.92 - 9.6j = 101.92 + 90.38j$$

$$r = \sqrt{(101.92^2 + 90.38^2)} = 136.22;$$

$$\theta = \tan^{-1}\left(\frac{90.38}{101.92}\right) = 41.56^\circ = \frac{41.56\pi}{180} = 0.7254^c \tag{2.31E}$$

$$Z = 136.221\angle 0.7254$$

The source current can now be computed as:

$$I = \frac{V_S}{Z} = \frac{5\angle\pi/2}{136.221\angle 0.7254} = \frac{5\angle 1.578}{136.221\angle 0.7254} = 36\angle 0.8453 \text{ mA} \tag{2.31F}$$
$$i(t) = 36 \ \cos(10\,000t + 0.8453)$$

EXAMPLE 2.8

See website for downloadable MATLAB code to solve this problem

Passive element circuit

Two fuses F_1 and F_2 (Figure 2.17), under normal conditions, are modeled as short circuits. However, if excess current flows through a fuse, it melts and consequently blows (becoming an open circuit).

Determine, using KVL, and mesh analysis, the voltages across R_1, R_2 and R_3 under normal condition (no blown fuses), when

$$V_{s1} = 115\,\text{V}; \quad V_{s2} = 115\,\text{V};$$
$$R_1 = R_2 = 5\,\Omega; \quad R_3 = 10\,\Omega; \quad R_4 = R_5 = 200\,\text{m}\Omega$$

Solution

Using KVL:

$$I_1(R_1 + R_4) - I_3R_1 = V_{s1}$$
$$I_2(R_2 + R_5) - I_3R_2 = V_{s2}$$
$$-I_1R_1 - I_2R_2 + I_3(R_1 + R_2 + R_3) = 0$$

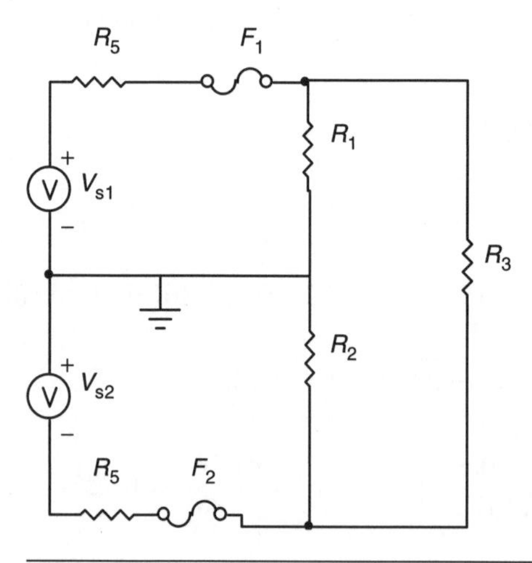

Figure 2.17 Fused circuit.

Substituting resistor values and rearranging:

$$\begin{array}{lll} 5.2I_1 & 0 & -5I_3 = 115 \\ 0 & 5.2I_2 & -5I_3 = 115 \\ 5I_1 & +5I_2 & -20I_3 = 0 \end{array}$$

$$I_1 = I_2 \qquad I_1 = 2I_3$$

$$\Delta = \begin{vmatrix} 5.2 & 0 & -5 \\ 0 & 5.2 & -5 \\ 5 & 5 & -20 \end{vmatrix} = 280.8$$

$$I_1 = I_2 = \frac{\begin{vmatrix} 115 & 0 & -5 \\ 115 & 5.2 & -5 \\ 0 & 5 & -20 \end{vmatrix}}{\Delta} = \frac{-119\,60}{-280.8} = 42.6\,\text{A} \tag{2.31G}$$

$$I_3 = I_1/2 = 21.3\,\text{A}$$

$$V_{R1} = R_1(I_1 - I_3) = 5 \times 21.3 = 106.5\,\text{V}$$

$$V_{R2} = R_2(I_3 - I_2) = -5 \times 21.3 = -106.5\,\text{V}$$

$$V_{R3} = I_3 R_3 = 10 \times 21.3 = 213\,\text{V}$$

EXAMPLE 2.9

See website for downloadable MATLAB code to solve this problem

Passive element circuit

Determine, using KVL, and mesh analysis, the voltages across R_1, R_2 and R_3 in Figure 2.17 (Example 2.8) for the following parameter values:

$$V_{s1} = 110\,\text{V}; \quad V_{s2} = 110\,\text{V};$$
$$R_1 = 100\,\Omega; \quad R_2 = 22\,\Omega; \quad R_3 = 70\,\Omega; \quad R_4 = R_5 = 13\,\Omega.$$

Solution

Using KVL:

$$\begin{aligned} I_1(R_1 + R_4) - I_3 R_1 &= V_{s1} \\ I_2(R_2 + R_5) - I_3 R_2 &= V_{s2} \\ -I_1 R_1 - I_2 R_2 + I_3(R_1 + R_2 + R_3) &= 0 \end{aligned} \tag{2.31H}$$

Substituting resistor values and rearranging:

$$
\begin{aligned}
&113I_1 \quad 0 \quad -100I_3 = 110\\
&0 \quad 35I_2 \quad -22I_3 = 110\\
&100I_1 + \quad 22I_2 \quad -192I_3 = 0\\
&\Delta = \begin{vmatrix} 113 & 0 & -100 \\ 0 & 35 & -22 \\ 100 & 22 & -192 \end{vmatrix} = -354\,668\\
&I_1 = \frac{\begin{vmatrix} 110 & 0 & -100 \\ 110 & 35 & -22 \\ 0 & 22 & -192 \end{vmatrix}}{\Delta} = \frac{-935\,660}{-354\,668} = 2.64\,\text{A} \qquad (2.31\,\text{I})\\
&I_2 = 4.31\,\text{A}\\
&I_3 = 1.86\,\text{A}\\
&V_{R1} = R_1(I_1 - I_3) = 100 \times 0.76 = 76\,\text{V}\\
&V_{R2} = R_2(I_3 - I_2) = -22 \times 2.43 = -53.46\,\text{V}\\
&V_{R3} = I_3R_3 = 70 \times 1.88 = 131.6\,\text{V}
\end{aligned}
$$

Problems

Node and mesh analysis methods

Q2.1 Determine the voltage across R_5 in Figure 2.18 when

$V_{s1} = 4\,\text{V}$; $V_{s2} = 2\,\text{V}$; $R_1 = 2\,\text{k}\Omega$; $R_2 = 4\,\text{k}\Omega$; $R_3 = 4\,\text{k}\Omega$; $R_4 = 2\,\text{k}\Omega$; $R_5 = 6\,\text{k}\Omega$; $R_6 = 2\,\text{k}\Omega$.

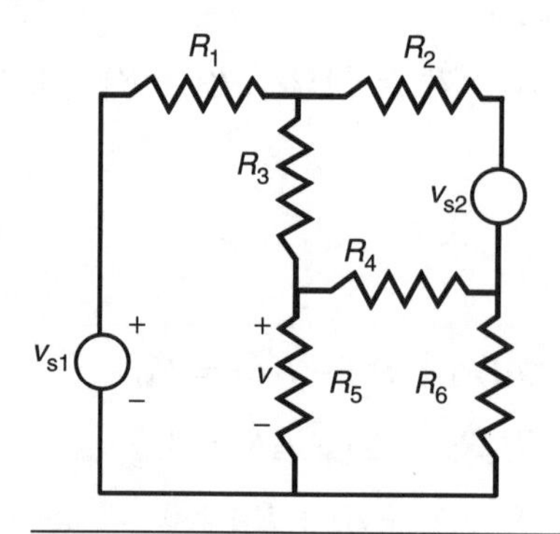

Figure 2.18 Circuit for Q2.1.

Q2.2 For the circuit in Figure 2.19, V_{S2} and R_S model a temperature sensor, and the voltage R_3 indicates the temperature. Determine the temperature.

$$V_{s1} = 24\,\text{V}; \quad V_{s2} = \text{kT}, \quad \text{where } k = 15\,\text{V/°C}; \quad V_{R3} = -4\,\text{V}$$
$$R_1 = R_s = 15\,\text{k}\Omega; \quad R_2 = 5\,\text{k}\Omega; \quad R_3 = 10\,\text{k}\Omega; \quad R_4 = 24\,\text{k}\Omega.$$

Q2.3 For the circuit in Figure 2.20 having $V_S = 10\,\text{V}$, $A_V = 50$, $R_1 = 3\,\text{k}\Omega$; $R_2 = 8\,\text{k}\Omega$; $R_3 = 2\,\text{k}\Omega$; $R_4 = 0.3\,\text{k}\Omega$, determine the voltage across R_4 using KCL and node analysis.

Q2.4 Determine (a) the current I, and (b) the voltage at node A, in Figure 2.21, where

$$V_1 = 5\,\text{V}; \quad V_2 = 10\,\text{V}$$
$$R_1 = 1\,\text{k}\Omega; \quad R_2 = 8\,\text{k}\Omega; \quad R_3 = 10\,\text{k}\Omega;$$
$$R_4 = 2\,\text{k}\Omega; \quad R_5 = 2\,\text{k}\Omega$$

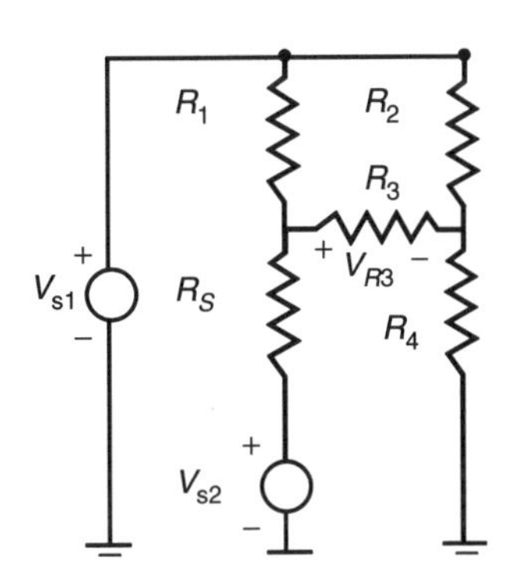

Figure 2.19 Circuit for Q2.2.

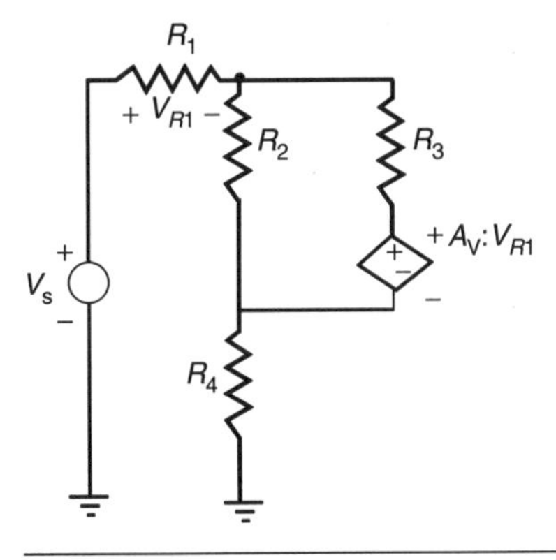

Figure 2.20 Circuit for Q2.3.

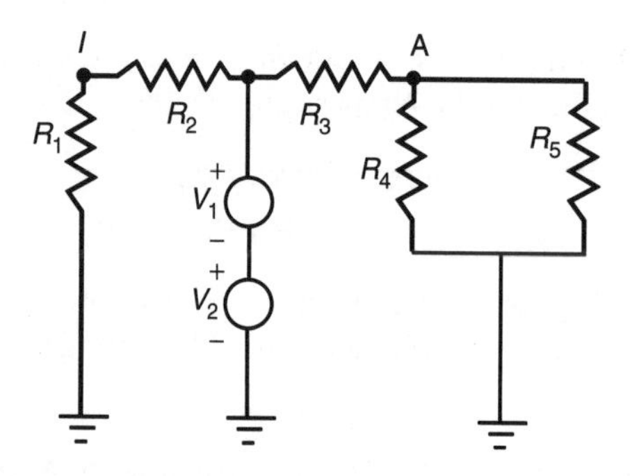

Figure 2.21 Circuit for Q2.4.

Q2.5 In Figure 2.22, F_1 and F_2 are fuses. Under normal conditions they are modeled as short circuits. However, if excess current flows through a fuse, it melts and consequently blows (becoming an open circuit). The component values are:

$$V_{s1} = 110\,\text{V}; \quad V_{s2} = 110\,\text{V};$$
$$R_1 = 100\,\Omega; \quad R_2 = 25\,\Omega; \quad R_3 = 75\,\Omega; \quad R_4 = R_5 = 15\,\Omega.$$

Normally, the voltages across R_1, R_2, and R_3 are 106.5 V, −106.5 V, and 213 V, respectively. If fuse F_1 now blows, or opens, determine, using KVL, and mesh analysis, the new voltages across R_1, R_2 and R_3.

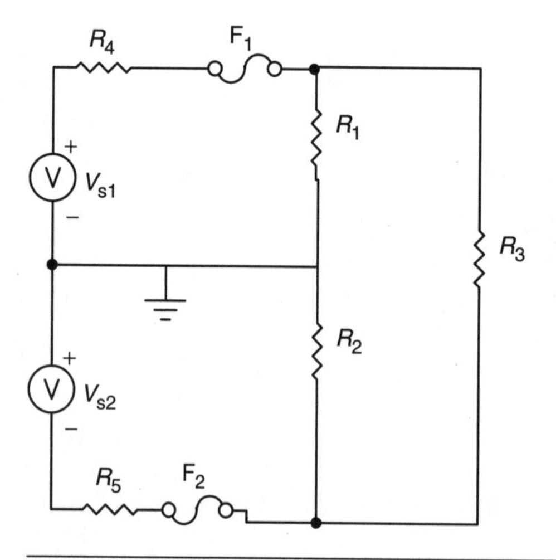

Figure 2.22 Circuit for Q2.5.

Thévenin and Norton equivalent circuits

Q2.6 In Figure 2.23, $V_S = 12\,V$; $R_1 = 7\,k\Omega$; $R_2 = 3\,k\Omega$; $R_3 = 8\,k\Omega$; $R_4 = 6\,k\Omega$. Determine:

(a) the Thévenin equivalent of the circuit to the left of *a*–*b*;

(b) the voltage across *a*–*b*.

Q2.7 In the circuit shown in the Figure 2.24:

$$V_1 = 15\,V;\quad V_2 = 12\,V;\quad I = 20\,mA;$$
$$R_1 = 10\,k\Omega;\quad R_2 = 2\,k\Omega;\quad R_3 = 4\,k\Omega;\quad R_4 = 5\,k\Omega;$$
$$R_5 = 8\,k\Omega;\quad R_6 = 6\,k\Omega;\quad R_7 = 3\,k\Omega;$$
$$C = 20\,\mu F.$$

Determine the Norton equivalent circuit with respect to *C*.

Sinusoidal sources

Q2.8 For the circuit shown in Figure 2.25, determine, for the values given,

(a) the equivalent impedance of the circuit the source current;

(b) the source current.

$$R_1 = 200\,\Omega;\quad R_2 = 100\,\Omega;\quad L = 50\,mH;\quad C = 50\,\mu F;$$
$$v_s(t) = 5\cos\left(5000t + \frac{\pi}{4}\right). \tag{2.31J}$$

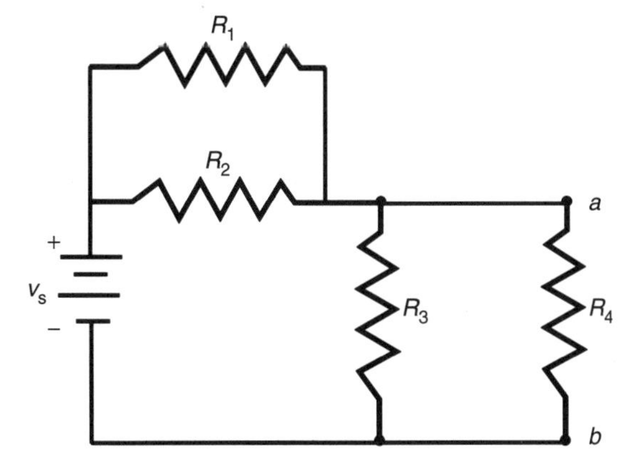

Figure 2.23 Circuit for Q2.6.

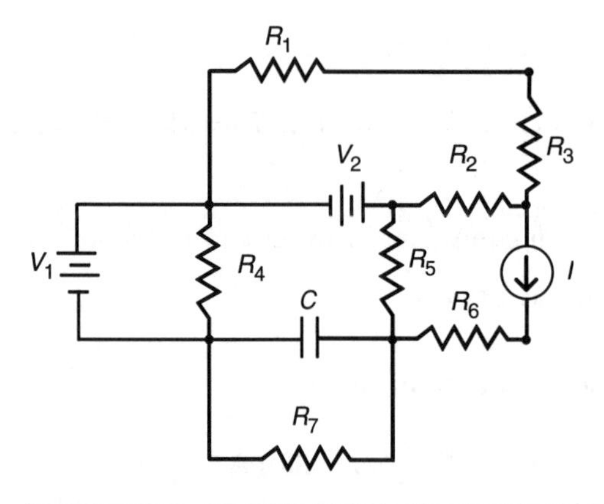

Figure 2.24 Circuit for Q2.7.

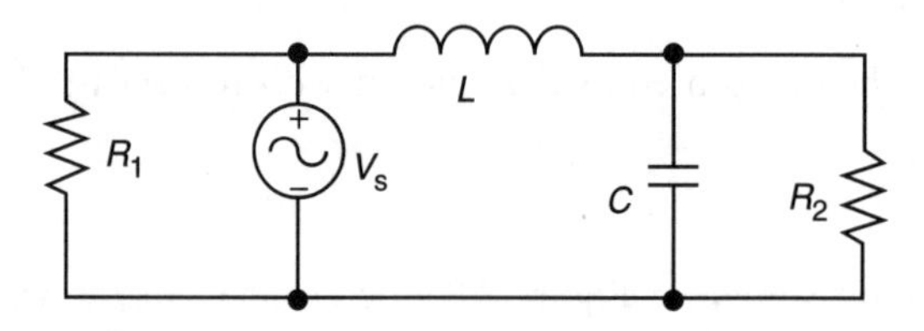

Figure 2.25 Circuit for Q2.8.

Further reading

[1] Horowitz, P. and Hill, W. (1989) *The Art of Electronics* (2nd. ed.), New York: Cambridge University Press.

[2] Rashid, M.H. (1996) *Power Electronics: Circuits, Devices, and Applications*, Prentice Hall.

[3] Rizzoni, G. (2003) *Principles and Applications of Electrical Engineering* (4th. ed.), McGraw-Hill.

CHAPTER 3

Semiconductor electronic devices

Chapter objectives

When you have finished this chapter you should be able to:

- understand how covalent bonds and doping materials impact on semiconductor electronic devices;
- understand the p–n junction and the diode effect;
- understand how a Zener diode works;
- understand bipolar junction transistors (BJTs);
- understand junction field-effect transistors (JFETs);
- understand metal-oxide semiconductor field-effect transistors (MOSFETs);
- understand transistor gates and switching circuits;
- understand complementary metal-oxide semiconductor (CMOS) field-effect transistor gates.

3.1 Introduction

In Rutherford's model of an atom electrons orbit a central nucleus. The limit to the number of electrons that can be included in each orbit or shell is $2n^2$, where n is the shell number (starting at 1 for the innermost shell). Moving out from the nucleus, therefore, there are 2, 8, 18, 32, . . ., electrons in an orbit. The energy at the innermost shell is least, and increases outwards (Figure 3.1). The outermost shell is referred to as the valence shell, and electrons in the valence shell are known as valence electrons.

Conductors such as copper, gold, and silver offer little resistance to current flow. On the other hand, insulators (non-conductors) such as plastics, glass, and

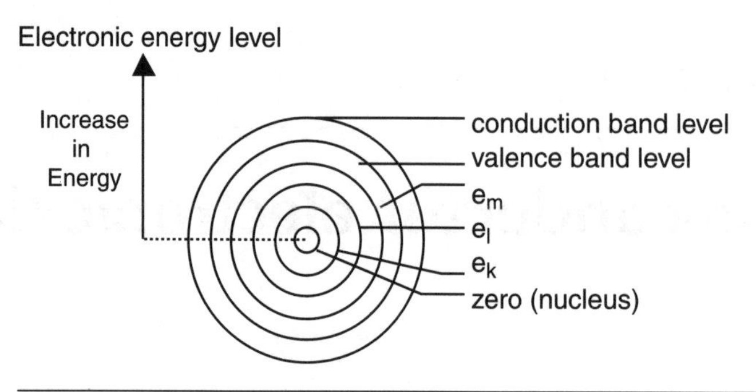

Figure 3.1 Energy levels in the atom.

Metals | Transition Zone | Nonmetals

	IA	IIA	IIIB	IVB	VB	VIB	VIIB	VIIIB			IB	IIB	IIIA	IVA	VA	VIA	VIIA	VIIIA
1	1 H																1 H	2 He
2	3 Li	4 Be											5 B	6 C	7 N	8 O	9 F	10 Ne
3	11 Na	12 Mg											13 Al	14 Si	15 P	16 S	17 Cl	18 Ar
4	19 K	20 Ca	21 Sc	22 Ti	23 V	24 Cr	25 Mn	26 Fe	27 Co	28 Ni	29 Cu	30 Zn	31 Ga	32 Ge	33 As	34 Se	35 Br	36 Kr
5	37 Rb	38 Sr	39 Y	40 Zr	41 Nb	42 Mo	43 Tc	44 Ru	45 Rh	46 Pd	47 Ag	48 Cd	49 In	50 Sn	51 Sb	52 Te	53 I	54 Xe
6	55 Cs	56 Ba	57 La	72 Hf	73 Ta	74 W	75 Re	76 Os	77 Ir	78 Pt	79 Au	80 Hg	81 Tl	82 Pb	83 Bi	84 Po	85 At	86 Rn
7	87 Fr	88 Ra	89 Ac															

58 Ce	59 Pr	60 Nd	61 Pm	62 Sm	63 Eu	64 Gd	65 Tb	66 Dy	67 Ho	68 Er	69 Tm	70 Yb	71 Lu
90 Th	91 Pa	92 U	93 Np	94 Pu	95 Am	96 Cm	97 Bk	98 Cf	99 Es	100 Fm	101 Md	102 No	103 Lw

Figure 3.2 The periodic table.

mica offer a great deal of resistance to current flow. Silicon is an example of a semiconductor, a material with properties between a conductor and an insulator. The most useful semiconductor materials are those made from atoms that have three, four, or five valence (outermost shell) electrons; these are trivalent atoms, tetravalent atoms, and pentavalent atoms, respectively. The periodic table (Figure 3.2) shows that trivalent atoms include boron, aluminum, gallium, and indium; tetravalent atoms include silicon and germanium; and pentavalent atoms include phosphorous, arsenic, and antimony.

Only electrons at the conduction-band level are free to take part in the process of electron current flow. Therefore, when valence electrons are exposed to outside sources of energy (e.g. electrical energy), they break away from their valence band energy level and migrate into the conduction band energy level. When the same electron falls back to the valence band energy level, its extra energy is given up, usually in the form of heat or light energy. This means that some medium has to be arranged to absorb the energy released (e.g. a heat sink).

Silicon has 14 electrons orbiting its nucleus in three specific shells containing two electrons in the innermost, 8 on the next orbit, and four (the valence electrons) in the next.

3.2 Covalent bonds and doping materials

In a covalent bond, two or more atoms share valence electrons. Semiconductor materials are materials in which the atoms exhibit covalent bonding, so that, for instance, silicon atoms (being tetravalent) form covalent bonds with four other atoms. The bonds of a tetravalent material are actually three-dimensional thus forming a tetrahedral lattice (a cubic crystal). Silicon is most often used as the semiconductor material as it is more abundant than alternatives such as germanium. A single grain of beach sand contains millions of tetravalent silicon atoms bonded in the form shown in Figure 3.3.

In practice, semiconductor materials are not made from simply one type of atom. Most semiconductor materials are produced by the addition of highly purified trivalent atoms (e.g. boron, aluminum, gallium, or indium) or pentavalent atoms (e.g. phosphorus, arsenic, or antimony). These are called the doping materials. As can be seen from Figures 3.4 and 3.5, when a tetravalent atom is doped with a trivalent atom, a 'hole' (resultant positive charge) is left, and when a tetravalent atom is doped with a pentavalent atom, that an excess electron (resultant negative charge) results. By controlling the amounts of doping materials, it is possible to affect the 'density' of holes and excess electrons. The doped

Figure 3.3 Covalent bonds for a tetravalent atom (e.g. silicon).

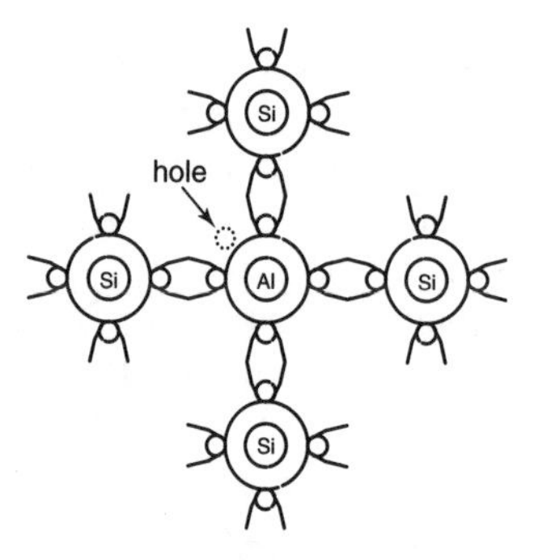

Figure 3.4 Covalent bond between a trivalent atom (aluminum) and a tetravalent atom (silicon).

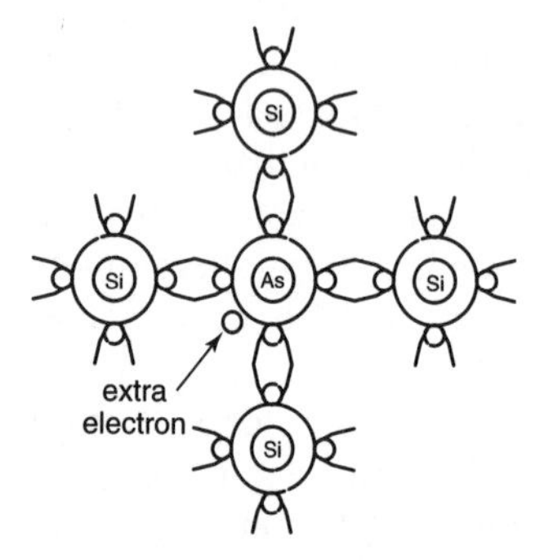

Figure 3.5 Covalent bond between a pentavalent atom (arsenic) and a tetravalent atom (silicon).

material is called an n-type semiconductor (excess of electrons) and a p-type (deficit of electrons) semiconductor.

When external energy (electrical, heat, or light) is applied to n-type semiconductor material, the excess electrons are made to wander through the material: the negative terminal of the applied potential repulses the free electrons and the positive terminal attracts the free electrons.

3.3 The p–n junction and the diode effect

A diode is formed by joining a p-type and n-type semiconductor together (Figure 3.6).

Initially both the p-type and n-type areas of the diode are totally neutral. However, the concentration of positive and negative carriers is quite different on opposite sides of the junction and a thermal energy-powered diffusion of positive carriers into the n-type material and negative carriers into the p-type material occurs. The n-type material acquires an excess of positive charge near

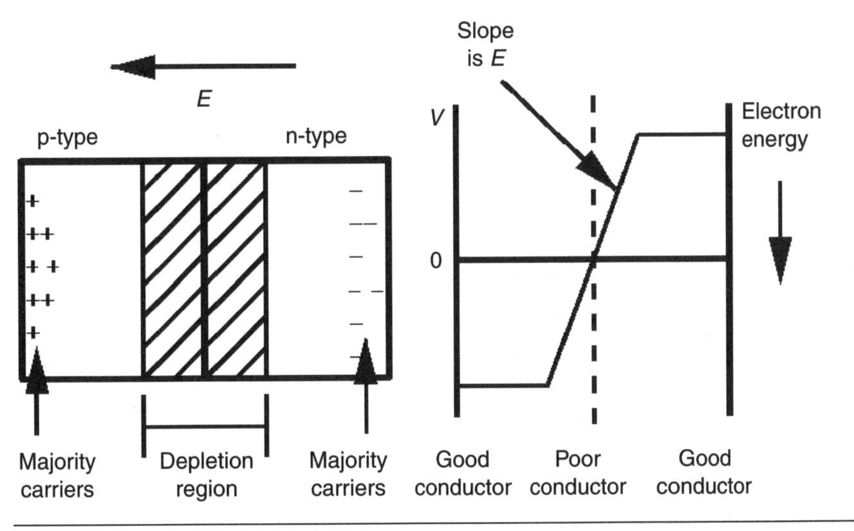

Figure 3.6 The p–n junction diode

the junction and the p-type material acquires an excess of negative charge. Eventually diffuse charges build up and an electric field is created which drives the minority charges and eventually equilibrium is reached. A region develops at the junction called the *depletion layer*. This region is essentially undoped (it is just intrinsic silicon).

3.3.1 Current though a diode

The behavior of a diode depends on the polarity of the source of potential energy connected to it (Figure 3.7). If the diode is *reverse biased* (positive potential on n-type material) the depletion layer increases. The only charge carriers able to support a net current across the p–n junction are the minority carriers and hence the reverse current is very small. A *forward-biased* diode (positive potential on p-type material) has a decreased depletion region; the majority carriers can diffuse across the junction. The voltage may become high enough to eliminate the depletion region entirely.

3.3.1.1 *Small-signal diode model*

An approximation to the current in the p–n junction region is given by

$$I = I_o\left(e^{V/V_T} - 1\right), \tag{3.1}$$

where both I_0 and V_T are temperature dependent (Figure 3.8(a)). This equation gives a reasonably accurate prediction of the current–voltage relationship of the

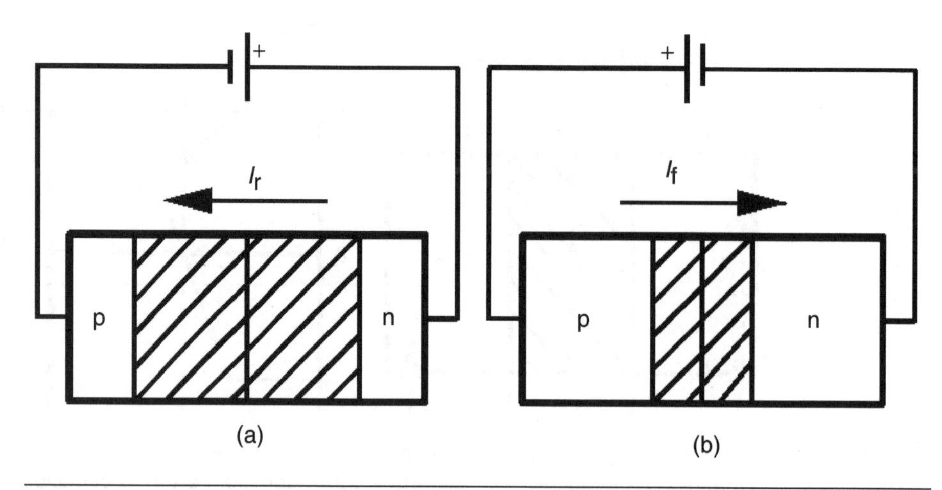

Figure 3.7 Diode circuit connections: (a) reversed bias; and (b) forward biased.

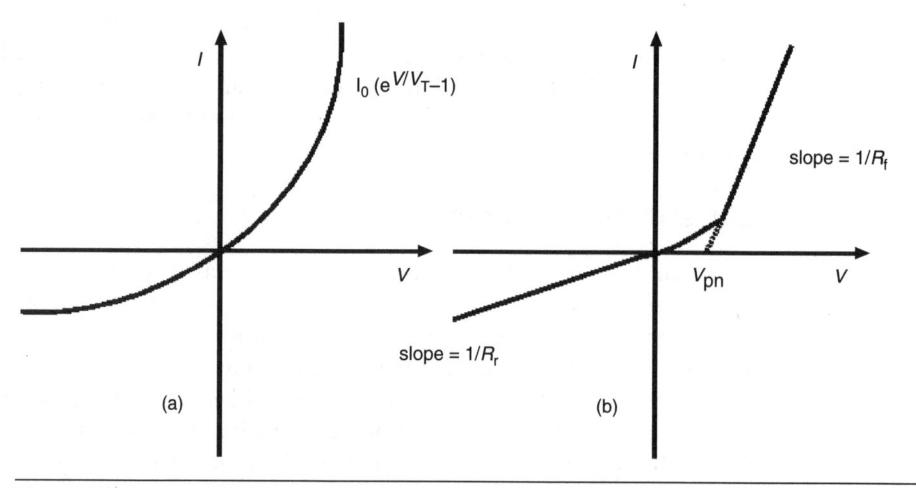

Figure 3.8 Current versus voltage: (a) in the p–n junction region; and (b) for an actual p–n diode.

p–n junction itself (especially the temperature variation) but can be improved somewhat by choosing I_0 and V_T empirically to fit a particular diode. However, for a real diode, other factors are also important: in particular, edge effects around the border of the junction cause the actual reverse current to increase slightly with reverse voltage, and the finite conductivity of the doped semiconductor ultimately restricts the forward current to a linear increase with increasing applied voltage.

3.3.1.2 *Piecewise linear diode model*

A better current–voltage curve for a real diode is shown in the Figure 3.8(b). Various regions of the curve can be identified: the linear region of forward-biasing, a non-linear transition region, a turn-on voltage, V_{pn}, and a reverse-biased region. We can assign a dynamic resistance to the diode in each of the linear regions: R_f in the forward-biased region and R_r in the reverse-biased region. These resistances are defined as the inverse slope of the curve: $1/R = \Delta I/\Delta V$. The voltage V_{pn}, represents the effective voltage drop across a forward-biased p–n junction (the turn-on voltage). For a germanium diode, V_{pn} is approximately 0.3 V, while for a silicon diode it is close to 0.6 V.

3.3.2 The p–n diode as a circuit element

Diodes are referred to as non-linear circuit elements because of the characteristic curve shown in Figure 3.8. For most applications the non-linear region can be avoided and the device can be modeled as the piecewise linear circuit elements: a conduction region of zero resistance; and an infinite resistance non-conduction region. For many circuit applications, this ideal diode model is an adequate representation of an actual diode and simply requires that the circuit analysis be separated into two parts: forward current and reverse current. Figure 3.9 shows a schematic symbol for a diode and the current–voltage curve for an ideal diode.

A diode can be described more accurately using the equivalent circuit model shown in Figure 3.10. If a diode is forward biased with a high voltage it acts like a resistor (R_f) in series with a voltage source (V_{pn}). When reverse biased it acts

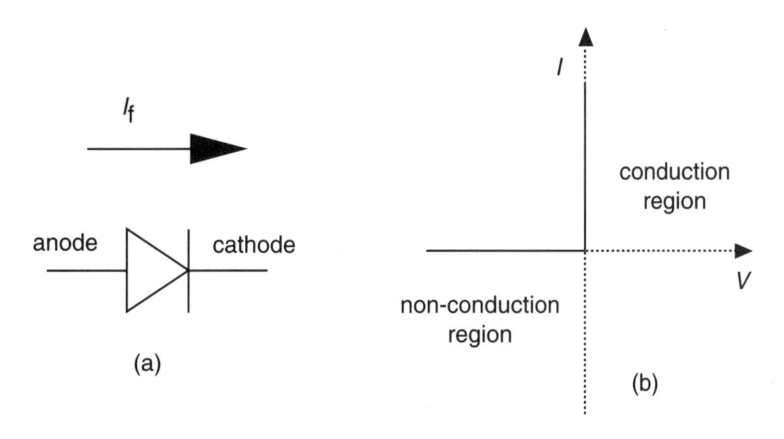

Figure 3.9 Diode: (a) the schematic symbol; and (b) the current versus voltage characteristic for an ideal diode.

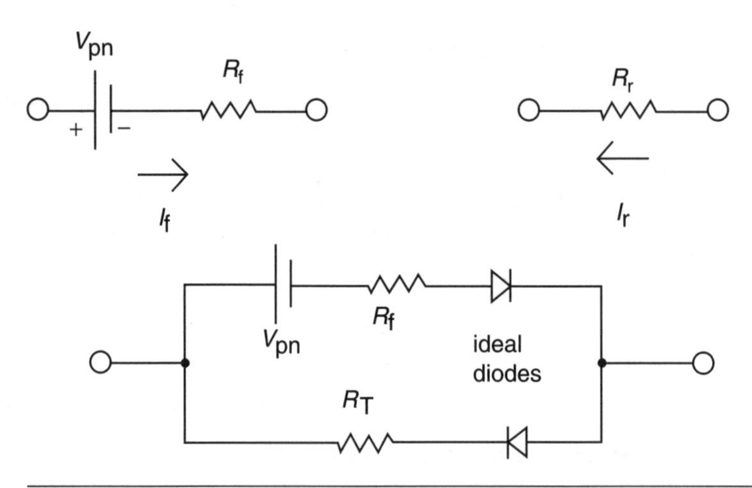

Figure 3.10 Equivalent circuit model of a junction diode.

simply as a resistor (R_r). These approximations are referred to as the linear element model of a diode.

3.4 The Zener diode

There are several other types of diodes beside the junction diode. In a Zener diode, as the reverse voltage increases, the diode can reach the avalanche-breakdown (Zener breakdown) condition. This causes an increase in current in the reverse direction. Zener breakdown occurs when the electric field near the junction becomes large enough to excite valence electrons directly into the conduction band. Avalanche-breakdown is when the minority carriers are accelerated in the electric field near the junction to sufficient energies that they can excite valence electrons through collisions. Figure 3.11 shows the current–voltage characteristic of a Zener diode, its schematic symbol and equivalent circuit model in the reverse-bias direction. The best Zener diodes have a breakdown voltage (V_Z) of 6–7 V.

3.4.1 The Zener diode voltage regulator

Figure 3.12 shows a simple voltage regulator circuit; note how the Zener diode terminals are connected. The Zener diode may be size rated for current or voltage, depending on the information available. R_L is the load resistance and V_S is an

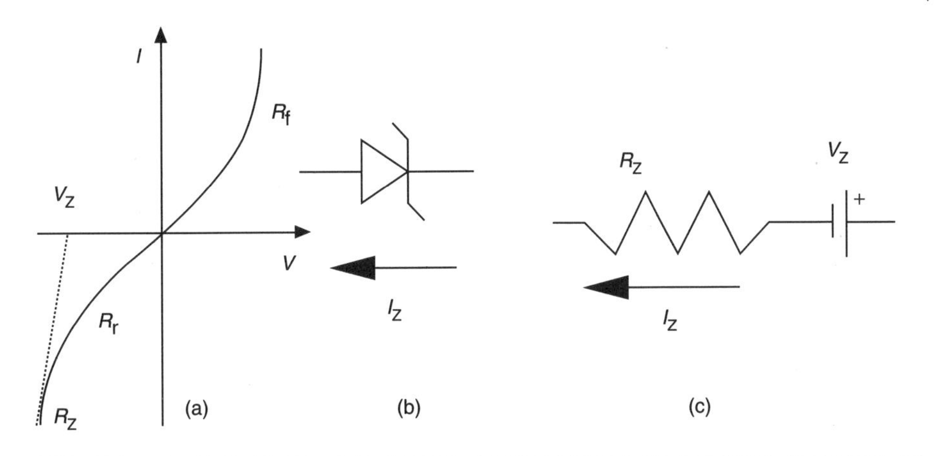

Figure 3.11 Zener diode: (a) current versus voltage characteristic; (b) the schematic symbol; and (c) the equivalent circuit model in the reverse-bias direction.

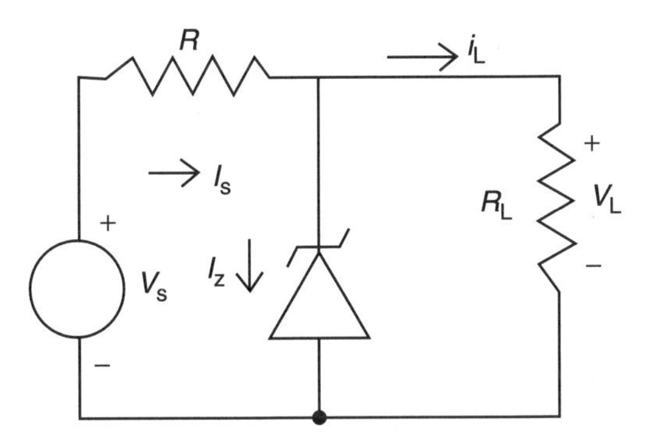

Figure 3.12 Zener diode voltage regulation.

unregulated source with a value exceeding the Zener voltage V_Z. The purpose of the circuit is to provide a constant d.c. voltage, V_L, across the load with a corresponding constant current through the load.

Resistance sizing

We analyze the circuit as follows, where V_Z and I_Z are the Zener voltage and current, respectively. The load voltage will be V_Z as long as the Zener diode is subject to reverse breakdown, leading to

$$I_Z = \frac{V_Z}{R_L}. \tag{3.2}$$

The load current is given by

$$I_L = I_S - I_Z. \tag{3.3}$$

The unregulated source current is given by

$$I_S = \frac{V_S - V_Z}{R}, \tag{3.4}$$

where R is referred to as a current-limiting resistor because it limits the power dissipated by the Zener diode.

The Zener diode current is given by

$$I_Z = \left(I_S - \frac{V_Z}{R_{L_{max}}}\right). \tag{3.5}$$

The power dissipated by the Zener diode is given by

$$P_{Z_{max}} = I_{Z_{max}} V_Z = \left(\frac{V_S - V_Z}{R}\right) V_Z - \frac{V_Z{}^2}{R_{L_{max}}}. \tag{3.6}$$

Current sizing

We analyze the circuit as follows:

$$P_{Z_{max}} = I_Z{}^2 R_Z + I_Z V_Z \tag{3.7}$$

$$I_Z{}^2 R_Z + I_Z V_Z - P_{Z_{max}} = 0 \tag{3.8}$$

Solving this quadratic equation, yields

$$I_Z = \frac{-V_Z \pm \sqrt{V_Z{}^2 - (-4R_Z P_{Z_{max}})}}{2R_Z} = \frac{1}{2}\left[-\frac{V_Z}{R_Z} \pm \sqrt{\left(\frac{V_Z}{R_Z}\right)^2 - \frac{4P_{Z_{max}}}{R_Z}}\right] \tag{3.9}$$

In reality, the Zener series resistance is very small compared to the series R, so $I_Z \approx I_S$.

Zener diodes are useful in electronic circuits where the requirement is to derive small regulated voltages from a single higher voltage source. The Zener diode regulator is an effective solution to power 5 V devices using a 9 V battery.

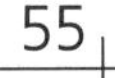

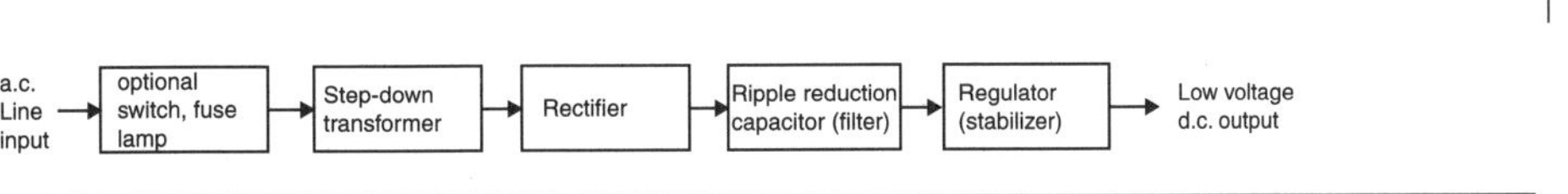

Figure 3.13 The main stages of a power supply system.

3.5 Power supplies

A power supply is required in virtually all mechatronic applications. No application is complete or fully operational without a reliable power supply and thus it is important to be able to design and maintain a power supply unit. Proprietary power supplies are available for almost all applications and these offer a wide range of voltages and current to choose from. However, for designs with voltage and current requirements, there is a need to custom build a power unit.

A power supply must be reliable and safe to use since it controls every component of the machine. An unconditioned power supply can damage the entire electronics of a product and even put the life of an operator at risk. Most power supplies available share common designs.

The d.c. voltage source is depicted on circuit diagrams as a battery but in reality the source can often be a d.c. supply derived from an a.c. line voltage which has been transformed, rectified, filtered and regulated (Figure 3.13). A d.c. power supply is often constructed using an inexpensive three-terminal regulator. These regulators are integrated circuits designed to provide the desirable attributes of temperature stability, output current limiting and thermal overload protection.

In power supply applications it is common to use a transformer to isolate the power supply from the a.c. line input. A rectifier can be connected to the transformer secondary to generate a d.c.voltage with little a.c. ripple. The object of any power supply is to reduce the *ripple* which is the periodic variation in voltage about the steady value.

3.5.1 Rectification

Figure 3.14 shows a half-wave rectifier circuit. The signal is almost exactly the top half of the input voltage sinusoidal signal, and, for an ideal diode, does not depend at all on the size of the load resistor. The peaks of the rectified signal are lower than the input a.c. waveform by an amount equal to the diode's forward voltage drop.

The full-wave (or bridge) rectifier circuit shown in Figure 3.15 produces an output similar to the half-wave rectifier but at twice the frequency of the line input and with the peaks less than input waveform peak amplitude by an amount equal to twice a diode's forward voltage drop. The diodes act to route the current from both

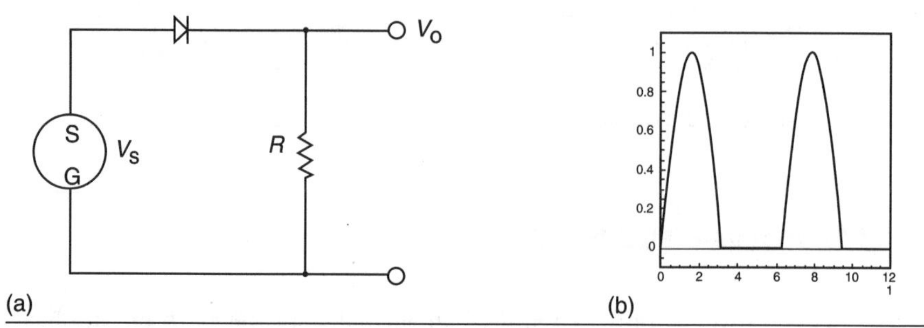

Figure 3.14 Half-wave rectifier: (a) circuit; and (b) output waveform.

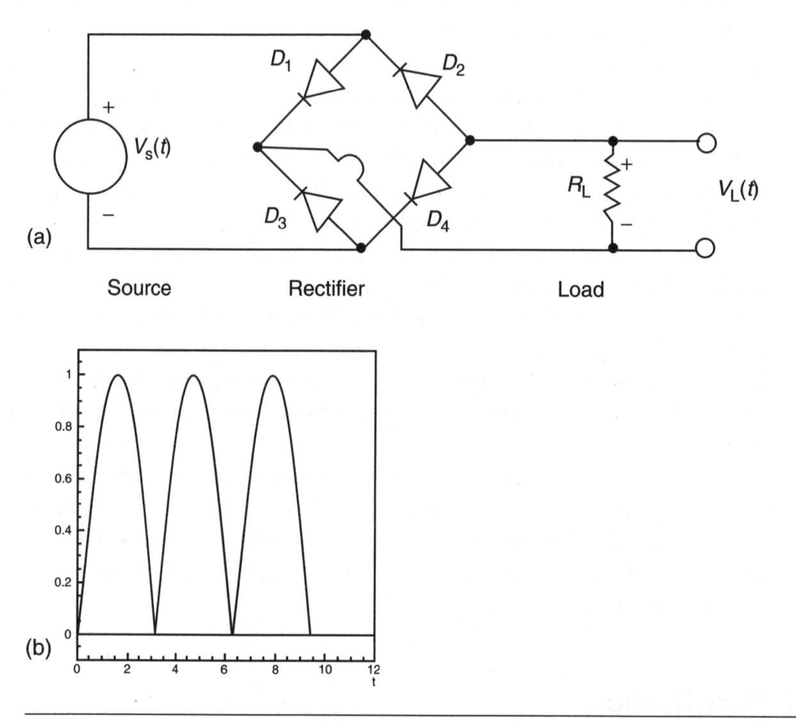

Figure 3.15 Full-wave rectifier: (a) circuit; and (b) output waveform.

halves of the a.c. suppply through the load resistor in the same direction, and the voltage developed across the load resistor becomes the rectified output signal. The diode bridge is a commonly used circuit and is available as a four-terminal component in a number of different power and voltage ratings.

After rectification, the supply signal is now a combination of an a.c. signal and a d.c. component, so for a d.c. supply the a.c. component needs to be filtered out (generally using a smoothing capactitor or other low-pass filter circuit). Any residual a.c. component after smoothing is called *ripple*.

3.6 Active components

The circuits we have encountered so far have been *passive* and dissipate power. Even a transformer that is capable of giving a voltage gain to a circuit is not an active element. *Active* elements in a circuit increase the power by controlling or modulating the flow of energy or power from an additional power supply into the circuit.

Transistors are active circuit elements and are typically made from silicon or germanium and come in two types. The bipolar junction transistor controls the current by varying the number of charge carriers. The field-effect transistor (FET) varies the current by varying the shape of the conducting volume.

We will discuss the general characteristics of these active devices, but since transistor gates and switches are more important in mechatronics than amplifiers, we will then concentrate on these applications.

Before going into details we will define some notation. The voltages that are with respect to ground are indicated by a single subscript. Voltages with repeated letters are power supply voltages, and voltages between two terminals are indicated by a double subscript.

3.6.1 The bipolar junction transistor (BJT)

By placing two p–n junctions together we can create a bipolar junction transistor. Germanium is favored for a pnp transistors, where the majority charge carriers are holes. Silicon is best for npn transistors where the majority charge carriers are electrons.

The thin and lightly doped central region is known as the *base* (B) which has majority charge carriers of opposite polarity to those in the surrounding material. The two outer regions are known as the *emitter* (E) and the *collector* (C). Under normal operating conditions the emitter will emit or inject majority charge carriers into the base region, and because the base is very thin, most will ultimately reach the collector. The emitter is highly doped to reduce resistance. The collector is lightly doped to reduce the junction capacitance of the collector–base junction.

The schematic circuit symbols for bipolar transistors are shown in Figure 3.16. The arrows on the schematic symbols indicate the direction of both I_B and I_C. The collector is usually at a higher voltage than the emitter. The emitter–base junction is forward biased while the collector–base junction is reversed biased.

3.6.1.1 *BJT operation (npn)*

If the collector, emitter, and base of an npn transistor are shorted together as shown in Figure 3.17(a), the diffusion process described earlier for diodes results in

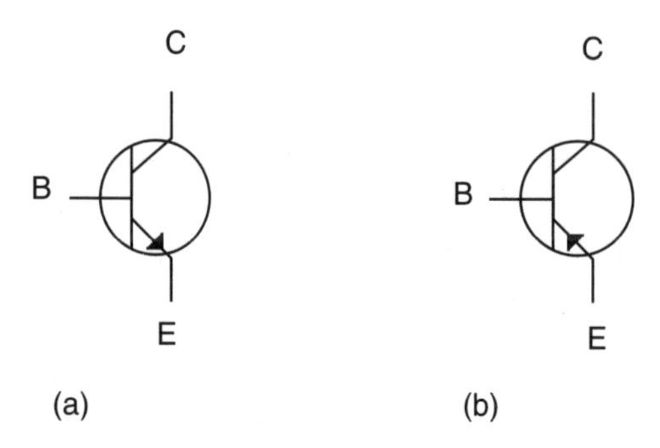

Figure 3.16 Bipolar junction transistor schematics: (a) npn; and (b) pnp.

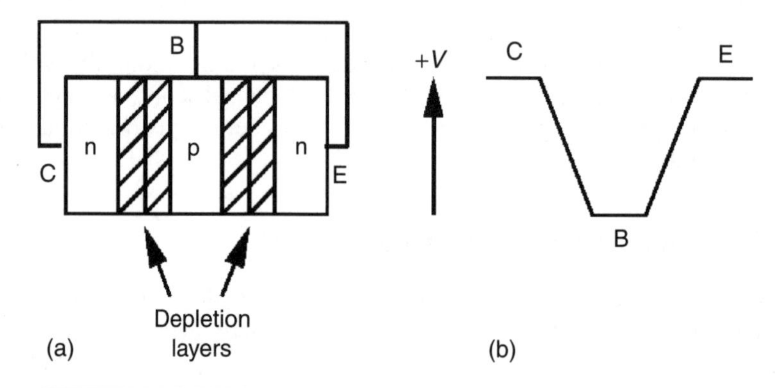

Figure 3.17 (a) npn BJT with collector, base, and emitter shorted; and (b) voltage levels developed within the shorted semiconductor.

the formation of two depletion regions that surround the base. The diffusion of negative carriers into the base and positive carriers out of the base results in a relative electric potential as shown in Figure 3.17(b).

When the transistor is biased for normal operation as in Figure 3.18(a), the base terminal is slightly positive with respect to the emitter (about 0.6 V for silicon), and the collector is positive by several volts. When properly biased, the transistor acts to make $I_C \gg I_B$. The depletion region at the reverse-biased base–collector junction grows and is able to support the increased electric potential change indicated in the Figure 3.18(b).

In a typical transistor, 95 to 99 percent of the charge carriers from the emitter make it to the collector and constitute almost all the collector current I_C. I_C is slightly less than I_E and we may write $\alpha = I_C/I_E$, where from above $\alpha = 0.95$ to 0.99.

The behavior of a transistor can be summarized by the characteristic curves shown in Figure 3.19. Each curve starts from zero in a non-linear fashion, rises

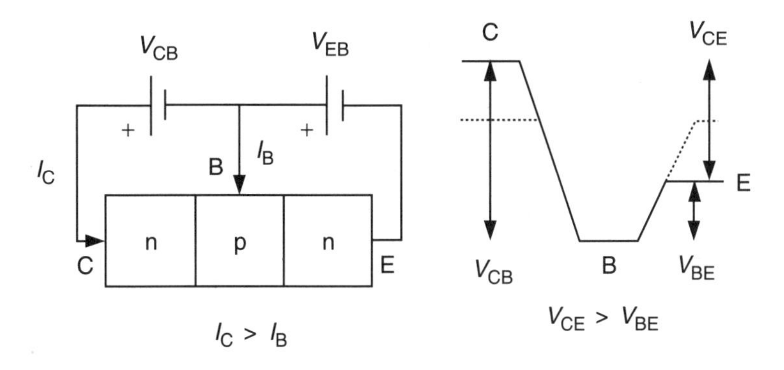

Figure 3.18 (a) npn BJT biased for operation; and (b) voltage levels developed within the biased semiconductor.

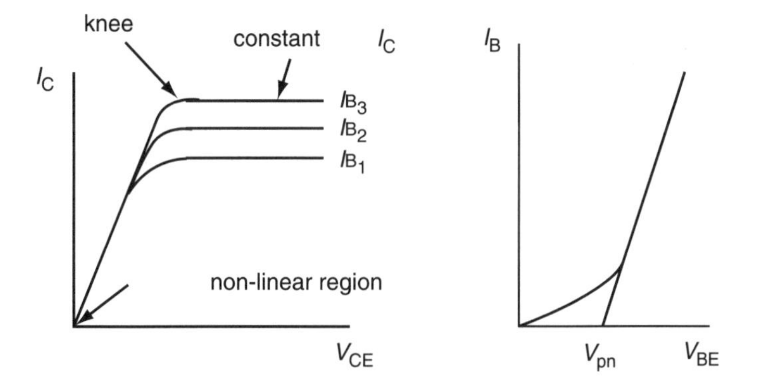

Figure 3.19 Characteristic curves for an npn BJT.

smoothly, then reaches a knee to enter a region of essentially constant I_C. This flat region corresponds to the condition where the depletion region at the base–emitter junction has essentially disappeared. To be useful as a linear amplifier, the transistor must be operated exclusively in the flat region, where the collector current is determined by the base current.

A small current flow into the base controls a much larger current flow into the collector. We can write

$$I_C = \beta I_B = h_{FE} I_B, \tag{3.10}$$

where β is the d.c. current gain and h_{FE} is called the static large signal forward-current transfer ratio, I_C/I_B. From the previous definition of α and the conservation of charge, $I_E = I_C + I_B$, we have

$$\beta = \frac{\alpha}{1 - \alpha}. \tag{3.11}$$

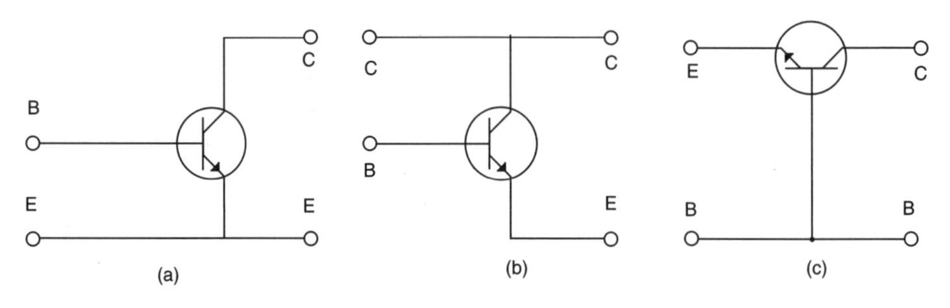

Figure 3.20 BJT basic configurations: (a) common emitter; (b) common collector, and (c) common base.

For example, if $\alpha = 0.99$ then $\beta = 99$ and the transistor is a current amplifying device.

3.6.1.2 *Basic circuit configurations*

For linear BJT operation, the design must ensure that the d.c. bias current and voltage allow the transistor to operate in the linear region of the characteristic curve. The d.c. *operating point* is defined by the values of I_B, I_C, V_{BE}, and V_{CE}. Correct a.c. operation must also be taken into account.

The BJT is a three-terminal device that we will use to form a four-terminal circuit. Small voltage changes in the base–emitter junction will produce large current changes in the collector and emitter, whereas small changes in the collector–emitter voltage have little effect on the base. The result is that the base is always part of the input to a four-terminal network. There are three common configurations: common emitter (CE), common collector (CC) and common base (CB), as shown in Figure 3.20.

3.6.1.3 *BJT self-bias d.c. circuit analysis*

A simple d.c. biasing procedure for a BJT is shown in Figure 3.21. Here two different supplies (V_{CB} and V_{BE}) are used to obtain the correct operating point on the characteristic curve.

In practice, the need for two different supplies is inconvenient and the resulting operating point is not very stable. A more realistic d.c. biasing procedure is shown in Figure 3.22; and overcomes both of these shortcomings.

The voltage supply, V_{CC}, appears across the pair of resistors R_1 and R_2. Consequently, the base terminal of the transistor will be determined by the voltage

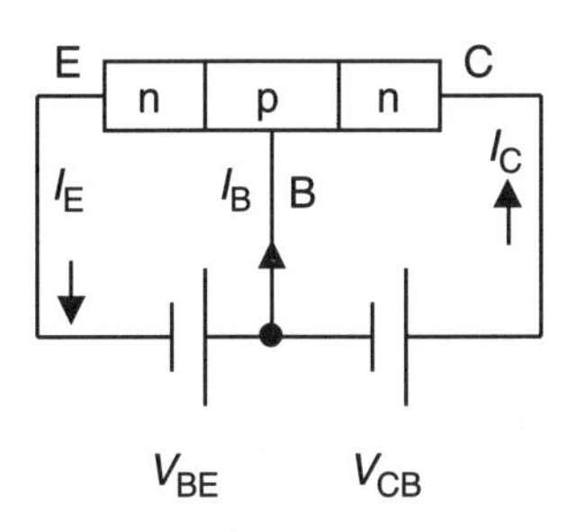

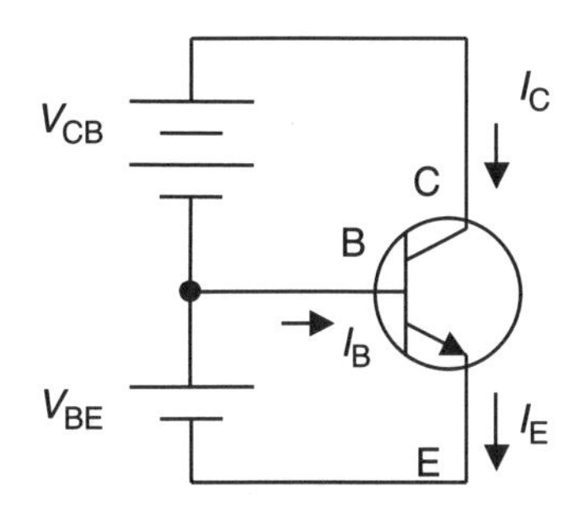

Figure 3.21 Self-biasing an npn BJT.

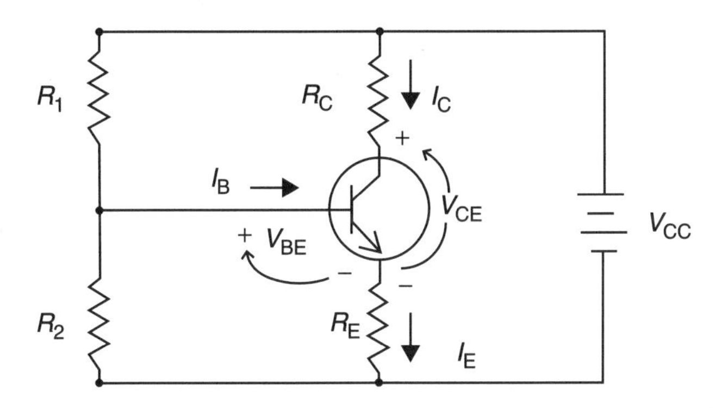

Figure 3.22 A practical self-biasing circuit for an npn BJT.

divider circuit formed by these two resistors:

$$V_{BB} = \frac{R_2}{R_1 + R_2} V_{CC}, \tag{3.12}$$

and the Thévenin equivalent resistance:

$$R_B = R_1 \| R_2. \tag{3.13}$$

This is shown in Figure 3.23.

Analysis:

Methodically, we can arrive at the equivalent voltage and resistance as follows.

Step 1: Thévenin equivalent resistance – disconnect load; zero all supplies

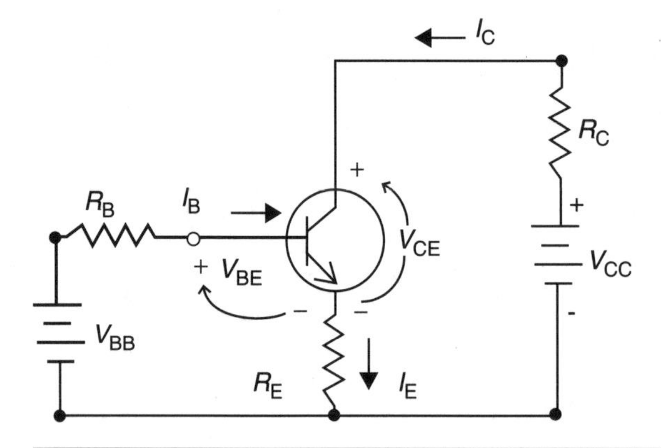

Figure 3.23 Equivalent self-biasing circuit to Figure 3.22.

- disconnect load R_{BE};
- zero all supplies (short voltage source V_{CC}, and open current source).

Step 2: Thévenin equivalent voltage

- disconnect load R_{BE}, leaving the load terminal open-circuit;
- define open-circuit voltage, V_{OC}, across the open load terminal;
- apply any method to solved the circuit problem;
- the Thévenin equivalent voltage is $V_{TH} = V_{OC}$.

Step 3: Thévenin equivalent circuit – reconnect the load
Around the base–emitter circuit

$$V_{BB} = I_B R_B + V_{BE} + I_E R_E$$

$$\text{But} \quad I_E = I_B + I_C \quad \text{and} \quad \frac{I_C}{I_B} = \beta; \quad \therefore I_E = I_B(1+\beta) \tag{3.14}$$

$$\text{So} \quad V_{BB} = I_B R_B + V_{BE} + I_B(1+\beta)R_E. \tag{3.15}$$

Around the collector–emitter circuit

$$V_{CC} = I_C R_C + V_{CE} + I_E R_E$$

$$\text{But} \quad I_E = I_B + I_C \quad \text{and} \quad \frac{I_C}{I_B} = \beta; \quad \therefore I_E = \frac{I_C}{\beta} + I_C = I_C \frac{(\beta+1)}{\beta} \tag{3.16}$$

$$\text{So} \quad V_{CC} = I_C R_C + V_{CE} + I_C \frac{(1+\beta)}{\beta} R_E. \tag{3.17}$$

From these equations

$$I_B = \frac{V_{BB} - V_{BE}}{R_B + I_B(1+\beta)} \tag{3.18}$$

$$\text{and} \quad I_C = \beta I_B$$

$$\text{So} \quad V_{CE} = V_{CC} - I_C\left\{R_C + \frac{(1+\beta)}{\beta}R_E\right\} \tag{3.19}$$

EXAMPLE 3.1

In Figure 3.24,

$$R_1 = 120\,\text{k}\Omega; \quad R_2 = 60\,\text{k}\Omega; \quad R_C = 8\,\text{k}\Omega; \quad R_E = 5\,\text{k}\Omega;$$

$$V_{CC} = 15\,\text{V}; \quad V_\gamma = 0.7\,\text{V}; \quad \beta = 100$$

Determine the d.c. bias point of the transistor in the circuit.

Solution

The supply voltage, V_{CC}, appears across the resistor divider comprising R_1 and R_2. Consequently, the base terminal of the transistor will see the Thévenin equivalent circuit composed of the Thévenin equivalent voltage (self-biasing):

$$V_{BB} = \frac{R_2}{R_1 + R_2}V_{CC} = \frac{60 \times 15}{120 + 60} = 5\,\text{V} \tag{3.19A}$$

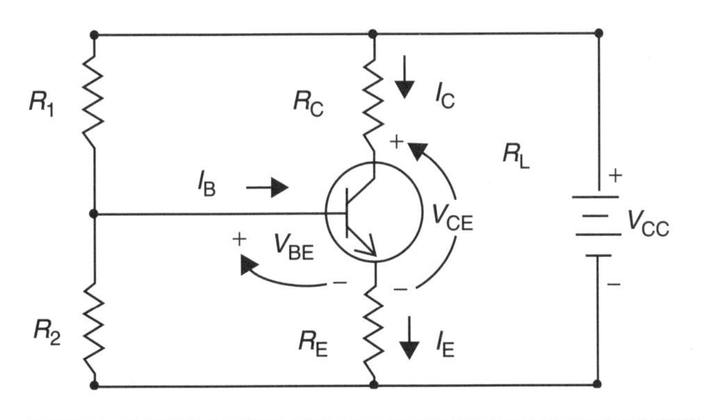

Figure 3.24 Circuit for Example 3.1.

and the Thevenin equivalent resistance:

$$R_B = R_1 \| R_2 = \frac{120\,\text{k} \times 60\,\text{k}}{120\,\text{k} + 60\,\text{k}} = \frac{7200}{180} = 40\,\text{k}\Omega$$

$$\therefore I_B = \frac{V_{BB} - V_{BE}}{R_B + (1+\beta)R_E} = \frac{5 - 0.7}{40\,\text{k} + (101 \times 5\,\text{k})} = \frac{4.3}{545\,\text{k}} = 7.89\,\mu\text{A}$$

$$\therefore I_C = 100 \times 7.89 \times 10^{-6} = 0.789\,\text{mA}$$

$$\therefore V_{CE} = V_{CC} - I_C\left\{R_C + \frac{(1+\beta)}{\beta}R_E\right\} \tag{3.19B}$$

$$= 15 - 0.789 \times 10^{-3}\left\{8 + \frac{101}{100} \times 5\right\} \times 10^3 = 4.704\,\text{V}$$

$$\therefore V_{CE}\big|Q = 4.704\,\text{V}; \quad I_C\big|Q = 0.789\,\text{mA}; \quad I_B\big|Q = 7.89\,\mu\text{A}$$

3.6.1.4 *Small-signal models of the BJT*

A simple transistor model is given by $I_C = h_{fe}I_B$. A more general transistor model capable of describing the family of characteristic curves is given by $I_C = f(I_B, V_{CE})$, where f is a transistor-dependent function.

For a.c. analysis only time changes are important and we may write

$$\frac{dI_C}{dt} = \frac{\partial I_C}{\partial I_B} \cdot \frac{dI_B}{dt} + \frac{\partial I_C}{\partial V_{CE}} \cdot \frac{dV_{CE}}{dt}, \tag{3.20}$$

where the partial derivatives are evaluated at a particular I_B and V_{CE}. The forward current transfer ratio is

$$h_{fe} = \frac{\partial I_C}{\partial I_B}|I_B, V_{CE} \tag{3.20A}$$

and describes the vertical spacing $\Delta I_C/\Delta I_B$ between the curves. The output admittance (inverse resistance) is

$$h_{oe} = \frac{\partial I_C}{\partial V_{CE}}|I_B, V_{CE} \tag{3.20B}$$

and describes the slope $\Delta I_C/\Delta V_{CE}$ of one of the curves as it passes through the operating point.

Using these definitions we may write

$$\frac{dI_C}{dt} = h_{fe} \times \frac{dI_B}{dt} + h_{oe} \times \frac{dV_{CE}}{dt}. \tag{3.21}$$

The input signal, V_{BE}, is also related to I_B and V_{CE}, and a similar argument to the above gives

$$V_{BE} = f(I_B,\ V_{CE})$$

For a.c. analysis only time changes are important and we may write

$$\frac{dV_{BE}}{dt} = \frac{\partial V_{BE}}{\partial I_B} \times \frac{dI_B}{dt} + \frac{\partial V_{BE}}{\partial V_{CE}} \times \frac{dV_{CE}}{dt}, \tag{3.22}$$

where the partial derivatives are evaluated at a particular I_B and V_{CE}. The reverse voltage transfer ratio is

$$h_{re} = \frac{\partial V_{BE}}{\partial V_{CE}} |I_B, V_{CE}. \tag{3.22A}$$

The input impedance is

$$h_{ie} = \frac{\partial V_{BE}}{\partial I_B} |I_B, V_{CE}. \tag{3.22B}$$

Using these definitions we may write

$$\frac{dV_{BE}}{dt} = h_{ie} \times \frac{dI_B}{dt} + h_{re} \times \frac{dV_{CE}}{dt}. \tag{3.23}$$

The differential equations are linear only in the limit of small a.c. signals, where the h-parameters are effectively constant. The h-parameters are in general functions of the variables I_B and V_{CE}. We arbitrarily picked I_B and V_{CE} as our independent variables. We could have picked any two of V_{BE}, V_{CE}, I_B, and I_C. The term h-parameter arises because the current and voltage variables are mixed, hence *hybrid* parameters.

In general the current and voltage signals will have both d.c. and a.c. components. The time derivatives involve only the a.c. component and if we restrict ourselves to sinusoidal a.c. signals, we may replace the time derivatives by

Table 3.1 Hybrid parameters

	Gain	*Resistance*	*Equivalence*
Input	$h_{re} = \frac{\partial V_{BE}}{\partial V_{CE}}$	$h_{ie} = \frac{\partial V_{BE}}{\partial i_B}$	Thévenin
Output	$h_{fe} = \frac{\partial i_C}{\partial i_B}$	$h_{oe} = \frac{\partial i_C}{\partial V_{CE}}$	Norton

the signals themselves (using complex notation). Our hybrid equations become

$$i_C = h_{fe} i_B + h_{oe} v_{CE} \tag{3.24}$$

$$v_{BE} = h_{ie} i_B + h_{re} v_{CE} \tag{3.25}$$

The parameters are summarized in Table 3.1.

The hybrid parameters are often used in manufacturers' specifications of transistors, but there are large variations between samples. Thus one should use the actual measured parameters in any detailed calculation based on this model. Table 3.2 shows typical values of h-parameters.

To illustrate the application of the h-parameter small-signal model discussed so far, consider the transistor amplifier circuit shown in Figure 3.25. The way to analyze this circuit is to treat the d.c. and a.c. equivalent circuits separately.

To obtain the d.c. circuit, the a.c. source is replaced with a short circuit (Figure 3.26) and the d.c. analysis method is used to find the operational Q point.

To obtain the a.c. circuit, the d.c. source is replaced with a short circuit (Figure 3.27). The transistor may now be replaced with its h-parameter small-signal model (Figure 3.28). Its analysis is simplified since the output impedance $h_{oe}{}^{-1}$ is very large and, if the load resistance R_L (in parallel with $h_{oe}{}^{-1}$) is considered to be small (that is, if $R_L h_{re}$ is less than or equal to 0.1), then the impedance $h_{oe}{}^{-1}$ in the model may be ignored.

Table 3.2 Typical values of hybrid parameters

	Parameter	*Minimum*	*Maximum*
Input	h_{ie} (kΩ)	2	4
	h_{re} ($\times 10^{-4}$)		8
	h_{fe}	50	300
Output	h_{oe} (μS)	5	35

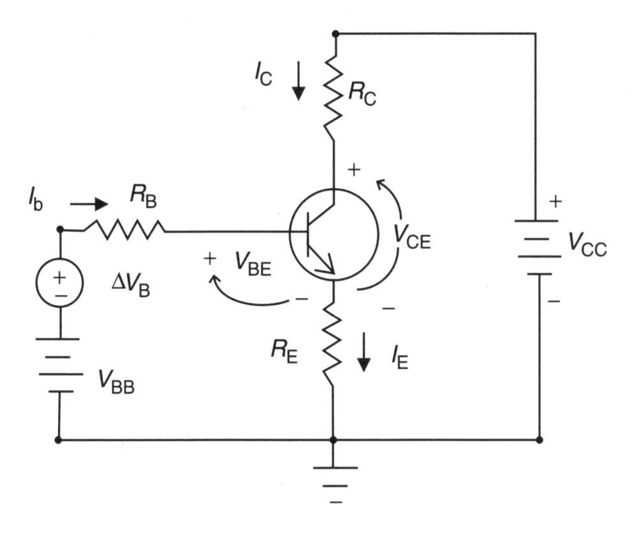

Figure 3.25 Transistor amplifier circuit.

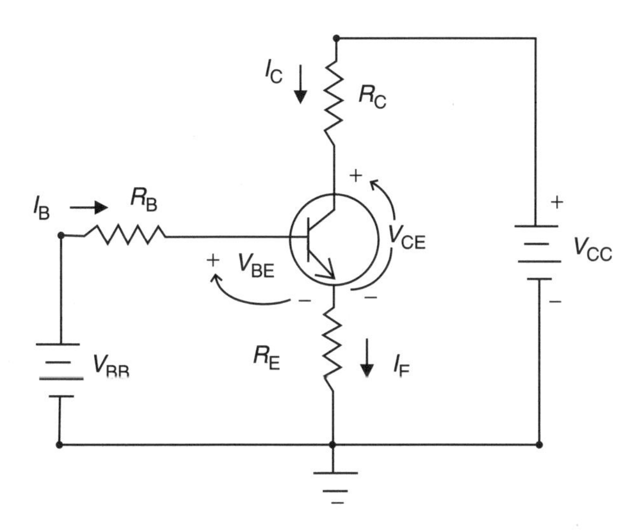

Figure 3.26 Equivalent d.c. circuit to Figure 3.25.

Applying the circuit analysis techniques already covered, noting that the emitter current is now the sum of the base and the collector currents, we can then find the change in the base voltage to be

$$\Delta V_B = \Delta I_B R_B + \Delta I_B h_{ie} + \Delta I_B(1 + h_{fe})R_E \tag{3.26}$$

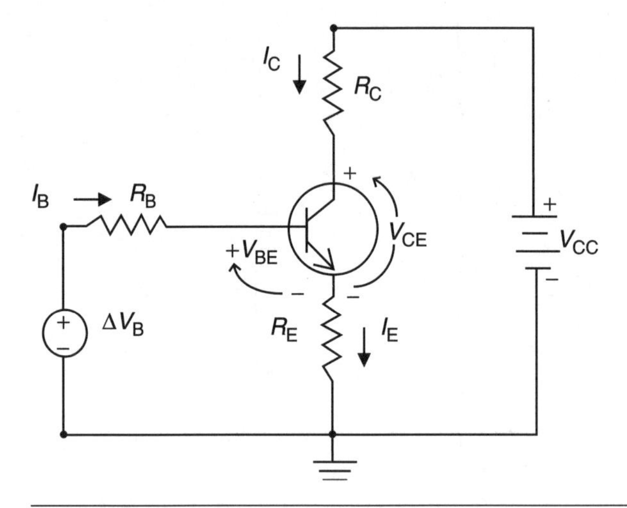

Figure 3.27 Equivalent a.c. circuit to Figure 3.25.

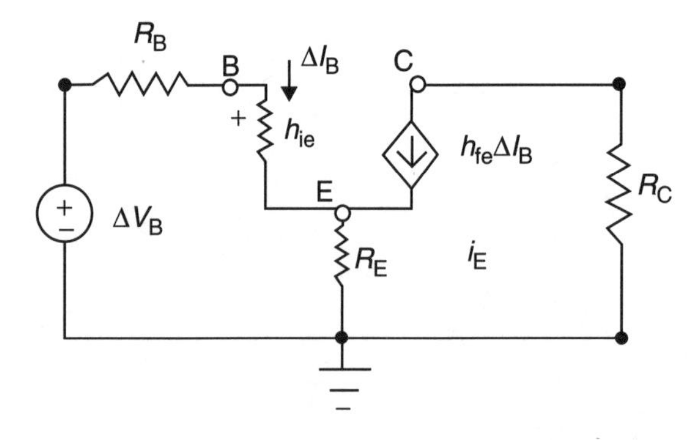

Figure 3.28 Equivalent a.c. circuit small-signal model to Figure 3.25.

which gives

$$\Delta I_B = \frac{\Delta V_B}{R_B + h_{ie} + (1 + h_{fe})R_E} \tag{3.27}$$

and

$$\Delta V_C = -\Delta I_C R_C. \tag{3.28}$$

We define $h_{fe} = \beta = I_C/I_B$

$$\therefore I_C = I_B h_{fe} \tag{3.29}$$

$$\Delta V_C = -\Delta I_C R_C = -h_{fe} \Delta I_B R_C = \frac{-h_{fe} R_C}{R_B + h_{ie} + (1 + h_{fe}) R_E} \Delta V_B. \qquad (3.30)$$

The open-loop gain is therefore given as

$$\mu = \frac{\Delta V_C}{\Delta V_B} = \frac{-h_{fe} R_C}{R_B + h_{ie} + (1 + h_{fe}) R_E} \qquad (3.31)$$

EXAMPLE 3.2

Determine the a.c. open-loop gain for the transistor amplifier shown in Figure 3.28, where,

$$R_B = 80\,\text{k}\Omega; \quad R_E = 80\,\Omega; \quad R_C = 1.5\,\text{k}\Omega;$$

$$h_{fe} = 200; \quad h_{fe} = 3;$$

Solution

$$\mu = \frac{-200 \times 1500}{80000 + 3 + (201 \times 80)} = -3.122 \qquad (3.31\text{A})$$

3.6.2 The junction field-effect transistor (JFET)

Junction field-effect transistors (JFETs) are unipolar semiconductor devices that control their current flow by means of electrical effects. Unipolar semiconductor devices conduct current by just one type of charge carrier: either holes or electrons depending on the type of semiconductor being used. The JFET has three external connections that are called the source (S), the gate (G), and the drain (D) as shown in Figure 3.29. A channel runs between the source and the drain connections as a solid piece of n- or p-type semiconductor material. The source connection provides the source of the charge carriers for the channel current. The drain connection provides the point where the charge carriers are drained or removed from the channel.

In an n-channel JFET, when a negative voltage is applied to the source terminal and a positive voltage is applied to the drain terminal, the device will conduct current by means of electron carriers flowing from source to drain as shown in Figure 3.30. In a p-channel JFET, when a positive voltage is applied to the source terminal and a negative voltage is applied to the drain terminal, holes drift from the source to drain but current flows in the opposite direction.

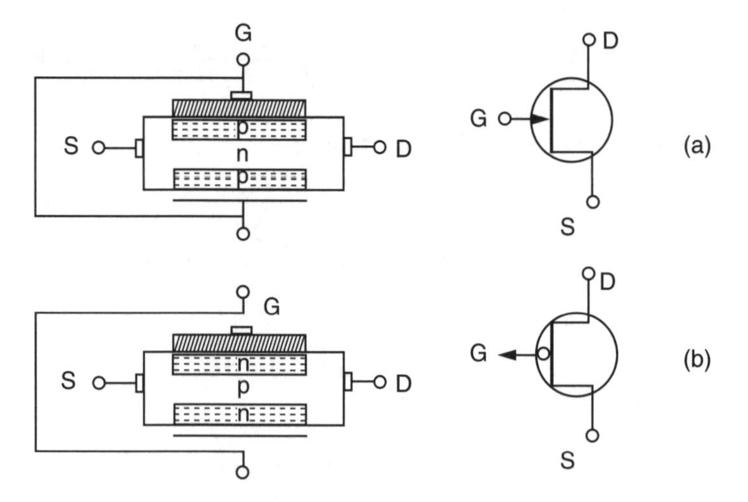

Figure 3.29 JFET construction and symbol: (a) n-channel; (b) p-channel.

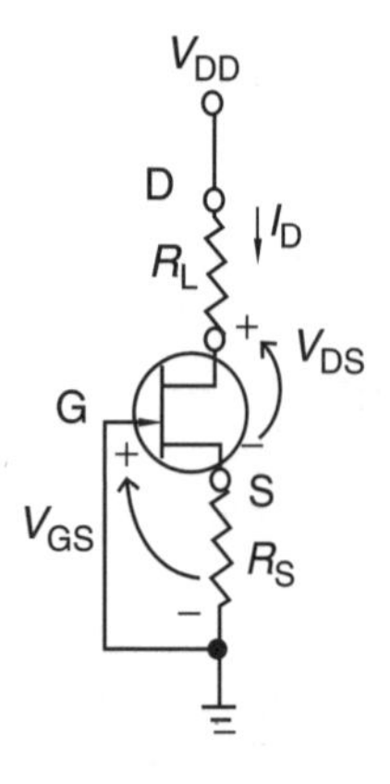

Figure 3.30 n-channel JFET.

3.6.2.1 *n-channel JFET bias voltages and currents*

The normal operating conditions for a JFET are:

- the gate terminal must be reverse-biased;
- the d.c. source, V_{DS}, is connected between the source and drain terminals in such a way that the source–gate junction must be reverse biased.

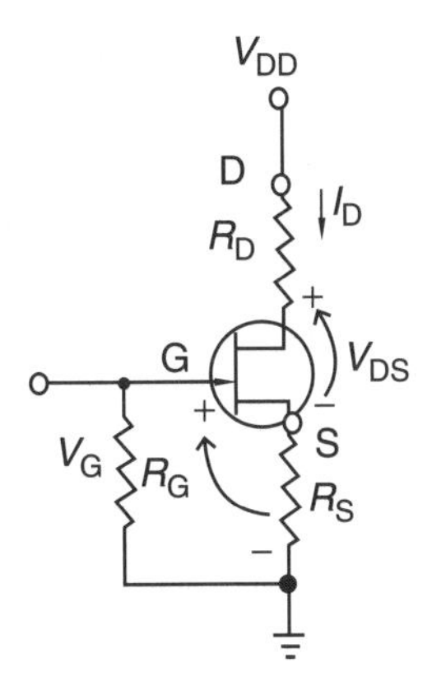

Figure 3.31 Biasing an n-channel JFET.

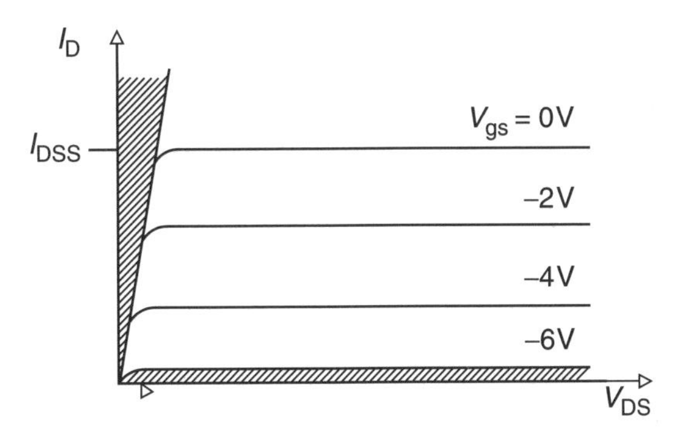

Figure 3.32 n-channel JFET drain characteristic.

Figure 3.31 shows the biasing of an n-channel JFET. A very important characteristic of the biasing of JFET is that maximum current, I_D flows from the source to the drain terminal if the reverse bias voltage, V_{GS}, is zero. By increasing the value of the reverse bias voltage, the depletion region increases accordingly. Figure 3.32 shows the drain characteristic curves for n-channel JFET.

3.6.3 The metal-oxide semiconductor field-effect transistor (MOSFET)

We have briefly mentioned that an FET is an active device in which the current is controlled by varying the shape of the conducting volume. Another FET technology utilizing metal-oxide semiconductor (MOS) technology is the metal-oxide semiconductor field effect transistor (MOSFET). Like the JFET, the MOSFET uses a reverse-biased voltage to control a current flow through a solid piece of semiconductor material.

The MOSFET, is similar to the JFET but exhibits an even larger resistive input impedance due to the thin layer of silicon dioxide that is used to insulate the gate from the semiconductor channel (hence the MOSFET's alternative name: the insulated gate FET). This insulating layer forms a capacitive coupling between the gate and the body of the transistor. The consequent lack of an internal d.c. connection to the gate makes the device more versatile than the JFET, but it also means that the insulating material of the capacitor can be easily damaged by the internal discharge of static charge developed during normal handling.

The MOSFET is widely used in large-scale digital integrated circuits where its high input impedance can result in very low power consumption per component. Many of these circuits feature bipolar transistor connections to the external terminals, thereby making the devices less susceptible to damage.

The MOSFET comes in four basic types: n-channel enhancement; n-channel depletion; p-channel enhancement; and p-channel depletion. The configuration of an n-channel depletion MOSFET is shown in Figure 3.33(a). Its operation is similar to the n-channel JFET discussed previously: a negative voltage placed on the gate generates a charge depleted region in the n-type material next to the gate, thereby reducing the area of the conduction channel between the drain and source. However, the mechanism by which the depletion region is formed is different from the JFET. As the gate is made negative with respect to the source, more positive carriers from the p-type material are drawn into the n-channel, where

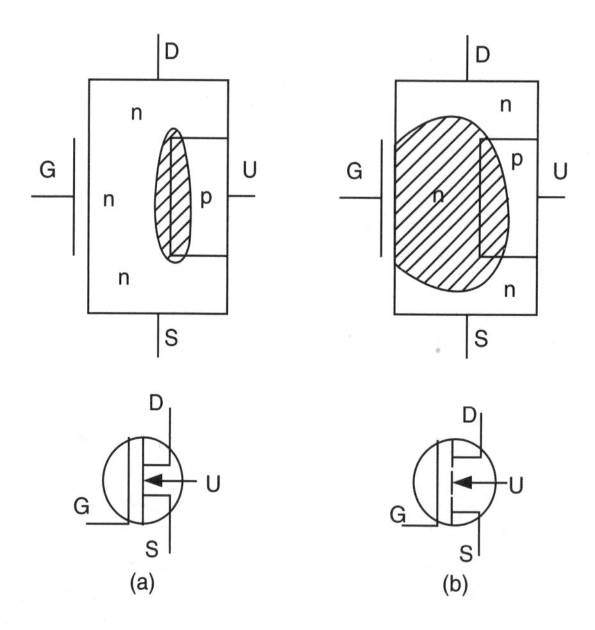

Figure 3.33 n-channel MOSFET operation and symbol: (a) depletion type; (b) enhancement type.

they combine with and eliminate the free negative charges. This action enlarges the depletion region towards the gate, reducing the area of the n-channel and thereby lowering the conductivity between the drain and source. For negative gate–source voltages, the observed effect is similar to a JFET, and g_m is also about the same size.

However, since the MOSFET gate is insulated from the channel, positive gate–source voltages may also be applied without losing the FET effect. Depending on the construction details, the application of a positive gate–source voltage to a depletion-type MOSFET can repel the minority positive carriers in the depleted portion of the n-channel back into the p-type material, thereby enlarging the channel and reducing the resistance. If the device exhibits this behaviour, it is known as an enhancement-depletion MOSFET.

A strictly enhancement MOSFET results from the configuration shown in Figure 3.33(b). Below some threshold of positive gate–source voltage, the connecting channel of n-type material between the drain and source is completely blocked by the depletion region generated by the p–n junction. As the gate–source voltage is made more positive, the minority positive carriers are repelled back into the p-type material, leaving free negative charges behind. The effect is to shrink the depletion region and increase the conductivity between the drain and source.

3.6.3.1 *The enhancement MOSFET*

Figure 3.34 shows the basic structure and symbol for enhancement-type MOSFET. The channel between the source and drain terminals is interrupted by a section of substrate material. A gate voltage is required on this device for current to flow from the source to drain terminal.

Figure 3.35 shows a biased n-channel enhancement MOSFET. Both d.c. voltages, V_{GS} and V_{DS}, must be forward biased in order to enhance the flow of current from the source to the drain terminal.

When $V_{GS} = 0$, no current flows from the source to the drain terminal and no enhancement region occurs. The greater V_{GS}, the more positive the gate terminal becomes, causing holes to move away from the gate area and/or electrons to move into the gate area. Consequently, an inversion layer is formed that completes a conductive channel between the source and the drain. The more positive the gate becomes, the wider the inversion layer becomes, which enhances the flow of current from the source to the drain terminal. Figure 3.36 shows the drain characteristic curves for an n-channel enhancement-type MOSFET.

The MOSFET transistor has four major regions (modes) of operation (Table 3.3): cutoff, ohmic (or triode), saturation and breakdown. In the non-saturated (or triode) region, the voltage drop across the drain–source terminals approaches zero volts as the magnitude of the voltage drop across the gate–source terminals approaches $V_{DD} - V_{SS}$. For example, in a 5 V system, the drain–source

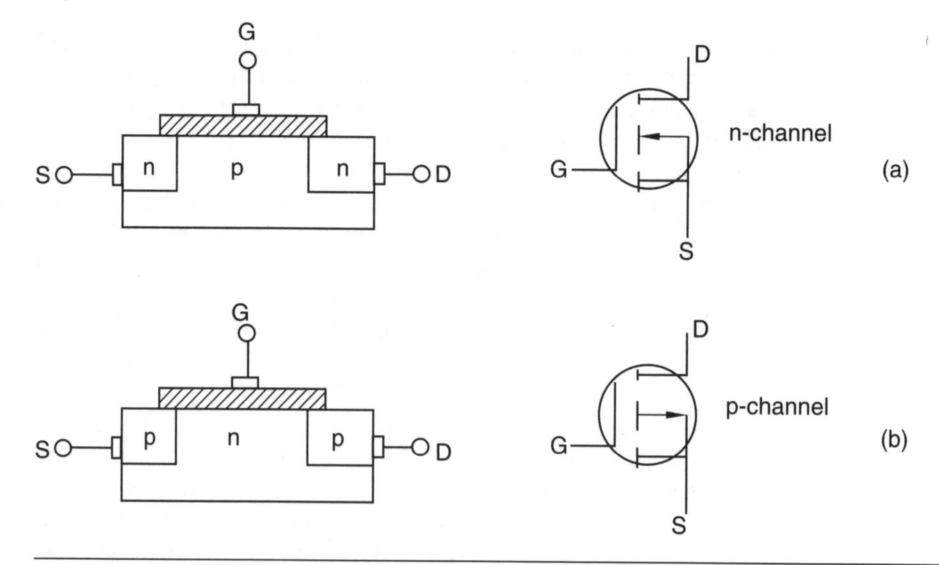

Figure 3.34 Enhancement-type MOSFET construction and symbol: (a) n-channel; (b) p-channel.

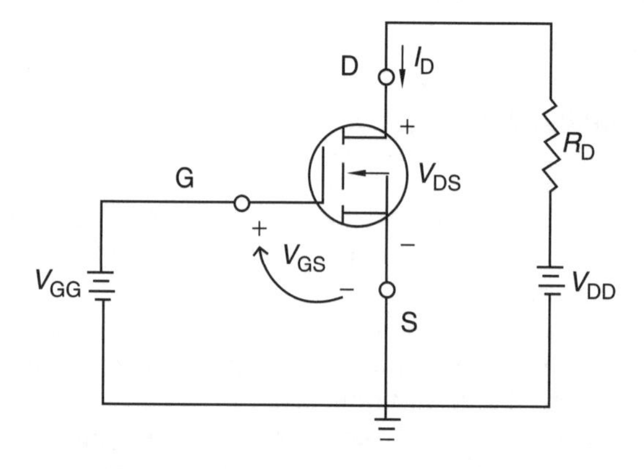

Figure 3.35 n-channel enhancement-type MOSFET biasing.

voltage approaches 0 V as the magnitude of the gate–source voltage drop approaches 5 V. In the cutoff region, the drain-to-source current, I_{DS}, approaches 0 A (i.e. the drain–source resistance approaches infinity: an open circuit). Hence, the drain and source terminals of a MOSFET transistor can be treated as an almost ideal switch alternating between the off (cutoff) and on (non-saturated) modes of operation.

In this case, we use the symbol V_T to represent the threshold voltage. The same equations are valid for p-channel devices if the subscript $_{SG}$ is substituted for $_{GS}$ and $_{SD}$ is substituted for $_{DS}$.

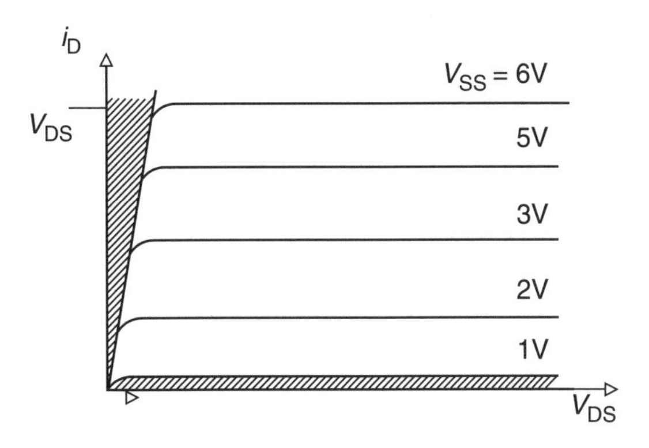

Figure 3.36 n-channel enhancement-type MOSFET drain characteristic.

Table 3.3 Regions for enhancement MOSFET Q-point calculation

Region	*Equation*
Cutoff	$V_{GS} < V_T$
Ohmic or triode	$V_{DS} < 0.25(v_{GS} - V_T), \quad v_{GS} > V_T$
	$R_{DS} = \dfrac{V_T^2}{2\, I_{DSS}(v_{GS} - V_T)}$
	$i_D \approx \dfrac{v_{DS}}{R_{DS}}$
Saturation	$V_{DS} \geq v_{GS} - V_T, \quad v_{GS} > V_T$
	$i_D = \dfrac{I_{DSS}}{V_T^2}(v_{GS} - V_T)^2 = k(v_{GS} - V_T)^2$
Breakdown	$V_{DS} > V_B$

EXAMPLE 3.3

See website for downloadable MATLAB code to solve this problem

Determine the resistance R_S in the circuit shown in Figure 3.37, where

$$R_1 = 3\,\text{M}\Omega; \quad R_2 = 2\,\text{M}\Omega; \quad R_D = 10\,\text{k}\Omega;$$

$$V_{DD} = 40\,\text{V}; \quad V_{DS} = 8\,\text{V}; \quad V_T = 4\,\text{V}; \quad I_{DSS} = 6\,\text{mA}$$

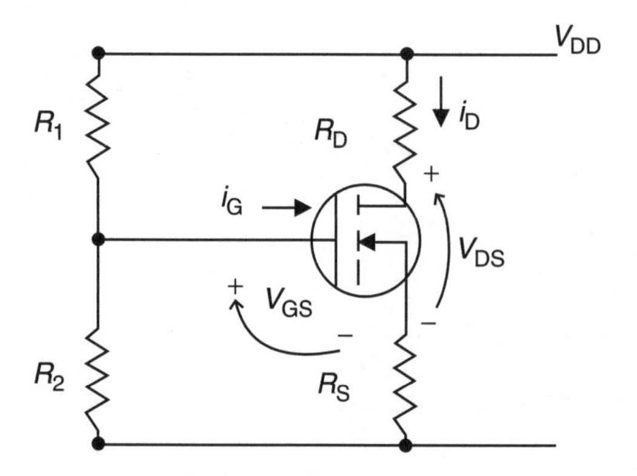

Figure 3.37 Self-biasing circuit for Example 3.3.

Solution

The voltage supply, V_{DD}, appears across the potential divider comprising R_1 and R_2. Consequently, the base terminal of the transistor will see the Thévenin equivalent circuit composed of the Thévenin equivalent voltage (self-biasing):

$$V_{GG} = \frac{R_2}{R_1 + R_2} V_{DD} = \frac{2 \times 40}{3 + 2} = 16\,\text{V} \tag{3.32}$$

and the Thévenin equivalent resistance:

$$R_G = R_1 || R_2 = \frac{3 \times 2}{3 + 2} = \frac{6}{5} = 1.2\,\text{M}\Omega \tag{3.33}$$

The gate circuit equation is:

$$V_{GG} = V_{GS} + i_G R_G + I_D R_S \tag{3.33A}$$

and since the MOSFET input resistance is infinite, $i_G = 0$, so

$$V_{GG} = V_{GS} + i_D R_S. \tag{3.34}$$

The drain circuit equation is:

$$V_{DD} = V_{DS} + I_D R_D + I_D R_S \tag{3.35}$$

$$I_D = \frac{I_{DSS}}{V_T^2}(v_{GS} - V_T)^2 = k(v_{GS} - V_T)^2. \tag{3.36}$$

From Equation 3.3A,

$$I_D R_S = V_{GG} - V_{GS}. \tag{3.37}$$

Substituting Equations 3.36 and 3.37 into Equation 3.35:

$$V_{DD} = i_D R_D + V_{DS} + V_{GG} - V_{GS}. \tag{3.37A}$$

Hence,

$$V_{DD} = I_{DSS}\left(\frac{v_{GS}}{V_r} - 1\right)^2 R_D + V_{DS} + V_{GG} - V_{GS}$$

$$40 = 6 \times 10 \times \left(\frac{v_{GS}}{4} - 1\right)^2 + 16 + 8 - V_{GS}$$

$$40 = 60\left(\frac{v_{GS}^2}{16} - \frac{v_{GS}}{2} + 1\right) + 24 - V_{GS}$$

$$40 = 3.75v_{GS}^2 - 30V_{GS} + 60 + 24 - V_{GS}$$

$$3.75v_{GS}^2 - 31V_{GS} + 44 = 0 \tag{3.37B}$$

$$0.375v_{GS}^2 - 3.1V_{GS} + 4.4 = 0$$

$$V_{GS} = \frac{3.1 \pm \sqrt{(3.1^2 - 4 \times 0.375 \times 4.4)}}{2 \times 0.375} = \frac{3.1 \pm 1.73}{0.7} = 6.4466 \text{ or } 1.96$$

$$V_{DS} \geq v_{GS} - V_T = 6.4466 - 4 = 2.9; \quad \text{hence} \quad 6.4466 \text{ is okay.}$$

$$i_D = I_{DSS}\left(\frac{v_{GS}}{V_T} - 1\right)^2 = 6 \times 10^{-3} \times \left(\frac{6.4466}{4} - 1\right)^2 = 2.245\,\text{mA}.$$

From Equation 3.35:

$$R_S = \frac{V_{DD} - i_D R_D - V_{DS}}{i_D} = \frac{40 - 2.225 \times 10^{-3} \times 10 \times 10^3 - 8}{2.225 \times 10^{-3}} = 4.25\ \text{k}\Omega. \tag{3.37C}$$

3.6.3.2 *The depletion MOSFET*

Figure 3.38 shows the basic structure and symbol for a depletion-type MOSFET. The channel between the source and drain terminals is a straight-through piece of semiconductor material.

In an n-channel depletion-type MOSFET, the electrons in the n-type material carry the charges, but in a p-channel depletion-type MOSFET, the charges are carried by holes through the p–n type material. The gate is separated from the channel by a very thin layer of silicon oxide, which is a very good insulating material. Consequently, no current flows between the gate terminal and the channel. However, as in JFET, applying a voltage to the gate of the MOSFET controls the amount of current that flows through the channel.

Figure 3.39 shows the n-type depletion MOSFET mode of operation. The normal operating conditions are as follows:

- the source–drain connection must be forward biased since the charge carriers flow from source to drain.
- the gate–source connection must be reverse biased to give the depletion mode operation.

When the bias voltage, V_{GS}, is zero, the maximum current I_D flows from the source to the drain. When the bias voltage, V_{GS}, is increased, the depletion region increased resulting in a smaller I_D flowing from the source. A very large

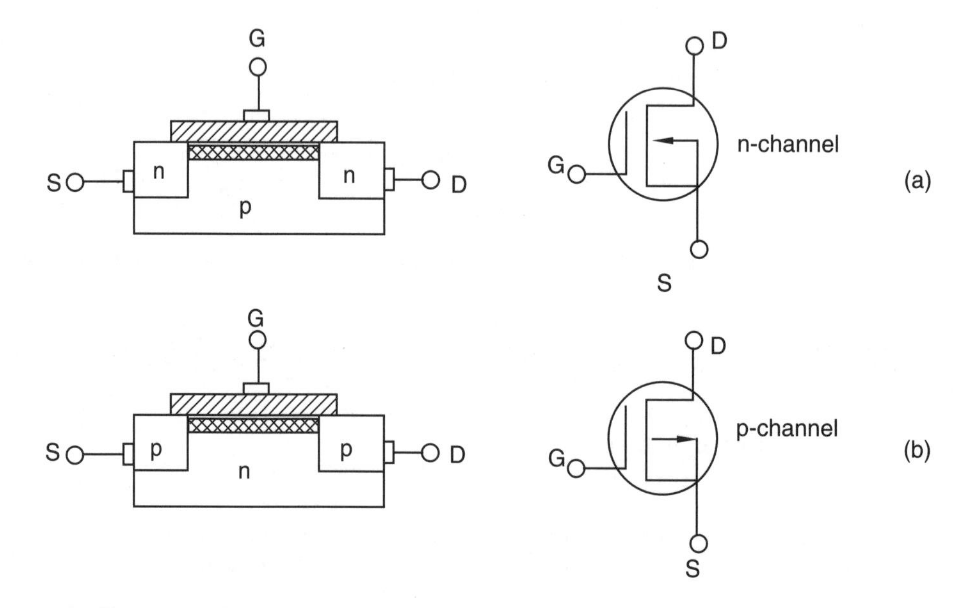

Figure 3.38 Depletion-type MOSFET construction and symbol: (a) n-channel; (b) p-channel.

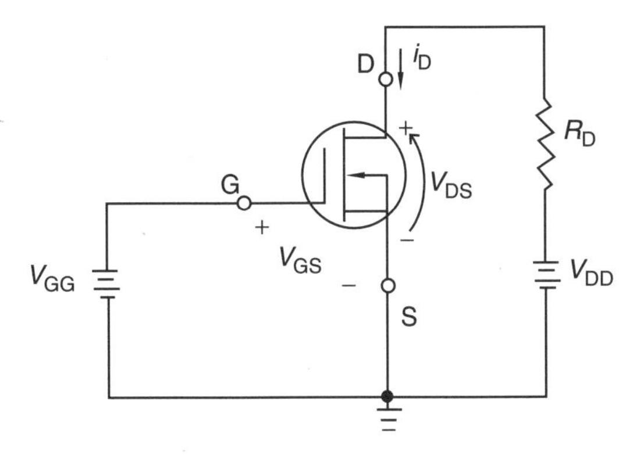

Figure 3.39 n-channel depletion-type MOSFET biasing.

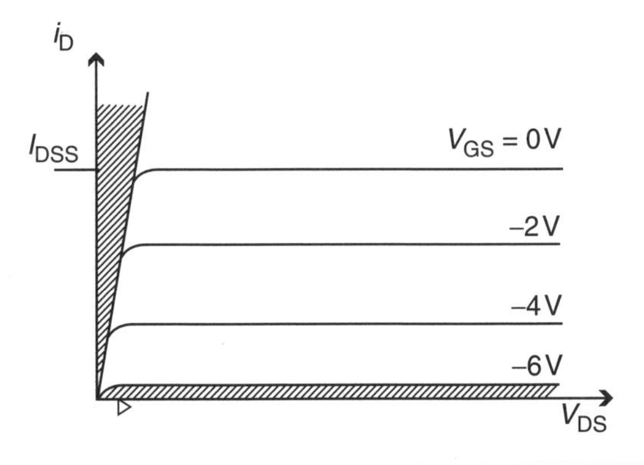

Figure 3.40 n-channel depletion-type MOSFET drain characteristic.

bias voltage, V_{GS}, can cause no current to flow from the source to the drain. Figure 3.40 shows the drain characteristic curves for an n-channel depletion-type MOSFET.

Again there are four regions of operation with the equations shown in Table 3.4. Here V_P represents the depletion type MOSFET threshold voltage. The same equations are valid for p-channel devices if the subscript $_{SG}$ is substituted for $_{GS}$ and $_{SD}$ is substituted for $_{DS}$.

3.6.3.3 *Rules for connections in MOSFETs*

Rule 1: The bulk (or substrate) connections for MOSFETs are normally connected to a power supply rail. p-channel bulk connections are typically tied to the V_{DD} rail and n-channel bulk connections are typically tied to the V_{SS} rail.

Table 3.4 Regions for depletion MOSFET Q-point calculation

Region	*Equation*
Cutoff	$V_{GS} < -V_P$
Ohmic or triode	$V_{DS} < 0.25(v_{GS} + V_P), \quad v_{GS} > -V_P$
	$R_{DS} = \dfrac{V_P{}^2}{2\,I_{DSS}(v_{GS} + V_P)}$
	$i_D \approx \dfrac{v_{DS}}{R_{DS}}$
Saturation	$V_{DS} \geq v_{GS} + V_P, \quad v_{GS} > -V_P$
	$i_D = \dfrac{I_{DSS}}{V_P{}^2}(v_{GS} + V_P)^2 = k(v_{GS} + V_P)^2$
Breakdown	$V_{DS} > V_B$

Rule 2: For proper operation as an ideal switch (see later), the p-channel MOSFET must be connected to the most positive voltage rail while the n-channel MOSFET must be connected to the most negative voltage rail.

Figure 3.41(a) shows the substrate connected to the power supply. Figure 3.41(b) shows the symbols for when the source–bulk connection has been shorted to ground. These symbols are most commonly used in documenting analog CMOS circuits. Figure 3.41(c) shows the schematic symbols for MOSFETs, and substrate connection is not indicated. Notice too that the

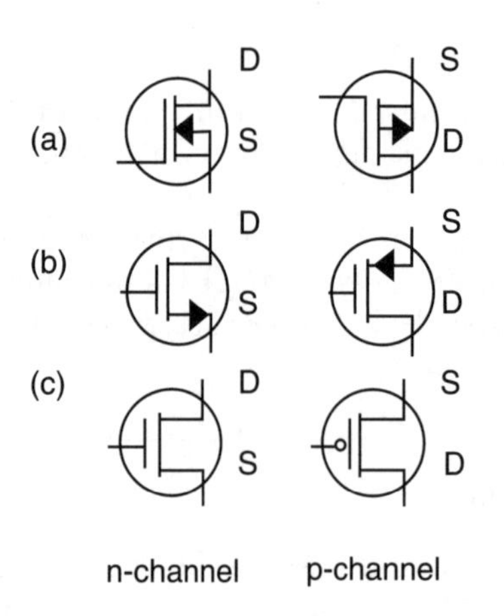

Figure 3.41 FET schematic symbols.

gates for the p- and n-channel devices differ. The p-channel device is identified by a 'bubble' on the gate input. The presence or absence of a 'bubble' on the gate input is used to signify what logic level is used to turn on the transistor. The presence of a 'bubble' on the p-channel device indicates that this device should have a logic low applied to the gate input to turn on the transistor while the absence of a 'bubble' on the n-channel device indicates that this device should have a logic high applied to the input to turn on the device. These schematic symbols are most commonly used when documenting CMOS logic circuits. The bulk-substrate connections are almost always connected to the power supply rails using MOSFET Rule 1.

3.6.4 The MOSFET small-signal model

The input signal I_D is related to V_{GS} and V_{DS}, and hence,

$$I_D = I_D(V_{GS}, V_{DS}) \tag{3.38}$$

For a.c. analysis, only time changes are important and we may write

$$\frac{dI_D}{dt} = \frac{\partial I_D}{\partial V_{GS}} \cdot \frac{dV_{GS}}{dt} + \frac{\partial I_D}{\partial V_{DS}} \cdot \frac{dV_{DS}}{dt} \tag{3.39}$$

The control current source is

$$g_m = \frac{\partial I_D}{\partial V_{GS}} |I_G, V_{DS}. \tag{3.39A}$$

The second term is small and can be neglected.
Using these definitions we may write

$$\frac{dI_D}{dt} = g_m \frac{dV_{GS}}{dt} \tag{3.40}$$

In general, the current and voltage signals will have both d.c. and a.c. components. The time derivatives involve only the a.c. component and if we restrict ourselves to sinusoidal a.c. signals, we may replace the time derivatives by the signals themselves (using complex notation). Our hybrid equations become

$$i_D = g_m \times v_{GS}, \tag{3.41}$$

where g_m is the controlled current source.

From the universal equation,

$$i_D = k(v_{GS} - V_T)^2; \quad k = \frac{I_{DSS}}{V_T^{\,2}} \tag{3.42}$$

$$g_m = \frac{\partial I_D}{\partial V_{GS}} |I_G, V_{DS} = \frac{\partial \lfloor k(V_{GS} - V_T)^2 \rfloor}{\partial V_{GS}} = 2k(V_{GS} - V_T). \tag{3.43}$$

From Equation 3.42:

$$(V_{GS} - V_T)^2 = \frac{I_D}{k} \text{ and } (V_{GS} - V_T) = \sqrt{\frac{I_D}{k}}, \tag{3.43A}$$

$$g_m = 2k\sqrt{\frac{I_D}{k}} = 2\sqrt{kI_D} = \frac{2\sqrt{I_{DSS} I_D}}{V_T}. \tag{3.44}$$

The MOSFET transistor small-signal model is therefore as shown in Figure 3.42.

EXAMPLE 3.4

See website for downloadable MATLAB code to solve this problem

Determine at the quiescent point, the value of g_m, for a MOSFET, where:

$$V_{DS} = 8\,\text{V}$$
$$V_{DS} = 6.9\,\text{V}$$
$$V_T = 4\,\text{V}$$
$$I_{DSS} = 6\,\text{mA}$$

Solution

$$i_D = k(v_{GS} - V_T)^2; \quad k = \frac{I_{DSS}}{V_T^{\,2}}$$

$$k = \frac{I_{DSS}}{V_T^{\,2}} = \frac{6 \times 10^{-3}}{4^2} = 0.375 \text{ mA/V}^2 \tag{3.44A}$$

$$i_D = k(v_{GS} - V_T)^2 = 0.375(6.9 - 4)^2 = 3.15 \text{ mA}$$

$$g_m = 2\sqrt{kI_D} = 2\sqrt{0.375 \times 3.15} = 2.173 \text{ mA/V}$$

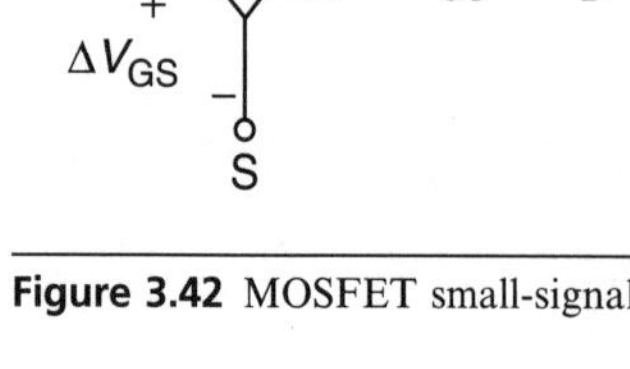

D

G

$g_m \; \Delta V_{GS} = \Delta I_D$

+

ΔV_{GS}

–

S

Figure 3.42 MOSFET small-signal model.

3.6.5 Transistor gate and switch circuits

In mechatronics applications, we are more interested in the transistor functioning as a gate or switch than as an amplifier (although the last application is also important). Here the device is either 'on' or 'off', and so circuits are easier to handle and do not involve the rigorous analyses that we have discussed in this chapter. We will first look at simple diode gates, then BJTs, followed by MOSFETs, and finish up with CMOS.

3.6.5.1 *Diode gates*

In the section on diodes, we noted that current flows through a diode device when it is forward biased and it is an open circuit when reverse biased. For the diode circuit shown in Figure 3.43, when $V_1 = 5\,\mathrm{V}$, diode D_1 will conduct and when $V_1 = 0\,\mathrm{V}$, it will not conduct. The same argument holds for diode D_2. If $V_1 = 5\,\mathrm{V}$, and $V_2 = 5\,\mathrm{V}$, both diodes will conduct. If $V_1 = 0\,\mathrm{V}$, and $V_2 = 0\,\mathrm{V}$, both diodes will not conduct. Consequently, the diodes are switching, but they are not used in practical mechatronics circuits; transistors are preferred in practical situations.

3.6.5.2 *BJT gates*

In analyzing small-signal models for the BJT, the cut-off region becomes important when dealing with the family of *I–V* characteristic curves. Within this region, virtually no current flows. However, when sufficient current is applied to the base of the transistor, saturation occurs and a reasonable amount of the collector current now flows. This characteristic is useful in realizing electronic gates and switches.

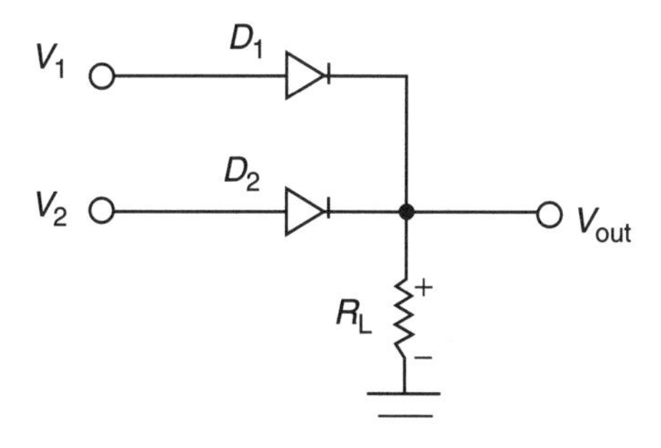

Figure 3.43 The diode OR gate.

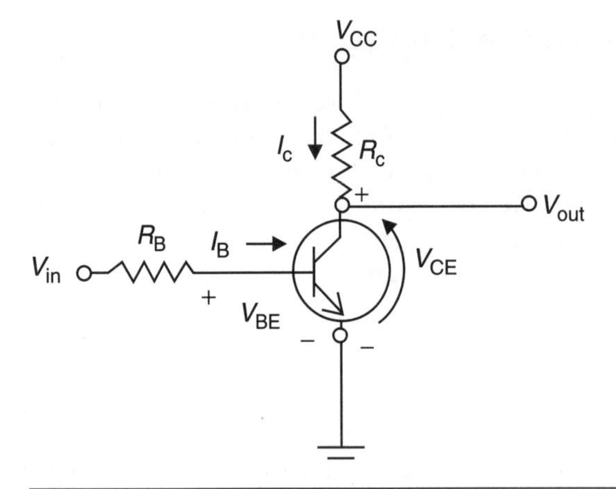

Figure 3.44 The BJT inverter.

A simple BJT switch (an inverter, shown in Figure 3.44) is analyzed by superimposing the load line equation on the *I–V* characteristic curves shown in Figure 3.45.

The base equation is:

$$V_{in} = i_B R_B + V_\gamma. \tag{3.45}$$

The collector equation is:

$$V_{CC} = i_C R_C + V_{CE} \tag{3.46}$$

and

$$V_{out} = V_{CE} \tag{3.47}$$

$$V_{out} = V_{CC} - i_C R_C \tag{3.47A}$$

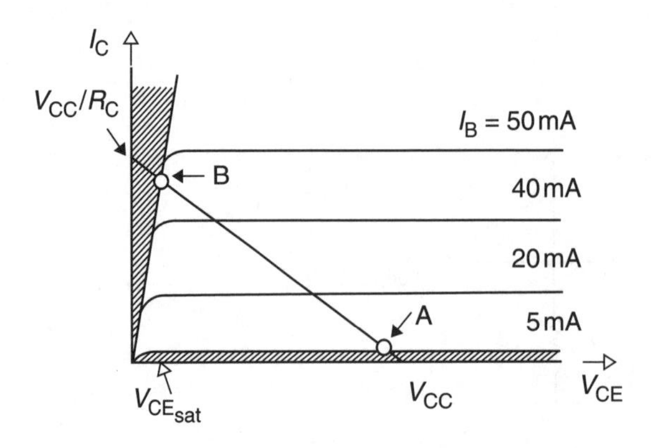

Figure 3.45 BJT characteristic curves with load line.

From the graph shown in Figure 3.45, when V_{in} is low (0 V), the transistor is in the cut-off region and little current flows; and consequently,

$$V_{out} = V_{CC} - i_C R_C \approx V_{CC}, \tag{3.47B}$$

since the second term is almost zero.

Hence, V_{out} is high (V_{CC}).

When V_{in} is high (V_{CC}), using the equation for the base circuit:

$$i_B = \frac{(5 - 0.2)}{8000} \approx 50 \text{ mA} \tag{3.47C}$$

In this case the transistor is in saturation region. Hence,

$$V_{out} = V_{CE/SAT} \approx 0.2 \text{ V or low.} \tag{3.47D}$$

We have now shown that the transistor circuit shown in Figure 3.44 is a BJT inverter (NOT gate).

3.6.5.3 *TTL gates*

The transistor–transistor logic (TTL) gate family is one that uses 5 V for logic level 1 (or high) and 0 V for logic level 0, as in the inverter example of the previous section. These types of logic gates can be combined in series or in parallel to form more complex circuits.

EXAMPLE 3.5

Complete the truth table (Table 3.5) for the TTL NAND gate shown in Figure 3.46 for the following parameters:

$$R_1 = 6\,\text{k}\Omega; \quad R_2 = R_3 = 3\,\text{k}\Omega; \quad R_4 = 2\,\text{k}\Omega;$$

$$V_{CC} = 5\,\text{V}; \quad V_{BEon} = 0.7\,\text{V}; \quad V_{TCEsat} = 0.2\,\text{V};$$

Table 3.5 Truth table for Example 3.5

V_1	V_2	Q_2	Q_3	V_{out}
0 V	0 V			
0 V	5 V			
5 V	0 V			
5 V	5 V			

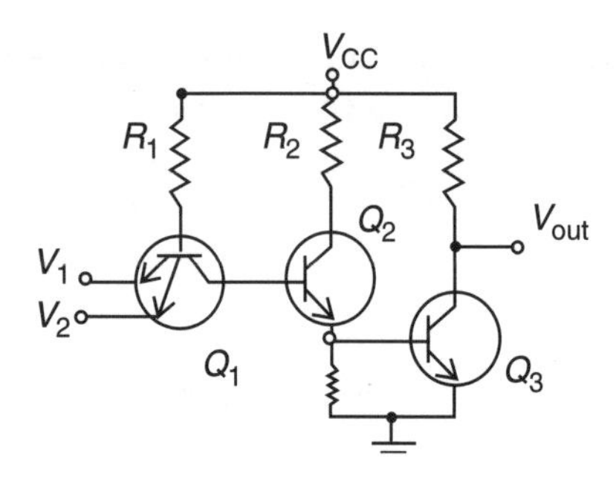

Figure 3.46 BJT NAND gate.

Solution

Case 1. When $V_1 = 0$; $V_2 = 0$: With the emitters of Q_1 connected to ground and the base of Q_1 at 5 V (high), then by the inverter (NOT) analysis, the collector of Q_1 is low, and consequently Q_1 is 'on'. The collector of Q_1 is the base of Q_2, which is now low. A low base of Q_2 means a high collector of Q_2 and a low emitter of Q_2. Since the emitter is high, Q_2 is 'off'. Since the base of Q_3 is low, the collector of Q_3 is high and a low base means Q_3 is 'off'. Since V_{out} is taken at the collector, which is high, V_{out} is high (5 V).

Case 2. When $V_1 = 0$; $V_2 = 5$ V: With the emitter of Q_1 connected to ground and the base of Q_1 at 5 V (high), then by the inverter (NOT) analysis, one base–emitter junction is forward-biased and the other is reverse-biased. The forward-biased junction means that current flows through Q_1, and hence Q_1 is 'on'. When Q_1 is 'on' its collector is low. The analysis proceeds in exactly the same way as for Case 1. Consequently, V_{out} is high (5 V).

Case 3. When $V_1 = 5$ V; $V_2 = 0$ V: This is symmetrical to Case 2, and hence, the same result is obtained, in which V_{out} is high (5 V).

Case 4. When $V_1 = 5$ V; $V_2 = 5$ V: In this case both base–emitter junctions in Q_1 are reverse-biased so Q_I is 'off'. Its collector is high and this corresponds to the base of Q_2. A high base of Q_2 means Q_2 is 'on'. The high emitter of Q_2, corresponds to a high base of Q_3. The collector of Q_3 is therefore low. This means V_{out} is low.

This last case is obtained by considering the current and output voltage as follows:

$$V_{CC} = I_{C3}R_3 + V_{CE3};$$

$$I_{C3} = \left(\frac{V_{CC} - V_{CE3}}{R_3}\right) = \left(\frac{5 - 0.2}{3000}\right) = 1.6\,\text{mA} \tag{3.47E}$$

$$V_{CC} = I_{C3}R_3 + V_{out};$$

$$V_{out} = V_{CC} - I_{C3}R_3 = 5 - 1.6 \times 10^{-3} \cdot 3 \times 10^3 = 0.2\,\text{V}$$

The completed truth table is shown in Table 3.6.

3.6.5.4 *MOSFET logic gates*

Creating AND and OR structures using MOSFETs is easily accomplished by placing the n-MOS and p-MOS transistors either in series (AND) (Figure 3.47) or parallel (OR) (Figure 3.48).

Table 3.6 Completed truth table for Example 3.5

V_1	V_2	Q_2	Q_3	V_{out}
0 V	0 V	off	off	5 V
0 V	5 V	off	off	5 V
5 V	0 V	off	off	5 V
5 V	5 V	on	on	0.2 V

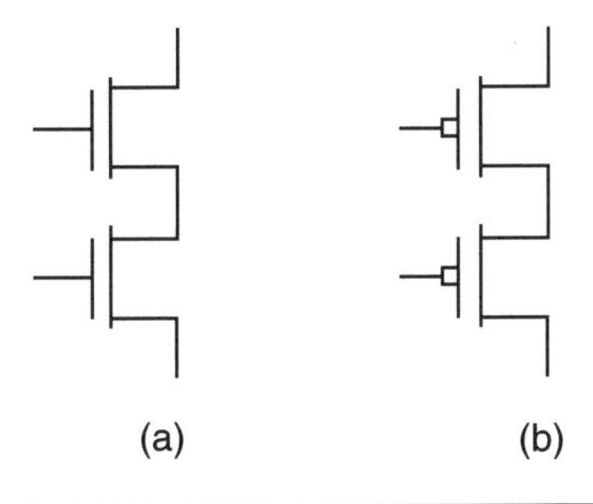

Figure 3.47 MOSFET AND structure: (a) n-type; (b) p-type.

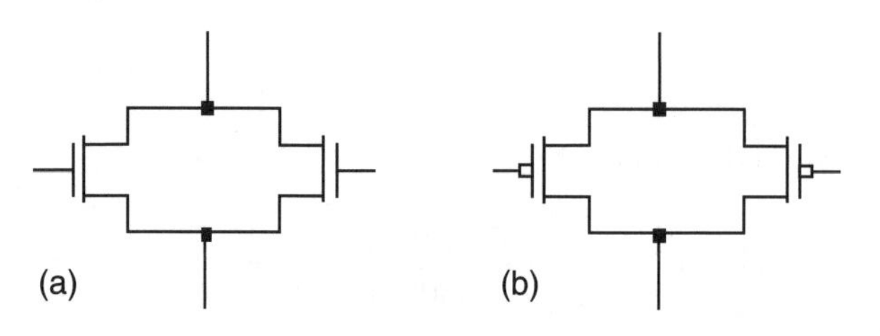

Figure 3.48 MOSFET OR structure: (a) n-type; (b) p-type.

3.6.5.5 *MOSFET logic gate analysis*

We now use the enhancement MOSFET circuit shown in Figure 3.49 to carry out the switching analysis (referring to the characteristics in Figure 3.50).

$$V_{DD} - i_D R_D - V_{out} = 0 \tag{3.48}$$

$$\begin{aligned} V_{out} &= V_{DD} - i_D R_D \\ \text{When} \quad i_D &= 0, V_{out} = V_{DD}(V_{in} = 0). \\ \text{When} \quad V_{out} &= V_{DSaat} \approx 0.5 \approx 0\ \text{V}(V_{in} = 5\ \text{V}); \quad i_D = \frac{V_{DD}}{R_D}. \end{aligned} \tag{3.49}$$

When $V_{in} = 0$, the MOSFET conducts virtually no current and $V_{out} = V_{DD} = 5\,\text{V}$.

When $V_{in} = 5\,\text{V}$, the MOSFET is in the saturation region, so $V_{out} = V_{DSaat} \approx 0.5$ or low (0 V).

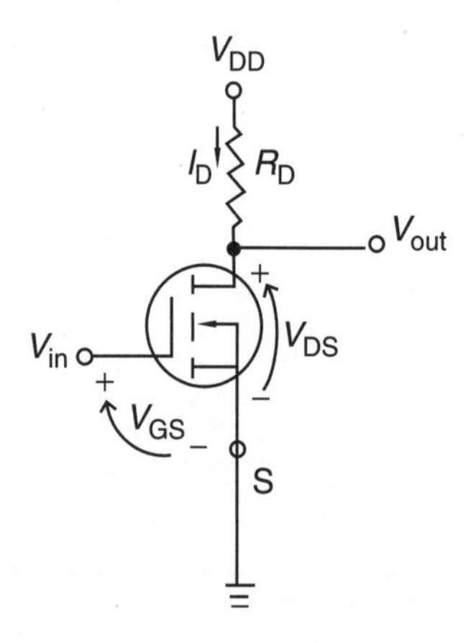

Figure 3.49 Logic gate analysis circuit.

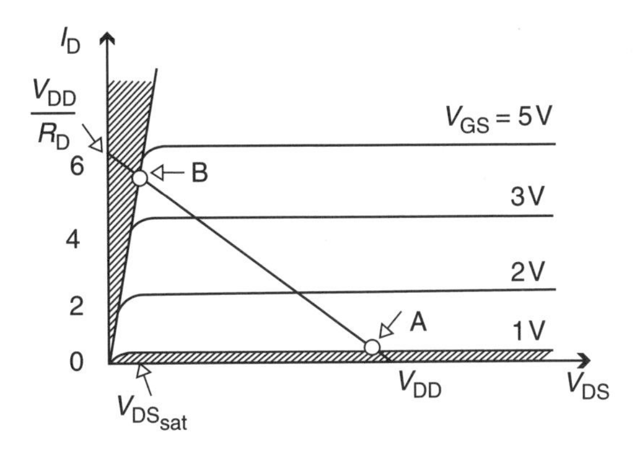

Figure 3.50 Logic gate analysis characteristic curves.

This analysis shows that the circuit is an inverter, which as in the case of the BJT, forms the basis of all MOS logic gates.

3.6.6 Complementary metal-oxide semiconductor (CMOS)

CMOS logic devices are the most common devices used today in the high density, multi-transistor circuits found in everything from complex microprocessor ICs to signal processing and communication circuits. The CMOS structure is popular because of its inherent low power requirements, high operating clock speed, and ease of implementation at the transistor level. Students of introductory electronic circuits can gain insight into the operation of these CMOS devices through a few exercises in constructing simple CMOS combinational logic circuits. CMOS logic circuits are created using both p- and n-channel MOSFETs connected in complementary configurations.

The complementary networks are used to connect the output of the logic device to either the power supply for a given input logic state. In a simplified view, the MOSFETs can be treated as simple switches. This is adequate for an introduction to simple CMOS circuits where switching speeds, propagation delays, drive capability, and rise and fall times are of little concern.

The p-channel MOSFET is a switch that is closed when the input voltage is low (0 V) and open when the input voltage is high (5 V). The n-channel MOSFET is modeled as a switch that is closed when the input voltage is high (5 V) and open when the input voltage is low (0 V). The basic idea behind CMOS logic circuits is to combine p-channel and n-channel devices such that there is never a conducting path from the supply voltage to ground. As a consequence, CMOS circuits consume very little power and produce low static dissipation.

3.6.6.1 *The CMOS inverter*

The most important CMOS gate is the CMOS inverter. It consists of only two transistors, an n-channel device and a p-channel device (Figure 3.51).

The n-channel transistor provides the switch connection to ground when the input is a logic high while the p-channel device provides the connection to the power supply when the input to the inverter circuit is a logic low. This is consistent with MOSFET Rule 2.

When $V_{in}=0$ (or low), transistor Q_n is 'off'.

However, for Q_p: $V_{in}-V_{GS}-V_{DD}=0$, or $V_{GS}=V_{in}-V_{DD}$. Since $V_{in}=0$, $V_{GS}=0-V_{DD}=-V_{DD}$. In a p-channel device, this is the same as $V_{GS}=V_{DD}$ ('on'). Hence, $V_{out}\approx V_{DD}$.

When $V_{in}=5\,V$ (or high), transistor Q_n is 'on'.

In this case, Q_p experiences $V_{GS}=V_{in}-V_{DD}=5-5=0$ ('off'). Hence, $V_{out}\approx 0$.

3.6.6.2 *The CMOS NOR gate*

We will illustrate the CMOS NOR gate using an example. It will then be easy to extend the concept learnt to other gates such as NAND and XOR. It is interesting to note CMOS gates do not necessarily need resistors and are indeed very simple compared to their counterpart switching devices already discussed.

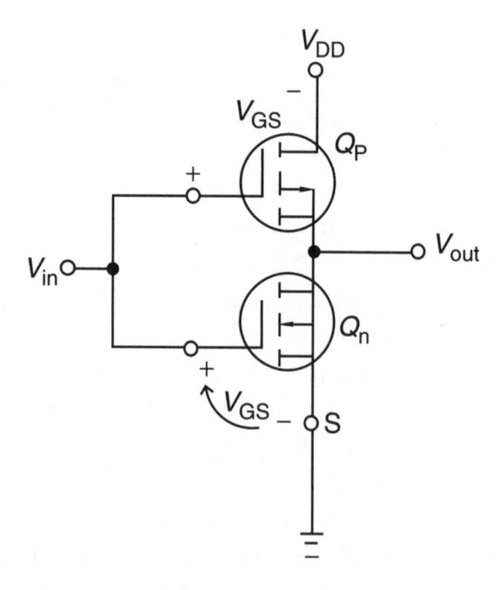

Figure 3.51 The CMOS inverter circuit.

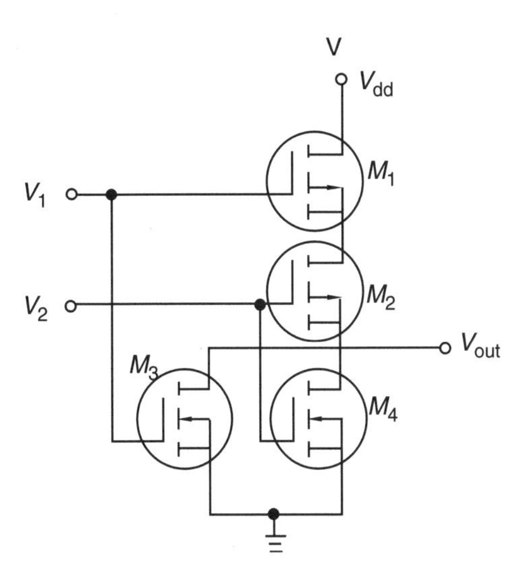

Figure 3.52 The CMOS NOR gate.

Table 3.7 Truth table for Example 3.6

V_1	V_2	M_1	M_2	M_3	M_4	V_{out}
0 V	0 V					
0 V	5 V					
5 V	0 V					
5 V	5 V					

EXAMPLE 3.6

Analyze the CMOS NOR gate shown in Figure 3.52 for its switching operations and complete the truth table in Table 3.7.

$$V_{CC} = 5\,V$$
$$V_T = 1.7\,V$$
$$V_{CEsat} = 0.2\,V$$
$$R_C = 2.2\,k\Omega$$

Solution

First redraw Figure 3.52 to make it easier to visualize (Figure 3.53). A guide to analysis is:

- treat the MOSFETS as *open* when *off*;
- treat the MOSFETS as linear *resistors* when *on*.

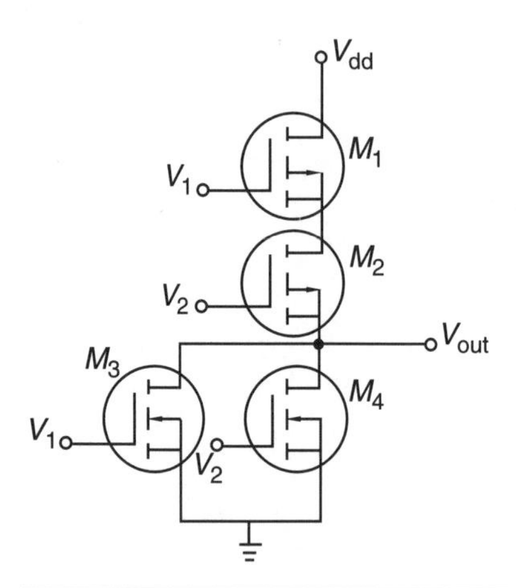

Figure 3.53 CMOS equivalent NOR gate.

Case 1: $V_1 = 0\,V$; $V_2 = 0\,V$: In this case, M_1 and M_2 are 'on', but M_3 and M_4 are 'off'. Thus, $V_{out} = V_D = 5\,V$. This condition is shown in Figure 3.54(a).

Case 2: $V_1 = 5\,V$; $V_2 = 0\,V$: In this case, M_2 and M_3 are 'on', but M_1 and M_4 are 'off'. Thus, $V_{out} = 0\,V$. This condition is shown in Figure 3.54(b).

Case 3: When $V_1 = 0\,V$; $V_2 = 5\,V$: This is symmetrical to Case 2, and hence, the same result is obtained, in which $V_{out} = 0\,V$. For completeness, we find that M_1 and M_4 are 'on', but M_2 and M_3 are 'off'. Thus, $V_{out} = 0\,V$. This condition is shown in Figure 3.54(c).

Case 4: When $V_1 = 5\,V$; $V_2 = 5\,V$: In this case, M_3 and M_4 are 'on', but M_1 and M_2 are 'off'. Thus, $V_{out} = 0.16\ V \approx 0\ V$. This condition is shown in Figure 3.54(d).

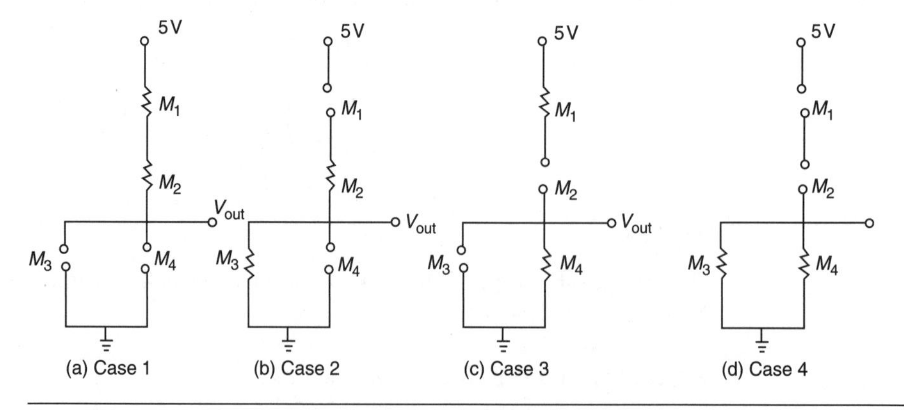

Figure 3.54 NOR gate analysis for Example 3.6.

Case 4 is obtained by considering the current and output voltage as follows:

$$V_{CC} = I_{C3}R_3 + V_{CE3};$$

$$I_{C3} = \left(\frac{V_{CC} - V_{CE3}}{R}\right) = \left(\frac{5 - 0.2}{2200}\right) = 2.2\,\text{mA} \tag{3.49A}$$

$$V_{CC} = I_{C3}R_3 + V_{out};$$

$$V_{out} = V_{CC} - I_{C3}R_3 = 5 - 2.2 \times 10^{-3} \times 2.2 \times 10^3 = 0.16\,\text{V}$$

The results of the analysis are summarized in Table 3.8.

Figure 3.55 shows an n-channel MOSFET AND structure with the source of $M1$ connected to ground (MOSFET Rule 2). This is turned on when a logic high is applied to the gate input. The logic expression for this circuit is $F = (A \cdot B)_{-L}$, meaning that the output F is low if A and B are high. This is

Table 3.8 Completed truth table for Example 3.6

V_1	V_2	M_1	M_2	M_3	M_4	V_{out}
0 V	0 V	on	on	off	off	5 V
0 V	5 V	on	off	off	on	0
5 V	0 V	off	on	on	off	0
5 V	5 V	off	off	on	on	0

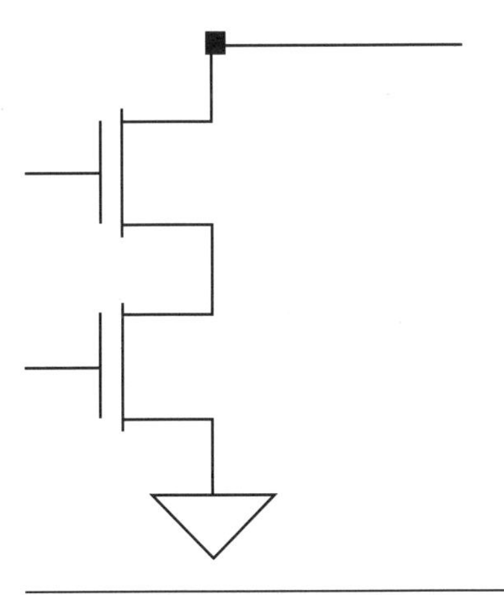

Figure 3.55 An n-channel transistor structure realizing the expression $F = (A \cdot B)_{-L}$.

called the analogous structure. If gate inputs A and B are a logic high, then the output node of the AND structure will be connected to ground (a logic low). If either input A or B is a logic low then there will not be a path to ground since both MOSFET transistors will not be turned on. In CMOS technology, a complementary transistor structure is required to connect the output node to the opposite power supply rail. The expression and transistor configuration for the complementary structure is obtained by applying DeMorgan's theorem.

Problems

Gates and switches

Q3.1 In the circuit shown in the Figure 3.56:

$$V_{CC} = 5\,\text{V}$$

$$R_1 = 6\,\text{k}\Omega$$

$$R_2 = 2\,\text{k}\Omega$$

$$R_3 = 3\,\text{k}\Omega$$

$$R_4 = 3.5\,\text{k}\Omega$$

$$R_5 = 1.5\,\text{k}\Omega$$

Show that the circuit is an AND gate and construct a truth table. (Carry out a detailed analysis as done in the chapter.)

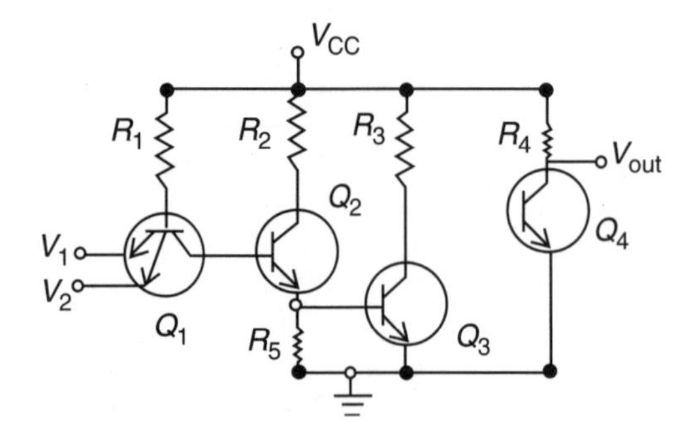

Figure 3.56 Circuit for Q3.1.

Q3.2 Show that the circuit in Figure 3.57 functions as an OR gate if the output is taken at V_{o2}. Construct a truth table.

Q3.3 Show that the circuit in Figure 3.57 functions as a NOR gate if the output is taken at V_{o1}. Construct a truth table.

Q3.4 Show that the circuit in Figure 3.58 functions as an AND gate if the output is taken at V_{o1}, and construct a truth table.

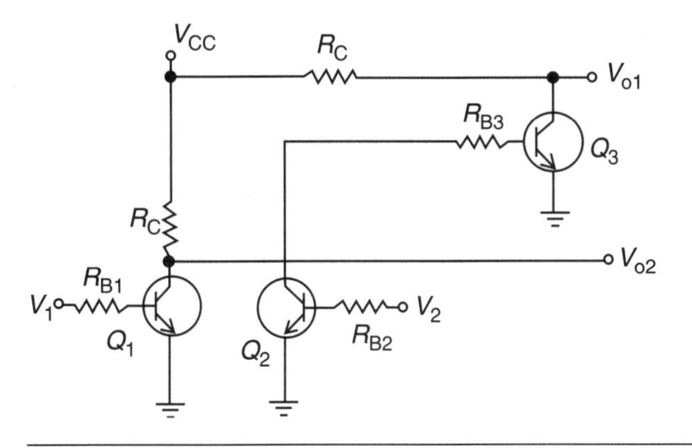

Figure 3.57 Circuit for Q3.2.

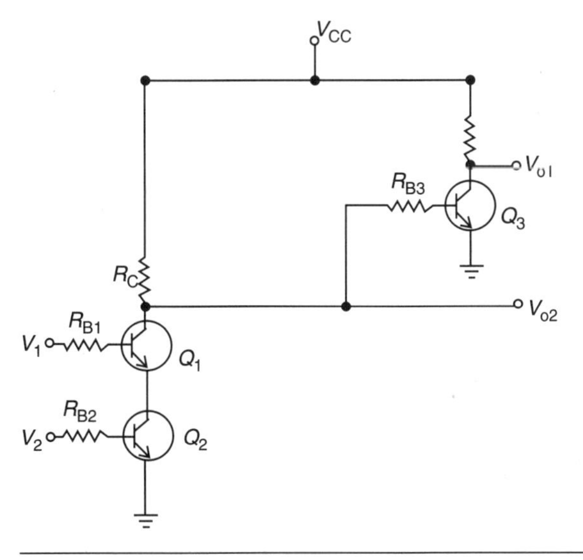

Figure 3.58 Circuit for Q3.4.

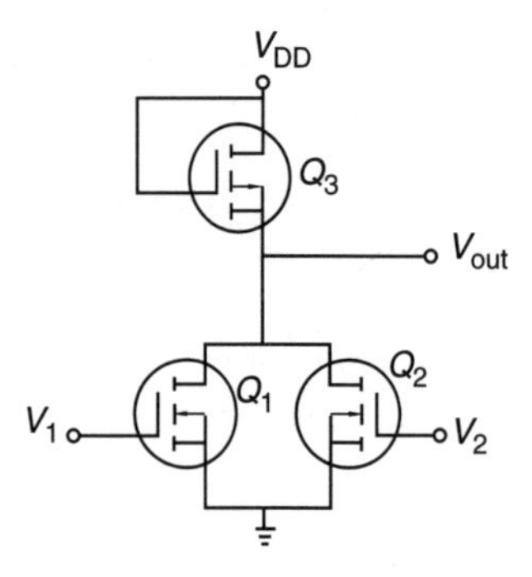

Figure 3.59 Circuit for Q3.6.

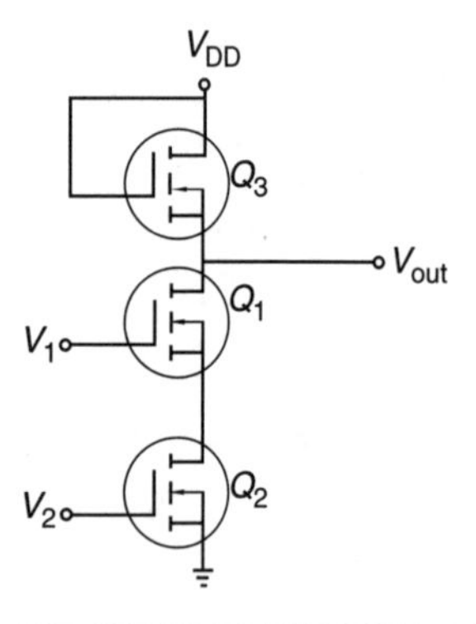

Figure 3.60 Circuit for Q3.7.

Q3.5 Show that the circuit in Figure 3.58 functions as a NAND gate if the output is taken at V_{o2}. Construct a truth table.

Q3.6 Show that the circuit in Figure 3.59 functions as a NOR gate and construct a truth table.

Q3.7 Show that the circuit in Figure 3.60 functions as a NAND gate, and construct a truth table.

Further reading

[1] Horowitz, P. and Hill, W. (1989) *The Art of Electronics* (2nd. ed.), New York: Cambridge University Press.

[2] Millman, J. and Gabrel, A. (1987) *Microelectronics* (2nd. ed.) New York: McGraw-Hill.
[3] Rashid, M.H. (1996) *Power Electronics: Circuits, Devices, and Applications*, Prentice Hall.
[4] Rizzoni, G. (2003) *Principles and Applications of Electrical Engineering* (4th. ed.), McGraw-Hill.

CHAPTER 4

Digital electronics

Chapter objectives

When you have finished this chapter you should be able to:

- handle combinational logic design using a truth table;
- understand Karnaugh maps and logic design;
- understand combinational logic modules such as the half adder, the full adder, multiplexers, and decoders;
- understand sequential logic modules such as the S-R flip-flop, the D flip-flop, and the J-K flip-flop;
- understand sequential logic design;
- understand data registers, counters, the Schmitt trigger, the 555 timer, the astable multivibrator, and the one-shot monostable multivibrator.

4.1 Introduction

Analog signals have a continuous range of values within some specified limits and can be associated with continuous physical phenomena. On the other hand, digital signals typically assume only two discrete values (states) and are appropriate for any phenomena involving counting or integer numbers. While we are mostly dealing with voltages and currents at specific points in analog circuits, we are interested in the information flow in digital circuits. The active elements in digital circuits are either BJTs or FETs (already discussed in Chapter 3). These transistors are designed to operate in only two states ('on' and 'off'), which normally correspond to two output voltages. Hence the transistors act as switches. The two digital states can be given various names: ON/OFF, true/false, high/low, or 1/0. The 1 and 0 notation naturally leads to the use of binary (base 2) numbers. Octal (base 8) and hexadecimal (base 16) numbers are also often used since they provide a condensed number notation. Decimal (base 10) numbers are not of much use in digital electronics.

Before going into the details of digital circuits and systems, we will first review number systems and Boolean algebra.

4.2 Number systems

We are accustomed to using a decimal system for most of our mathematical computations. This is a base ten system, in which each digit of a number represents a power of 10. Consider the decimal number *abc*. We can write this as

$$abc_{10} = (a \times 10^2) + (b \times 10^1) + (c \times 10^0) \tag{4.1}$$

For example, the number 789.45 can be expressed as:

$$\left(7 \times 10^2\right) + \left(8 \times 10^1\right) + \left(9 \times 10^0\right) + \left(4 \times 10^{-1}\right) + \left(5 \times 10^{-2}\right).$$

This format can be rewritten more generally by assigning the digits 7, 8, 9, 4, and 5 to the expressions D(2), D(1), D(0), D(−1), and D(−2), respectively. The numbers in parentheses are the same as the exponents of the powers of ten they correspond to. We can also express the base, in this case, 10, as r. These substitutions give us the following expression:

$$\left[\mathrm{D}(2) \times \mathrm{r}^2\right] + \left[\mathrm{D}(1) \times \mathrm{r}^1\right] + \left[\mathrm{D}(0) \times \mathrm{r}^0\right] + \left[\mathrm{D}(-1) \times \mathrm{r}^{-1}\right] + \left[\mathrm{D}(-2) \times \mathrm{r}^{-2}\right].$$

We now have a generalization for expressing numbers of any base r in terms of a *power series*. Using this generalization, we can convert from any base to decimal (or any other if you can readily add and take exponents in that base).

For example, we can find the decimal value of 325.02_3 as follows:

$$\begin{aligned}&\left[3 \times 3^2\right] + \left[2 \times 3^1\right] + \left[5 \times 3^0\right] + \left[0 \times 3^{-1}\right] + \left[2 \times 3^{-2}\right]\\&= (3 \times 9) + (2 \times 3) + (5 \times 1) + (0 \times 1/3) + (2 \times 1/9)\\&= 27 + 6 + 5 + 2/9 = 38.22\end{aligned}$$

4.2.1 Binary numbers

The binary number system is a base 2 number system, using only the digits 0 and 1. It is commonly used when dealing with computers because it is well suited to representing logical expressions, which have only 2 values: TRUE (1) and FALSE (0). Single binary digits are often referred to as bits. In the binary system a number *abc* can be written as

$$abc_2 = (a \times 2^2) + (b \times 2^1) + (c \times 2^0) \tag{4.2}$$

The left most bit is the highest-order bit and represents the most significant bit (MSB), while right most bit, which is the lowest-order bit is the least significant bit (LSB).

4.2.2 Octal numbers

Octal numbers are base 8 numbers, using only the digits 0 through 7. The eight octal numbers are represented with the symbols 0,…, 7. In the octal system a number *abc* can be written as

$$abc_8 = (a \times 8^2) + (b \times 8^1) + (c \times 8^0) \tag{4.3}$$

Since 8 is power of 2, each of its digits can be represented as a group of bits. The number of bits is the same as the power of 2 that the base is. In other words, since $8 = 2^3$, a base 8 digit can be represented as three bits, as in the following examples:

$$237.44_8 = 010|011|111|.|100|100_2$$
$$7372.01_8 = 111|011|111|010|.|000|001_2$$

4.2.3 Hexadecimal numbers

Hexadecimal numbers (often referred to as hex) are base 16 numbers, using the digits 0 through 9, and the letters A through F (A representing 10_{10}, B representing 11_{10}, and so on). Since 16 is power of 2, each of its digits can be represented as a group of bits, the number of bits being the same as the power of 2 that the base is. In other words, since $16 = 2^4$, a base 16 digit can be represented as four bits as in the following examples:

$$\mathrm{A443.4CB}_{16} = 1010|0100|0100|0011|.|0100|1100|1011_2$$
$$\mathrm{1AA.03}_{16} = 0001|1010|1010|.|0000|0011_2$$

Our number example *abc* can be written as

$$abc_{16} = (a \times 16^2) + (b \times 16^1) + (c \times 16^0) \tag{4.4}$$

4.2.4 Base conversion

4.2.4.1 *Conversion from binary, octal, or hex to decimal*

Notice that in our examples, each group of bits on the right corresponds to a digit in the higher based number on the left. It is also easy to convert the other way, from binary to hex or octal. One simply starts at the decimal point and counts out groups of threes or fours, depending on the base to which one is converting, and add leading and following zeros to fill the outer groups. Octal to hexadecimal conversion, or vice versa, is most easily performed by first converting to binary first. This ease of conversion makes octal and hex a good shorthand for binary numbers.

Conversion from binary, octal, or hex to decimal can be done using a set of rules, but it is much easier to use a calculator or tables (see Table 4.1).

4.2.4.2 *Conversion from decimal to other bases*

We have seen how to convert to decimal from other bases using a power series. One can use the same principle to convert from decimal to other bases. In this case, one can make a power series in the base one wants to change to, with coefficients of 1. We find the first number larger than the decimal number. The next step is to divide the decimal number by the next smallest number in the power series, and take an

Table 4.1 Decimal, binary, octal, and hexadecimal equivalents

Decimal	*Binary*	*Octal*	*Hex*
00	00000	00	00
01	00001	01	01
02	00010	02	02
03	00011	03	03
04	00100	04	04
05	00101	05	05
06	00110	06	06
07	00111	07	07
08	01000	10	08
09	01001	11	09
10	01010	12	0A
11	01011	13	0B
12	01100	14	0C
13	01101	15	0D
14	01110	16	0E
15	01111	17	0F
16	10000	20	10

integer result. Then the process is repeated with the remainder. The number in the new base is the result of the divides, lined up in order of exponent from left to right, while we make sure to remember to put zeros in places where the divide result was zero.

For example, let us find the value of 175_{10} in base 3. First we take a power series of 3:

$$3^0 = 1 \quad 3^1 = 3 \quad 3^2 = 9 \quad 3^3 = 27 \quad 3^4 = 81 \quad 3^5 = 243$$

Since 81 is the next smallest number, we start by dividing 175 by 81, and continue through the power series as follows:

$$\begin{aligned} 175/81 &= \mathbf{2} \quad \text{Remainder } 13 \\ 13/27 &= \mathbf{0} \quad \text{Remainder } 13 \\ 13/9 &= \mathbf{1} \quad \text{Remainder } 4 \\ 4/3 &= \mathbf{1} \quad \text{Remainder } 1 \\ 1/1 &= \mathbf{1} \quad \text{Remainder } 0 \end{aligned} \tag{4.4A}$$

MODing the decimal number can simplify this method by the new base, and repeating the process with the integer result of dividing the decimal number by the new base until the division result is 0. The result is then the results of the MODs taken from bottom to top. For example, let us repeat the example from above:

$$\begin{aligned} 175 \text{ MOD } 3 &= \mathbf{1} \quad 175/3 = 58 \\ 58 \text{ MOD } 3 &= \mathbf{1} \quad 58/3 = 19 \\ 19 \text{ MOD } 3 &= \mathbf{1} \quad 19/3 = 6 \\ 6 \text{ MOD } 3 &= \mathbf{0} \quad 6/3 = 2 \\ 2 \text{ MOD } 3 &= \mathbf{2} \quad 2/3 = 0 \end{aligned} \tag{4.4B}$$

Thus, the result again is 20111_3.

4.2.4.3 *Fractions in different bases*

Base conversions are handled similarly for fractional parts of numbers. We can take a power series for the base we are converting to and subtract. Alternatively, we can use a method similar to the one we just described, multiplying by the new base instead of taking MODs, and taking off the integer parts of the results from top to bottom to obtain the number in the new base.

As an illustration, let us use both methods to convert the fraction 0.59376_{10} into its binary equivalent. We will use two methods: power series and multiplication methods.

Power series method

First, we will take a power series of 8, with increasing negative exponents as follows:

$$2^{-1} = 0.5 \quad 2^{-2} = 0.25 \quad 2^{-3} = 0.125 \quad 2^{-4} = 0.0625 \quad 2^{-5} = 0.031\,25 \quad 2^{-6} = 0.015\,625$$

We will stop at six decimal places. Unlike base conversions between integers, fractional parts of numbers are not limited to a set number of digits, and in some cases they can even go on infinitely. For example, the number 1/3, expressed as 0.1 in base three, is infinitely long in base 10. Thus it is often necessary to decide on a set precision to expand the number to. Next we will subtract:

$$0.593\,76 = \mathbf{1} \times 0.500\,000 + 0.093\,76$$

$$0.093\,42 = \mathbf{0} \times 0.250\,000 + 0.093\,76$$

$$0.093\,42 = \mathbf{0} \times 0.125\,000 + 0.093\,76$$

$$0.093\,42 = \mathbf{1} \times 0.062\,500 + 0.031\,26$$

$$0.031\,26 = \mathbf{1} \times 0.031\,250 + 0.000\,01$$

$$0.000\,01 = \mathbf{0} \times 0.016\,525 + 0.00\,001$$

Thus our result is $0.100\,11_2$.

Multiplication method

$$0.593\,76 \times 2 = \mathbf{1} + 0.187\,52$$

$$0.187\,52 \times 2 = \mathbf{0} + 0.375\,04$$

$$0.375\,04 \times 2 = \mathbf{0} + 0.750\,08$$

$$0.750\,08 \times 2 = \mathbf{1} + 0.500\,16$$

$$0.500\,16 \times 2 = \mathbf{1} + 0.000\,32$$

$$0.000\,32 \times 2 = \mathbf{0} + 0.000\,64$$

Again, we get the result 0.10011_2.

EXAMPLE 4.1

(a) Convert the binary number 1001 1110 to hexadecimal and to decimal.
(b) Convert the octal number 175_8 to hexadecimal.
(c) Convert the number 146 to binary by repeated subtraction of the largest power of 2 contained in the remaining number.
(d) Devise a method similar to that used in the previous problem and convert 785 to hexadecimal by subtracting powers of 16.

Solution

(a) $10011110_2 = 9E_{16}$
$= 2^7 + 2^4 + 2^3 + 2^2 + 2^1 = 158_{10}$

(b) $175^8 = 001111101_2 = 07D_{16}$

(c) $146_{10} = 2^7 + 2^4 + 2^1 = 10010010_2$

(d) $785_{10} = 3 \times 16^2 + 16 + 1 = 311_{16}$

4.2.5 Number representation

There is some computer terminology that need to be defined.

Word: is a binary number consisting of an arbitrary number of bits.

Nibble: is a 4-bit word (one hexadecimal digit).

Byte: is an 8-bit word.

We often use the expressions 16-bit word (short word) or 32-bit word (long word) depending on the type of computer being used. Most fast computers today actually employ a 64-bit word at the hardware level.

If a word has n bits it can represent 2^n different numbers in the range 0 to 2^{n-1}. Negative numbers are usually represented by *two's complement* notation. To obtain the two's complement of a number, take the complement (invert each bit) and then add 1. All the negative numbers will have a 1 in the MSB position, and the numbers will now range from -2^n to -2^{n-1}. The electronic advantages of the two's complement notation becomes evident when addition is performed. Convince yourself of this advantage.

4.3 Combinational logic design using truth tables

We will design some useful circuits using basic logic gates, and use these circuits later on as the building blocks for more complicated circuits. We describe the basic AND, NAND, OR or NOR gates as being *satisfied* when the inputs are such that a change in any one input will change the output. A satisfied AND or OR gate has a true output, whereas a satisfied NAND or NOR gate has a false output. We sometimes identify the input logic variables A, B, C, etc. with an *n*-bit number ABC. . . .

The following steps are a useful formal approach to combinational problems:

1. Devise a truth table of the independent input variables and the resulting output quantities.
2. Write Boolean algebra statements that describe the truth table.
3. Reduce the Boolean algebra.
4. Implement the Boolean statements using the appropriate logic gates.

4.3.1 Boolean algebra

The binary 0 and 1 states are naturally related to the true and false logic variables. We will find the following Boolean algebra useful. Consider two logic variables A and B and the result of some Boolean logic operation Q. We can define

$$\mathrm{Q} \equiv \mathrm{A\ AND\ B} \equiv \mathrm{A} \cdot \mathrm{B} \tag{4.5}$$

Q is true if and only if A is true AND B is true.

$$\mathrm{Q} \equiv \mathrm{A\ OR\ B} \equiv \mathrm{A} + \mathrm{B} \tag{4.6}$$

Q is true if A is true OR B is true.

$$\mathrm{Q} \equiv \mathrm{A\ NOT\ B} \equiv \mathrm{A} \cdot \overline{\mathrm{B}} \tag{4.7}$$

Q is true if A is true and B is false.

A useful way of displaying the results of a Boolean operation is with a truth table. We summarize a number of basic Boolean rules in Table 4.2, and present de Morgan's theorems, which are useful in Boolean operations.

4.3.2 de Morgan's theorems

de Morgan's theorems state that:

$$\overline{(\mathrm{X} \cdot \mathrm{Y})} = \overline{\mathrm{X}} + \overline{\mathrm{Y}} \tag{4.8}$$

$$\overline{(\mathrm{X} + \mathrm{Y})} = \overline{\mathrm{X}} \cdot \overline{\mathrm{Y}} \tag{4.9}$$

Table 4.2 Boolean identities

(a)	*(b)*	*(c)*
$\overline{0} = 1$	$\overline{1} = 0$	Fundamental laws
$X+0=X$	$X \cdot 1=X$	
$X+1=1$	$X \cdot 0=0$	
$X+X=X$	$XX=X$	Idempotent law
$\overline{\overline{X}} = X$		Involution law
$X+Y=Y+X$	$XY=YX$	Commutative law
$X+XY=X$	$X(X+Y)=X$	Absorption law
$X+\overline{X}Y = X+Y$	$X(\overline{X}+Y) = XY$	
$\overline{(X+Y)} = \overline{X}\,\overline{Y}$	$\overline{(XY)} = \overline{X}+\overline{Y}$	De Morgan law
$X+YZ=(X+Y)(X+Z)$	$X(Y+Z)=XY+XZ$	Distributive law
$X+(Y+Z)=(X+Y)+Z$	$X(YZ)=(XY)Z$	Associative law
$=X+Y+Z$	$=XYZ$	

Consider the truth table that defines the OR gate. Using the lines in this table that yield a true result gives.

$$Q = \overline{A} \cdot B + A \cdot \overline{B} + A \cdot B \tag{4.10}$$

$$= \overline{A} \cdot B + A \cdot \overline{B} + A \cdot B + A \cdot B \tag{4.11}$$

$$= A \cdot (B + \overline{B}) + B \cdot (A + \overline{A}) \tag{4.12}$$

$$= A + B \tag{4.13}$$

Since Q is a two-state variable all other input state combinations must yield a false. If the truth table had more than a single output result, each such result would require a separate equation. An alternative is to write an expression for the false condition.

$$\overline{Q} = \overline{A} \cdot \overline{B} \tag{4.14}$$

$$\overline{\overline{Q}} = \overline{\overline{A}} + \overline{\overline{B}} \tag{4.15}$$

$$Q = \overline{\overline{A} \cdot \overline{B}} \tag{4.16}$$

$$Q = A + B \tag{4.17}$$

4.3.3 Logic gates

Electronic circuits which combine digital signals according to the Boolean algebra are referred to as *logic gates* because they control the flow of information. *Positive logic* is an electronic representation in which the true state is at a higher voltage, while *negative logic* has the true state at a lower voltage. We will use the positive logic type in this chapter.

In digital circuits no inputs must be left unconnected. Logic circuits are grouped into families, each with their own set of detailed operating rules. Some common logic families are:

- RTL: resistor–transistor logic;
- DTL: diode–transistor logic;
- TTL: transistor–transistor logic;
- NMOS: n-channel metal-oxide silicon;
- CMOS: complementary metal-oxide silicon;
- ECL: emitter–coupled logic.

ECL is very fast. MOS has very low power consumption and hence is often used in large scale integration (LSI) technology. TTL is normally used for small-scale integrated circuit units.

Any logic operation can be formed from NAND or NOR gates or a combination of both. Gates also often have more than two inputs. Inverter gates can be formed by applying the same signal to both inputs of a NOR or NAND gate.

4.3.3.1 *The AND gate*

AND gates are used to determine when *both* inputs are true. The schematic symbol is shown in Figure 4.1, and the truth table is shown in Table 4.3.

4.3.3.2 *The NAND gate*

NAND gates are negated AND gates. They are true when *at least one* input is *not* true. As a side note, it is often easier (and cheaper) to buy NAND gates instead of AND gates. This is due to the fact that on the transistor level, the number of transistors required to construct a NAND gate is less than the number needed for

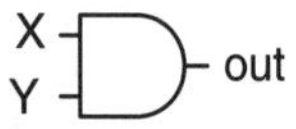

Figure 4.1 Two-input AND gate.

Table 4.3 AND truth table.

X	*Y*	*X AND Y*
0	0	0
0	1	0
1	0	0
1	1	1

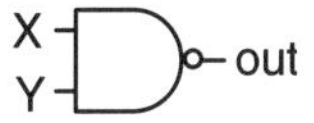

Figure 4.2 Two-input NAND gate.

Table 4.4 NAND truth table

X	*Y*	*X NAND Y*
0	0	1
0	1	1
1	0	1
1	1	0

an AND gate. The schematic symbol is shown in Figure 4.2, and the truth table is shown in Table 4.4.

4.3.3.3 *The OR gate*

OR gates are used to determine when *at least one* input is true. The schematic symbol is shown in Figure 4.3, and the truth table is shown in Table 4.5.

4.3.3.4 *The NOR gate*

NOR gates are negated OR gates. They are true when *all* inputs are *not* true. The schematic symbol is shown in Figure 4.4, and the truth table is shown in Table 4.6.

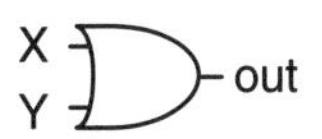

Figure 4.3 Two-input OR gate.

Table 4.5 OR truth table

X	*Y*	*X OR Y*
0	0	0
0	1	1
1	0	1
1	1	1

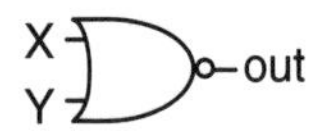

Figure 4.4 Two-input NOR gate

Table 4.6 NOR truth table.

X	*Y*	*X NOR Y*
0	0	1
0	1	0
1	0	0
1	1	0

4.3.3.5 *The NOT gate*

NOT gates return the *opposite* of the input. The schematic symbol is shown in Figure 4.5, and the truth table is shown in Table 4.7. The open circle is used to indicate the NOT or negation function and can be replaced by an inverter in any circuit. A signal is negated if it passes through the circle.

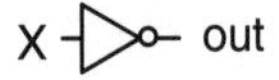

Figure 4.5 The NOT gate.

Table 4.7 NOT truth table

X	*NOT*
0	1
1	0

4.3.3.6 *The buffer gate*

Buffers return a delayed output which is the *same* as the input. The schematic symbol is shown in Figure 4.6, and the truth table is shown in Table 4.8.

4.3.3.7 *The tri-state buffer (TSB) gate*

Tri-state buffers only have output when current is applied to an enabling input (Y in the diagram and table). When an enable signal is present, the output is true when the primary input (X in the diagram and table) is true, and false when the primary input is false. When the enable is off, the TSB effectively has infinite resistance, preventing any signal whatsoever from flowing through the gate. The symbol is shown in Figure 4.7, and the truth table is shown in Table 4.9.

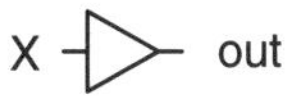

X out

Figure 4.6 The buffer.

Table 4.8 Buffer truth table

X	*NOT*
0	0
1	1

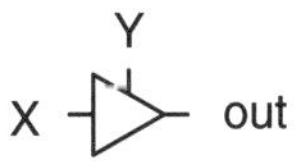

Y
X out

Figure 4.7 The tri-state buffer.

Table 4.9 Tri-state buffer truth table

X	*Y*	*TSB out*
0	0	–
0	1	0
1	0	–
1	1	1

4.3.3.8 *The AND-OR-INVERT gate*

Some logic families provide a gate known as an AND-OR-INVERT (AOI) gate (see Figure 4.8).

Here, we note that

$$Q = \overline{A \cdot B + C \cdot D} \tag{4.18}$$

4.3.3.9 *The exclusive-OR gate*

The exclusive-OR gate (EOR or XOR) is a very useful two-input gate. XOR gates are true when *an odd number* of inputs are true. (Note that an XNOR gate is true only when *an even number* of inputs are true.) The schematic symbol is shown in Figure 4.9, and the truth table is shown in Table 4.10.

$$Q = \overline{A} \cdot B + A \cdot \overline{B} \tag{4.19}$$

We can draw the implementation directly from the truth table (Figure 4.10).

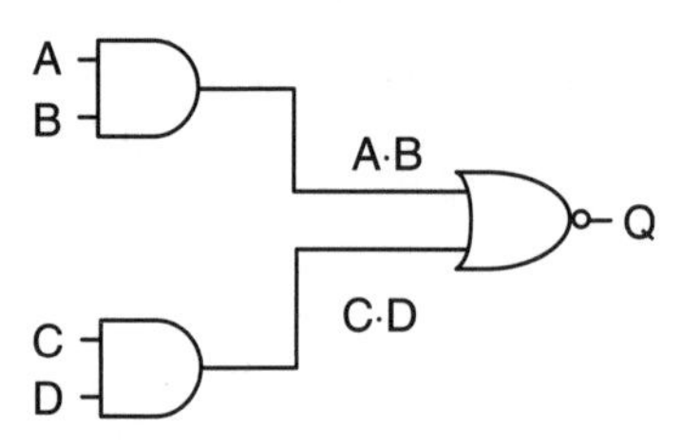

Figure 4.8 The AOI gate.

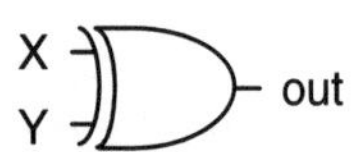

Figure 4.9 The XOR gate.

Table 4.10 XOR truth table

X	*Y*	*XOR*
0	0	0
0	1	1
1	0	1
1	1	0

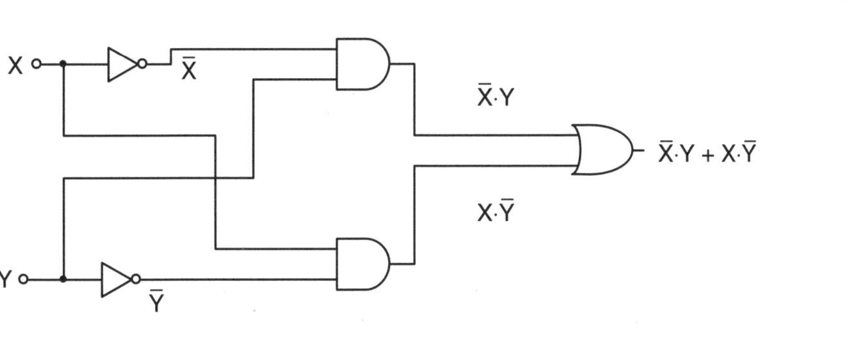

Figure 4.10 An XOR implementation.

4.4 Karnaugh maps and logic design

Karnaugh maps consist of a grid of 2^N squares where N is the number of variables in the Boolean expression being minimized (Figure 4.11). These maps are useful for minimizing expressions with six or fewer variables. Two, three, and four variables are very easily dealt with. Five and six variables are more difficult, but possible. Seven or more variables are extremely difficult, if not impossible. The main objective of using Karnaugh maps is to minimize Boolean expressions without having to use Boolean algebra theorems and equation manipulations.

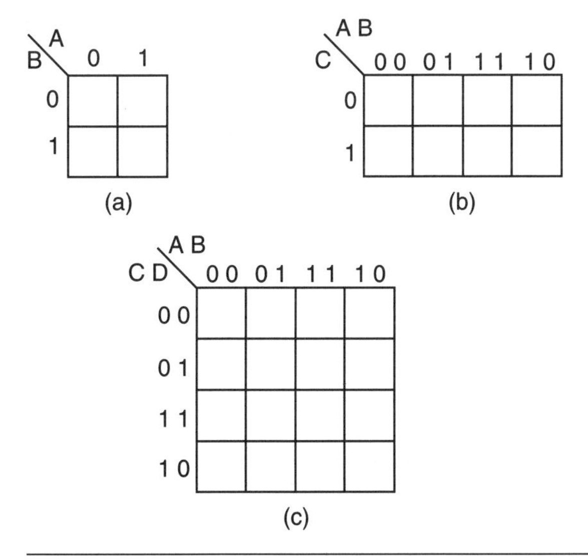

Figure 4.11 Karnaugh map grids for (a) 2; (b) 3, and (c) 4 variables.

The first step in using a Karnaugh map to minimize a Boolean expression is to map the expression onto the grid. A Boolean expression is mapped in the same way as a truth table is constructed. Where a one would be found in the output of the truth table a one is placed on the Karnaugh map. In other words, for a particular combination of input values where the output is true, the same combination of input values on the borders of the Karnaugh map indicate that that cell should be marked as true. Where a zero would be found in the output of the truth table a zero is placed in the corresponding position of the Karnaugh map. We notice that the variable values on the Karnaugh map differ from their neighbor by only one bit. This is *logical adjacency*. Physically adjacent blocks on a Karnaugh map are also logically adjacent. Physically adjacent locations on a truth table, such as the rows with input values of binary three and four, are not logically adjacent.

For example, the equation $W = A\overline{B} + \overline{A}B$ would require a 2×2 Karnaugh map with 1s entered in the upper right and lower left cells. Other examples include $W = A\overline{B} + AB$, $W = ABC + \overline{A}B\overline{C} + A\overline{B}\,\overline{C} + AB\overline{C}$, and $W = ABCD + ABC\overline{D} + \overline{A}\,\overline{B}\,\overline{C}\,\overline{D} + \overline{A}BC\overline{D}$. These examples are shown in Figure 4.12.

Consider the expression $Z = AB\overline{C}\,\overline{D} + A\overline{B}\,\overline{C}\,\overline{D}$. This can be simplified using Boolean algebra as follows:

$$A\overline{B}\,\overline{C}\,\overline{D} + AB\overline{C}\,\overline{D} = A\overline{C}\,\overline{D}(\overline{B} + B) = A\overline{C}\,\overline{D}(1) = A\overline{C}\,\overline{D}$$

This result can be arrived at by plotting the expression on a Karnaugh map (Figure 4.13).The two adjacent 1s are ringed. Notice that the variable B changes as

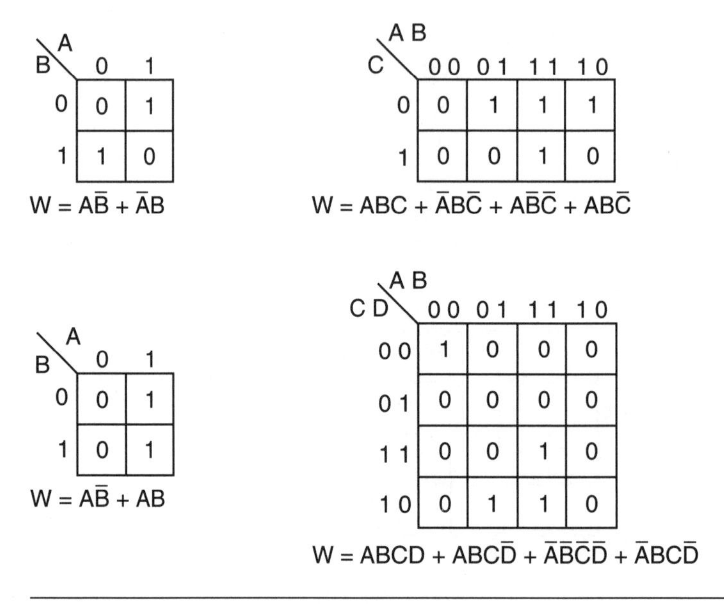

Figure 4.12 Further examples of Karnaugh maps.

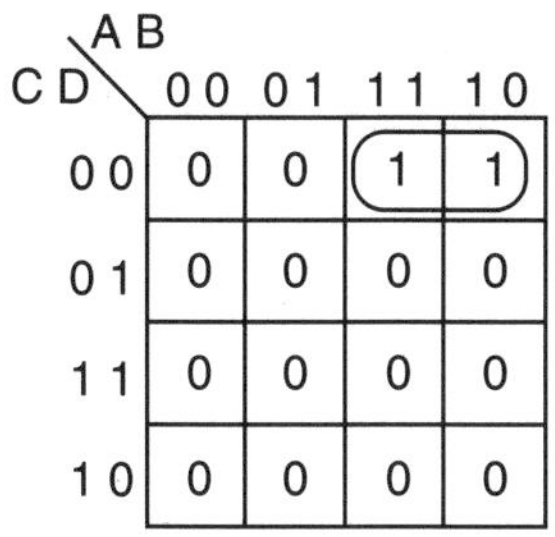

CD \ AB	00	01	11	10
00	0	0	1	1
01	0	0	0	0
11	0	0	0	0
10	0	0	0	0

$Z = AB\overline{C}\overline{D} + A\overline{B}\overline{C}\overline{D}$
$= A\overline{C}\overline{D}$

Figure 4.13 Using a Karnaugh map to simplify an expression.

we move across the ring, but all other variables maintain their value. The variable(s) that change are removed, in this case variable B changes and therefore is removed. This results in the solution, $A\overline{C}\overline{D}$.

Consider the expression $Z = AB\overline{C}\overline{D} + A\overline{B}\overline{C}\overline{D} + AB\overline{C}D + A\overline{B}\overline{C}D$. Again using Boolean algebra we can simplify it:

$$= A\overline{C}[\overline{D}(B + \overline{B}) + D(B + \overline{B})]$$

$$= A\overline{C}(D + \overline{D})$$

$$= A\overline{C}$$

This is arrived at using the Kanaugh by encircling the four 1s in the upper right corner with one large ring and noting that the only variable that changes is D (see Figure 4.14).

On a 4×2 Karnaugh map the cells on the left and right edge are logically adjacent. On a 4×4 map the cells on the left and right edges are adjacent as well as the cells on the top and bottom. This also implies that the four corners are logically adjacent.

CD \ AB	00	01	11	10
00	0	0	1	1
01	0	0	1	1
11	0	0	0	0
10	0	0	0	0

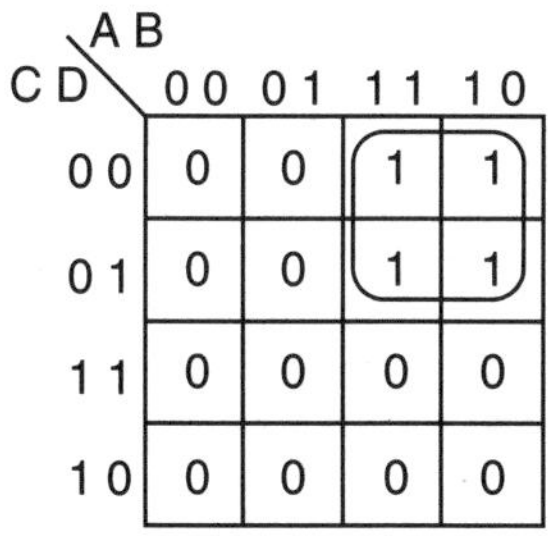

Figure 4.14 Using a Karnaugh map to simplify an expression.

The following describes the use of Karnaugh maps to minimize Boolean expressions so the physical gate count is minimized. The process has six steps:

1. Plot the expression on the Karnaugh map by placing 1s and 0s in the appropriate cells.
2. Form groups of adjacent 1s. Make groups as large as possible, but the group size must be a power of two.
3. Ensure that every 1 is in a group. 1s can be part of more than one group.
4. Select the least number of groups that covers all the 1s.
5. Translate each group into a product term by including each variable or its complement if the variable does not change value over the group. If the input value of the variable is 0 and does not change over the group, the complemented variable is included in the product term; if its value is a 1 and does not change then include the variable itself.
6. OR each product term together.

If a truth table has *don't care* outputs they can be represented on the Karnaugh map by either 1s or 0s. If we select wisely we can minimize our logic even further. An example of this situation is an encoder. For example, a 4-to-2 binary encoder (see Figure 4.15) takes a 4-bit input and generates a 2-bit binary output. The only input combinations we are interested in are those with only a single input asserted. If more than one input is asserted we may not care what the output is and therefore a *don't care* label is satisfactory. A partial truth table for this encoder is shown in Table 4.11.

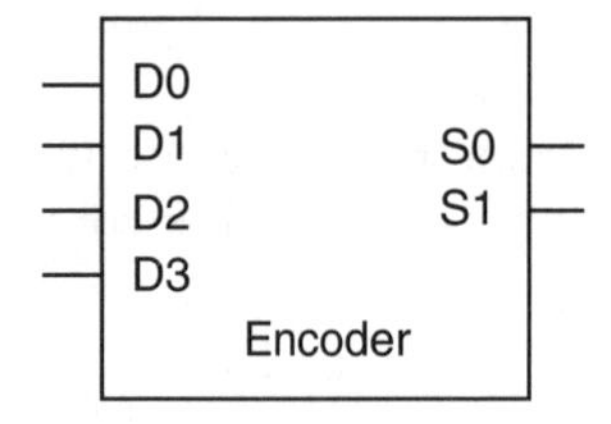

Figure 4.15 A 4-to-2 binary encoder schematic.

Table 4.11 Partial truth table for 4-to-2 encoder

D_0	D_1	D_2	D_3	S_1	S_0
1	0	0	0	0	0
0	1	0	0	0	1
0	0	1	0	1	0
0	0	0	1	1	1

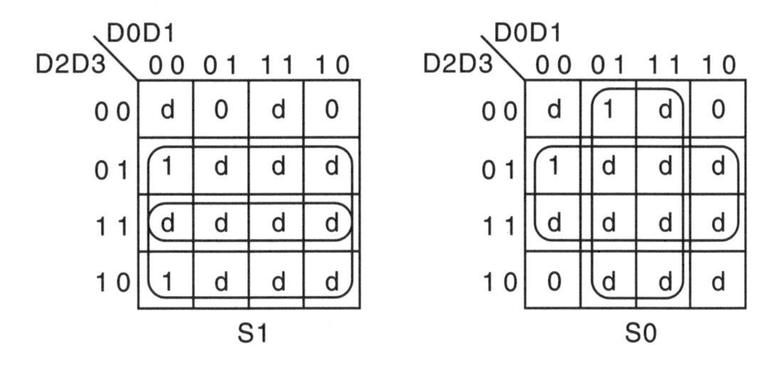

Figure 4.16 Karnaugh maps showing *don't care* conditions.

Don't care outputs are represented on a Karnaugh map by a *d* (Figure 4.16), which can be set to either 1 or 0 to help minimization.

EXAMPLE 4.2

Minimize the Boolean expression $X = A\,B\,\overline{C} + \overline{A}\,\overline{B}\,\overline{C} + \overline{A}\,B\,\overline{C} + A\,B\,C + A\,\overline{B}\,\overline{C}$ using a Karnaugh map.

Solution

Following the steps given above, first plot the expression on the Karnaugh map. In other words, where X is true for some combination of input values place a 1 on the map corresponding to the same input values (Figure 4.17(a)).

The second, third, and fourth steps are to group all of the 1s. The groups must be as large as possible and their size must be a power of two. Use as few groups as possible, but make sure every one is covered (Figure 4.17(b)).

The fifth step is to translate each group into a product term. This results in two product terms: AB and $\overline{C}$.

The final step is to OR each of these product terms together. This yields, $X = AB + \overline{C}$.

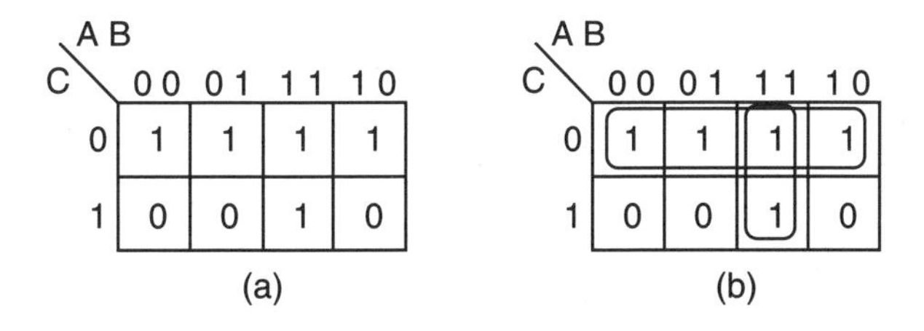

Figure 4.17 Karnaugh map for Example 4.2: (a) plotted expression; (b) solution.

We have shown in this section that Karnaugh maps are useful for minimizing Boolean expressions, which in turn minimizes the number of gates required in a practical realization of the logic expression. As already mentioned, four or less variables are easy to work with, but five and six variables are difficult to plot and minimize.

4.5 Combinational logic modules

4.5.1 The half adder

A half adder that adds two binary numbers can be made from basic logic gates. Consider adding two binary numbers X and Y to give a sum bit (S) and carry bit (C). The truth table for all combinations of X and Y is shown in Table 4.12.

From the truth table, it can be seen that:

$$S = \overline{X}Y + X\overline{Y} = X \oplus Y \tag{4.21}$$

$$C = XY \tag{4.22}$$

The implementation of these two equations is shown in Figure 4.18.

Table 4.12 The half adder truth table

X_0	Y_0	S	C_1
0	0	0	0
0	1	1	0
1	0	1	0
1	1	0	1

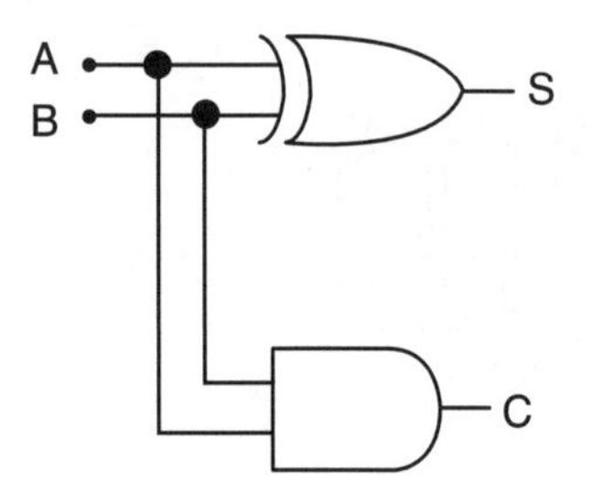

Figure 4.18 The half adder.

4.5.2 The full adder

The half adder cannot handle the addition of any two arbitrary numbers because it does not allow the input of a carry bit from the addition of two previous digits. A circuit that can handle three inputs can perform the addition of any two binary numbers (Table 4.13 and Figure 4.19).

From the truth table

$$\begin{aligned} C_2 &= \overline{C}_1XY + C_1X\overline{Y} + C_1\overline{X}Y + C_1XY \\ &= XY + C_1X + C_1Y \end{aligned} \tag{4.23}$$

C_2 shows majority logic and a majority detector is shown in Figure 4.20.

$$\begin{aligned} S &= \overline{C}_1\overline{X}Y + \overline{C}_1X\overline{Y} + C_1\overline{X}\,\overline{Y} + C_1XY \\ &= C_1 \oplus (X \oplus Y) \end{aligned} \tag{4.24}$$

Figure 4.21 shows the full adder (with majority detection) which is able to add three single bits of information and return the sum and carry bits.

Table 4.13 The full adder truth table

X	*Y*	C_1	C_2	S_B
0	0	0	0	0
1	0	0	0	1
0	1	0	0	1
1	1	0	1	0
0	0	1	0	1
1	0	1	1	0
0	1	1	1	0
1	1	1	1	1

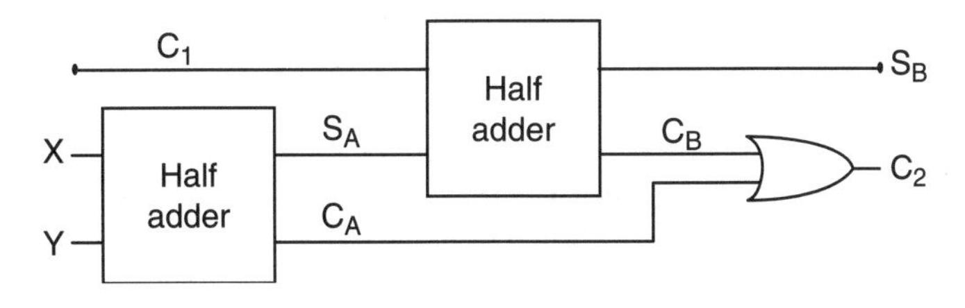

Figure 4.19 The full adder.

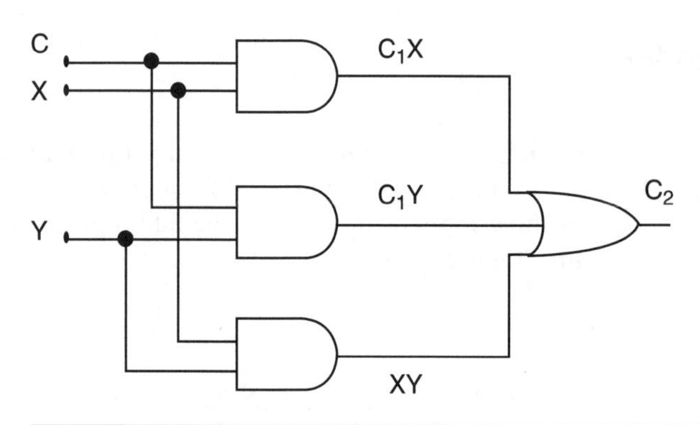

Figure 4.20 An implementation of the majority detector.

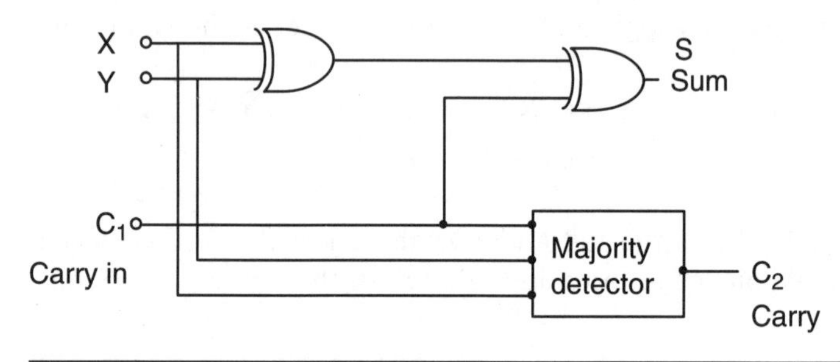

Figure 4.21 The full adder with majority detector.

The circuit shown in Figure 4.22 is able to add any two numbers of any size. The inputs are $X_2X_1X_0$ and $Y_2Y_1Y_0$, and the output is $C_3Z_2Z_1Z_0$.

EXAMPLE 4.3

If the input to the circuit in Figure 4.23 is written as a number ABCD, write the nine numbers that will yield a true Q.

Solution

The output of the circuit gate is:

$$Q = AB + CD + \overline{A}C\overline{D}$$

The truth table is shown in Table 4.14, from which we can deduce that ABCD = (2, 3, 6, 7, 11, 12, 13, 14, 15) gives Q true.

EXAMPLE 4.4

Using the two's complement notation, the 3-bit number ABC can represent the numbers from −3 to 3 as shown in Table 4.15

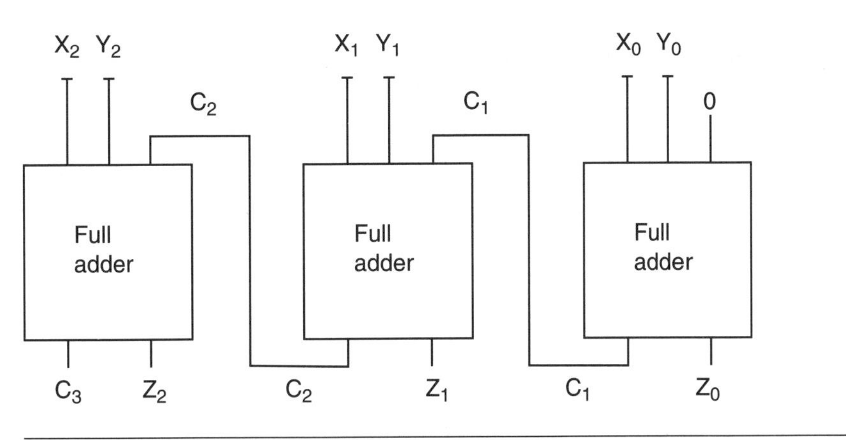

Figure 4.22 A circuit capable of adding two 3-bit numbers.

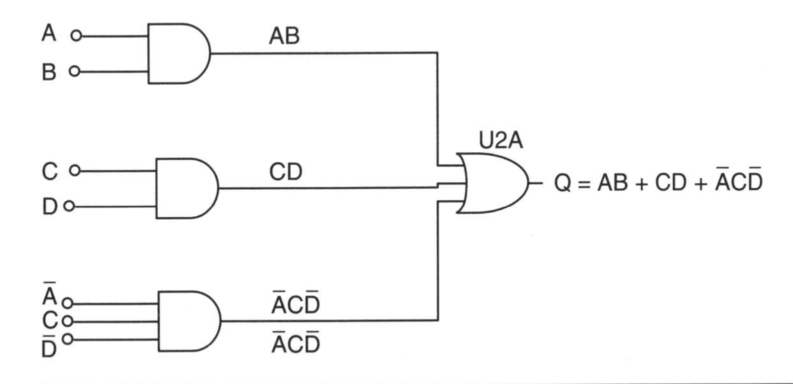

Figure 4.23 Circuit for Example 4.3.

(ignore −4). Assuming that A, B, C and $\overline{A}\,\overline{B}\,\overline{C}$ are available as inputs, devise a circuit that will yield a 2-bit output EF that is the absolute value of the ABC number, using only two- and three-input AND and OR gates.

Solution

1. Fill a truth table with the ABC and EF bits.
2. Write a Boolean algebra expression for E and for F.
3. Implement these expressions.

The truth table is shown in Table 4.15.

Table 4.14 Truth table for Example 4.3

A	*B*	*C*	*D*	*AB*	*CD*	$\overline{A}C\overline{D}$	*Q*
0	0	0	0	0	0	0	0
0	0	0	1	0	0	0	0
0	0	1	0	0	0	1	1
0	0	1	1	0	1	0	1
0	1	0	0	0	0	0	0
0	1	0	1	0	0	0	0
0	1	1	0	0	0	1	1
0	1	1	1	0	1	0	1
1	0	0	0	0	0	0	0
1	0	0	1	0	0	0	0
1	0	1	0	0	0	0	0
1	0	1	1	0	1	0	1
1	1	0	0	1	0	0	1
1	1	0	1	1	0	0	1
1	1	1	0	1	0	0	1
1	1	1	1	1	1	0	1

Table 4.15 Truth table for Example 4.4

Value	*A*	*B*	*C*	*E*	*F*
0	0	0	0	0	0
1	0	0	1	0	1
2	0	1	0	1	0
3	0	1	1	1	1
−1	1	1	1	0	1
−2	1	1	0	1	0
−3	1	0	1	1	1
−4	1	0	0		

The Boolean expressions are as follows:

$$
\begin{aligned}
E &= \overline{A}B\overline{C} + \overline{A}BC + AB\overline{C} + A\overline{B}C \\
&= \overline{A}B + A(B \oplus C) \\
F &= \overline{A}\,\overline{B}C + \overline{A}BC + ABC + A\overline{B}C \\
&= \overline{A}(\overline{B}C + BC) + A(\overline{B}C + BC) \\
&= \overline{A}C + AC \\
&= (\overline{A} + A)C \\
&= C
\end{aligned}
\tag{4.24A}
$$

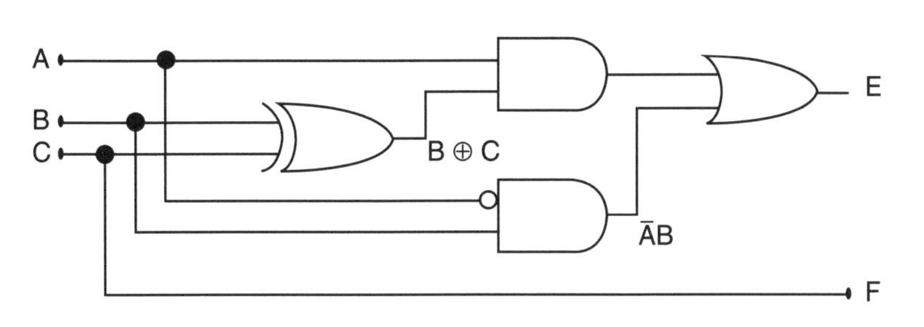

Figure 4.24 Circuit for Example 4.4.

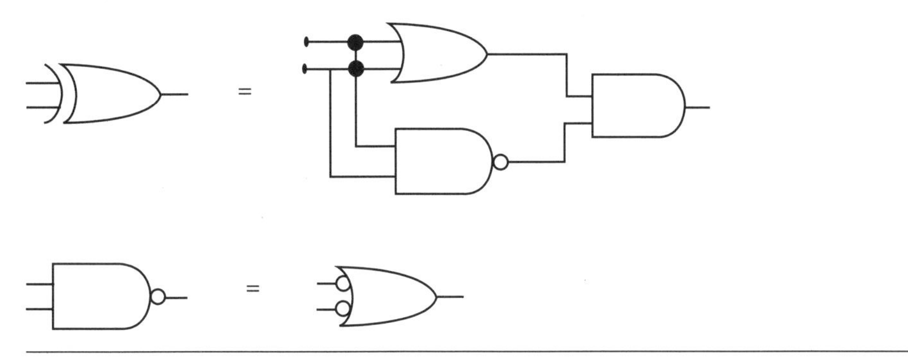

Figure 4.25 Gate equivalency for Example 4.4.

The implemented expressions are shown in Figure 4.24. The gate equivalency is shown in Figure 4.25.

EXAMPLE 4.5

Suppose that a 2-bit binary number AB is to be transmitted between devices in a noisy environment. To reduce undetected errors introduced by the transmission, an extra bit, a parity bit P, is often included in the transmission to add redundancy to the information. Assume that P is set true or false as needed to make an odd number of true bits in the resulting 3-bit number ABP. When the number is received, logic circuits are required to generate an error signal E whenever the odd number of bits condition is not met. Implement the logic gate to carry out the functions described.

Solution

Develop a truth table of E in terms of A, B and P.
The truth table is shown in Table 4.16.
The Boolean expression for E as determined directly from the truth table.

$$E = \overline{A}\,\overline{B}\,\overline{P} + \overline{A}\,B\,P + A\,\overline{B}\,P + AB\,\overline{P} \tag{4.25}$$

Table 4.16 Truth table for Example 4.5

A	B	P	E
0	0	0	1
0	0	1	0
0	1	0	0
0	1	1	1
1	0	0	0
1	0	1	1
1	1	0	1
1	1	1	0

Using de Morgan's theorem twice, this expression can be reduced to one EOR and one NEOR operation. (Note: this is very similar to the half adder problem).

$$\begin{aligned} E &= \overline{A}(\overline{B}\,\overline{P} + BP) + A(\overline{B}P + B\overline{P}) \\ &= \overline{A}\,\overline{(B \oplus P)} + A(B \oplus P) \\ &= \overline{[A \oplus \overline{(B \oplus P)}]} \end{aligned} \tag{4.26}$$

The implemented expressions are shown in Figure 4.26.

4.5.3 Multiplexers

Multiplexers and decoders are used when many lines of information are being gated and passed from one part of a circuit to another.

A multiplexer (MUX), also known as a data selector, is a combinational network containing up to 2^n data inputs, n control inputs, and an output (see Figure 4.27). The MUX allows one of the 2^n to be selected as the output. The control lines are used to make this selection. A MUX with 2^n input lines and n selection lines (referred to as a 1-of-2^n MUX) may be wired to realize any Boolean function of $n+1$ variables.

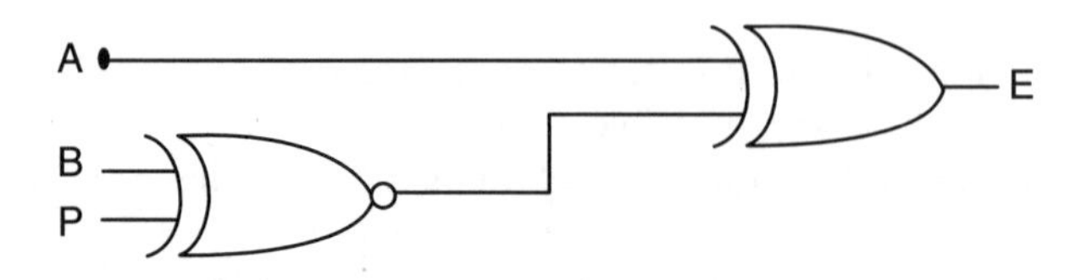

Figure 4.26 Gate implementation for Example 4.5.

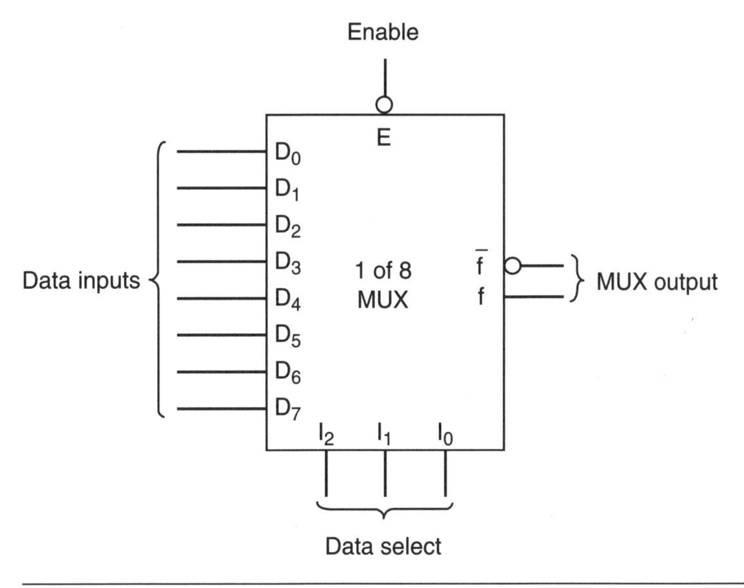

Figure 4.27 Multiplexer schematic.

Multiplexing is when multiple data signals share a common propagation path. Time multiplexing is when different signals travel along the same wire but at different times. These devices have data and address lines, and usually include an enable/disable input. When the device is disabled the output is locked into some particular state and is not affected by the inputs.

Figure 4.28 shows the gate arrangement for the multiplexer in Figure 4.27.

Multiplexers provide the designer with numerous choices without the need for minimizing logic for a particular application. They are also quite fast in operation, but are more expensive than basic logic gates.

4.5.4 Decoders

A decoder de-multiplexes signals back to several different lines. Figure 4.29 is a binary-to-octal decoder (3-line to 8-line decoder). Hexadecimal decoders are 4-line to 16-line devices. When the decoder is disabled the outputs will be high. A decoder is normally disabled while the address lines are changing to avoid glitches on the output lines.

4.5.5 Read and write memory

There are basically two main types of memory (read and write) from which other types are derived. These are now briefly described.

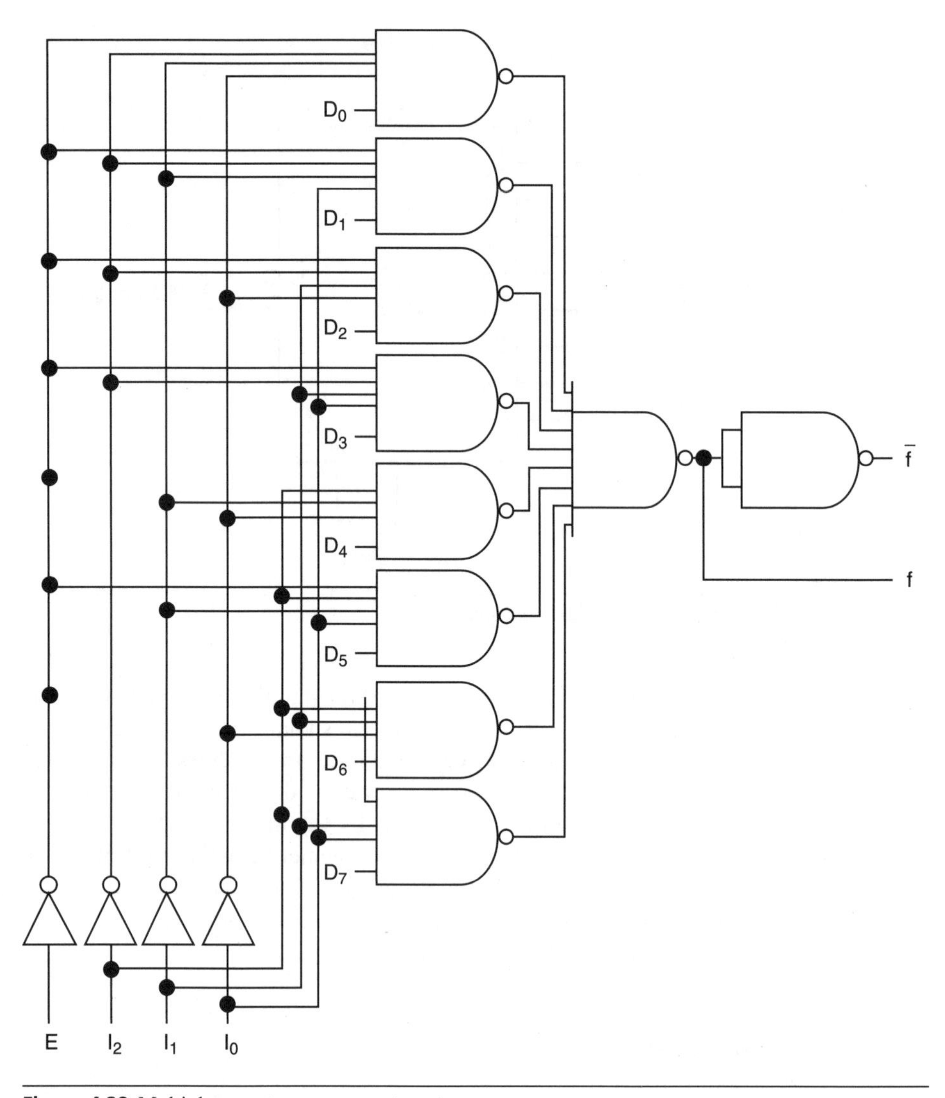

Figure 4.28 Multiplexer gate arrangement.

4.5.5.1 *RAM*

In random access memory (RAM), a particular memory location can be specified, and after a set amount of time, the contents of that location are available for reading or writing. This delay time is very expensive. Caches are much faster than regular RAM, and therefore more expensive (in terms of money). Modern RAM requires a refresh signal to regenerate the data that it contains. Older RAM (especially core memory made from little cores of iron that looked like miniature

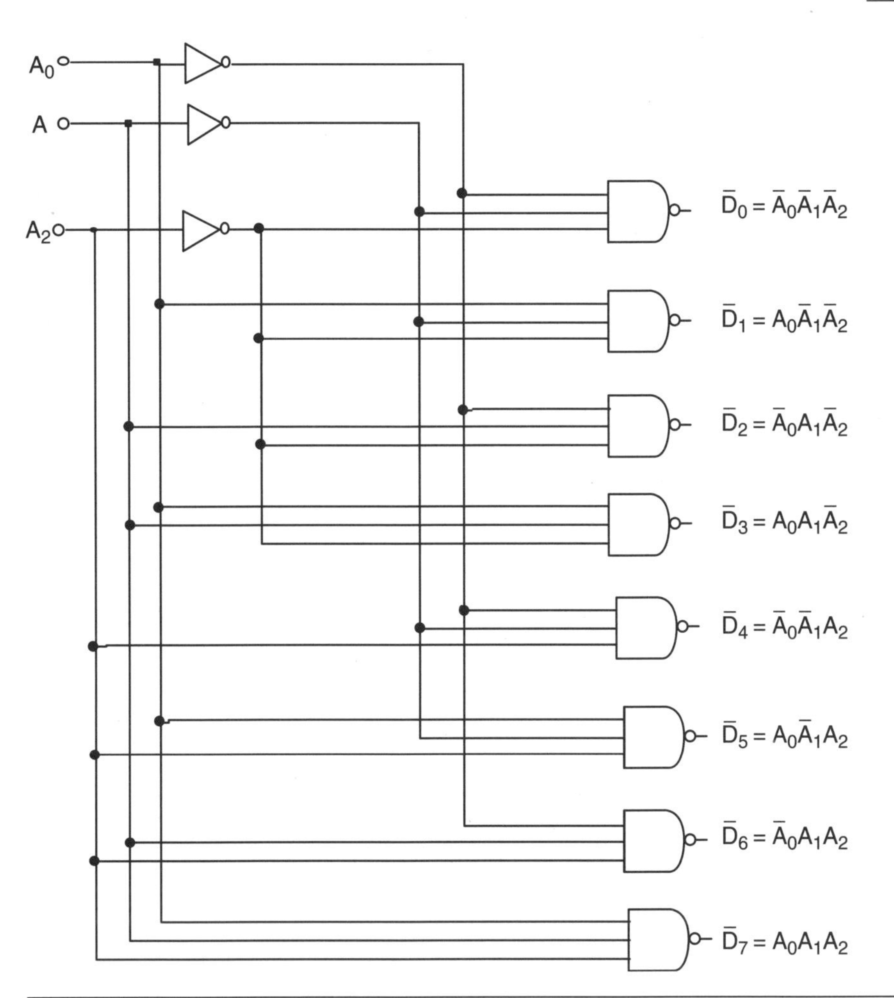

Figure 4.29 Octal decoder.

donuts) retained this signal until it was explicitly changed. That meant that if a computer lost power and then was powered back up, the memory still held all its data. In this configuration, core memory was set and read by induced current. A charge moving through a wire affects charge in a nearby wire.

In modern computer memory technology, we come across terms such as DRAM, SIMM, SDRAM, and DIMM. These are types of RAM that differ in implementation, but can be treated identically. CMOS (Figure 4.30) is a type of RAM implementation that uses a small battery to keep the refresh signal running even when the computer is off. CMOS is usually used for configurations that need to be present at boot time but might change. The important thing to remember about RAM is that you can use any part of it you want, by specifying an address in memory, waiting a little while, and then doing whatever you were planning on.

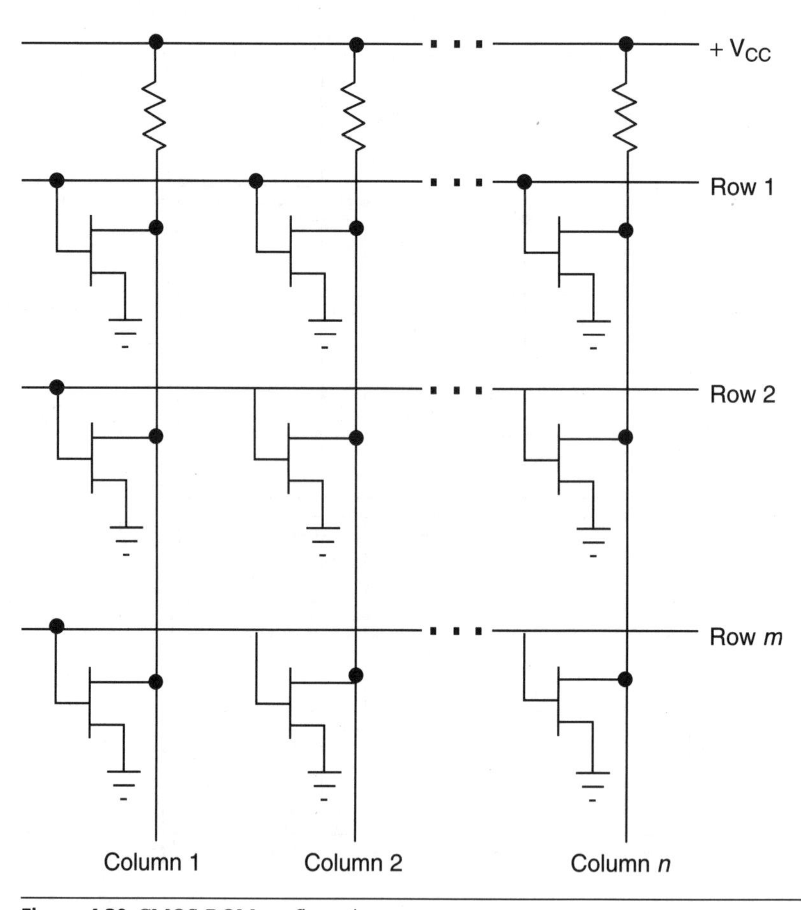

Figure 4.30 CMOS ROM configuration.

4.5.5.2 *ROM*

ROM stands for read-only memory, which means that it is basically binary digits etched in silicon. Note that this makes it very suitable for information that is needed to boot a computer, but very inappropriate for boot-time configurations, since that kind of information changes often.

The information held in ROM is specified by the designer and embedded in the unit to form the required interconnection pattern. The field programmable ROM (FPROM) is a variant of normal ROM in which the pattern may be defined by the user. Another ROM type is erasable PROM (EPROM). Here the program inside the chip can be erased using strong ultraviolet light. ROM has address inputs to select a particular memory location, data outputs carrying the information from the selected location, and enable inputs.

To use the device as a combinational logic circuit implementing a Boolean expression, the address inputs of the memory become the Boolean input variables and the data outputs become the required functions. The functions need to be expanded into canonical form (every variable appearing in each term) for ROM implementation (where every term represents an address).

The biggest problem with ROM is that if you make a mistake 'burning' the data in, you can not easily correct it later. However, software is available that allows you to emulate a program before you finally commit it to the chip.

The only important difference between ROM and RAM is that ROM is permanent (for example, the BIOS of a desktop PC), whereas RAM is volatile.

4.5.5.3 *Programmable logic array (PLA)*

Another type of ROM is the PLA. This comprises an array of many fuses which can be blown during programming to configure the device to simulate logical expressions. This development is much more economical than ordering huge batches of ROM to be burnt with a possible faulty design, but is still so expensive that it is primarily used only for prototypes.

A combinational circuit may have 'don't care' conditions. In a ROM implementation, these conditions become addresses that will never occur and because not all the bit patterns available are used, it is considered a waste of resource. For cases where the number of 'don't care' conditions is excessive, it is more economical to use PLAs (Figure 4.31). It is similar to a ROM in concept, but does not provide full decoding of the variables and does not generate all the 'min-terms' as in the ROM. Field-programmable and erasable variants are also available.

4.5.5.4 *Two-state storage elements*

Analog voltage storage times are limited since the charge on a capacitor will eventually leak away. The problem of discrete storage reduces to the need to store a large number of two-state variables. The four commonly used methods are:

1. magnetic domain orientation;
2. presence or absence of charge (not amount of charge) on a capacitor;
3. presence or absence of an electrical connection; and
4. the d.c. current path through the latches and flip-flops of a digital circuit.

The only one of these to concern us is the last which we will discuss in a later section.

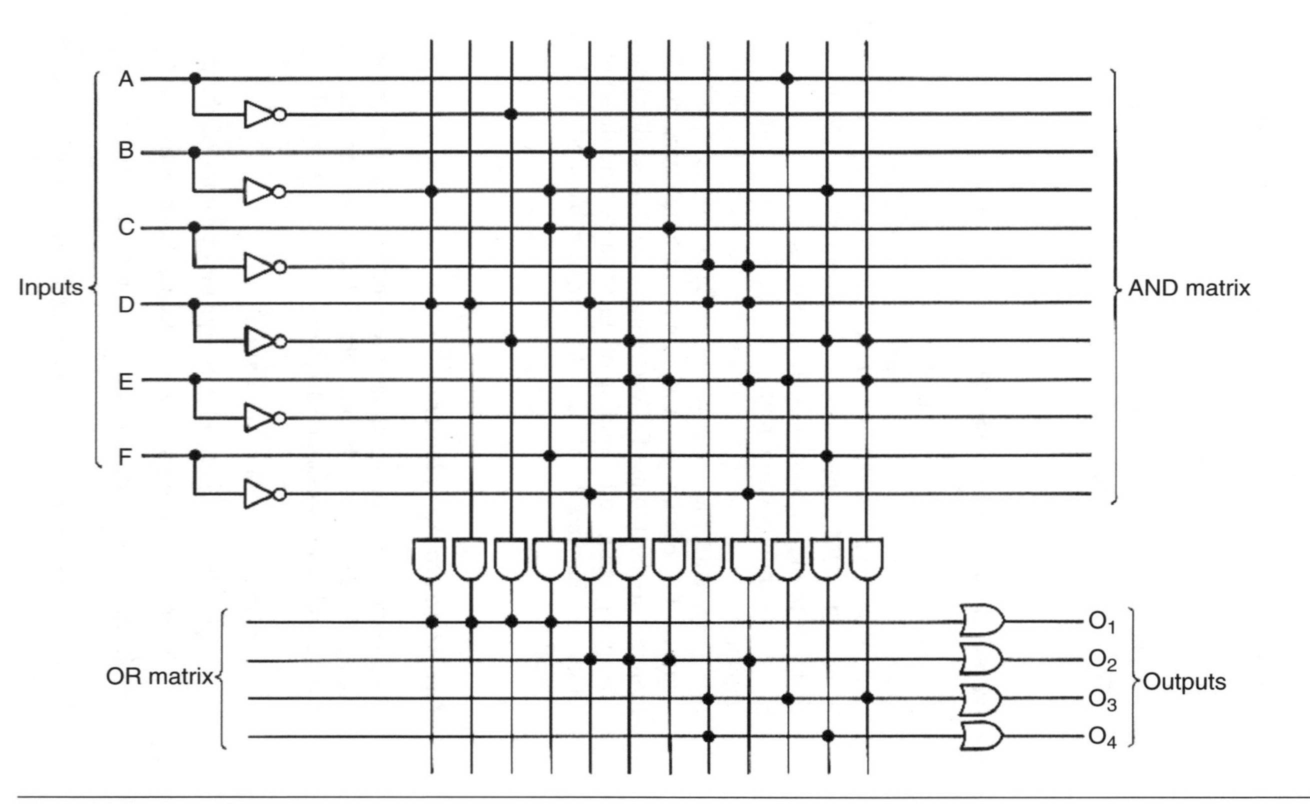

Figure 4.31 PLA configuration.

4.6 Timing diagrams

We discuss timing diagrams before discussing flip-flops because they are useful in understanding how flip-flops behave.

Normally signals flip from one logic state to another. The time it takes the signal to move between states is the *transition time* (t_t), which is measured between 10 percent and 90 percent of the peak-to-peak signal levels. Delays within the logic elements result in a *propagation (pulse) delay* (t_{pd}), where the time is measured between 50 percent of the input signal and 50 percent of the output response. Definitions of the transition time and propagation delay are shown in Figure 4.32.

Signal racing is the condition when two or more signals change almost simultaneously. The condition may cause glitches or spikes in the output signal as shown in Figure 4.33. The effects of these glitches can be eliminated by using synchronous timing techniques. In synchronous timing the glitches are allowed to come and go, and the logic state changes are initiated by a timing pulse (clock pulse).

4.7 Sequential logic components

Unlike combinational logic devices which are based on a combination of present inputs only, sequential logic devices provide outputs that depend on present and past input values. As a result of this memory property, sequential logic devices can store information and consequently they are extremely important in digital logic circuits. We will first consider different types of flip-flops and then their applications in counters, registers, etc.

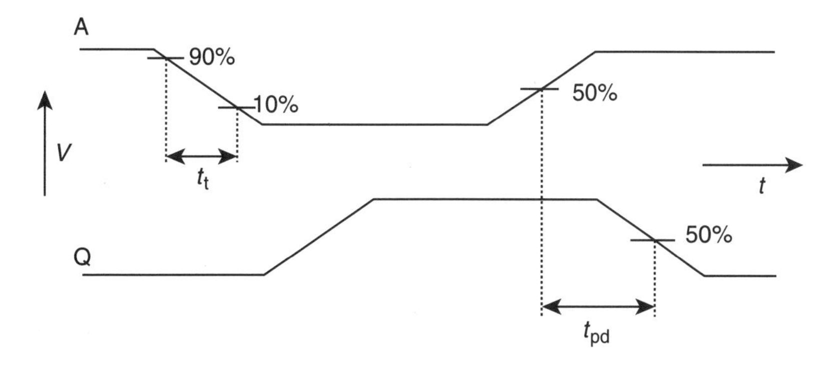

Figure 4.32 Transition time and propagation delay.

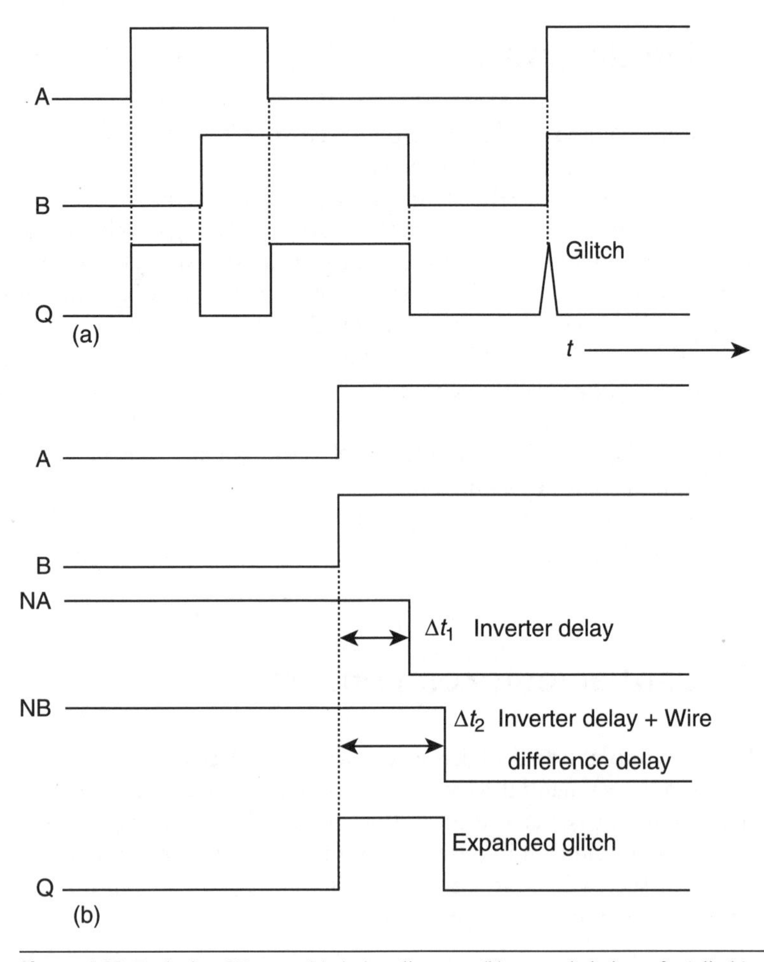

Figure 4.33 Exclusive OR gate: (a) timing diagram; (b) expanded view of a 'glitch' caused by signal racing.

4.7.1 Latches and flip-flops

It is possible using basic logic gates to build a circuit that remembers its present condition. It is also possible to build counting circuits. The basic counting unit is the flip-flop or latch. All flip-flops have two inputs: data and enable/disable, and typically Q and $\overline{Q}$ outputs. A *ones-catching* latch can be built as shown in Figure 4.34.

When the control input C is false, the output Q follows the input D, but when the control input goes true, the output latches true as soon as D goes true and then stays there independent of further changes in D.

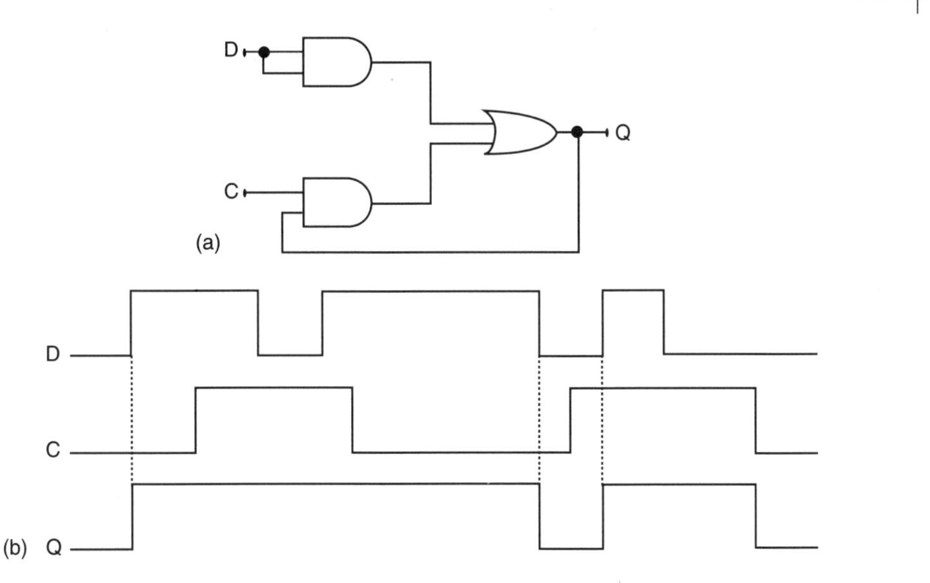

Figure 4.34 (a) AND and OR gates used as a *ones-catching* latch; (b) the associated timing diagram.

One of the most useful latches is known as the *transparent* latch or D-type latch. The transparent latch is like the ones-catching latch but the input D is frozen when the latch is disabled. The operation of this latch is the same as that of the statically triggered D-type flip-flop discussed later.

4.7.2 The reset-set (R-S) flip-flop

The R-S flip-flop can be built by cross-connecting two NOR gates as shown in Figure 4.35. The truth table is shown in Table 4.17.

The ideal flip-flop has only two rest states, set and reset, defined by $Q\overline{Q} = 10$ and $Q\overline{Q} = 01$, respectively.

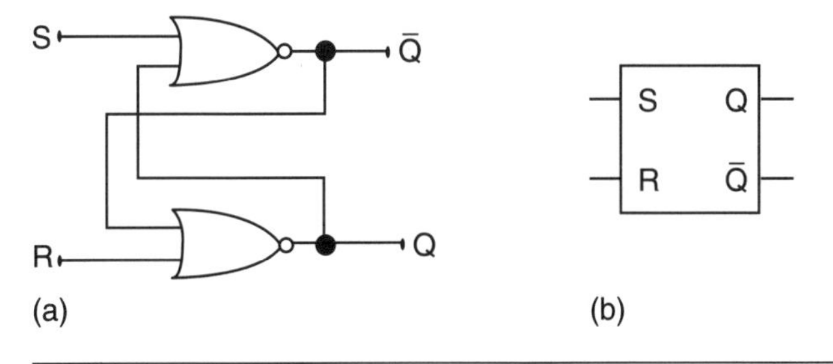

Figure 4.35 R-S flip-flop: (a) using NOR gates; (b) schematic.

Table 4.17 Truth table for R-S flip-flip.

S	R	Q	$\overline{Q}$
0	0	no change	
0	1	0	1
1	0	1	0
1	1	undefined	

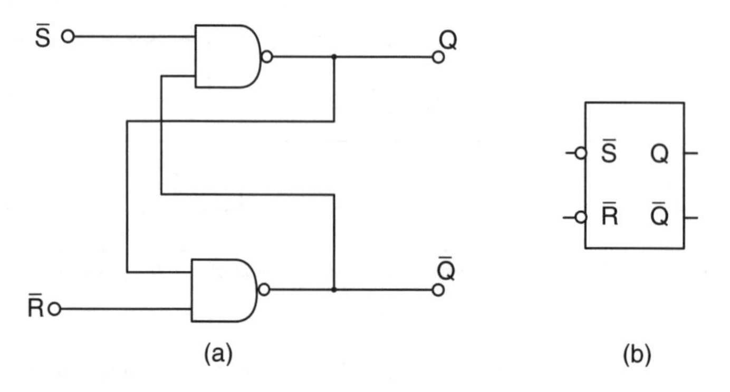

Figure 4.36 R-S flip-flop: (a) using NAND gates; (b) schematic.

Table 4.18 Alternative truth table for R-S flip-flop

S	R	Q	$\overline{Q}$
0	0	no change	
0	1	1	0
1	0	0	1
1	1	undefined	

A very similar flip-flop can be constructed using two NAND gates as shown in Figure 4.36 and Table 4.18.

4.7.3 The clocked flip-flop

A clocked flip-flop has an additional input that allows output state changes to be synchronized to a clock pulse.

We distinguish two types of clock inputs. These are (a) *static clock input* in which the clock input is sensitive to the signal level, and (b) *dynamic clock input* in which a clock input is sensitive to the signal edges.

4.7.3.1 *The clocked R-S flip-flop*

The static clocked (level-sensitive) R-S flip-flop is shown in Figure 4.37 and its truth table is given in Table 4.19. The symbol X represents either the binary state 0 or 1.

The first five rows in the truth table give the static input and output states. The last four rows show the state of the outputs after a complete clock pulse *p*.

4.7.3.2 *The D-type flip-flop*

The D-type flip-flop avoids the undefined states that occur in the R-S flip-flop by reducing the number of input options. Statically triggered D-type flip-flops (transparent latch) are implemented with clocked R-S flip-flop (see Figure 4.38 and Table 4.20).

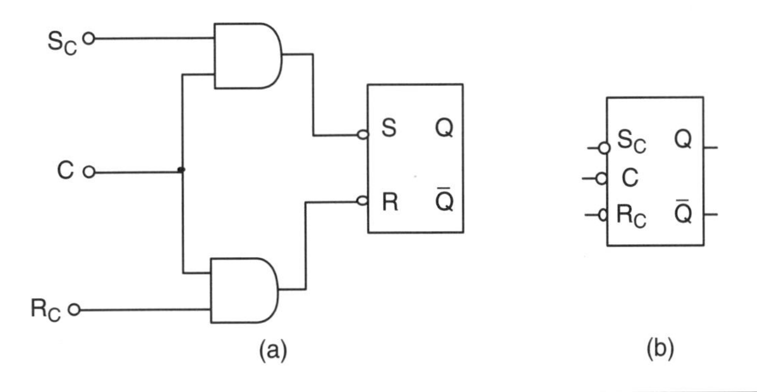

Figure 4.37 Clocked R-S flip-flop: (a) construction; (b) schematic.

Table 4.19 Truth table for clocked R-S flip-flop

S_c	R_c	C	Q	$\bar{Q}$
X	X	0	no change	
0	0	1	no change	
0	1	1	0	1
1	0	1	1	0
1	1	1	undefined	
0	0	*p*	no change	
0	1	*p*	0	1
1	0	*p*	1	0
1	1	*p*	undefined	

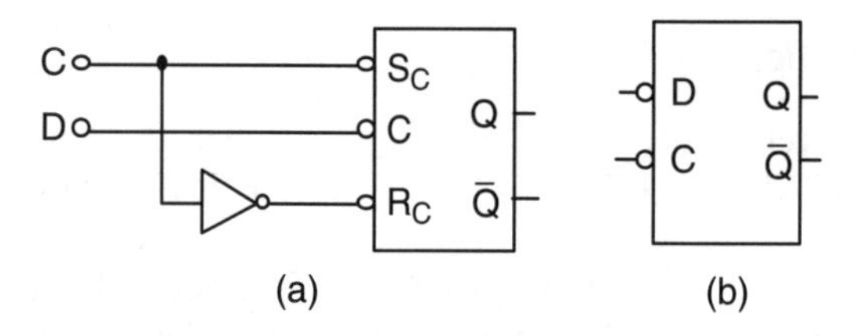

Figure 4.38 D-type flip-flop: (a) construction; (b) schematic.

Table 4.20 Truth table for D-type flip-flop

S_c	R_c	C	Q	$\overline{Q}$
X	0	no change		
0	p	0	1	
1	p	1	0	

4.7.3.3 *The J-K flip-flop*

The J-K flip-flop simplifies the R-S flip-flop truth table but keeps two inputs (Table 4.21). The basic J-K flip-flop is constructed from an R-S flip-flop with the addition of the gates shown in Figure 4.39. The toggle state is useful in counting circuits. If the C pulse is too long this state is undefined and hence the J-K flip-flop can only be used with rigidly defined short clock pulses.

Table 4.21 Truth table for J-K flip-flop

J	K	C	Q	$\overline{Q}$
0	0	p	no change	
0	1	p	0	1
1	0	p	1	0
1	1	p	toggle	

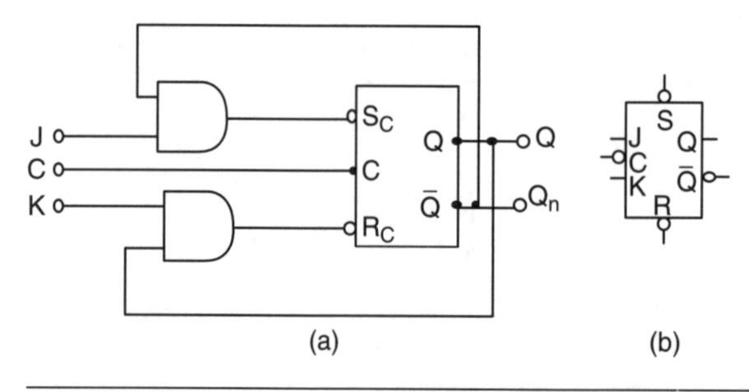

Figure 4.39 J-K flip-flop: (a) construction; (b) schematic.

4.7.4 The master–slave flip-flop

We can simulate a dynamic clock input by putting two flip-flops in tandem, one driving the other, in a master–slave arrangement (as shown in Figure 4.40). The slave is clocked in a complementary fashion to the master.

This arrangement is *pulse triggered*. The data inputs are written to the master flip-flop while the clock is true and transferred to the slave when the clock becomes false. The arrangement guarantees that the $Q\overline{Q}$ outputs of the slave can never be connected to the slave's own RS inputs. The design overcomes *signal racing* (i.e. the input signals never catch up with the signals already in the flip-flop). There are however a few special states when a transition can occur in the master and be transferred to the slave when the clock is high. These are known as *ones catching* and are common in master–slave designs.

4.7.5 Edge triggering

Edge triggering is when the flip-flop state is changed as the rising or falling edge of a clock signal passes through a threshold voltage (Figure 4.41). This true dynamic clock input is insensitive to the slope or time spent in the high or low state.

Both types of dynamic triggering are represented on a schematic diagram by a special symbol (>) near the clock input (Figure 4.41 and Table 4.22). In addition to the clock and data inputs most IC flip-flop packages will also include *set* and *reset* (or mark and erase) inputs. The additional inputs allow the flip-flop to be *preset* to an initial state without using the clocked logic inputs.

EXAMPLE 4.6

Figure 4.43 shows a 3-bit binary ripple counter. Draw the timing diagram for the clock input, Q_1, Q_2, and Q_3.

Solution

We assume negative-edge-triggered devices. The timing diagram is shown in Figure 4.44.

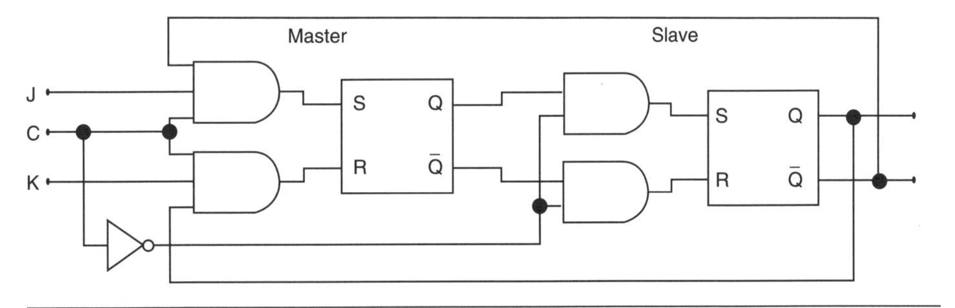

Figure 4.40 Master–slave flip-flop.

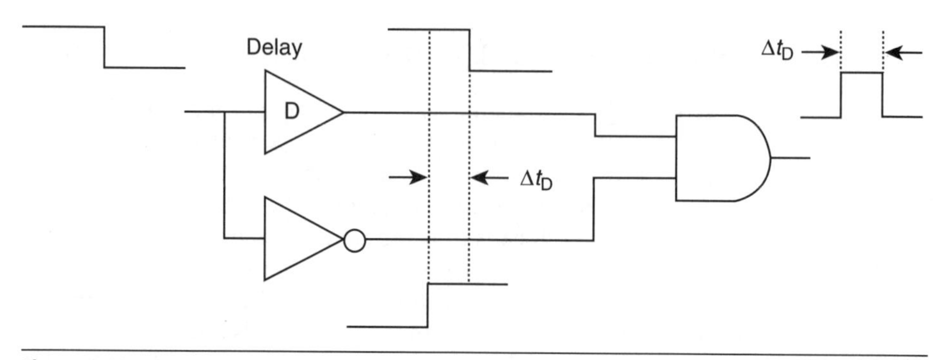

Figure 4.41 Edge triggering.

Table 4.22 Truth table indicating edge triggering

J	K	C	S	R	Q	$\bar{Q}$
0	0	↓	1	1	no change	
0	1	↓	1	1	0	1
1	0	↓	1	1	1	0
1	1	↓	1	1	toggle	
X	X	X	0	1	1	0
X	X	X	1	0	0	1

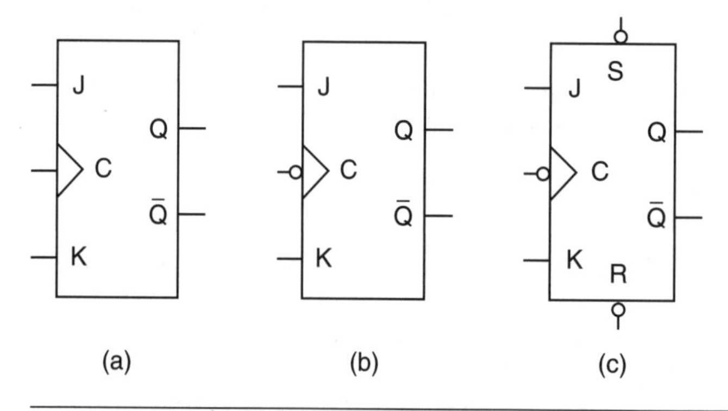

Figure 4.42 J-K flip flop: (a) positive edge-triggered; (b) negative edge-triggered; (c) negative edge-triggered with set and reset inputs.

4.8 Sequential logic design

Combinational logic outputs depend only on the current inputs. There is no 'memory'. Sequential logic outputs depend on the current and past sequence of

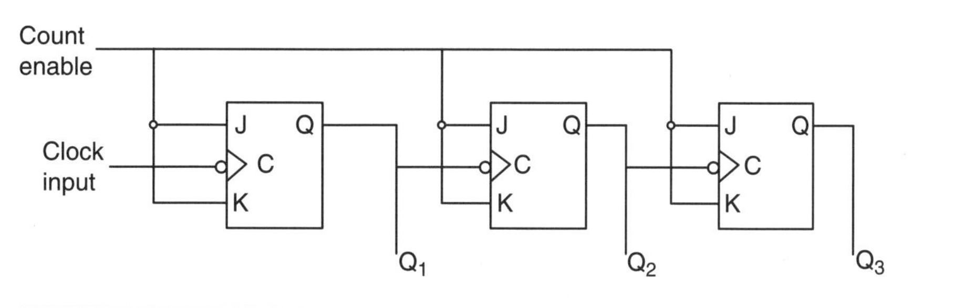

Figure 4.43 Circuit for Example 4.6.

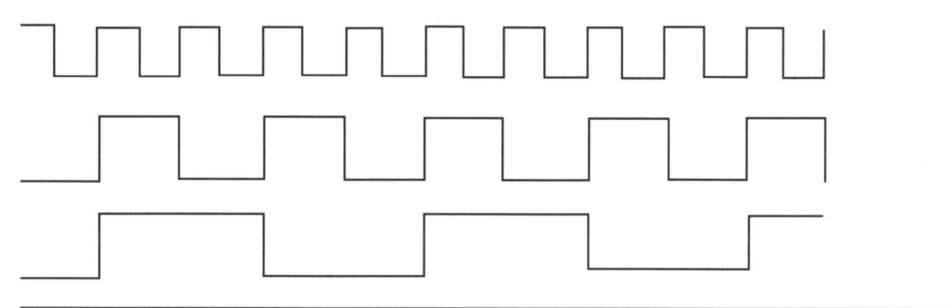

Figure 4.44 Timing diagram for Example 4.6.

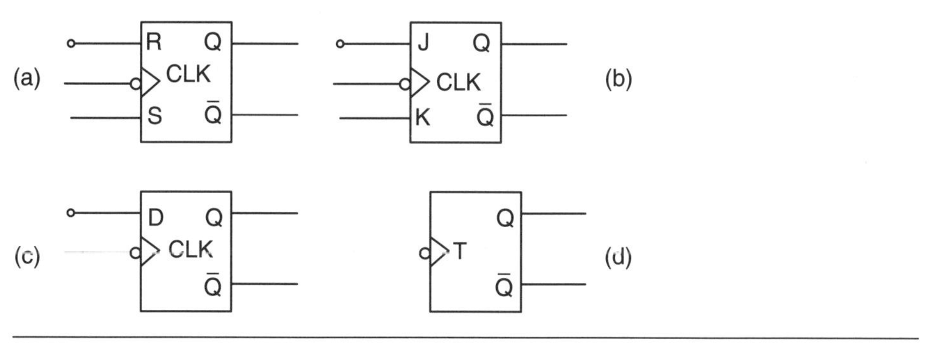

Figure 4.45 Flip-flops: (a) R-S; (b) J-K; (c) D-type; and (d) T-type.

inputs. There is 'memory'. Figure 4.45 shows four flip-flops commonly used: (a) R-S; (b) J-K; (c) D-type; and (d) T-type.

4.8.1 The D-type flip-flop

The D-type flip-flop (Table 4.23) is a common memory component. It is normally positive edge-triggered.

Table 4.23 Truth table for D-type flip-flop indicating edge triggering

D	*CLK*	*Q*	$\overline{Q}$
0	$\uparrow$	0	1
1	$\uparrow$	1	0
X	0	last Q	last $\overline{Q}$
X	1	last Q	last $\overline{Q}$

4.8.2 The J-K flip-flop

The J-K flip-flop (Table 4.24) is a two input memory component. It is normally positive or negative edge-triggered. Its use may result in a simpler circuit under some circumstances.

4.8.3 The T-type flip-flop

As the name implies, a toggle flip-flop (Table 4.25) toggles between two states.

4.8.4 Clocked synchronous state machine

The clocked synchronous state-machine is also referred to as synchronous finite state machine in which all the flip-flops are simultaneously clocked. This is a more realistic structure unlike the asynchronous state-machine in which the output of one flip-flop clocks the subsequent flip-flop.

Table 4.24 Truth table for J-K flip-flop indicating edge triggering

J	*K*	*CLK*	*Q*	$\overline{Q}$
X	X	0	last Q	last $\overline{Q}$
X	X	1	last Q	last $\overline{Q}$
0	0	$\uparrow$	last Q	last $\overline{Q}$
0	1	$\uparrow$	0	1
1	0	$\uparrow$	1	0
1	1	$\uparrow$	last $\overline{Q}$	last Q

Table 4.25 Truth table for T-type flip-flop

T	*Q*	$\overline{Q}$
$\uparrow$	change state	change state

4.8.5 General state machine structures

There are two commonly used structures: (a) the Mealy model; and (b) the Moore model. In the Mealy model (Figure 4.46), the outputs depend on the current inputs and state.

In the Moore model (Figure 4.47), the outputs depend on the current state only.

4.8.5.1 *Analysis of a state machine circuit*

The analysis of a state machine circuit starts with a circuit diagram and finishes with a state diagram. The next states and outputs are functions of current state and inputs.

The three general steps are:

1. Determine the next state and output functions as Boolean expressions;
2. Use the Boolean expressions to construct the state/output table. This table holds information about the next state and output for every possible state and input combination.
3. Draw a state diagram that presents Step 2 in a graphical form.

4.8.5.2 *Design of a state machine circuit*

The general design procedures for state machine circuits are:

1. Construct a state diagram corresponding to the word description or specification.

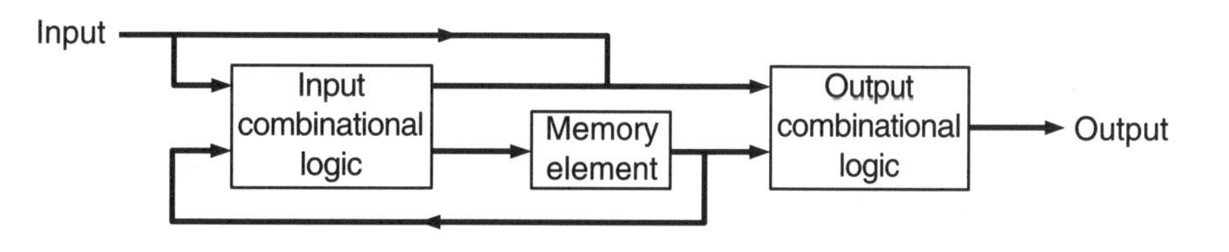

Figure 4.46 The Mealy model state machine.

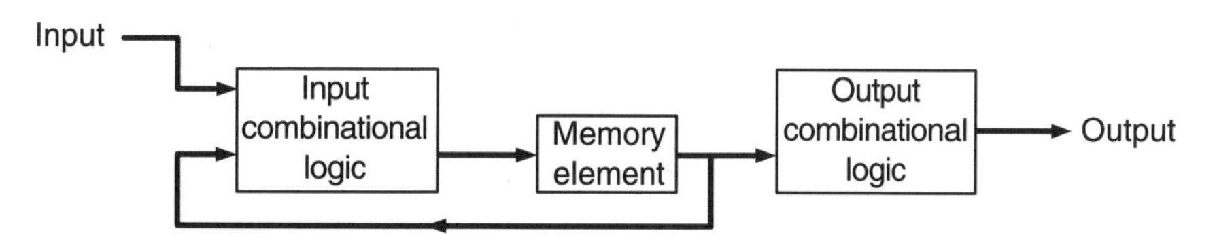

Figure 4.47 The Moore model state machine.

2. Construct a next state/output table from the state diagram.
3. Minimize the number of states in the state/output table.
4. Assign binary bits to the named states to produce a fully assigned state/output table.
5. Choose a flip-flop type to construct the excitation table. The number of flip-flops required is related to the number of states.
6. Derive excitation equations and output equations.
7. Draw a circuit diagram that shows the complete circuit.

The design of sequential circuits can be carried out by a systematic procedure. A state diagram, with its associated state transition table, is the equivalent of a Karnaugh map. The initial specification for a logic circuit is usually in the form of either a transition table or a state diagram. The design will differ depending on the type of flip-flop chosen. Therefore, the first step in sequential logic design is to choose a flip-flop and define its behavior in the form of an excitation table.

4.8.5.3 *Characteristic equations*

The characteristic equations for the R-S, D-type, T-type, and J-K flip-flops are shown in Table 4.26.

4.8.5.4 *Truth tables and excitation tables*

Truth tables and excitation tables for the R-S, D-type, T-type, and J-K flip-flops are shown in Tables 4.27–4.30.

4.8.6 Modulo-*N* binary counters

There are several modulo-N binary counters. The modulus of a binary counter is 2^N, where N is the number of flip-flops in the counter. This follows from the fact

Table 4.26 Characteristic equations for the R-S, D-type, T-type, and J-K flip-flops

Flip-flop types	*Characteristic equations*
R-S	$Q_{t+1} = S + \overline{R}Q_t$ (SR = 0)
J-K	$Q_{t+1} = J\overline{Q_t} + \overline{K}Q_t$
D	$Q_{t+1} = D$
T	$Q_{t+1} = T\overline{Q_t} + \overline{T}Q_t = T \oplus Q_t$

Table 4.27 Truth table and excitation table for the R-S flip-flop

Truth table for R-S flip-flop				*Excitation table for R-S flip-flop*			
S	*R*	Q_t	Q_{t+1}	Q_t	Q_{t+1}	*S*	*R*
0	0	0	0	0	0	0	d[b]
0	0	1	1	0	1	1	0
0	1	0	0	1	0	0	1
0	1	1	0	1	1	d	0
1	0	0	1				
1	0	1	1				
1	1	X[a]	X				
1	1	X	X				

[a]: not allowed; [b]: don't care

Table 4.28 Truth table and excitation table for the D-type flip-flop

Truth table for D-type flip-flop			*Excitation table for D-type flip-flop*		
D	Q_t	Q_{t+1}	Q_t	Q_{t+1}	*D*
0	0	0	0	0	0
0	1	0	0	1	1
1	0	1	1	0	0
1	1	1	1	1	1

Table 4.29 Truth table and excitation table for the T-type flip-flop

Truth table for T-type flip-flop			*Excitation table for T-type flip-flop*		
T	Q_t	Q_{t+1}	Q_t	Q_{t+1}	*T*
0	0	0	0	0	0
0	1	1	0	1	1
1	0	1	1	0	1
1	1	0	1	1	0

that there are 2^N combinations of 0s and 1s consisting of *N* bits. For example, in the case of binary up counter, the counting sequence is $00\ldots0_2$ to $11\ldots1_2$, which is equivalent to $(2^N - 1)_{10}$. The counting sequence for a binary down counter is the opposite $11\ldots1_2$ to $00\ldots0_2$. We will examine how to design a modulo-*N* counter. The state transition diagram for an up–down counter is shown in Figure 4.48.

Table 4.30 Truth table and excitation table for the J-K flip-flop

Truth table for J-K flip-flop				*Excitation table for J-K flip-flop*			
J	*K*	Q_t	Q_{t+1}	Q_t	Q_{t+1}	*J*	*K*
0	0	0	0	0	0	0	d
0	0	1	1	0	1	1	d
0	1	0	0	1	0	d	1
0	1	1	0	1	1	d	0
1	0	0	1				
1	0	1	1				
1	1	0	1				
1	1	1	0				

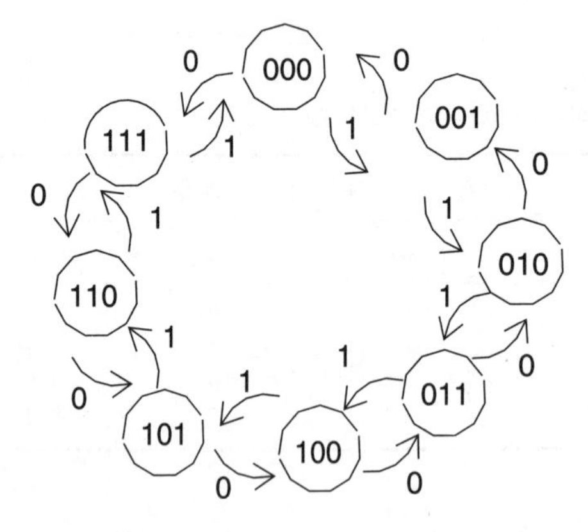

Figure 4.48 State transition diagram for an up–down counter.

4.8.6.1 *Design of modulo-4 binary up–down counter*

The state transition diagram for a modulo-4 binary up–down counter is shown in Figure 4.49. For example, if the current state of the counter is 2, an addition of 1 will cause the counter to change up to 3 while an input of 0 will cause the counter to count down to 1. The states of a modulo-4 binary up–down counter are given in Table 4.31.

The first and second columns of the first row of Table 4.31 show the first current state 0 (00) which when a 0 is added counts down to 3 (11) in the third and fourth columns of the first row. The first and second columns of the second row show the second current state 1 (01) which when a 0 is added counts down to 0 (00) in the third and fourth columns of the second row. The first and second columns of the third row show the third current state 2 (10) which when a 0 is added

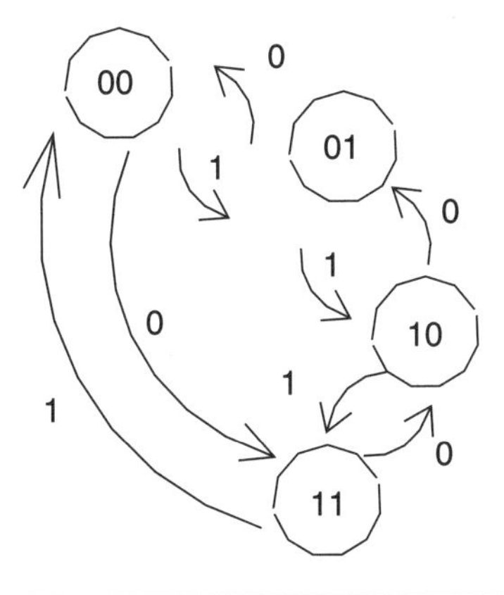

Figure 4.49 State transition diagram for a modulo-4 up–down counter.

Table 4.31 States of a modulo-4 binary up–down counter

Current state q_1	*Current state q_2*	*Next state Q_1*	*Next state Q_2*
0	0	1	1
0	1	0	0
1	0	0	1
1	1	1	0
0	0	0	1
0	1	1	0
1	0	1	1
1	1	0	0

counts down to 1 (01) in thc third and fourth columns of thc third row. Thc first and second columns of the fourth row show the fourth current state 3 (11) which when a 0 is added counts down to 2 (10) in the third and fourth columns of the third row.

The table also shows the next state for additions of 1s to the current state.

Step 1: There are two current states q_1 and q_2, and one input X. There are two given next states Q_1 and Q_2. We fill up the first five columns of Table 4.32 and the information is used to count up or down.

Step 2: Using q_1 and Q_1, find S_1 and R_1 from the excitation table (Table 4.27) and use these to fill in columns 6 and 7. For example, in the first row, q_1 and Q_1 (Q_t and Q_{t+1}) have values of 01, so S and R have values of 1 and 0, respectively.

Table 4.32 State transition table for a modulo-4 up–down counter

X	*Current State q_1*	*Current State q_2*	*Next State Q_1*	*Next State Q_2*	*S_1*	*R_1*	*S_2*	*R_2*	*Y*
0	0	0	1	1	1	0	1	0	1
0	0	1	0	0	0	d	0	1	0
0	1	0	0	1	0	1	1	0	1
0	1	1	1	0	d	0	0	1	0
1	0	0	0	1	0	d	1	0	1
1	0	1	1	0	1	0	0	1	0
1	1	0	1	1	d	0	1	0	1
1	1	1	0	0	0	1	0	1	0

Step 3: Using q_2 and Q_2, find S_2 and R_2 from the excitation table and use these to fill in columns 8 and 9. For example, in the first row, q_2 and Q_2 (Q_t and Q_{t+1}) have values of 01, so S and R have values of 1 and 0, respectively. Other rows can similarly be filled.

Step 4: Represent the state transition table using a Karnaugh map (X against q_1 and q_2); X is along the ordinate (0 and 1) and q_1 and q_2 are along the abscissa (00, 01, 11, and 10) as shown in Figure 4.50.

$$
\begin{aligned}
S_1 &= \bar{x}\,\overline{q_1}\,\overline{q_2} + x\overline{q_1}q_2 = (\bar{x}\,\overline{q_2} + xq_2)\overline{q_1} \\
S_2 &= \overline{q_2} \\
R_1 &= \bar{x}q_1\overline{q_2} + xq_1q_2 = (\bar{x}\,\overline{q_2} + xq_2)q_1 \\
R_2 &= q_2
\end{aligned}
\qquad (4.26A)
$$

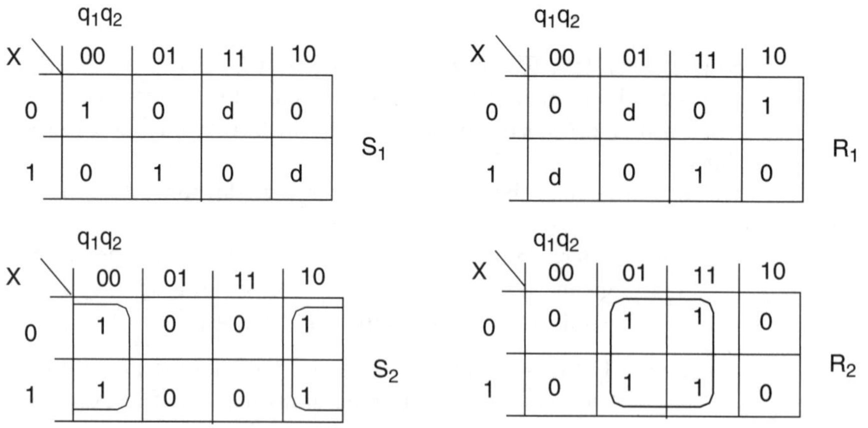

S_1

X \ q_1q_2	00	01	11	10
0	1	0	d	0
1	0	1	0	d

R_1

X \ q_1q_2	00	01	11	10
0	0	d	0	1
1	d	0	1	0

S_2

X \ q_1q_2	00	01	11	10
0	1	0	0	1
1	1	0	0	1

R_2

X \ q_1q_2	00	01	11	10
0	0	1	1	0
1	0	1	1	0

Figure 4.50 Karnaugh maps for modulo-4 design.

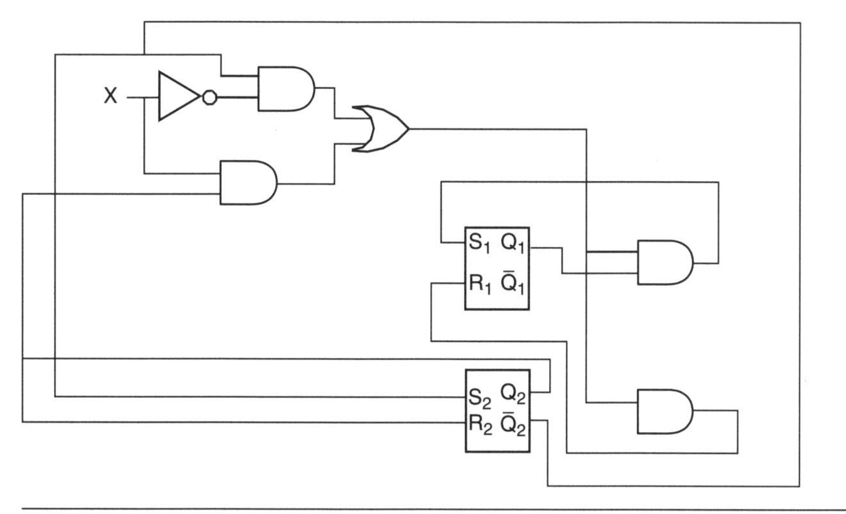

Figure 4.51 Modulo-4 counter implementation.

Step 5: Complete the design and draw the required circuit as shown in Figure 4.51.

EXAMPLE 4.7

Complete the circuit design for the Moore machine state diagram shown in Figure 4.52 using D-type flip-flops.

Solution

The next state (S*)/output is obtained from the present state (S). There are four states (A, B, C, D).

- A is connected to A and B with an output of 0.
- B is connected to B and C with an output of 0.
- C is connected to A and D with an output of 0.
- D is connected to A and B with an output of 1.

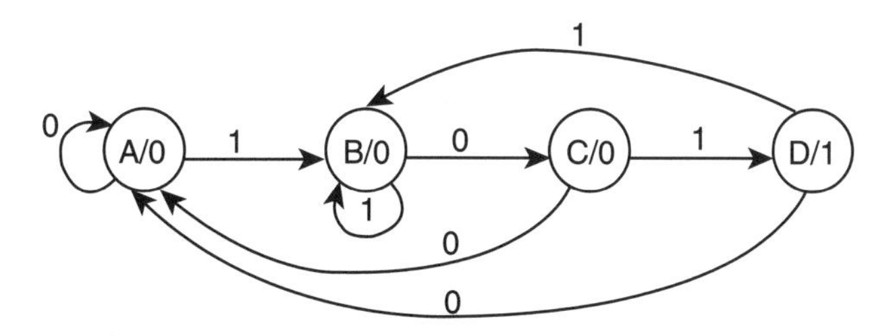

Figure 4.52 Moore machine state diagram.

Putting these together, we obtain the next state/output shown in Table 4.33. The input (0/1) is the next state (S*).

This table is expanded using the values on the arrow arcs as the inputs. The extended next state/output is shown in Table 4.34.

We now assign binary bits to named states as follows:

$$A = 00$$
$$B = 01$$
$$C = 11$$
$$D = 01$$

These assigned binary bits are now used to obtain Table 4.35.

We now create a fully assigned table with excitation values as shown in Table 4.36.

We now derive the excitation and output equations by considering conditions where D_1 and D_0 are equal to 1 and then writing the conditions of Q_1 and Q_0:

$$\begin{aligned} D1 &= (\overline{Q1}^{*}Q0^{*}\overline{In}) + (Q1^{*}Q0^{*}In) \\ D0 &= (\overline{Q1}^{*}\overline{Q0}^{*}In) + (\overline{Q1}^{*}Q0^{*}\overline{In}) + (\overline{Q1}^{*}Q0^{*}In) + (Q1^{*}\overline{Q0}^{*}In) \end{aligned} \tag{4.26B}$$

Table 4.33 Next state/output table

Present state (S)	*Input*		
	0	*1*	*Output*
A	A	B	0
B	C	B	0
C	A	D	0
D	A	B	1

Table 4.34 Expanded next state/output table

Present state (S)	*Input*	*Next state (S*)*	*Output*
A	0	A	0
A	1	B	0
B	0	C	0
B	1	B	0
C	0	A	0
C	1	D	0
D	0	A	1
D	1	B	1

Table 4.35 Binary bit assignment for next state/output

Present state (S)	*Input*	*Next state (S*)*	*Output*
0 0	0	0 0	0
0 0	1	0 1	0
0 1	0	1 1	0
0 1	1	0 1	0
1 1	0	0 0	0
1 1	1	1 0	0
1 0	0	0 0	1
1 0	1	0 1	1

Table 4.36 Excitation values

S				*S**			*Excitation*		
Q1	*Q0*		*Input*	*Q1**	*Q0**		*D1*	*D0*	*Output*
0	0	A	0	0	0	A	0	0	0
0	0	A	1	0	1	B	0	1	0
0	1	B	0	1	1	C	1	1	0
0	1	B	1	0	1	B	0	1	0
1	0	D	0	0	0	A	0	0	1
1	0	D	1	0	1	B	0	1	1
1	1	C	0	0	0	A	0	0	0
1	1	C	1	1	0	D	1	0	0

Rearranging D0 equation:

$$\begin{aligned} \mathrm{D0} &= (\overline{\mathrm{Q1}}^*\overline{\mathrm{Q0}}^*\mathrm{In}) + (\mathrm{Q1}^*\overline{\mathrm{Q0}}^*\mathrm{In}) + (\overline{\mathrm{Q1}}^*\mathrm{Q0}^*\overline{\mathrm{In}}) + (\overline{\mathrm{Q1}}^*\mathrm{Q0}^*\mathrm{In}) \\ &= \overline{\mathrm{Q1}}^*(\overline{\mathrm{Q0}}^*\mathrm{In}) + \mathrm{Q1}^*(\overline{\mathrm{Q0}}^*\mathrm{In}) + (\overline{\mathrm{Q1}}^*\mathrm{Q0})^*\overline{\mathrm{In}} + (\overline{\mathrm{Q1}}^*\mathrm{Q0})^*\mathrm{In} \\ &= (\overline{\mathrm{Q0}}^*\mathrm{In}) + (\overline{\mathrm{Q1}}^*\mathrm{Q0}) \end{aligned}$$

$$\mathrm{Output} = \mathrm{Q1}^*\overline{\mathrm{Q0}} \quad \text{Moore machine – ignore input (In)} \tag{4.26C}$$

Finally we draw the circuit diagram as shown in Figure 4.53 using the following:

$$\begin{aligned} \mathrm{D1} &= (\overline{\mathrm{Q1}}^*\mathrm{Q0}^*\overline{\mathrm{In}}) + (\mathrm{Q1}^*\mathrm{Q0}^*\mathrm{In}) \\ \mathrm{D0} &= (\overline{\mathrm{Q0}}^*\mathrm{In}) + (\overline{\mathrm{Q1}}^*\mathrm{Q0}) \\ \mathrm{Output} &= \mathrm{Q1}^*\overline{\mathrm{Q0}} \end{aligned} \tag{4.26D}$$

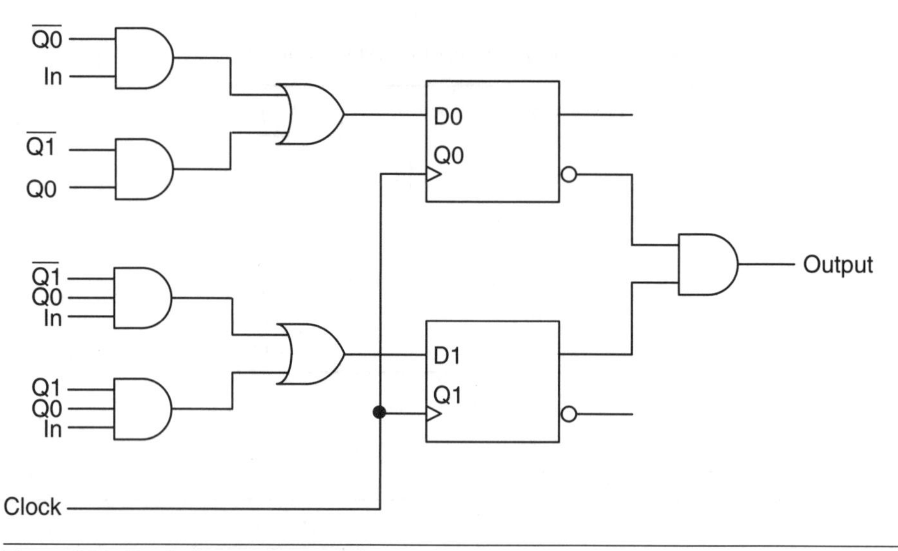

Figure 4.53 Circuit diagram for Example 4.7.

4.9 Applications of flip-flops

4.9.1 Registers

A data register consists of a group or cascade of N negative edge-triggered D-type flip-flops arranged to hold and manipulate a data word using some common circuitry, one bit in each flip-flop. The mechanism for storage consists of first transferring data values D_i, fron N data lines to the outputs Q of the flip-flops on the negative edge of the load signal. Then, a pulse is read on the read line to present the data D_i at the register outputs R_i of the AND gates. Data registers are frequently used in micro-controllers to hold data for arithmetic calculations in the ALU. The value of N determines the size of bits that can be stored. We will consider data registers, shift registers, counters and divide-by-N counters.

4.9.1.1 *Data registers*

The circuit shown in Figure 4.54 uses the clocked inputs of D-type flip-flops to load data into the register on the rising edge of a LOAD pulse.

It is also possible to load data and still leave the clock inputs free (Figure 4.55). The loading process requires a two-step sequence. First the register must be cleared, then it can be loaded.

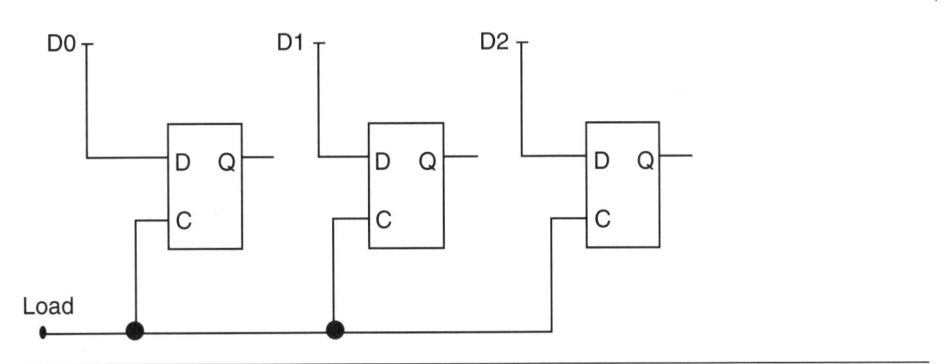

Figure 4.54 A data register using the clocked inputs to D-type flip-flops.

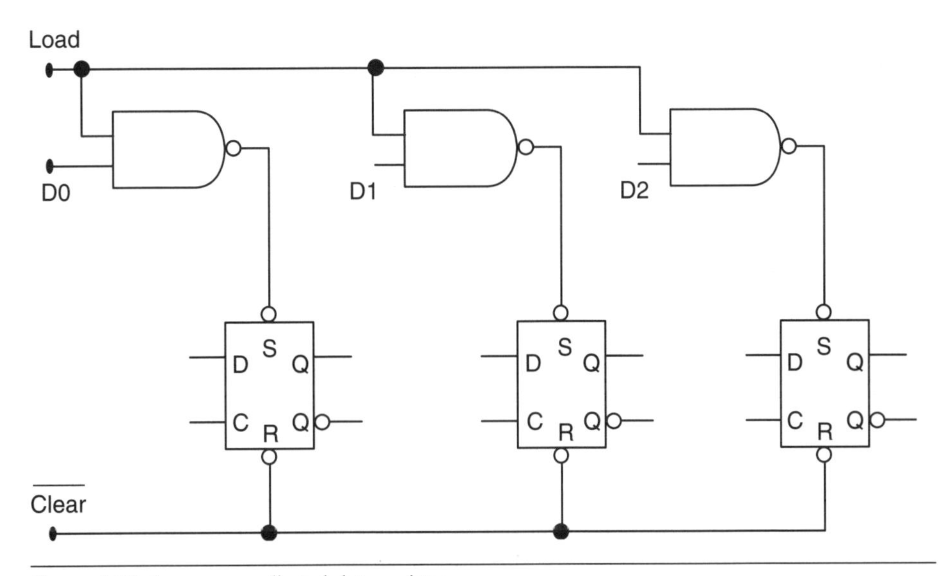

Figure 4.55 A more complicated data register.

4.9.1.2 *Shift registers*

A simple shift register is shown in Figure 4.56. A register of this type can move 3-bit parallel data words to a serial-bit stream. It could also receive a 3-bit serial-bit stream and save it for parallel use.

If A is connected back to D the device is known as a circular shift register or ring counter. A circular shift register can be preloaded with a number and then used to provide a repeated pattern at Q.

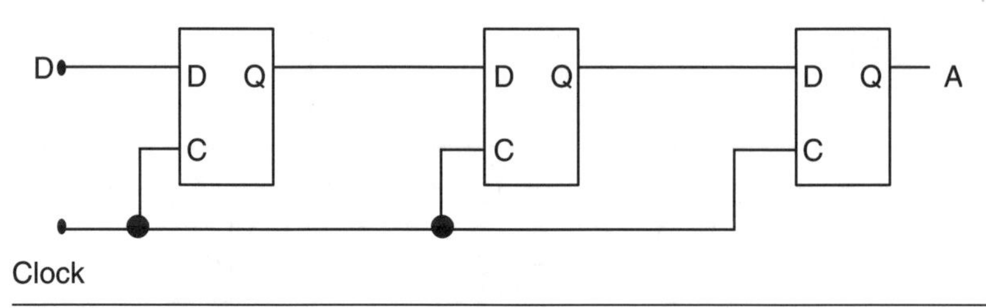

Figure 4.56 A D-type flip-flop 3-bit shift register.

4.9.2 Counters

Counters are composed of N negative edge-triggered toggle (T-type) flip-flops connected in sequence or cascaded. The output of each is the clock input to the next stage, while each flip-flop is held in toggle mode. A simple square wave is used as the clock.

There are several different ways of categorizing counters:

- binary-coded decimal (BCD);
- binary;
- one direction;
- up–down;
- asynchronous ripple-through;
- synchronous.

Counters are also classified by their clearing and preloading abilities. The BCD-type counter is decimal, and is most often used for displays. In the synchronous counter each clock pulse is fed simultaneously or synchronously to all flip-flops. For the ripple counter, the clock pulse is applied only to the first flip-flop in the array and its output is the clock to the second flip-flop, etc. The clock is said to ripple through the flip-flop array.

4.9.2.1 *The binary counter*

Figure 4.57 shows a 3-bit binary, ripple-through, up counter. Table 4.37 shows the associated truth table. The truth table for a 4-bit binary counter is shown in Table 4.38.

Because of pulse delays, the counter will show a transient and incorrect result for short time periods. If the result is used to drive additional logic elements, these transient states may lead to a spurious pulse. This problem is avoided by the synchronous clocking scheme shown in Figure 4.58. All output signals will change state at essentially the same time.

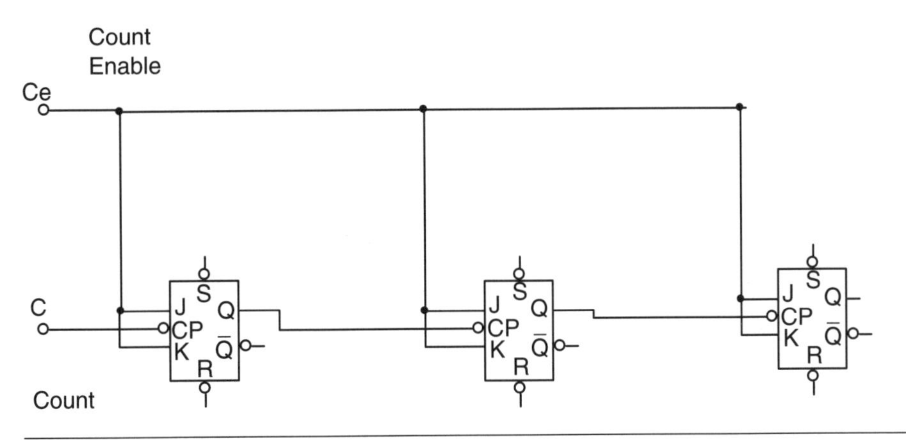

Figure 4.57 A J-K flip-flop 3-bit ripple counter.

Table 4.37 Truth table for a 3-bit binary counter

0	*0*	*0*
0	0	1
0	1	0
0	1	1
1	0	0
1	0	1
1	1	0
1	1	1

Table 4.38 Truth table for a 4-bit binary counter

B_3	B_2	B_1	B_0
0	0	0	0
0	0	0	1
0	0	1	0
0	0	1	1
0	1	0	0
0	1	0	1
0	1	1	0
0	1	1	1
1	0	0	0
1	0	0	1
1	0	1	0
1	0	1	1
1	1	0	0
1	1	0	1
1	1	1	0
1	1	1	1

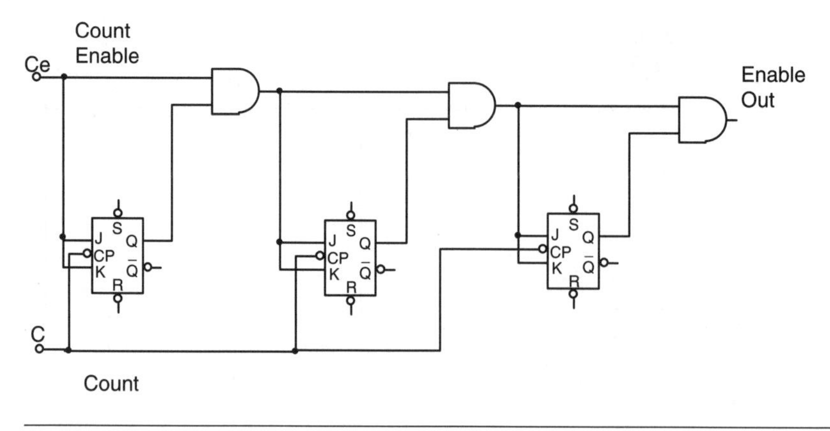

Figure 4.58 A 3-bit synchronous counter.

4.9.2.2 *The decade counter*

A decade counter is a negative edge-triggered counter whose output is binary-coded decimal (BCD) from 0 to 9 (the binary equivalents give a pattern of four bits). The counter resets to 0000 after the count of 9 (1001). The truth table of BCD counter is shown in Table 4.39.

4.9.2.3 *Divide-by-N counters*

A common feature of many digital circuits is a high-frequency clock with a square wave output. If this signal of frequency, *f*, drives the clock input of a J-K flip-flop wired to toggle on each trigger, the output of the flip-flop will be a square wave of

Table 4.39 Truth table for a 7490 decade BCD counter

D	*C*	B_1	*A*
0	0	0	0
0	0	0	1
0	0	1	0
0	0	1	1
0	1	0	0
0	1	0	1
0	1	1	0
0	1	1	1
1	0	0	0
1	0	0	1
0	0	0	0

frequency $f/2$. This single flip-flop is a divide-by-2 counter. In a similar manner an n flip-flop binary counter will yield an output frequency that is f divided by 2^n.

4.9.3 The Schmitt trigger

A noisy input signal to a logic gate could cause unwanted state changes near the voltage threshold. Schmitt trigger logic, modeled as shown in Figure 4.59, reduces this problem by using two voltage thresholds: a high threshold to switch the circuit during low-to-high transitions and a lower threshold to switch the circuit during high-to-low transitions. Such a trigger scheme is immune to noise as long as the peak-to-peak amplitude of the noise is less than the difference between the threshold voltages. A gate symbol with the Schmitt trigger feature has a small hysteresis curve drawn inside the gate symbol. Schmitt triggers are mostly used in inverters or simple gates to condition slow or noisy signals before passing them to more critical parts of the logic circuit.

The Schmitt trigger inverter can be analysed in two steps:

Step 1:

$$\textit{Case } 1\text{: } V_{\text{out}} = +V_{\text{sat}} \tag{4.27}$$

From Figure 4.59(a),

$$V^+ = \frac{R_2}{R_1 + R_2} V_{\text{sat}} \tag{4.28}$$

$$V^- = V_{\text{in}} \tag{4.29}$$

$$\varepsilon = V^+ - V^- \tag{4.30}$$

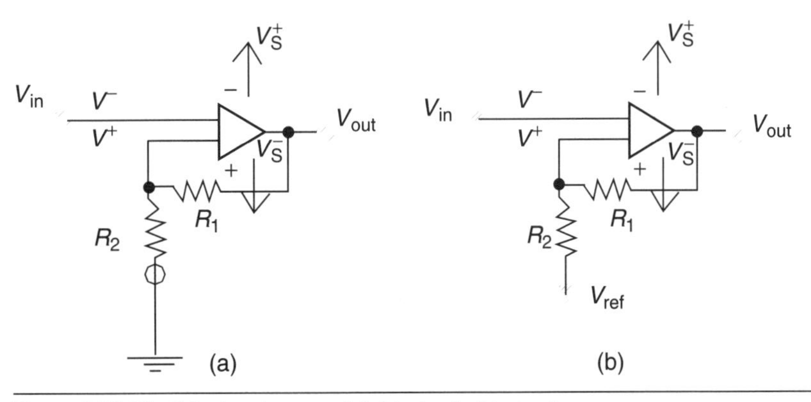

Figure 4.59 The Schmitt trigger model: (a) step 1; (b) step 2.

For the comparator to switch from positive to negative,

$$\varepsilon < 0 \text{ or } V_{in} > \frac{R_2}{R_1 + R_2} V_{sat} \tag{4.31}$$

Step 2:

$$V_{out} = -V_{sat} \tag{4.32}$$

From Figure 4.59(b),

$$V^+ = \frac{-R_2}{R_1 + R_2} V_{sat} \tag{4.33}$$

$$V^- = V_{in} \tag{4.34}$$

$$\varepsilon = V^+ - V^- = \frac{-R_2}{R_2 + R_1} V_{sat} - V_{in} \tag{4.35}$$

For the comparator to switch from positive to negative,

$$\varepsilon > 0 \text{ or } V_{in} < \frac{-R_2}{R_1 + R_2} V_{sat} \tag{4.36}$$

The non-inverting terminal is connected to a reference voltage, hence the non-inverting voltage is:

$$\frac{V_{out} - V^+}{R_1} - \frac{V^+ - V_{ref}}{R_2} = 0 \text{ or } \frac{V^+}{R_1} + \frac{V^+}{R_1} = \frac{V_{out}}{R_1} + \frac{V_{ref}}{R_2} \tag{4.37}$$

Yielding,

$$V^+ = \frac{R_2}{R_1 + R_2} V_{out} + \frac{R_1}{R_1 + R_2} V_{ref} \tag{4.38}$$

Hence the switching level from positive to negative is

$$V_{in} > V^+ \text{ or } V_{in} > \frac{R_2}{R_1 + R_2} V_{sat} + \frac{R_1}{R_1 + R_2} V_{ref}. \tag{4.39}$$

and the switching level from negative to positive is

$$V_{in} < \frac{-R_2}{R_1 + R_2} V_{sat} + \frac{R_1}{R_1 + R_2} V_{ref}. \tag{4.40}$$

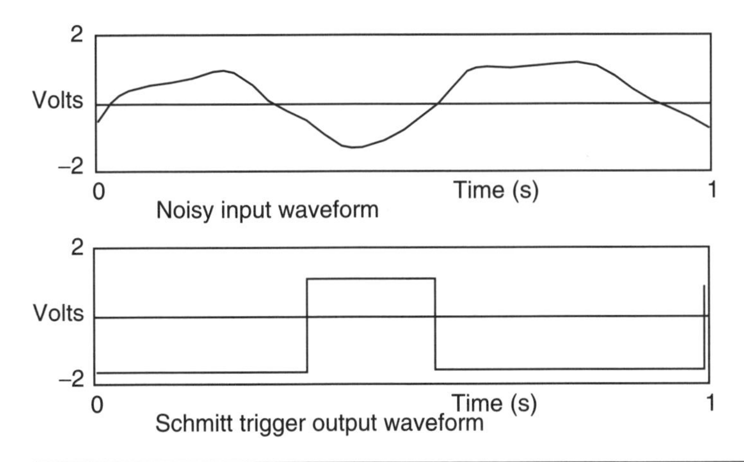

Figure 4.60 Schmitt trigger output for a noisy input.

Figure 4.60 shows the response of a Schmitt trigger with appropriate switching thresholds to a noisy waveform. Accordingly, the Schmitt trigger is able to transform a corrupt signal into a square impulse.

4.9.4 The 555 timer

The 555 timer (Figure 4.61) is a general purpose timer useful for a variety of functions. The concept of this timer is developed from the S-R flip-flop. This circuit works on the principle of alternately charging and discharging a capacitor. The 555 begins to discharge the capacitor by grounding the *discharge* terminal when the voltage detected by the *threshold* terminal exceeds 2/3 the power supply voltage (V_{cc}). It stops discharging the capacitor when the voltage detected by the *trigger* terminal falls below 1/3 the power supply voltage. Thus, when both the *threshold* and *trigger* terminals are connected to the capacitor's positive terminal, the capacitor voltage will cycle between 1/3 and 2/3 of the power supply voltage in a 'saw tooth' pattern.

During the charging cycle, the capacitor receives charging current through the series combination of the R_A and R_B resistors (Figure 4.62). As soon as the *discharge* terminal on the 555 timer goes to ground potential (a transistor inside the 555 connected between that terminal and ground turns on), the capacitor's discharging current only has to go through the R_B resistor. The result is an *RC* time constant that is much longer for charging than for discharging, resulting in a charging time greatly exceeding the discharging time.

The 555's *output* terminal produces a square wave voltage signal that is 'high' (nearly V_{cc}) when the capacitor is charging, and 'low' (nearly 0 V) when the capacitor is discharging. Because the capacitor's charging and discharging times

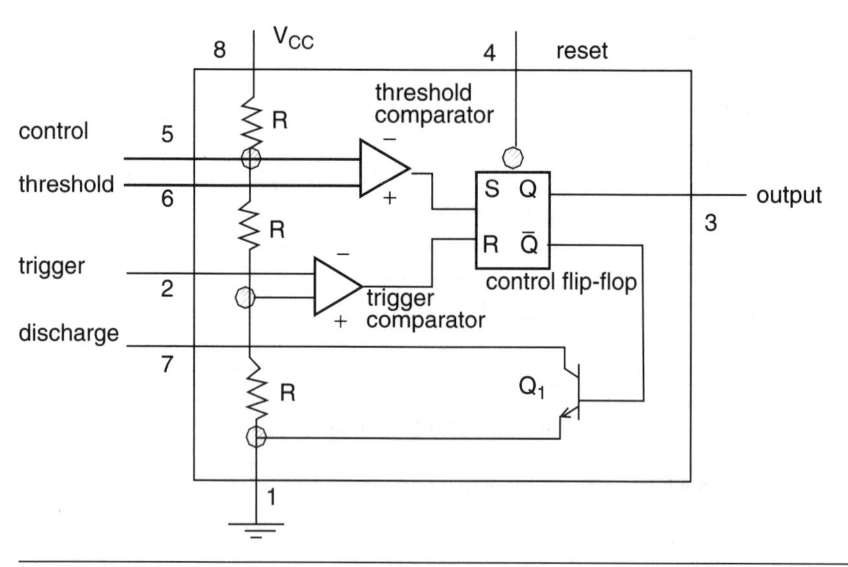

Figure 4.61 The 555 timer.

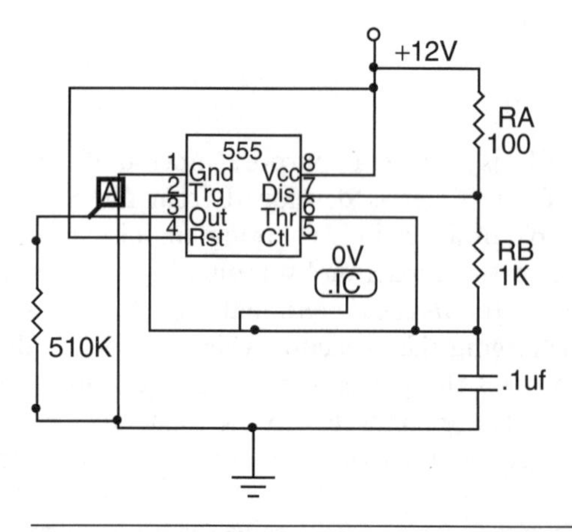

Figure 4.62 The 555 timer analysis.

are unequal, the 'high' and 'low' times of the output's square waveform will be unequal as well.

The proportion between 'high' and 'low' times of a square wave is expressed as that wave's *duty cycle*. A square wave with a 50 percent duty cycle is perfectly symmetrical: its 'high' time is precisely equal to its 'low' time. A square wave that is 'high' 10 percent of the time and 'low' 90 percent of the time is said to have a 10 percent duty cycle.

The operation and principles of the 555 make it useful in a number of applications. The NE555 astable multivibrator is very useful in mechatronics applications since it can be applied in timing processes, driving stepper motors, etc.

4.9.5 The astable multivibrator

The astable multivibrator is useful in the generation of timing or clock waveforms which are a square wave signal of fixed period and amplitude (Figure 4.63). The circuit periodically switches between two states without ever settling at a stable state.

An analysis of the circuit is similar to that of the Schmitt trigger except for the absence of both negative and positive feedback connections. We model the Schmitt trigger as shown in Figure 4.64.

$$V^{+} = \frac{R_2}{R_3 + R_2} V_{\text{out}} = \frac{R_2}{R_3 + R_2} V_{\text{sat}} \tag{4.41}$$

Using the transient method,

$$V^{-}(t) = V_{\text{c}}(t) = V_{\text{sat}}(1 - \text{e}^{-t/\tau});\ \tau = R_1 C \tag{4.42}$$

$$\varepsilon = V^{+} - V^{-} \tag{4.43}$$

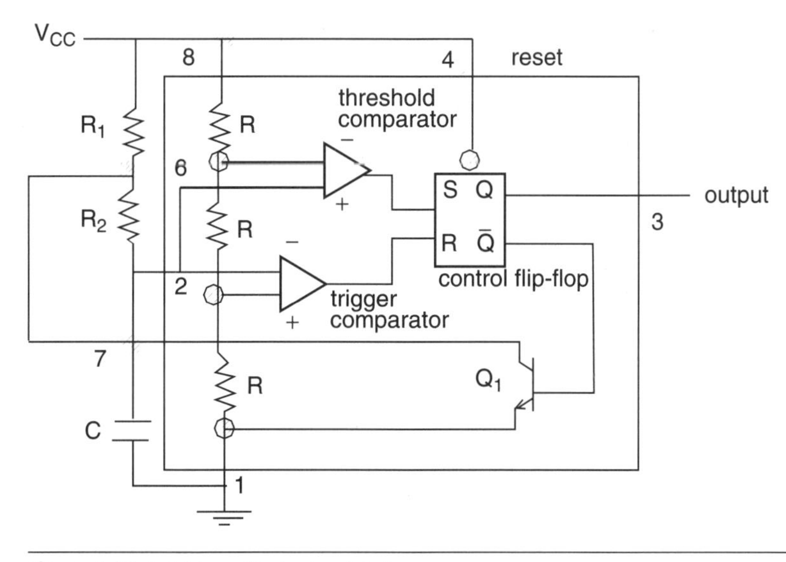

Figure 4.63 Astable multivibrator detail.

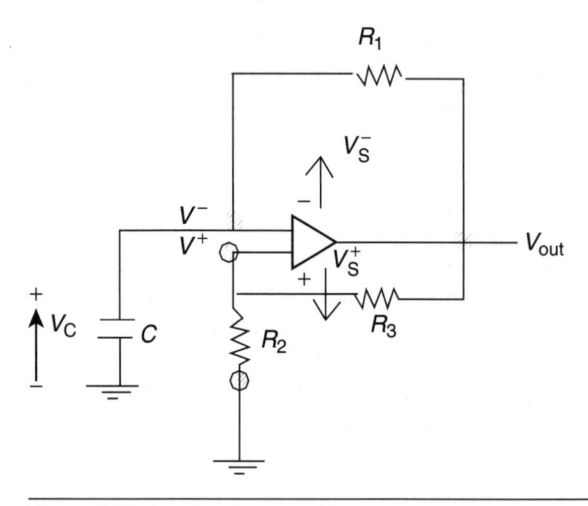

Figure 4.64 Schmitt trigger model.

As the inverting terminal voltage exceeds a threshold voltage, ε becomes negative and

$$V^- > \frac{R_2}{R_1 + R_2} V_{\text{sat}} \tag{4.44}$$

$$V^+ = \frac{-R_2}{R_1 + R_2} V_{\text{sat}} \tag{4.45}$$

The capacitor, which has charged towards $+V_{sat}$ now perceives a negative voltage, $-V_{sat}$ and consequently begins to discharge torwards the new value of V_{sat}, which is $-V_{sat}$ according to the function

$$V_c(t) = [V_c(t_0) + V_{\text{sat}}]e^{-(t-t_0)/\tau} - V_{\text{sat}} \tag{4.46}$$

The period, T, of the waveform is determined by the charge and discharge time of the capacitor, and it can be shown that the period is the square waveform given by

$$T = 2R_1C \quad \log_e\left(\frac{2R_2}{R_3} + 1\right). \tag{4.47}$$

However, for an astable 555 timer:

$$T_t = 0.69(R_1 + R_2)C \text{ and } T_- = 0.69R_2C. \tag{4.48}$$

4.9.6 The monostable multivibrator

The monostable multivibrator is essentially an unstable flip-flop (Figure 4.65). When a monostable multivibrator is set by an input clock or trigger pulse, it will return to the reset state on its own accord after a fixed time delay. They are often used in pairs with the output of the first used to trigger the second. Unfortunately the time relationship between the signals becomes excessively interdependent and it is better to generate signal transitions synchronized with the circuit clock.

4.9.7 The data bus

A bus is a common wire connecting various points in a circuit; examples are the ground bus and power bus. A data bus carries digital information and is usually a group of parallel wires connecting different parts of a circuit with each individual wire carrying a different logic signal. A data bus is connected to the inputs of several gates and to the outputs of several gates. You cannot directly connect the outputs of normal gates. For this purpose three-state output logic is commonly used but will not be discussed here.

A data bus line may be time-multiplexed to serve different functions at different times. At any time only one gate may drive information onto the bus line but several gates may receive it. In general, information may flow on the bus wires in both directions. This type of bus is referred to as a bidirectional data bus.

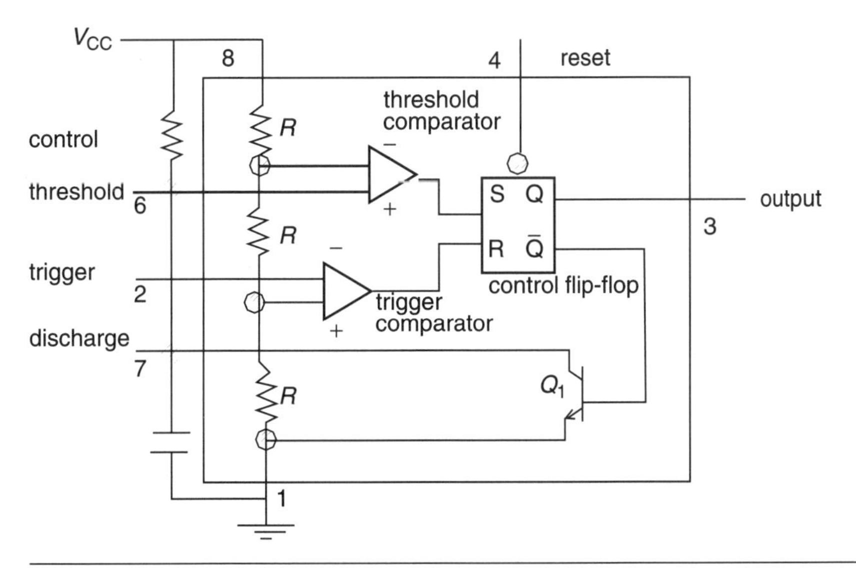

Figure 4.65 Monostable multivibrator.

4.9.8 Standard integrated circuits

Many of the circuit configurations discussed in this chapter are available as standard integrated circuit packages. For a comprehensive listing reference should be made to catalogs from the many IC manufacturers.

Problems

Number system

Q4.1 Convert the following base 10 numbers to their binary and hex numbers: (a) 400; (b) 250; (c) 175; (d) 80.

Q4.2 Convert the following hex numbers to binary and base 10: (a) 72; (b) 56; (c) 24; (d) 13.

Q4.3 Convert the following binary numbers to hex and base 10: (a) 10101010; (b) 1010101; (c) 101010; (d) 10101.

Q4.4 Perform the following binary additions. (Check your answers by converting to their decimal equivalents.) (a) 10101010 + 1010101; (b) 1010101 + 101010; (c) 101010 + 10101; (d) 10101010 + 10101010.

Q4.5 Perform the following binary subtractions. (Check your answers by converting to their decimal equivalents.) (a) 10101010 − 1010101; (b) 1010101 − 101010; (c) 101010 − 10101.

Q4.6 Find the two's complement of the following binary numbers: (a) 10101010; (b) 1010101; (c) 101010; (d) 10101.

Combinational logic

Q4.7 Use a truth table to prove that $X+\overline{X}Y=X+Y$.

Q4.8 Use a truth table to prove that $X+XY=X$.

Q4.9 Prove that the following Boolean identity $XY+XZ+\overline{Y}Z=XY+\overline{Y}Z$ is valid.

Q4.10 Find the logic function defined by the truth table given in Table 4.40.

Table 4.40 Truth table for Q4.10

X	*Y*	*Z*	*f*
0	0	0	0
0	0	1	0
0	1	0	1
0	1	1	1
1	0	0	0
1	0	1	1
1	1	0	1
1	1	1	1

Q4.11 Find the logic function corresponding to the truth table shown in Table 4.41 in the simplest sum-of-products form.

Q4.12 Complete a Karnaugh map for the function described by the truth table in Table 4.42.

Table 4.41 Truth table for Q4.11

X	*Y*	*Z*	*f*
0	0	0	1
0	0	1	1
0	1	0	0
0	1	1	0
1	0	0	0
1	0	1	0
1	1	0	1
1	1	1	1

Table 4.42 Truth table for Q4.12

X	*Y*	*Z*	*f* (*X*, *Y*, *Z*)
0	0	0	1
0	0	1	1
0	1	0	0
0	1	1	1
1	0	0	1
1	0	1	1
1	1	0	1
1	1	1	0

Table 4.43 Truth table for Q4.13

W	*X*	*Y*	*Z*	*f* (*W*, *X*, *Y*, *Z*)
0	0	0	0	1
0	0	0	1	0
0	0	1	0	1
0	0	1	1	0
0	1	0	0	0
0	1	0	1	0
0	1	1	0	0
0	1	1	1	0
1	0	0	0	1
1	0	0	1	0
1	0	1	0	1
1	0	1	1	0
1	1	0	0	1
1	1	0	1	1
1	1	1	0	1
1	1	1	1	0

(a) What is the minimum expression for the function? (b) Draw the appropriate circuit.

Q4.13 Table 4.43 shows the truth table for a logic function *f*.
(a) Complete a Karnaugh map for the logic function; (b) What is the minimum expression for the function?

Sequential logic

Q4.14 What is the relationship between the circuit shown in Figure 4.66 and a D-type flip-flop?

Q4.15 Figure 4.67 shows a 2-bit synchronous binary up–down counter. Sketch a timing diagram for this circuit.

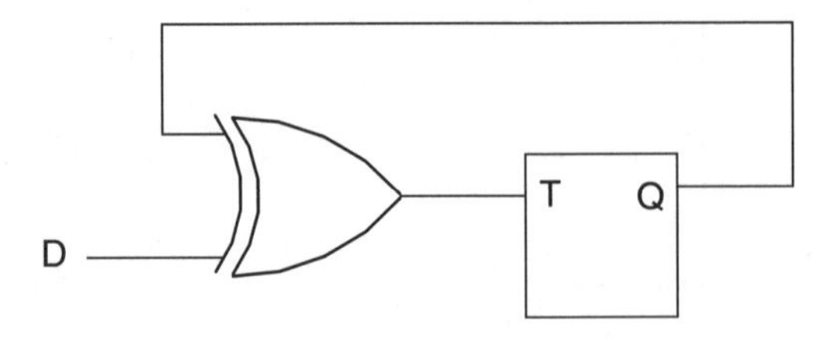

Figure 4.66 Circuit for Q4.14.

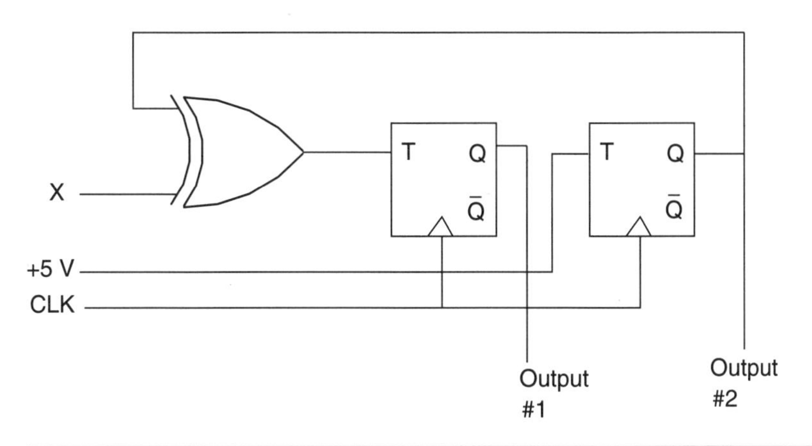

Figure 4.67 Circuit for Q4.15.

Q4.16 Figure 4.68 shows a circuit in which the clock input signal is a square wave having a period of 4 s, a maximum value of 5 V, and a minimum value of 0 V. Assume that initially all the flip-flops are in the *reset* state. (a) What does the circuit do? (b) Sketch the timing diagram, including all outputs.

Q4.17 A binary pulse counter can be constructed by appropriately interconnecting T-type flip-flops. It is desired to construct a counter of this type which can count up to 100_{10}.
(a) How many flip-flops are needed? (b) Sketch the circuit needed to implement this counter.

Q4.18 Complete the timing diagram for the circuit shown in Figure 4.69.

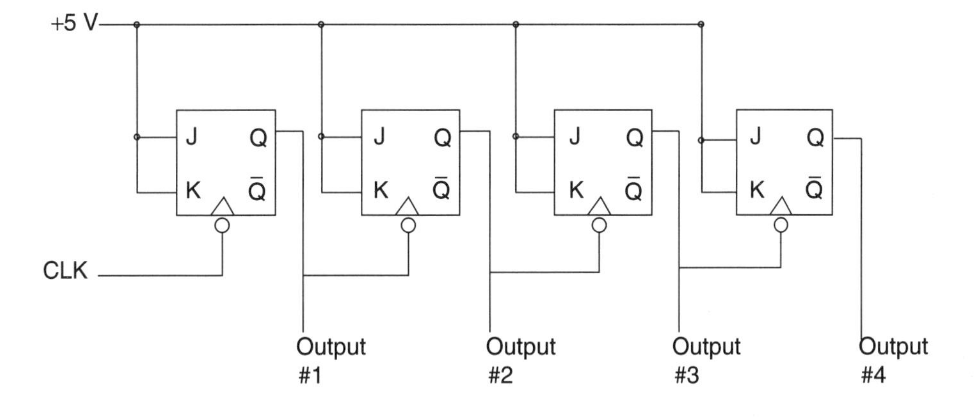

Figure 4.68 Circuit for Q4.16.

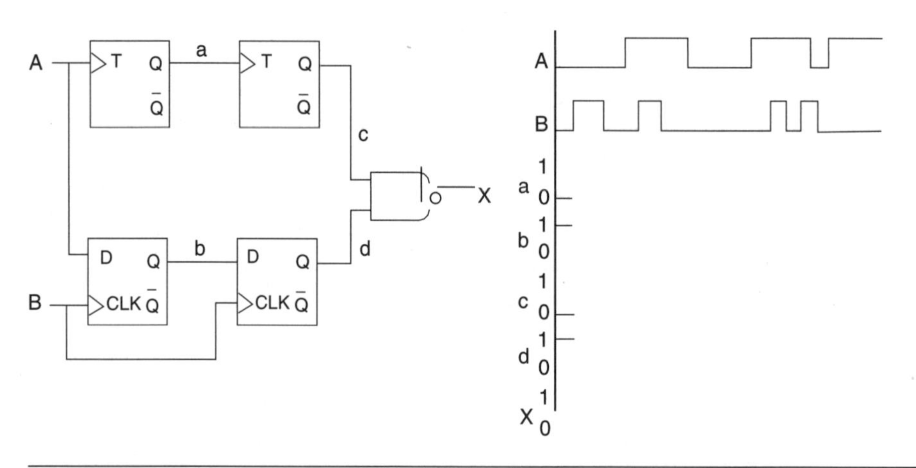

Figure 4.69 Circuit for Q4.18.

Q4.19 Repeat question Q4.18 but with the output taken from $\overline{Q}$ of the last flip-flops.

Q4.20 Complete the timing diagram for the circuit shown in Figure 4.70.

Q4.21 Complete the timing diagram for the circuit shown in Figure 4.71.

Q4.22 Design a modulo-N up–down counter using J-K flip-flops for $N=$(a) 4; (b) 6; (c) 8.

Q4.23 Design a modulo-N up–down counter using D-type flip-flops for $N=$(a) 4; (b) 6; (c) 8.

Q4.24 Design a modulo-N up–down counter using T-type flip-flops for $N=$(a) 4; (b) 6; (c) 8.

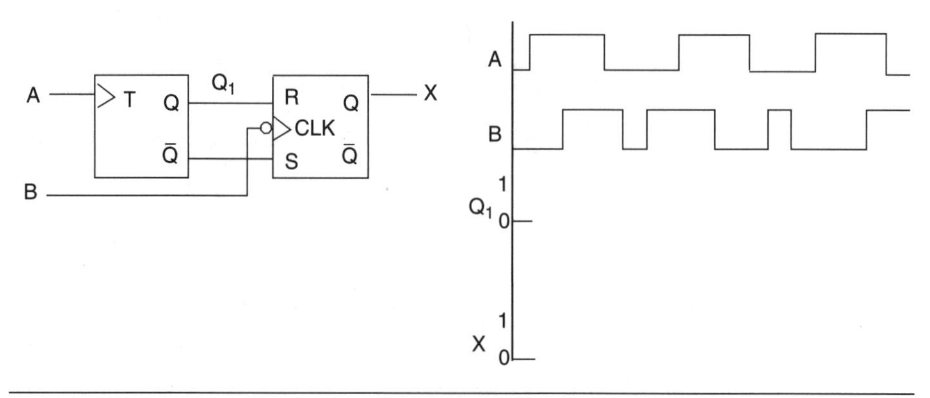

Figure 4.70 Circuit for Q4.20.

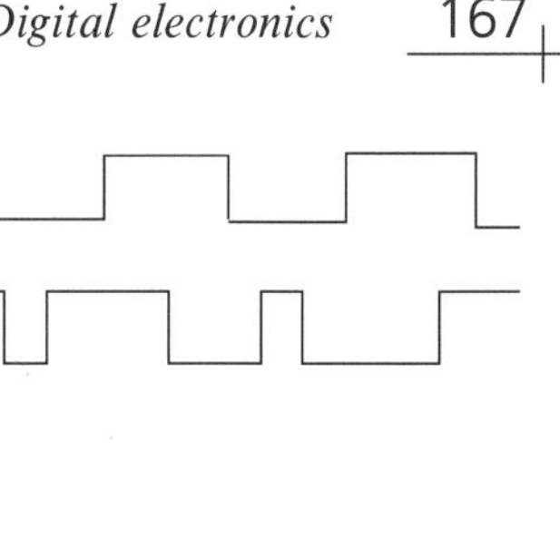

Figure 4.71 Circuit for Q4.21.

Q4.25 Design a modulo-N up–down counter using R-S flip-flops for $N=$(a) 6; (b) 8.

Q4.26 Design a Moore machine using
(a) R-S flip-flops; (b) J-K flip-flops; (c) T-type flip-flops.

Computing

Q4.27 A PC has 42 MB of standard memory. Determine
(a) the number of words available; (b) the number of nibbles available; (c) the number of bits available.

Q4.28 For a microprocessor having n registers, determine
(a) the number of control lines required to connect each register to all other registers; (b) the number of control lines required if a bus is used.

Q4.29 It is planned to implement a 4 kB 16-bit memory. Determine the number of bits required for
(a) the memory address register; (b) the memory data register.

Further reading

[1] Fletcher, W.I. (1980) *An Engineering Approach to Digital Design*, Prentice-Hall.
[2] Givone, D.D. (2002) *Digital Principles and Design* (1st. ed.), McGraw-Hill.
[3] Horowitz, P. and Hill, W. (1989) *The Art of Electronics* (2nd. ed.), New York: Cambridge University Press.

[4] Histand, M.B. and Alciatore, D.G. (2002) *Introduction to Mechatronics and Measurement Systems* (2nd. ed.), McGraw-Hill.
[5] Rashid, M.H. (1996) *Power Electronics: Circuits, Devices, and Applications*, Prentice Hall.
[6] Rizzoni, G. (2003) *Principles and Applications of Electrical Engineering* (4th. ed.), McGraw-Hill.
[7] Stiffler, A.K. (1992) *Design with Microprocessors for Mechanical Engineers*, McGraw-Hill.

CHAPTER 5

Analog electronics

Chapter objectives

When you have finished this chapter you should be able to:

- understand the basics of amplifiers;
- understand and apply different types of amplifier such as inverting, non-inverting, unity-gain buffer, summer, difference, instrumentation, integrator, differentiator, comparator, and sample and hold amplifiers;
- understand and apply active filters such as low-pass active filters, high-pass active filters, and active band-pass filters.

5.1 Introduction

Usually electrical signals in mechatronic and measuring systems come from transducers, which convert physical quantities (displacement, temperature, strain, flow, etc.) into voltages; the output of which is usually in analog signal form since it is continuous and varies with time. Often the signals from transducers need to be cleaned up because they could be distorted (noisy, too small, corrupted, or d.c. offset).

The primary purpose for the analog signal conditioning circuitry is to modify the transducer or sensor output into a form that can be optimally converted to a discrete time digital data stream by the data acquisition system. Some important input requirements of most data acquisition systems are:

- The input signal must be a voltage waveform. The process of converting the sensor output to a voltage can also be used to reduce unwanted signals, that is noise.

- The dynamic range of the input signal should be at or near the dynamic range of the data acquisition system (usually equal to the voltage reference level, V_{ref}, or $2V_{ref}$). This is important in maximizing the resolution of the analog to digital converter (ADC).
- The source impedance, Z_S, of the input signal should be low enough so that changes in the input impedance, Z_{in}, of the data acquisition system do not affect the input signal.
- The bandwidth of the input signal must be limited to less than half of the sampling rate of the analog to digital conversion.

Analog signal processing comprises of the following issues: signal isolation, signal preprocessing, and removal of undesirable signals. These are now briefly discussed.

5.1.1 Signal isolation

In many data acquisition applications it is necessary to isolate the sensor from the power supply of the computer. This is done in one of two ways: magnetic isolation or optical isolation. Magnetic isolation by means of a transformer is primarily used for coupling power from the computer or the wall outlet to the sensor. Optical isolation is used for coupling the sensor signal to the data acquisition input. This is usually done through the use of a light emitting diode and a photodetector often integrated into a single IC package.

5.1.2 Signal preprocessing

Many times it is desirable to perform preprocessing on the sensor signal before data acquisition. Depending on the application, this can help lower the required computer processing time, lower the necessary system sampling rate, or even perform functions that will enable the use of a much simpler data acquisition system entirely. For example, while an accelerometer system can output a voltage proportional to acceleration, it may be desired to only tell the computer when the acceleration is greater than a certain amount. This can be accomplished in the analog signal conditioning circuitry. Thus, the data acquisition system is reduced to only having a single binary input (and thus no need for an ADC).

5.1.3 Removal of undesired signals

Many sensors output signals that have many different components to them or other signals may corrupt the signal. It may be desirable or even necessary to remove such components before the signal is digitized. This *noise* can also be removed using analog circuitry. For example, 60 Hz interference can distort the output of low output sensors. The signal conditioning circuitry can remove this before it is amplified and digitized.

The simplest and the most common form of signal processing is amplification where the magnitude of the voltage signal is increased. Other forms of signal processing include inversion, addition, subtraction, comparison, differentiation, and integration. The operational amplifier is an integrated circuit that can perform these operations. Accordingly, in the remaining part of this section, simple circuit models of the operational amplifier (op amp) will be introduced. The simplicity of the models will permit the use of the op amp as a circuit element, or building block, without the need to describe its internal workings in detail. For the purpose of many instrumentation applications, the op amp can be treated as an ideal device.

5.2 Amplifiers

The op amp was designed to perform mathematical operations and is a common feature of modern analog electronics. The op amp is constructed from several transistor stages, which usually include a differential input stage, an intermediate gain stage and a push-pull output stage. The differential amplifier consists of a matched pair of bipolar transistors or FETs. The push-pull amplifier transmits a large current to the load and hence has a small output impedance. An op amp can perform a great number of operations such as addition, filtering, or integration, which are based on the properties of ideal amplifiers and of ideal circuit elements.

The op amp is a linear amplifier with $V_{out} \propto V_{in}$. The d.c. open-loop voltage gain of a typical op amp is 10^2 to 10^6. The gain is so large that most often feedback is used to obtain a specific transfer function and control the stability. Inexpensive IC versions of op amps are readily available, making their use popular in any analog circuit. These operate from d.c. to about 20 kHz, while the high-performance models operate up to 50 MHz. A popular device is the 741 op-amp which drops off 6 dB/octave above 5 Hz. Op amps are usually available as an IC in an 8-pin dual, in-line package (DIP). Some op amp ICs have more than one op amp on the same chip. Figure 5.1 shows the internal design of a 741 op amp IC.

Many sensors output a voltage waveform so no signal conditioning circuitry is needed to perform the conversion to a voltage. However, dynamic range modification, impedance transformation, and bandwidth reduction may all be necessary in the signal conditioning system depending on the amplitude and bandwidth of the signal and the impedance of the sensor. It is especially important to review the analysis of ideal op amp circuits.

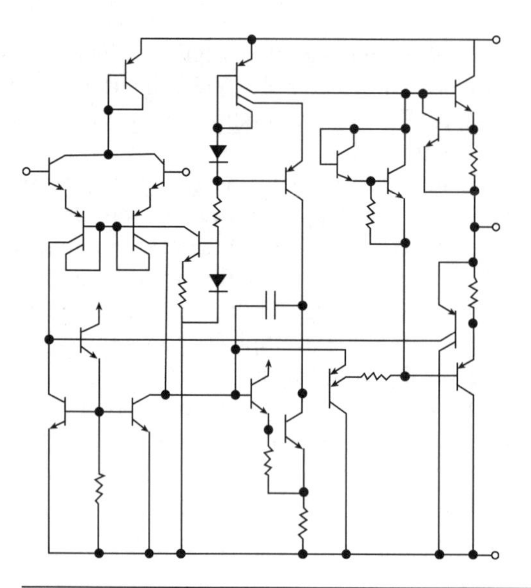

Figure 5.1 A 741 op amp internal design. (Courtesy of National Semiconductor Inc.)

5.3 The ideal operational amplifier model

The ideal op amp model is shown in Figure 5.2. Before continuing we define some terminology:

- **linear amplifier:** the output is directly proportional to the amplitude of input signal.
- **open-loop gain, A:** the voltage gain without feedback $(\approx 10^6)$ as shown in Figure 5.2(a).
- **closed-loop gain, G:** the voltage gain with negative feedback (approximation to $H(j\omega)$) as shown in Figure 5.2(b).
- **negative feedback:** the output is connected to the inverting input forming a feedback loop (usually through a *feedback resistor* R_F) as shown in Figure 5.2(b).

Figure 5.2(c) shows an ideal model for analyzing circuits containing op amps. The ideal op am model is based on the following assumptions:

- it has infinite impedance at both inputs, consequently there is no current drawn from the input circuits;

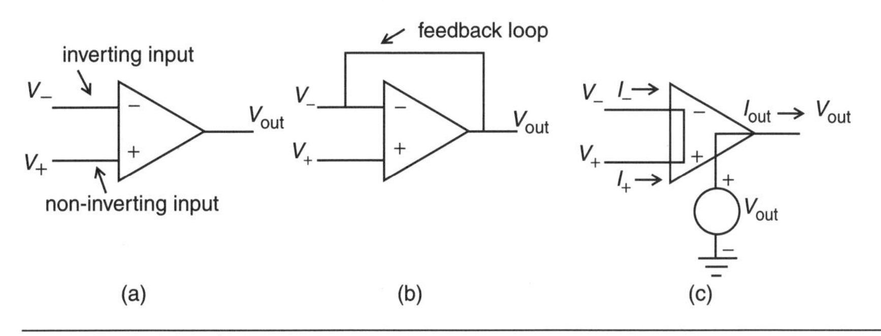

Figure 5.2 The op amp model: (a) open loop; (b) closed loop; (c) ideal.

- it has infinite gain, hence the difference between the input and output voltages is zero. This is denoted by short circuiting the two inputs;
- it has zero output impedance, so that the output voltage is independent of the output current.

All the signals are referenced to ground and feedback exists between the output and the inverting input in order to achieve stable linear behavior.

5.4 The inverting amplifier

The most common circuit used for signal conditioning is the inverting amplifier circuit as shown in Figure 5.3. This amplifier was first used when op amps only had one input, the inverting (–) input. The analysis is done by noting that at the inverting input node, Kirchoff's current law requires that

$$i_{\mathrm{I}} + i_{\mathrm{out}} = 0 \tag{5.1}$$

$$\frac{V_{\mathrm{i}}}{R_{\mathrm{I}}} + \frac{V_{\mathrm{out}}}{R_{\mathrm{F}}} = 0. \tag{5.2}$$

Leading to the voltage gain

$$\frac{V_{\mathrm{out}}}{V_{\mathrm{i}}} = -\frac{R_{\mathrm{F}}}{R_{\mathrm{I}}}. \tag{5.3}$$

Thus the level of sensor outputs can be matched to the level necessary for the data acquisition system. The input impedance is approximately R_{I} and

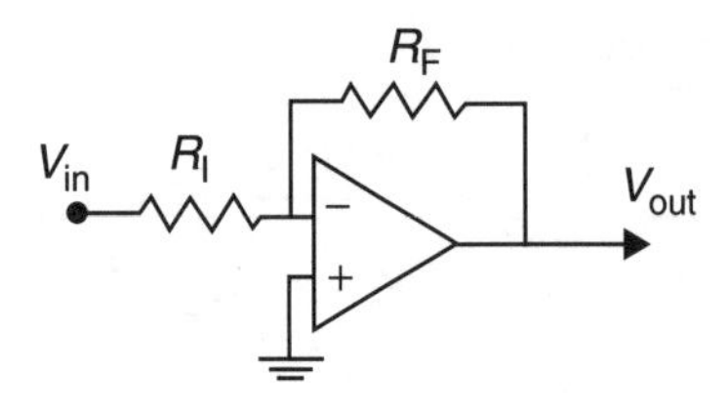

Figure 5.3 The inverting amplifier.

the output impedance is nearly zero, so this circuit provides impedance transformation between the sensor and the data acquisition system.

5.5 The non-inverting amplifier

To avoid the negative gain (i.e. phase inversion) introduced by the inverting amplifier, a non-inverting amplifier configuration is commonly used as shown in Figure 5.4. The analysis of the non-inverting amplifier is done in the same way as the inverting amplifier by noting that at the inverting input node, Kirchoff's current law requires that

$$i_f + i_n - i_i = 0; \qquad (i_n \approx 0) \tag{5.4}$$

$$\frac{V_{out} - V_s}{R_f} - \frac{V_s - 0}{R_i} = 0 \tag{5.5}$$

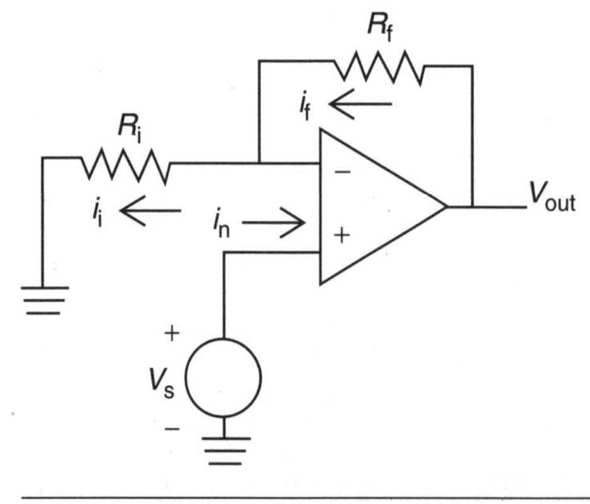

Figure 5.4 The non-inverting amplifier.

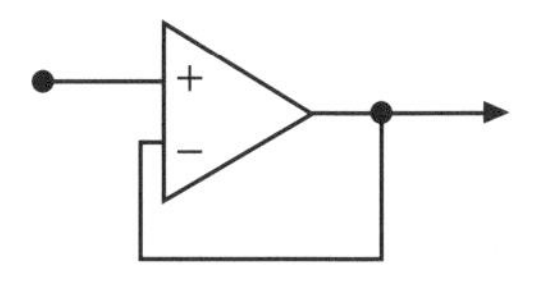

Figure 5.5 The unity-gain buffer.

Leading to the voltage gain

$$\frac{V_{out}}{V_s} = 1 + \frac{R_f}{R_i} \tag{5.6}$$

The input impedance is nearly infinite (limited only by the op amp's input impedance) and the output impedance is nearly zero. The circuit is ideal for sensors that have a high source impedance and thus would be affected by the current draw of the data acquisition system.

5.6 The unity-gain buffer

If $R_f = 0$ and $R_i = 0$ and is open (removed), then the gain of the non-inverting amplifier is unity. This circuit, as shown in Figure 5.5 is commonly referred to as a unity-gain buffer or simply a buffer.

5.7 The summing amplifier

The op amp can be used to add two or more signals together as shown in Figure 5.6. The analysis of the summing amplifier is done by noting that at the inverting input node, Kirchoff's current law requires that

$$i_1 + i_2 + ... + i_i + i_f - i_n = 0; \qquad (i_n \approx 0) \tag{5.7}$$

$$\frac{V_1}{R_1} + \frac{V_2}{R_2} + ... + \frac{V_i}{R_i} + \frac{V_{out} - 0}{R_f} = 0. \tag{5.8}$$

Leading to

$$V_{out} = -R_f \sum_{i=1}^{N} \frac{V_i}{R_i} \tag{5.9}$$

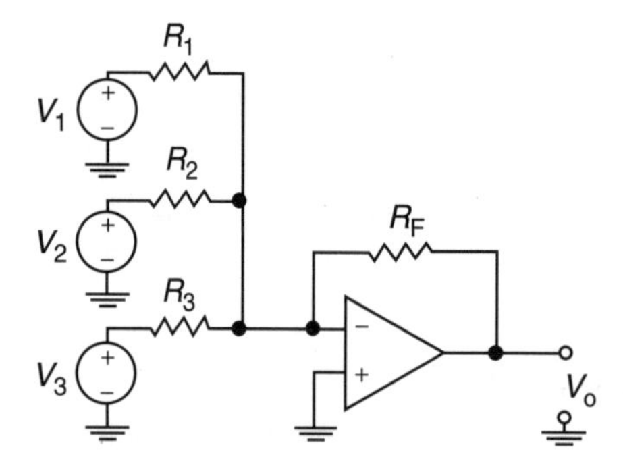

Figure 5.6 The summing amplifier.

$$V_{\text{out}} = -\sum_{i=1}^{N} V_{\text{i}} \quad \text{if} \quad R_{\text{f}} = R_{\text{i}}. \tag{5.10}$$

This circuit can be used to combine the outputs of many sensors such as a microphone array.

5.8 The difference amplifier

The op amp can also be used to subtract two signals as shown in Figure 5.7. The priniciple of superposition is used to analyze the difference amplifier.

In the first step, we short V_2 so that we have an inverter, and that the output to V_1 is

$$V_{\text{o1}} = -\frac{R_2}{R_1} V_1 \tag{5.11}$$

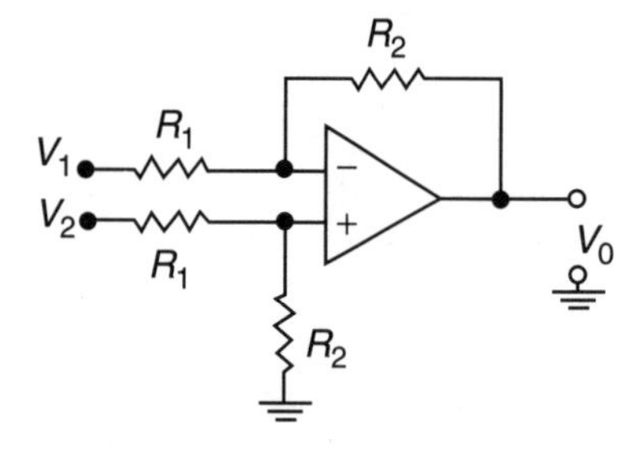

Figure 5.7 The difference amplifier.

In the second step, we short we short V_1 so that we have a non-inverter. We then use the voltage divider principle to obtain the non-inverting voltage as

$$V_3 = \frac{R_2}{R_1 + R_2} V_2 \tag{5.12}$$

$$V_{o2} = \left(1 + \frac{R_2}{R_1}\right) V_3 = \left(1 + \frac{R_2}{R_1}\right)\left(\frac{R_2}{R_1 + R_2}\right) V_2 \tag{5.13}$$

The priniciple of superposition means that the output voltage is the sum of the input voltages:

$$V_o = V_{o1} + V_{o2} = -\left(\frac{R_2}{R_1}\right) V_1 + \left(1 + \frac{R_2}{R_1}\right)\left(\frac{R_2}{R_1 + R_2}\right) V_2. \tag{5.14}$$

or

$$V_o = V_{o1} + V_{o2} = \left(\frac{R_2}{R_1}\right)(V_2 - V_1). \tag{5.15}$$

Thus the difference amplifier magnifies the difference between the two input signals by the closed-loop gain (R_2/R_1). This circuit is commonly used to remove unwanted d.c. offset. It can also be used to remove differences in the ground potential of the sensor and the ground potential of the data acquisition circuitry (so-called ground loops). In this case V_2 can be the output of the sensor and V_1 can be the signal that is to be removed.

5.9 The instrumentation amplifier

When the input signals are very low level and also have noise, the difference amplifier is not able to extract a satisfactory difference signal. Possibly the most important circuit configuration for amplifying sensor output when the input signals are very low level is the instrumentation amplifier (IA). The requirements for an instrumentation amplifier are as follows:

- Finite, accurate and stable gain, usually between 1 and 1000.
- Extremely high input impedance.
- Extremely low output impedance.
- Extremely high common mode rejection ratio (CMRR).

The CMRR is defined as:

$$\text{CMRR} = \left(\frac{A_{\text{vd}}}{A_{\text{vc}}}\right) \tag{5.16}$$

where

$$A_{\text{vd}} = \frac{V_{\text{out}}}{V^{+} - V^{-}} = \text{differential-mode gain}$$
$$A_{\text{vc}} = \frac{V_{\text{out}}}{\frac{V^{+}+V^{-}}{2}} = \text{common-mode gain} \tag{5.17}$$

That is, CMRR is the ratio of the gain of the amplifier for differential-mode signals (signals that are different between the two inputs) to the gain of the amplifier for common-mode signals (signals that are the same at both inputs). The difference amplifier described above, clearly does not satisfy the second requirement of high input impedance. To solve this problem, a non-inverting amplifier is placed at each one of the inputs to the difference amplifier as shown in Figure 5.8. Remember that a non-inverting amplifier has a nearly infinite input impedance. Notice that instead of grounding the resistors, the two resistors are connected together to create one common resistor, R_G.

The analysis of the instrumentation amplifier is simplified by taking advantage of symmetry: we recognize two symmetric halves at the input (draw a vertical line between R_1 and R_3, and a horizontal line across R_G). This results in an equivalent circuit shown in Figure 5.9.

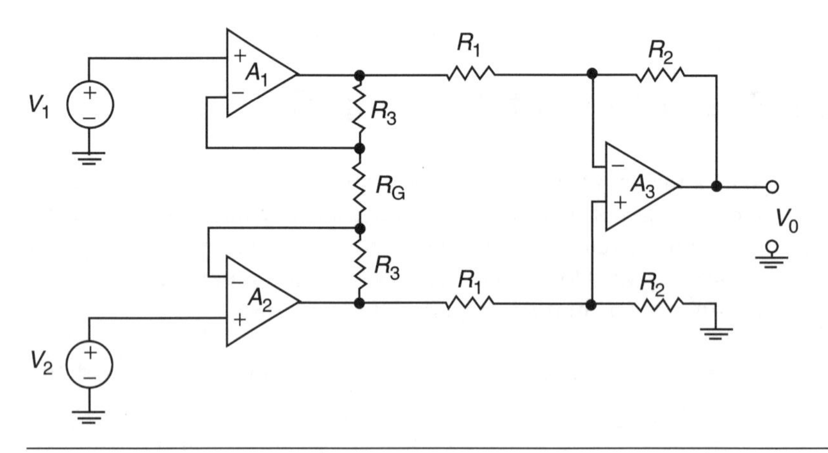

Figure 5.8 The instrumentation amplifier.

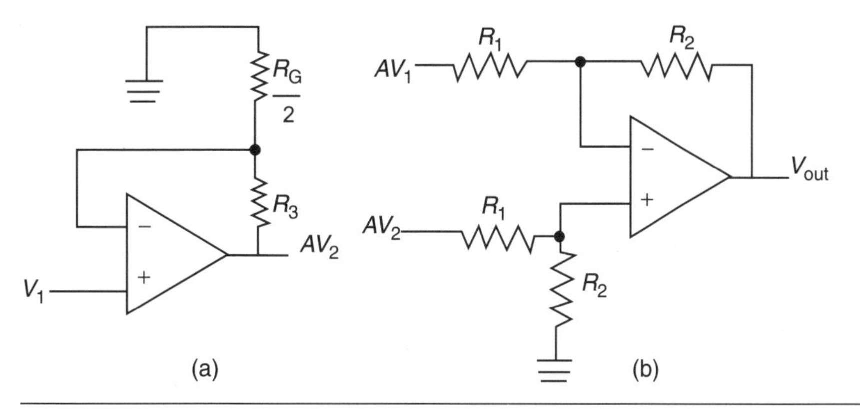

Figure 5.9 Equivalent instrumentation amplifier for analysis: (a) step 1; (b) step 2.

The analysis is done in two steps:

- Step 1: Using Figure 5.9(a), which is a non-inverting amplifier,

$$A = 1 + \frac{R_3}{R_G/2} = 1 + \frac{2R_3}{R_G}. \tag{5.18}$$

- Step 2: Using Figure 5.9(b), which is a difference amplifier,

$$V_{out} = \left(\frac{R_2}{R_1}\right)(AV_2 - AV_1) = \left(\frac{R_2}{R_1}\right)\left(1 + \frac{2R_3}{R_G}\right)(V_2 - V_1). \tag{5.19}$$

- The overall differential gain of the circuit is

$$A_{vd} = \left(\frac{R_2}{R_1}\right)\left(1 + \frac{2R_3}{R_G}\right). \tag{5.20}$$

5.10 The integrator amplifier

The op amp in Figure 5.10 provides an output that is proportional to the integral of $v_{in}(t)$, an arbitrary function of time (e.g. a pulse train, a triangular wave, or a square wave).

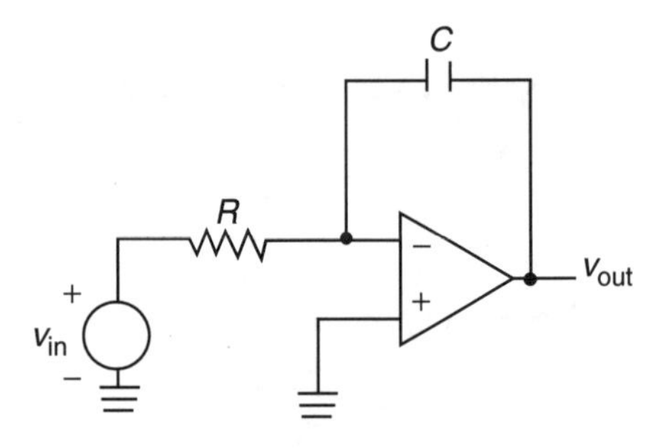

Figure 5.10 The integrating amplifier.

The analysis of the differential circuit is based on the observation that

$$i_{in}(t) = -i_f(t) \tag{5.21}$$

$$i_{in}(t) = \frac{v_{in}(t)}{R} \tag{5.22}$$

and

$$i_f(t) = C\frac{dv_{out}(t)}{dt} \tag{5.23}$$

$$\frac{1}{RC}v_{in}(t) = -\frac{dv_{out}(t)}{dt} \tag{5.24}$$

$$v_{out}(t) = -\frac{1}{RC}\int_{-\infty}^{t} v_s(t)\,dt. \tag{5.25}$$

5.11 The differentiator amplifier

The op amp in Figure 5.11 provides an output that is proportional to the integral of $v_{in}(t)$, an arbitrary function of time (e.g. a pulse train, a triangular wave, or a square wave).

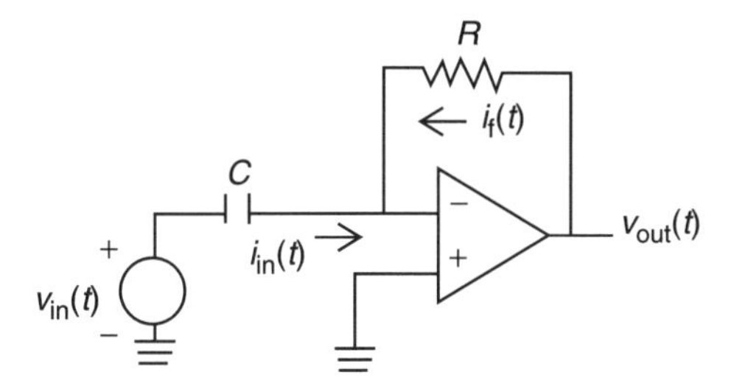

Figure 5.11 The differentiating amplifier.

The analysis of the differential circuit is based on the observation that

$$i_{\text{in}}(t) = -i_{\text{f}}(t) \tag{5.26}$$

$$i_{\text{in}}(t) = C\frac{\mathrm{d}v_{\text{in}}(t)}{\mathrm{d}t} \tag{5.27}$$

and

$$i_{\text{f}}(t) = \frac{v_{\text{out}}(t)}{R} \tag{5.28}$$

$$v_{\text{out}}(t) = -\text{RC}\frac{\mathrm{d}v_{\text{in}}(t)}{\mathrm{d}t}. \tag{5.29}$$

5.12 The comparator

Sometimes there is no need to send the entire range of voltages from a sensor to an analog-to-digital converter (ADC). Instead, often a sensor is used simply as a switch. Figure 5.12 functions as a *comparator* which takes an analog sensor

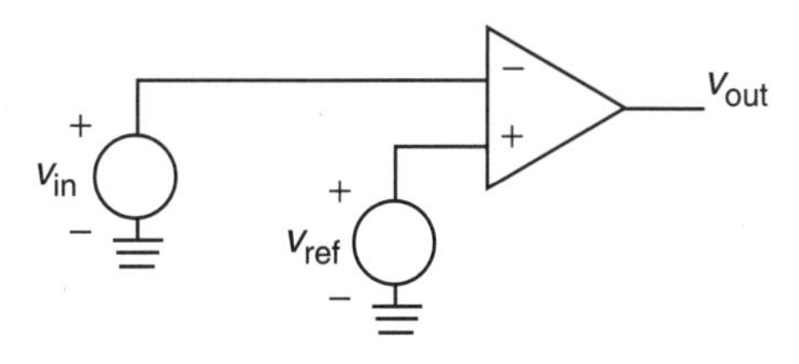

Figure 5.12 The comparator.

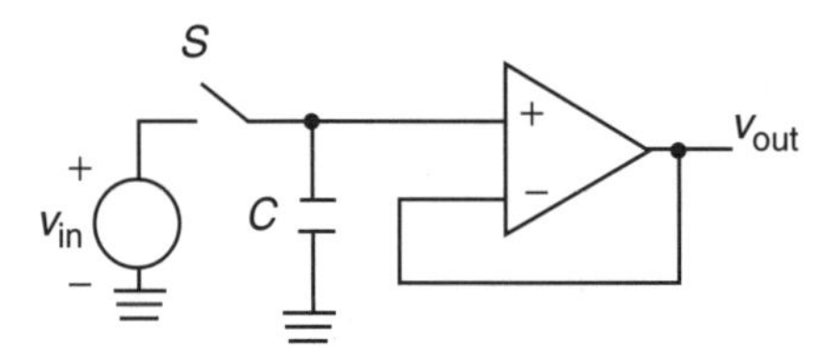

Figure 5.13 The sample and hold amplifier.

voltage and compares it to a threshold voltage, V_{ref}. If the sensor's voltage is greater than the threshold, the output of the circuit is maximum (typically 5 V). If the sensor's output is less than the threshold, the output of the circuit is minimum (usually 0 V).

5.13 The sample and hold amplifier

The purpose of sample and hold circuitry is to take a snapshot of the sensor signal and hold the value. An ADC must have a stable signal in order to accurately perform a conversion. An equivalent circuit for the sample and hold is shown in Figure 5.13. The switch connects the capacitor to the signal conditioning circuit once every sample period. The capacitor then holds the measured voltage until a new sample is acquired. Often, the sample and hold circuit is incorporated in the same integrated circuit package as the amplifier.

5.13.1 Problems with sample and hold amplifiers

- **Finite aperture time:** The sample and hold takes a period of time to capture a sample of the sensor signal. This is called the aperture time. Since the signal will vary during this time, so the sampled signal can be slightly inaccurate.
- **Signal feedthrough:** When the sample and hold is not connected to the signal, the value being held should remain constant. Unfortunately, some signal does bleed through the switch to the capacitor, causing the voltage being held to change slightly.
- **Signal droop:** The voltage being held on the capacitor starts to slowly decrease over time if the signal is not sampled often enough.

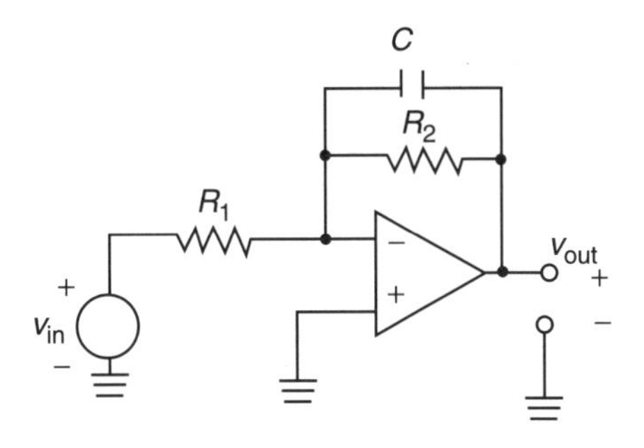

Figure 5.14 Low-pass active filter.

5.14 Active filters

There are many practical applications that involve filters of one kind or another; for example, filters to eliminate impurities from drinking water, sunglasses to reduce the light intensity reaching the eye, etc. Similarly, in electric circuits, it is possible to attenuate or reduce/eliminate the amplitude of unwanted frequencies caused by electric noise or other forms of interference. This section treats the analysis of electric filters.

5.14.1 The low-pass active filter

The inverting amplifier configuration can be modified to limit the bandwidth of the incoming signal. For example, the feedback resistor can be replaced with a resistor/capacitor combination as shown in Figure 5.14.

We now analyze the filter as follows:

$$Z_F = Z_R \Big\| Z_C = R_2 \Big\| \frac{1}{j\omega C} \tag{5.30}$$

$$Z_F = \frac{R_2 \frac{1}{j\omega C}}{R_2 + \frac{1}{j\omega C}} = \frac{R_2}{1 + j\omega C R_2} \tag{5.31}$$

$$Z_i = Z_R = R_1. \tag{5.32}$$

The gain of this filter is given by:

$$A_{LP}(j\omega) = -\frac{Z_F}{Z_i} = -\frac{R_2/R_1}{1 + j\omega CR_2} = \left(\frac{-R_2}{R_1}\right)\frac{1}{1 + j\frac{\omega}{\omega_0}} \tag{5.33}$$

$$A_{LP}(j\omega) = H_0 \frac{1}{1 + j\frac{\omega}{\omega_0}}, \tag{5.34}$$

where

$$H_0 = \left(\frac{-R_2}{R_1}\right); \ \omega_0 = \frac{1}{CR_2}; f_0 = \frac{1}{2\pi R_2 C} \tag{5.35}$$

$$\text{As } \omega \to 0, \ A_{LP}(j\omega) \to H_O$$

$$\text{As } \omega \to \infty, \ A_{LP}(j\omega) \to 0. \tag{5.35A}$$

$$\text{When } \omega = \omega_0, \ A_{LP}(j\omega) = H_0 \frac{1}{1 + j}.$$

Under the last condition, the filter *rolls off* and

$$A_{LP}(j\omega) = H_0 \frac{1}{\sqrt{2}}; \ |A_{LP}(j\omega)|_{dB} = 20 \ \log_{10} H_0 - 20 \ \log_{10} \sqrt{2} \tag{5.35B}$$

$$\therefore \ |A_{LP}(j\omega)|_{dB} = 20 \ \log_{10} H_o - 3 \text{ dB}.$$

This means that the filter rolls off at 20 dB per 10-times increase in frequency (20 dB/decade) times the order of the filter. That is

the rate of attenuation = (order of filter) × (20 dB/decade).

Thus a first order filter rolls off at 20 dB/decade as shown in Figure 5.15.

This is sometimes expressed as a roll off of 6 dB/octave (where an octave is a doubling of frequency).

5.14.2 The high-pass active filter

The input resistor of the inverting amplifier is replaced by a resistor/capacitor pair to create a high-pass filter as shown in Figure 5.16.

We now analyze the filter as follows:

$$Z_F = Z_R = R_2 \tag{5.36}$$

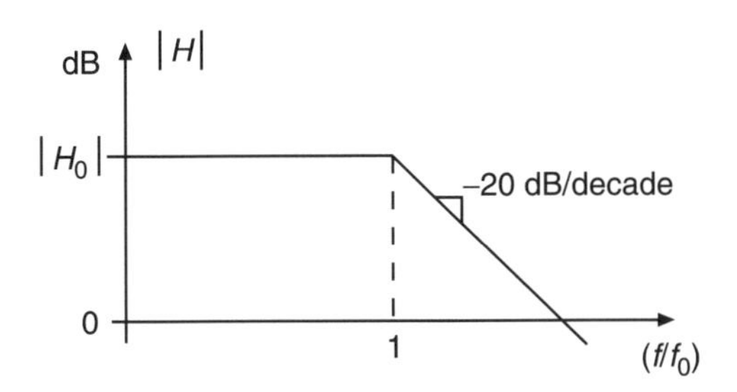

Figure 5.15 Frequency response of a single pole low-pass filter.

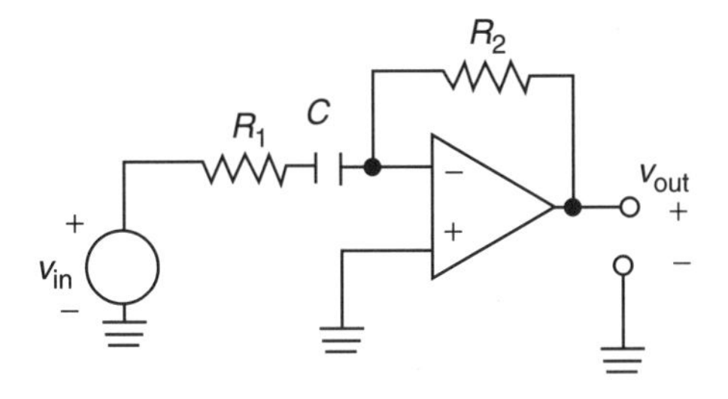

Figure 5.16 High-pass active filter.

$$Z_i = Z_R + Z_C = R_1 + \frac{1}{j\omega C} \tag{5.37}$$

$$A_{HP}(j\omega) = -\frac{Z_F}{Z_i} = \frac{-R_2}{R_1 + \frac{1}{j\omega C}} = \frac{-j\omega C R_2}{1 + j\omega C R_1} \tag{5.38}$$

$$A_{HP}(j\omega) - \frac{-j\omega C R_2}{1 + j\frac{\omega}{\omega_0}}. \tag{5.39}$$

The gain of this filter is given by:

$$\therefore\ A_{HP}(j\omega) = H_0 \frac{j\omega C R_2}{1 + j\frac{\omega}{\omega_0}}, \tag{5.40}$$

where

$$H_0 = \left(\frac{-R_2}{R_1}\right); \quad \omega_0 = \frac{1}{CR_1}; \quad f_0 = \frac{1}{2\pi R_2 C} \tag{5.41}$$

$$\begin{aligned} &\text{As } \omega \to 0,\ A_{HP}(j\omega) \to 0 \\ &\text{As } \omega \to \infty,\ A_{HP}(j\omega) \to H_0. \end{aligned} \tag{5.41A}$$

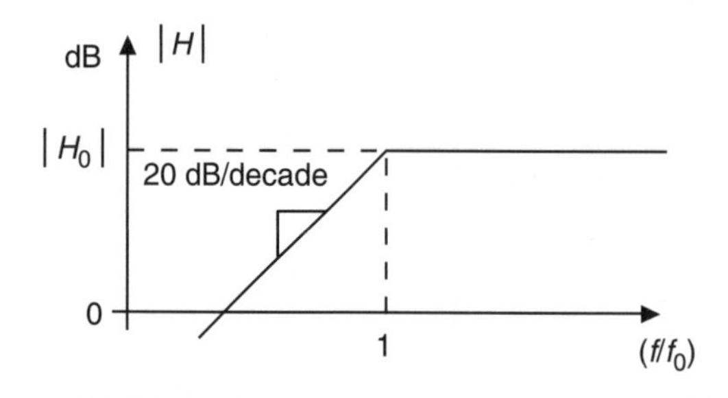

Figure 5.17 Frequency response of a single pole high-pass filter.

Under the last condition, the filter acts as a linear amplifier. The frequency response of this filter is shown in Figure 5.17. The frequency ω_0 is called the 'cutoff frequency'; this is the point where the filter begins to filter out the higher-frequency signal.

5.14.3 The band-pass active filter

The band-pass active filter is a combination of the low-pass active filter and the high-pass active filter as shown in Figure 5.18.

We analyze the circuit as follows:

$$Z_F = Z_R \Big\| Z_C = R_2 \Big\| \frac{1}{j\omega C_2} \tag{5.42}$$

$$Z_F = \frac{R_2 \frac{1}{j\omega C_2}}{R_2 + \frac{1}{j\omega C_2}} = \frac{R_2}{1 + j\omega C_2 R_2} \tag{5.43}$$

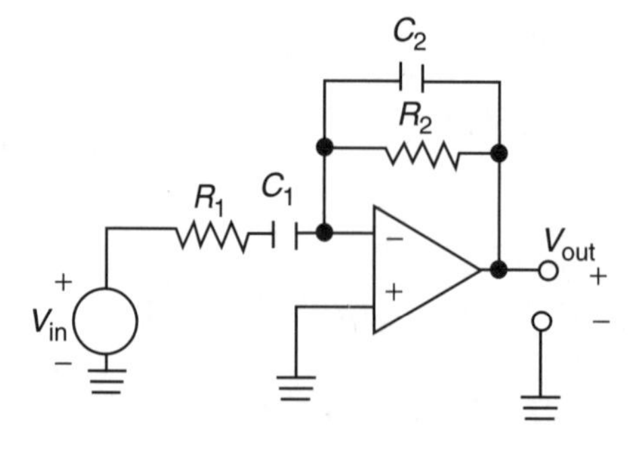

Figure 5.18 Band-pass active filter.

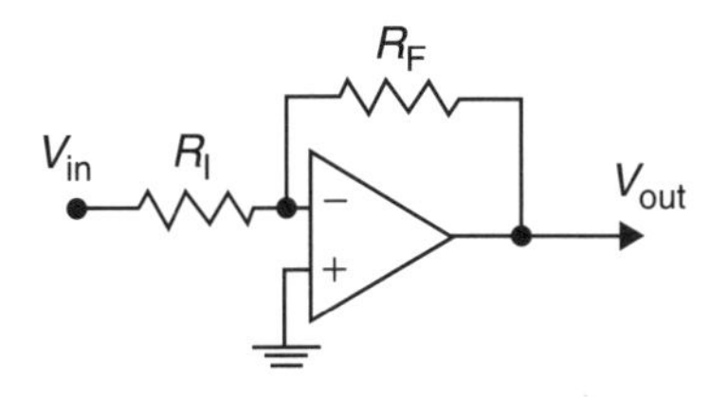

Figure 5.19 Circuit for Example 5.1.

$$Z_i = Z_R + Z_C = R_1 + \frac{1}{j\omega C_1} = \frac{1 + j\omega C_1 R_1}{j\omega C_1}. \tag{5.44}$$

The gain of this filter is given by:

$$A_{LP}(j\omega) = -\frac{Z_F}{Z_i} = -\frac{j\omega C_1 R_2}{(1 + j\omega C_2 R_2)(1 + j\omega C_1 R_1)}, \tag{5.45}$$

where

$$\omega_l = \frac{1}{C_1 R_2}; \quad \omega_{LP} = \frac{1}{C_2 R_2}; \quad \omega_{HP} = \frac{1}{C_1 R_1}; \quad \omega_{HP} > \omega_{LP} \tag{5.46}$$

EXAMPLE 5.1

See website for downloadable MATLAB code to solve this problem

In the circuit shown in Figure 5.19,

$$R_F = 8\,\text{k}\Omega; \quad R_I = 4\,\text{k}\Omega; \quad V_i = 5\,\text{mV}.$$

Determine the output voltage.

Solution

This is an inverting amplifier, so from Equation 5.3:

$$\frac{V_{out}}{V_i} = -\frac{R_F}{R_I}$$
$$V_{out} = -\frac{R_F}{R_I} V_i = -\frac{8}{4} \times 5\text{ mV} = -10\text{ mV} \tag{5.46A}$$

EXAMPLE 5.2

See website for downloadable MATLAB code to solve this problem

In the circuit shown in Figure 5.20,

$$R_F = 8\,\text{k}\Omega; \quad R_I = 4\,\text{k}\Omega; \quad V_s = 10\,\text{mV}.$$

Determine (a) the voltage gain of this amplifier; (b) the output voltage.

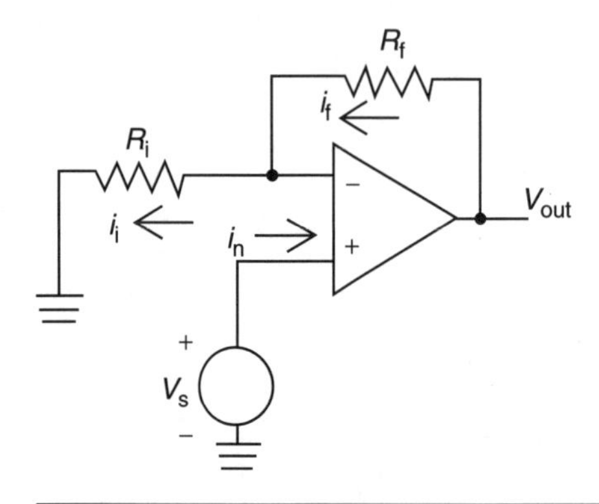

Figure 5.20 Circuit for Example 5.2.

Solution

This is a non-inverting amplifier, so from Equation 5.6:

(a) The voltage gain of this amplifier is

$$1 + \frac{R_f}{R_i} = 1 + \frac{8}{4} = 3. \quad (5.46B)$$

(b) The output voltage is

$$V_{out} = \left(1 + \frac{R_f}{R_i}\right) V_s = 3 \times 10\,\text{mV} = 30\,\text{mV}. \quad (5.46C)$$

EXAMPLE 5.3

See website for downloadable MATLAB code to solve this problem

In the circuit shown in Figure 5.21,

$$R_F = 10\,\text{k}\Omega; \quad R_1 = 6\,\text{k}\Omega; \quad R_2 = 6\,\text{k}\Omega; \quad R_3 = 6\,\text{k}\Omega;$$

$$V_1 = V_2 = V_3 = 5\,\text{V}.$$

Determine the output voltage.

Solution

This is a summing amplifier, so from Equation 5.9:

$$V_{out} = -R_f \sum_{i=1}^{N} \frac{V_i}{R_i} = -10\left(\frac{1}{6} + \frac{1}{6} + \frac{1}{6}\right) \times 5\,\text{V} = -25\,\text{V}. \quad (5.46D)$$

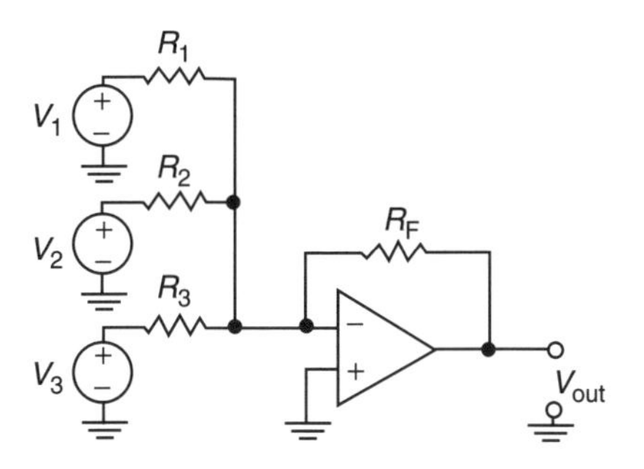

Figure 5.21 Circuit for Example 5.3.

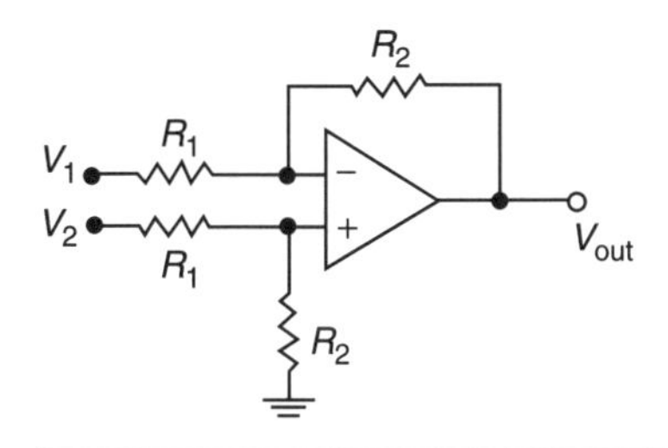

Figure 5.22 Circuit for Example 5.4.

EXAMPLE 5.4

See website for downloadable MATLAB code to solve this problem

In the circuit shown in Figure 5.22,

$$R_1 = 5\,\text{k}\Omega; \quad R_2 = 15\,\text{k}\Omega;$$
$$V_1 = 10\,\text{V}; \quad V_2 = 15\,\text{V}.$$

Determine the output voltage for these values and for when $R_1 = R_2$.

Solution

This is a difference amplifier, so from Equation 5.15:

$$V_{\text{out}} = \left(\frac{R_2}{R_1}\right)(V_2 - V_1) = \left(\frac{15}{5}\right)(15 - 10) = 15\,\text{V}. \tag{5.46E}$$

When $R_1 = R_2$, the output voltage is 5 V.

EXAMPLE 5.5

See website for downloadable MATLAB code to solve this problem

For the instrumentation amplifier circuit shown in Figure 5.8,

$$R_1 = 5\,\text{k}\Omega; \quad R_2 = 20\,\text{k}\Omega; \quad R_3 = 15\,\text{k}\Omega; \quad R_G = 3\,\text{k}\Omega;$$
$$V_1 = 10\,\text{V}; \quad V_2 = 15\,\text{V}.$$

Determine (a) the gain; (b) the output voltage.

Solution

$$\text{(a)}\quad G = \left(\frac{20}{5}\right) \times \left(1 + \frac{2 \times 15}{3}\right) = 44. \tag{5.46F}$$

$$\text{(b)}\quad V_{out} = \left(\frac{20}{5}\right)\left(1 + \frac{2 \times 15}{3}\right)(15 - 10) = 220\,\text{V}. \tag{5.46G}$$

EXAMPLE 5.6

See website for downloadable MATLAB code to solve this problem

For the low-pass active filter circuit shown in Figure 5.14,

$$R_1 = 5\,\text{k}\Omega; \quad R_2 = 20\,\text{k}\Omega;$$
$$C = 2\,\mu\text{F}.$$

Determine the closed-loop gain and the decibel value at which the filter rolls off.

Solution

$$\begin{aligned} H_0 &= \left|\frac{-R_2}{R_1}\right| = \frac{20}{5} = 4 \\ \omega_0 &= \frac{1}{2 \times 10^{-6} \times 20 \times 10^3} = 25 \\ f_0 &= \frac{25}{2\pi} = 3.98\,\text{Hz} \end{aligned} \tag{5.46H}$$

The filter roll off is given as

$$\begin{aligned} &A_{LP}(j\omega) = H_0\frac{1}{\sqrt{2}}; \quad \left|A_{LP}(j\omega)\right|_{dB} = 20\,\log_{10} H_0 - 20\,\log_{10}\sqrt{2} \\ &\therefore\ \left|A_{LP}(j\omega)\right|_{dB} = 20\,\log_{10} H_0 - 3\,\text{dB} = 20\,\log_{10}(4) - 3\,\text{dB} = 9.04\,\text{dB} \end{aligned} \tag{5.46I}$$

Problems

Signal processing

Q5.1 Find the current i_i in the circuit shown in Figure 5.23, where

$$R_1 = 3\,\Omega; \quad R_2 = 2\,\Omega; \quad R_3 = 4\,\Omega;$$
$$V_s = 12\,\text{V}.$$

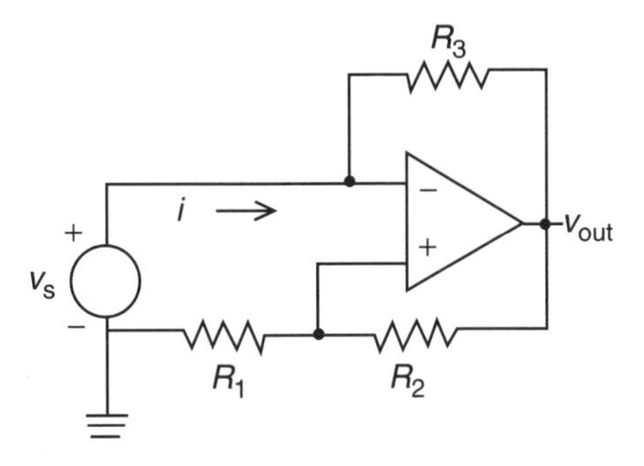

Figure 5.23 Circuit for Q5.1.

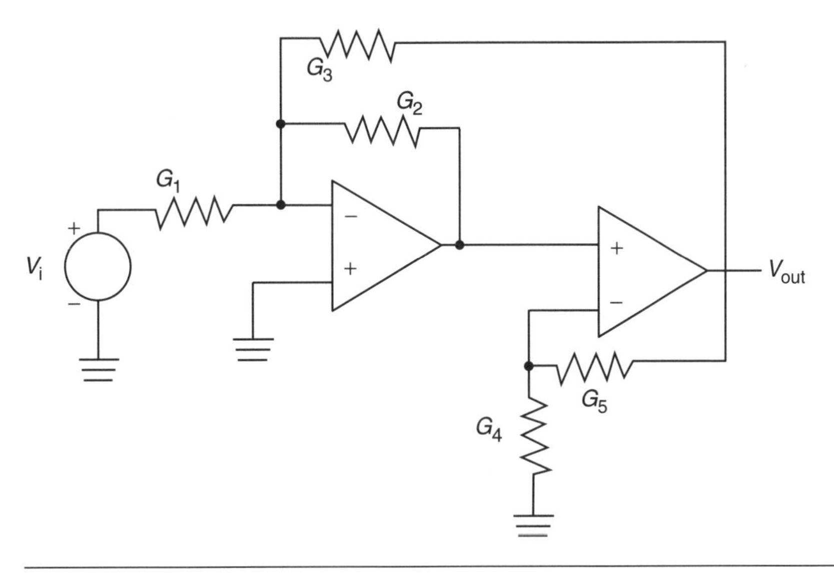

Figure 5.24 Circuit for Q5.2.

Q5.2 In the circuit shown in Figure 5.24, assume the op amps are ideal, and that

$$G_1 = 16\text{s}; \quad G_2 = 8\text{s}; \quad G_3 = 2\text{s}; \quad G_4 = 6\text{s}; \quad G_5 = 2\text{s}.$$

Determine:

(a) the expression for the overall gain $A_v = V_{out}/V_{in}$;

(b) the conductance $G = i_{in}/V_{in}$ that the voltage source, V_i sees.

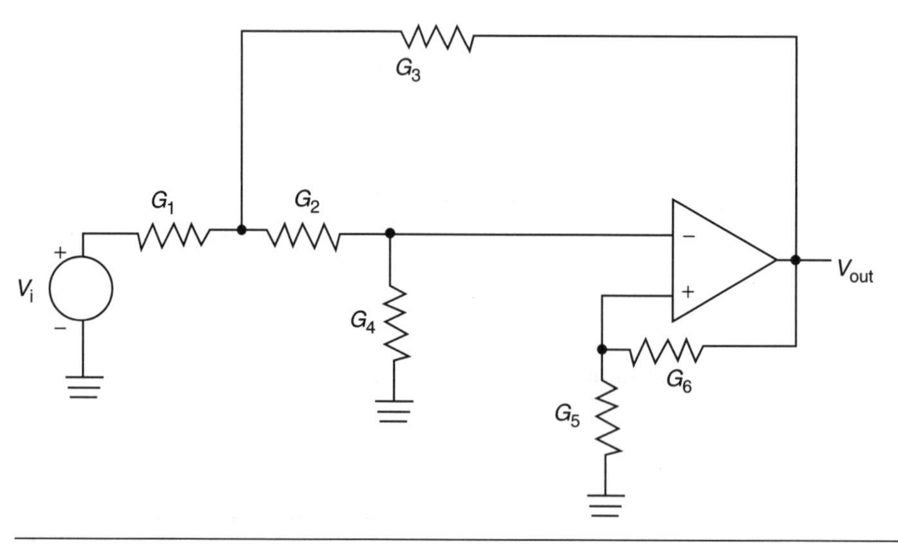

Figure 5.25 Circuit for Q5.3.

Q5.3 In the circuit shown in Figure 5.25, assume the op amps are ideal, and that

$$G_1 = 2\text{s}; \quad G_2 = 4\text{s}; \quad G_3 = 4\text{s}; \quad G_4 = 4\text{s}; \quad G_5 = 6\text{s}; \quad G_6 = 6\text{s}.$$

Determine:

(a) the expression for the overall gain $A_v = V_{out}/V_{in}$;

(b) the conductance $G = i_{in}/V_{in}$ that the voltage source, V_i sees.

Q5.4 Figure 5.26 shows a circuit that will remove the d.c. component of the input voltage, $V_1(t)$, and simultaneously amplify the a.c. portion.

$$V_1(t) = 18 + 10^{-3} \sin \omega t \, \text{V}; \quad V_{battery} = 24\,\text{V}; \quad R_F = 10\,\text{k}\Omega.$$

Determine:

(a) R_S such that no d.c. voltage appears at the output;

(b) $V_{out}(t)$, using the calculated value of R_S.

Q5.5 In Figure 5.27, $R_1 = 30\,\text{k}\Omega$; $R_F = 300\,\text{k}\Omega$; $R_2 = 3\,\text{k}\Omega$; and

$$V_1(t) = 10^{-3} + 10^{-3} \cos \omega t \, \text{V}.$$

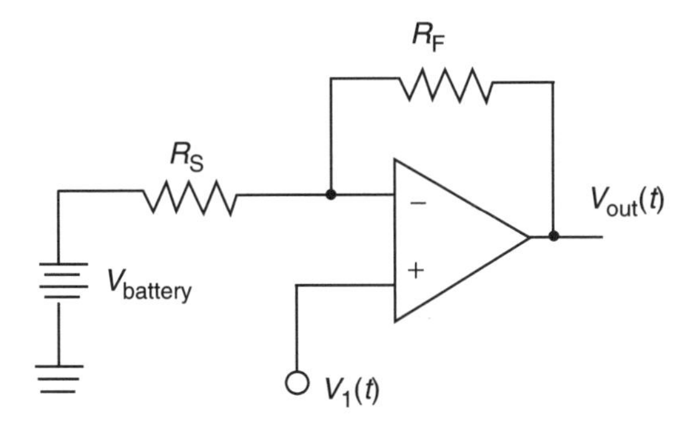

Figure 5.26 Circuit for Q5.4.

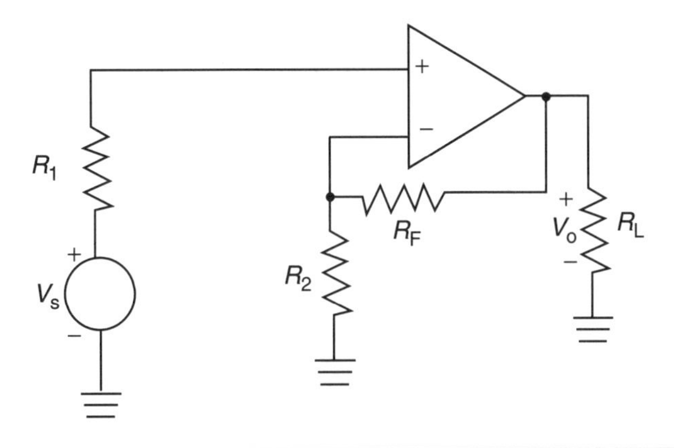

Figure 5.27 Circuit for Q5.5 and Q5.6.

Determine:

(a) the expression for the output voltage;

(b) the value of the output voltage.

Q5.6 Repeat Q5.5 with $R_1 = 50\,\text{k}\Omega$; $R_F = 300\,\text{k}\Omega$; $R_2 = 5\,\text{k}\Omega$; and

$$V_1(t) = 10^{-3} + 10^{-3} \cos \omega t\,\text{V}.$$

Q5.7 In Figure 5.28, $R_S = 60\,\text{k}\Omega$; $R_L = 240\,\text{k}\Omega$; and

$$V_S(t) = 20 \times 10^{-3} + 5 \times 10^{-3} \cos \omega t\,\text{V}.$$

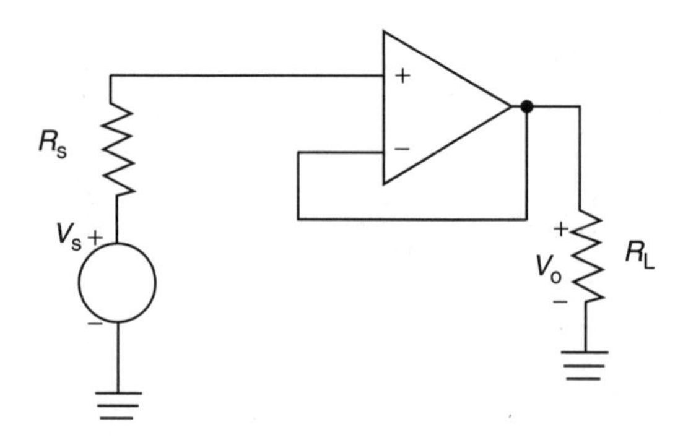

Figure 5.28 Circuit for Q5.7.

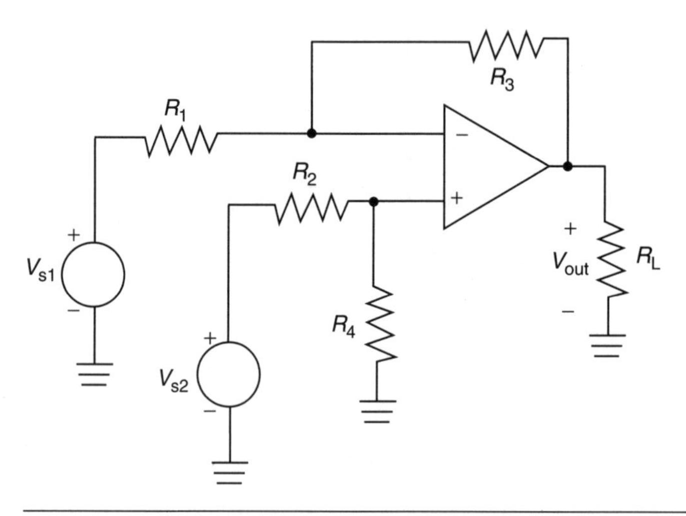

Figure 5.29 Circuit for Q5.8, Q5.9, Q5.11, and Q5.12.

Determine the output voltage, V_{out}.

Q5.8 In the circuit shown in Figure 5.29,

$$V_{S1}(t) = V_{S2}(t) = 20 \times 10^{-3} + 5 \times 10^{-3} \cos \omega t \,\text{V};$$
$$R_1 = 3\,\text{k}\Omega; \quad R_2 = 6\,\text{k}\Omega; \quad R_3 = 15\,\text{k}\Omega; \quad R_4 = 12\,\text{k}\Omega.$$

Determine the output voltage, V_{out}.

Q5.9 In the circuit shown in Figure 5.29,

$$V_{S1} = 10 \times 10^{-3}\,\text{V}; \quad V_{S2} = 21 \times 10^{-3}\,\text{V};$$
$$R_1 = 2\,\text{k}\Omega; \quad R_2 = 78\,\text{k}\Omega; \quad R_3 = 20\,\text{k}\Omega; \quad R_4 = 65\,\text{k}\Omega.$$

Determine the output voltage.

Q5.10 In a differential amplifier, $A_{v1} = -16$ and $A_{v2} = +20$. Derive expressions for and then determine the value of (a) the common-mode gain; (b) the differential-mode gain.

Q5.11 In the circuit shown in Figure 5.29,

$$V_{S1} = 1\,\text{V}; \quad V_{S2} = 2\,\text{V};$$

$$R_1 = 5\,\text{k}\Omega; \quad R_2 = 2\,\text{k}\Omega; \quad R_3 = 15\,\text{k}\Omega; \quad R_4 = 3\,\text{k}\Omega.$$

Determine:

(a) the output voltage;

(b) the common-mode component of the output voltage;

(c) the differential-mode component of the output voltage.

Q5.12 In the circuit shown in Figure 5.29,

$$V_{S1} = 12\,\text{V}; \quad V_{S2} = 20\,\text{V};$$

$$R_1 = 1\,\text{k}\Omega; \quad R_2 = 15\,\text{k}\Omega; \quad R_3 = 80\,\text{k}\Omega; \quad R_4 = 70\,\text{k}\Omega.$$

Determine:

(a) the output voltage;

(b) the common-mode component of the output voltage;

(c) the differential-mode component of the output voltage.

Filters

Q5.13 Figure Q5.30 shows an active filter in which,

$$C = 1\,\mu\text{F}; \quad R_1 = 10\,\text{k}\Omega; \quad R_2 = 2\,\text{k}\Omega.$$

Determine:

(a) the gain (in dB) in the pass band;

(b) the cut-off frequency;

(c) if the circuit is a low- or high-pass filter.

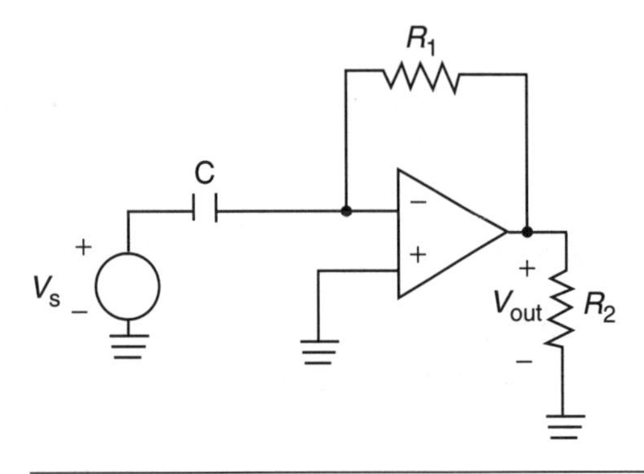

Figure 5.30 Circuit for Q5.13.

Q5.14 Figure Q5.31 shows an op amp that is used as a filter, with component values of

$$C = 1\,\mu\text{F};\quad R_1 = 10\,\text{k}\Omega;\quad R_2 = 25\,\text{k}\Omega;\quad R_L = 15\,\text{k}\Omega.$$

Determine:

(a) the gain V_{out}/V_s (in dB) in the pass band;

(b) the cut-off frequency;

(c) if the circuit is a low- or high-pass filter.

Q5.15 The op amp circuit shown in Figure 5.31 is used as a filter, with component values of

$$C = 100\,\mu\text{F};\quad R_1 = 10\,\text{k}\Omega;\quad R_2 = 1505\,\text{k}\Omega;\quad R_L = 15\,\text{k}\Omega.$$

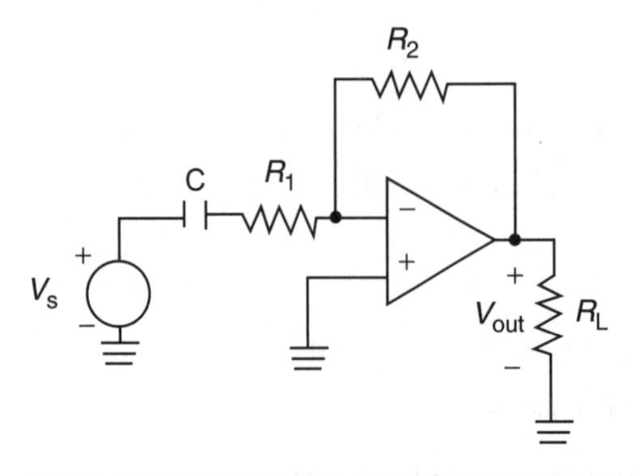

Figure 5.31 Circuit for Q5.14 and Q5.15.

Determine:

(a) the gain V_{out}/V_s (in dB) in the pass band;

(b) the cut-off frequency;

(c) if the circuit is a low- or high-pass filter.

Q5.16 Figure 5.32 shows an active filter with

$$C = 10\,\text{pF}; \quad R_1 = 5\,\text{k}\Omega; \quad R_2 = 75\,\text{k}\Omega; \quad R_L = 40\,\text{k}\Omega.$$

Determine:

(a) the magnitude of the voltage transfer function at very low and at very high frequencies;

(b) the cut-off frequency.

Q5.17 Figure 5.33 shows an active filter with

$$C = 10\,\text{nF}; \quad R_1 = 1\,\text{k}\Omega; \quad R_2 = 5\,\text{k}\Omega; \quad R_3 = 95\,\Omega; \quad R_L = 90\,\text{k}\Omega.$$

Determine:

(a) an expression for the voltage transfer function in the form $H_v(j\omega) = V_{out}(j\omega)/V_{in}(j\omega)$;

(b) the cut-off frequency;

(c) the pass band gain.

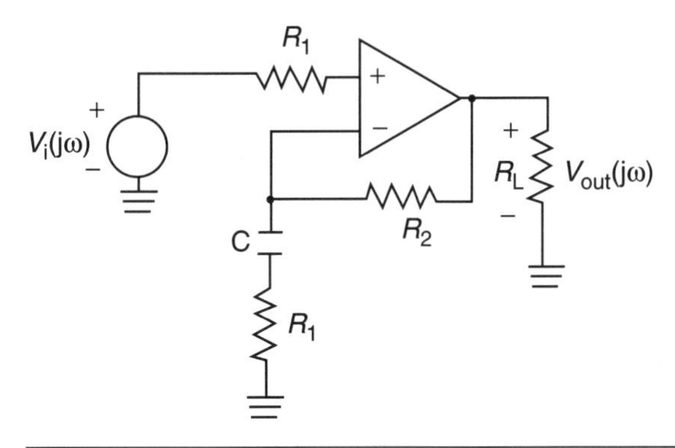

Figure 5.32 Circuit for Q5.16.

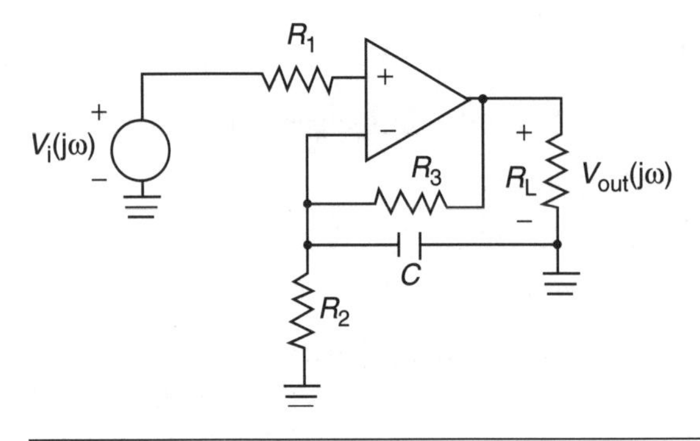

Figure 5.33 Circuit for Q5.17.

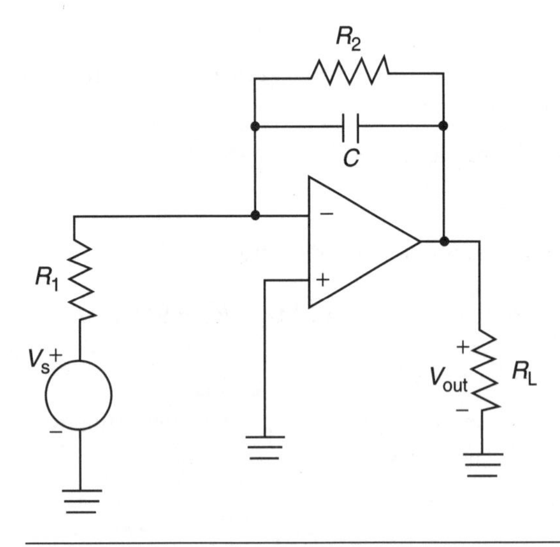

Figure 5.34 Circuit for Q5.18, Q5.19, and Q5.20.

Q5.18 The op amp circuit shown in Figure 5.34 is an active filter with component values of

$$C = 1\,\mu\text{F}; \quad R_1 = 10\,\text{k}\Omega; \quad R_2 = 8\,\text{k}\Omega; \quad R_L = 10\,\Omega.$$

Determine:

(a) an expression for the voltage transfer function in the standard form;

(b) the gain in dB in the pass band;

(c) the cut-off frequency;

(d) if the circuit is a low- or high-pass filter.

Q5.19 Solve Q5.18 with:

$$C = 0.5\,\text{nF}; \quad R_1 = 2500\,\text{k}\Omega; \quad R_2 = 75\,\text{k}\Omega; \quad R_L = 15\,\Omega.$$

Q5.20 Solve Q5.18 with:

$$C = 1\,\text{nF}; \quad R_1 = 10500\,\text{k}\Omega; \quad R_2 = 35\,\text{k}\Omega; \quad R_L = 20\,\Omega.$$

Further reading

[1] Alciatore, D.G. and Histand, M.B. (2002) *Introduction to Mechatronics and Measurement Systems* (2nd. ed.), McGraw-Hill.

[2] Horowitz, P. and Hill, W. (1989) *The Art of Electronics* (2nd. ed.), New York: Cambridge University Press.

[3] Rashid, M.H. (1996) *Power Electronics: Circuits, Devices, and Applications*, Prentice Hall.

[4] Rizzoni, G. (2003) *Principles and Applications of Electrical Engineering* (4th. ed.), McGraw-Hill.

[5] Stiffler, A.K. (1992) *Design with Microprocessors for Mechanical Engineers*, McGraw-Hill.

CHAPTER 6

Microcomputers and microcontrollers

Chapter objectives

When you have finished this chapter you should be able to:

- understand microprocessors' and microcomputers' fundamentals;
- understand the architectures of the PIC 16F84 and 16F877 microcontrollers, including their main features;
- understand programming a PIC using assembly language;
- understand programming a PIC using C;
- interface common PIC peripherals using the PIC millennium board, for numeric keyboard, LCD display applications;
- interface the PIC to some other practical mechatronics systems.

6.1 Introduction

The digital circuits presented in Chapter 4 allow the implementation of combinational and sequential logic operations by interconnecting ICs containing gates and flip-flops. This is considered a hardware solution because it consists of a selection of specific ICs, which when hardwired on a circuit board, carry out predefined functions. To make a change in functionality, the hardware circuitry must be modified and may require a redesign. However, in many mechatronics systems, the control tasks may involve complex relationships among many inputs and outputs, making a strictly hardware solution impractical. A more

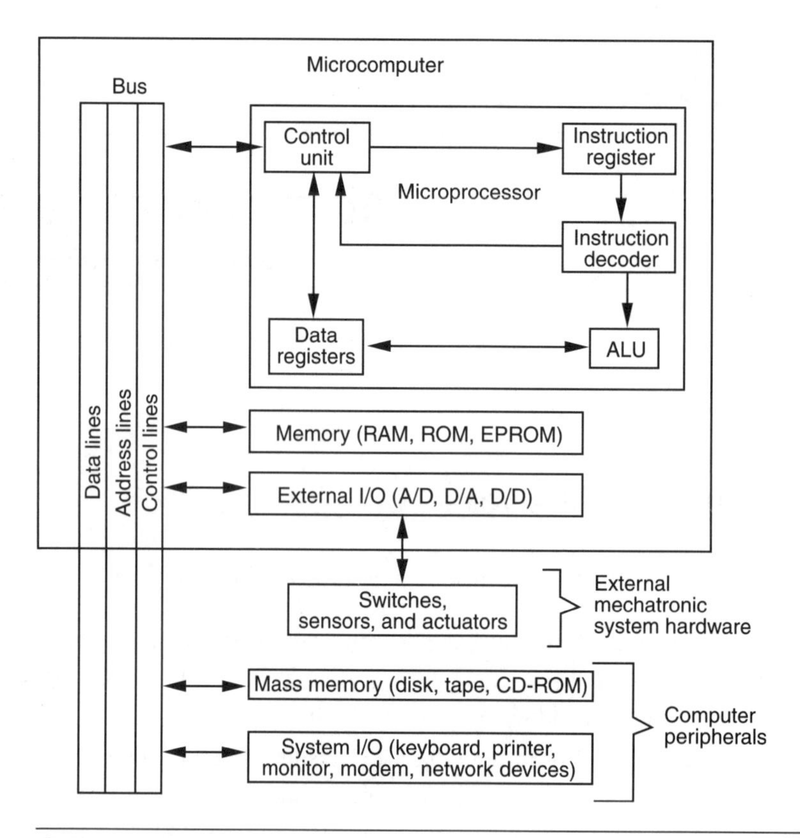

Figure 6.1 Microcomputer architecture.

satisfactory approach in complex digital design involves the use of a microprocessor-based system to implement a software solution. Software is a procedural program consisting of a set of instructions to execute logic and arithmetic functions and to access input signals and control output signals. An advantage of a software solution is that, without making changes in hardware, the program can be easily modified to alter the mechatronics system's functionality.

A *microprocessor* is a single, very-large-scale-integration (VLSI) chip that contains many digital circuits that perform arithmetic, logic, communication, and control functions.

When a microprocessor is packaged on a printed circuit board with other components, such as interface and memory chips, the resulting assembly is referred to as a *microcomputer* or *single-board computer*. Figure 6.1 illustrates the overall architecture of a typical microcomputer system using a microprocessor.

The microprocessor, also called the *central processing unit* (CPU) or *microprocessor unit* (MPU), is where the primary computation and system control operations occur. The *arithmetic logic unit* (ALU) within the CPU executes mathematical functions on data structured as binary words.

The following define some terms that are key to a CPU's operation in the storage and retrieval of data.

- A *word* is an ordered set of bits, usually 8, 16, 32, or 64 bits long.
- The *instruction decoder* interprets instructions fetched sequentially from memory by the control unit and stored in the instruction register. Each instruction is a set of coded bits that commands the ALU to perform bit manipulation, such as binary addition and logic functions, on words stored in the CPU data registers. The ALU results are also stored in data registers and then transferred to memory by the control unit.
- The *bus* is a set of shared communication lines that serves as the 'central nervous system' of the computer. Data, address, and control signals are shared by all system components via the bus. Each component connected to the bus communicates information to and from the bus via its own bus controller. The data lines, address lines, and control lines allow a specific component to access data addressed to that component.
- The *data lines* are used to communicate words to and from data registers in the various system components such as memory, CPU, and input/output (I/O) peripherals.
- The *address lines* are used to select devices on the bus or specific data locations within memory. Devices usually have a combinational logic address decoder circuit that identifies the address code and activates the device.
- The *control lines* transmit read and write signals, the system clock signal, and other control signals such as system interrupts, which are described in subsequent sections.
- The *read-only memory* (ROM) is used for permanent storage of data that the CPU can read, but the CPU cannot write data to ROM. ROM does not require a power supply to retain its data and therefore is called non-volatile memory.
- *Random access memory* (RAM) is used for storing data that is used during the program run-time: Data can be read from or written to RAM at any time, provided power is maintained. The data in RAM is considered volatile because it is lost when power is removed. There are two main types of RAM: *static* RAM (SRAM), which retains its data in flip-flops as long as the memory is powered, and *dynamic* RAM (DRAM), which consists of capacitive storage of data that must be refreshed (rewritten) periodically because of charge leakage.
- *Erasable-programmable* ROM (EPROM): Data stored in an EPROM can be erased with ultraviolet light applied through a transparent quartz

window on top of the EPROM IC. Then new data can be stored on the EPROM. Another type of EPROM is *electrically erasable* (EEPROM). Data in EEPROM can be erased electrically and rewritten through its data lines without the need for ultraviolet light. Since data in RAM are volatile, ROM, EPROM, EEPROM, and peripheral mass memory storage devices such as magnetic disks and tapes and optical CD-ROMs are sometimes needed to provide permanent data storage.

- *Reduced instruction-set computer* (RISC): When the set of instructions is small, the microprocessor is known as a RISC microprocessor. RISC microprocessors are cheaper to design and manufacture, and are usually faster than the conventional microprocessor. However, more programming steps may be required for complex algorithms, due to the limited set of instructions.
- *Machine code:* Communication to and from the microprocessor occurs through in/out (I/O) devices connected to the bus. External computer peripheral I/O devices include keyboards, printers, displays, modems, and network devices. For mechatronics applications, analog-to-digital (A/D), digital-to-analog (D/A) and digital-to-digital(D/D) I/O devices provide interfaces between the microcomputer and switches, sensors, and actuators. The instructions that can be executed by the CPU are defined by a binary code called *machine code*. The instructions and corresponding codes are microprocessor dependent. A unique binary string represents each instruction that causes the microprocessor to perform a low-level function (e.g. add a number to a register or move a register's value to a memory location).
- *Assembly language:* Microprocessors can be programmed using *assembly language*, which comprises mnemonic commands corresponding to each instruction (e.g. 'ADD' to add a number to a register and 'MOV' to move a register's contents to a memory location).
- *Assembler:* Assembly language must be converted to machine code using the assembler, so that the microprocessor is able to understand and execute the instructions.
- *High-level languages:* Programs can also be written in a higher-level language such as BASIC or C, provided that a compiler is available that can generate machine code for the specific microprocessor being used. The advantages of using a high-level language are:

  - ease of learning and use;
  - ease of debugging programs (the process of finding and removing errors);
  - ease of comprehension of programs;

- availability of programming techniques, such as variable and array management, assignment statements with complex calculations, logical comparison expressions, iteration, interrupts, pauses, and special purpose functions.

Disadvantages include

- the resulting machine code may be less efficient (i.e. slower and require more memory) than a corresponding well-written assembly language program;
- consumption of more EEPROM space.

Now that we have presented the architecture of microprocessors and microcomputers, and defined useful terminology, we now delve into microcontrollers, which are instrumental in achieving flexible mechatronics systems.

6.2 Microcontrollers

Recently, there have been two directions in the ongoing advances of microprocessor technology. One direction supports CPUs for the personal computer and workstation industry, where the main constraints are high speed and large word size (32 and 64 bits). The other direction includes development of the microcontroller, which is a single IC containing specialized circuits and functions that are applicable to mechatronics system design.

The microcontroller contains a microprocessor, memory, I/O capabilities, and other on-chip resources. It is basically a microcomputer on a single IC.

Popular microcontrollers that have being in great demand for realizing mechatronics systems are:

- Microchip's PIC;
- Motorola's 68HC11; and
- Intel's 8096.

Factors that have driven the development of the microcontroller are low cost, versatility, ease of programming, and small size. Microcontrollers are attractive in mechatronics system design since their small size and broad functionality allow them to be physically embedded in a system to perform all of the necessary control functions.

Microcontrollers are used in a wide variety of applications including home appliances, entertainment equipment, telecommunication equipment, automobiles, trucks, airplanes, toys, and office equipment. All these products involve devices that require some sort of intelligent control based on various inputs. For example, the microcontroller in a microwave oven monitors the control panel for user input, updates the graphical displays when necessary, and controls the timing and cooking functions. In an automobile, there are many microcontrollers to control various subsystems, including cruise control, antilock braking, ignition control, keyless entry, environmental control, and air and fuel flow. An office fax machine controls actuators to feed paper, uses photosensors to scan a page, sends or receives data on a phone line, and provides a user interface complete with menu-driven controls. A toy robot dog has various sensors to detect inputs from its environment (e.g. bumping into obstacles, being patted on the head, light and dark, voice commands), and an onboard microcontroller actuates motors to mimic actual dog behavior (e.g. bark, sit, and walk) based on this input. Microcontrollers and the software running on them control all of these powerful and interesting devices.

Figure 6.2 shows a block diagram for a typical full-featured microcontroller. Included in the figure are lists of typical external devices that might interface to the microcontroller. The components of a microcontroller are the:

- CPU
- RAM

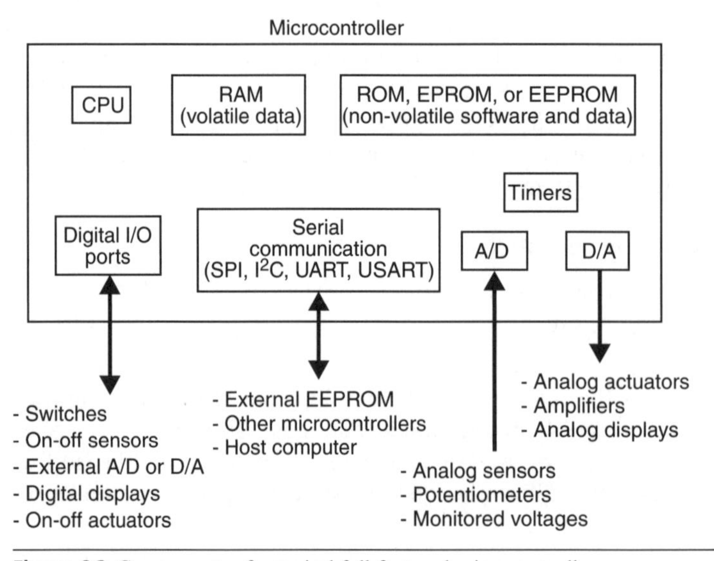

Figure 6.2 Components of a typical full-featured microcontroller.

- ROM
- Digital I/O ports
- A serial communication interface
- Timers
- Analog-to-digital (A/D) converters, and digital-to-analog (D/A) converters

The CPU executes the software stored in ROM and controls all the microcontroller components. The RAM is used to store settings and values used by an executing program. The ROM is used to store the program and any permanent data. A designer can have a program and data permanently stored in ROM by the chip manufacturer, or the ROM can be in the form of EPROM or EEPROM, which can be reprogrammed by the user. Software permanently stored in ROM is referred to as *firmware*. Microcontroller manufacturers offer programming devices that can download a compiled machine code file from a PC directly to the EEPROM of the microcontroller, usually via the PC serial port and special-purpose pins on the microcontroller. These pins can usually be used for other purposes once the device is programmed. Additional EEPROM may also be available and used by the program to store settings and parameters generated or modified during execution. The data in EEPROM is non-volatile, which means the program can access the data when the microcontroller power is turned off and back on again.

The digital I/O *ports* allow binary data to be transferred to and from the microcontroller using external pins on the IC. These pins can be used to read the state of switches and on–off sensors, to interface to external analog-to-digital (ADC) and digital-to-analog (DAC) converters, to control digital displays, and to control on–off actuators. The I/O ports can also be used to transmit signals to and from other microcontrollers to coordinate various functions. The microcontroller can also use a serial port to transmit data to and from external devices, provided these devices support the same serial communication protocol. Examples of such devices include external EEPROM memory ICs that might store a large block of data for the microcontroller, other microcontrollers that need to share data, and a host computer that might download a program into the microcontroller's onboard EEPROM.

There are various standards or protocols for serial communication including SPI (serial peripheral interface), I^2C (inter-integrated circuit), UART (universal asynchronous receiver-transmitter), and USART (universal synchronous asynchronous receiver-transmitter).

The ADC allows the microcontroller to convert an external analog voltage (e.g. from a sensor) to a digital value that can be processed or stored by the CPU. The DAC allows the microcontroller to output an analog voltage to a non-digital device (e.g. a motor amplifier). ADCs and DACs and their applications are discussed in Chapter 7. Onboard timers are usually provided to help create delays or ensure events occur at precise time intervals (e.g. reading the value of a sensor).

Microcontrollers typically have less than 1 kB to several tens of kilobytes of program memory, compared with microcomputers where RAM memory is measured in megabytes or gigabytes. Also, microcontroller clock speeds are slower than those used for microcomputers. For some applications, a selected microcontroller may not have enough speed or memory to satisfy the needs of the application. Fortunately, microcontroller manufacturers usually provide a wide range of products to accommodate different applications. Also, when more memory, or I/O capability, is required, the functionality of the microcontroller can be expanded with additional external components (e.g. RAM or EEPROM chips, external ADCs and DACs, and other microcontrollers).

In the remainder of this chapter, we focus on the PIC microcontroller due to its wide acceptance in industry, abundant information resources, low cost, and ease of use. PIC is a large and diverse family of low-cost microcontrollers manufactured by Microchip Technology. They vary in physical size, the number of I/O pins available, the size of the EEPROM and RAM space for storing programs and data, and the availability of ADCs and DACs. Obviously, the more features and capacity a microcontroller has, the higher the cost. We first focus specifically on the PIC16F84 device, which is a low-cost 8-bit microcontroller with EEPROM flash memory for program and data storage. It does not have a built-in ADC, DAC or serial communication capability, but it supports 13 digital I/O lines and serves as a good learning platform because of its low cost and ease of programming. The PIC16F877, which has more enhanced capabilities and was used for most of the applications presented in the chapter for the case studies of Chapter 20 is discussed later in this chapter. Once a user knows how to interface and program one microcontroller, it is easy to extend that knowledge to other microcontrollers with different features and programming options.

6.3 The PIC16F84 microcontroller

The block diagram for the PIC16F84 microcontroller is shown in Figure 6.3. This diagram, along with complete documentation of all of the microcontroller's features and capabilities, can be found in the manufacturer's data sheets. The PIC16F8X data sheets are contained in a book available from Microchip and as a PDF file on its website (www.microchip.com).

The PIC16F84 is a low cost, high performance, CMOS, fully static 8-bit microcontroller. Similar to all the PIC microcontrollers, the PIC16F84 employs an advanced reduced instruction set computer (RISC) architecture. The separate

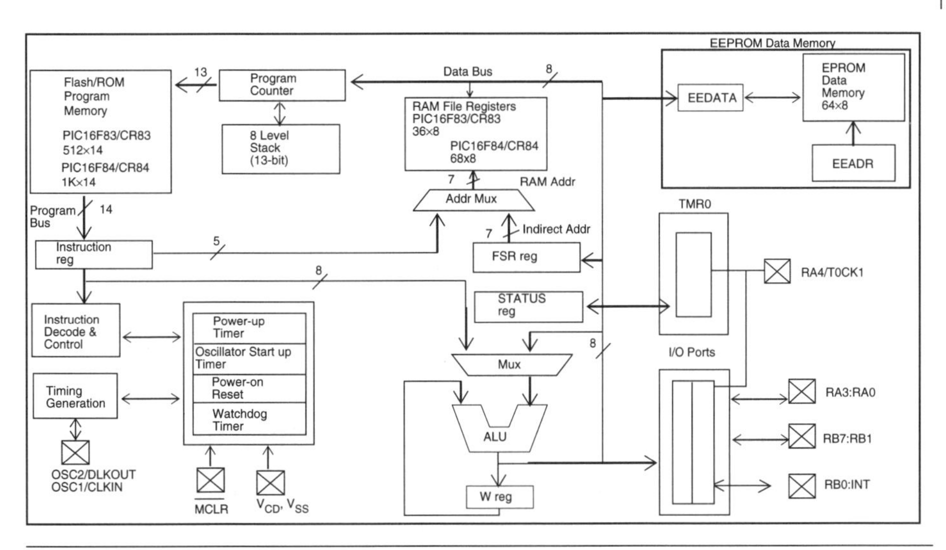

Figure 6.3 The PIC16F84 microcontroller block diagram. (Courtesy of Microchip Technology Inc.)

instruction and data buses of the Harvard architecture allow a 14-bit wide instruction word with a separate 8-bit wide data bus.

The high performance RISC CPU features make PIC very versatile and useful in many mechatronics projects. It has 35 single word instructions. It operates at a maximum frequency of 10 MHz. The program memory allows 1 kB of flash memory, 68 bytes of data RAM and 64 bytes of data EEPROM. It also has four interrupt sources. The PIC16F84 provides 1000 erase/write cycles to the flash memory. The EEPROM can retain data for more than 40 years.

Equally unique are the peripheral features of the PIC16F84. It has 13 I/O pins for individual direction control. It provides a 25 mA sink per pin and 20 mA source per pin; sufficient to drive LEDs. An 8-bit timer/counter makes the PIC suited for timer-based mechatronic projects.

The PIC16F84 generally operates on a regulated 5 V supply but has an operable range from 2 V to 6 V. Its power consumption is minimal and hence can be operated with dry cells for long hours. The PIC16F84 is available as an 18-pin dual-in-line package. These features are now summarized in the next subsection.

6.3.1 Features of the PIC16F84 microcontroller

- 8-bit wide data bus CMOS microcontroller;
- 18-pin DIP, SOIC;
- 1792 bytes of flash EEPROM program memory subdivided into 14-bit words (0h–3Fh);

- 68 bytes of RAM data memory;
- 64 bytes of non-volatile EEPROM data memory;
- 1024 (1 kB) instructions capability;
- 4 MHz clock speed (maximum 10 MHz);
- 15 special function hardware registers;
- 36 general purpose registers (SRAM);
- 13 I/O pins.

6.3.2 The PIC16F84 microcontroller architecture

The PIC16F84 block diagram is shown in Figure 6.3; it uses the Harvard architecture since its program memory and data memory are accessed from different memories. This offers an improvement over the von Neumann architecture in which the program and data are accessed simultaneously from the same memory (accessed over the same bus). All peripheral interface controllers (PIC) developed by Microchip utilize the Harvard architecture. It means that the registers are separate from the program memory. The PIC16F84's processor is its *arithmetic logic unit* (ALU). It receives, processes and stores data to and retrieves data from the *data registers*.

The 13-bit wide *program counter* (PC) provides the addresses to the program memory, which are then read out and stored in the *instruction register*. It is then decoded by the *instruction decode and control* circuitry. The *program memory* contains the executable code that is run as the application.

To implement two parameters in the operation of the ALU, a temporary holding register is used. This is referred to as the *accumulator* in most microprocessors but referred to as the *W register* in the PIC. Every arithmetic operation goes through the W register. As an example, suppose the contents of two registers are to be added. The contents of one of the registers are moved to the W register and then the contents of the second are added to it. Solutions can be retained in W register for immediate use or sent back for storage to the source register.

The *multiplexer* (MUX) before the ALU selects the operation that the ALU is required to perform. Some of these operations include addition, subtraction, left shifting, right shifting, etc.

The *status register* is the primary *central processing unit* (CPU) execution control register. It is used to control the execution of the application code from the *program memory* and monitor the status of the arithmetic and bit-wise operation.

There can be cases when a 7-bit register address is specified within an instruction. This is known as direct addressing and any register in the address bank can be accessed. Arithmetic and bit-wise operations that access a register can store the result in either the W register or the source register. Even further, there can be instances in some applications where the direct addressing of a register or explicitly specifying a value may not be sufficient. In such cases, a method of arithmetically specifying an address will be required. PIC does this by loading the *file select register* (FSR) with the address that needs to be accessed. The contents of the FSR are then multiplexed with seven immediate address bits. Indexed addressing is described in high-level programming as array variables.

In the PIC, instructions take one word or address. This is a characteristic of the RISC philosophy. This means that there is not enough space in a *goto* or *call* instruction for the entire address of the new location of the program counter while implementing tables. A table is a code artifact in which the program counter is written to force a jump to a specific location in the program memory and is an important feature of the PIC.

The *stack* is solely devoted to the program counter and cannot be accessed by the application code. In most other microcontrollers, the stack can be accessed by the application code. However, only the higher series of PIC allow this.

TMR0 is an 8-bit basic timer comprising an incrementing counter that can be preset (loaded) by the application code with a specific value. An external source or an instruction clock can clock the counter. Each TMR0 input is matched to two instruction clocks for synchronization. This feature limits the maximum speed of the timer to one half of the instruction clock speed. *TOCK* and *TOCE* bits are used to select the clock source and clock edge, respectively, that increments TMR0. These bits are located in the *option* register.

EEPROM and flash memory access is very important. There are four registers required for EEPROM access and these are EECON1, EECON2, EEADR and EEDATA. These registers are used to control the access to the EEPROM. EEADR and EEDATA provide the address and data, respectively, to interface the 256-byte data EEPROM memory. EECON1 and EECON2 are used to indicate that the operation has completed. EECON2 is pseudo-register that cannot be read from but data can be written to it. EECON1 contains data bits for controlling the access to the EEPROM.

6.3.3 Memory organization of the PIC16F84 microcontroller

There are two memory blocks in the PIC16F8X series of microcontrollers: program memory and data memory. Each memory block has its own bus to allow simultaneous accessing during the same oscillator cycle. This is the Harvard architecture, which improves bandwidth over traditional von Neumann architecture.

6.3.3.1 *Program memory (flash) organization*

The PIC16F8X has eight stack levels that can page each program memory space of 1 k so that the total memory space of 8 k can be utilized as shown in Figure 6.4. The first $1\,k \times 14$ (0000h–03FFh) are physically implemented as shown. Conceptually, the configuration is similar to layers used in commercial drafting packages.

The reset vector is at 0000h and the interrupt vector is at 0004h location.

Accessing a location above the physically implemented address will result in wrap around. For example, locations 10h, 410h, 810h, C10h, 1010h, 1410h, 1810h, 1C10h will be the same instructions in each of the eight pages, each being 1 k different from the other. This is because

$$400\text{h} = 1024\,(1\,\text{k})$$

$$800\text{h} = 2048\,(2\,\text{k})$$

$$\text{C00h} = 3072\,(3\,\text{k})$$

$$1000\text{h} = 4096\,(4\,\text{k})$$

$$1400\text{h} = 5120\,(5\,\text{k})$$

$$1800\text{h} = 6144\,(6\,\text{k})$$

$$1\text{C00h} = 7164\,(7\,\text{k})$$

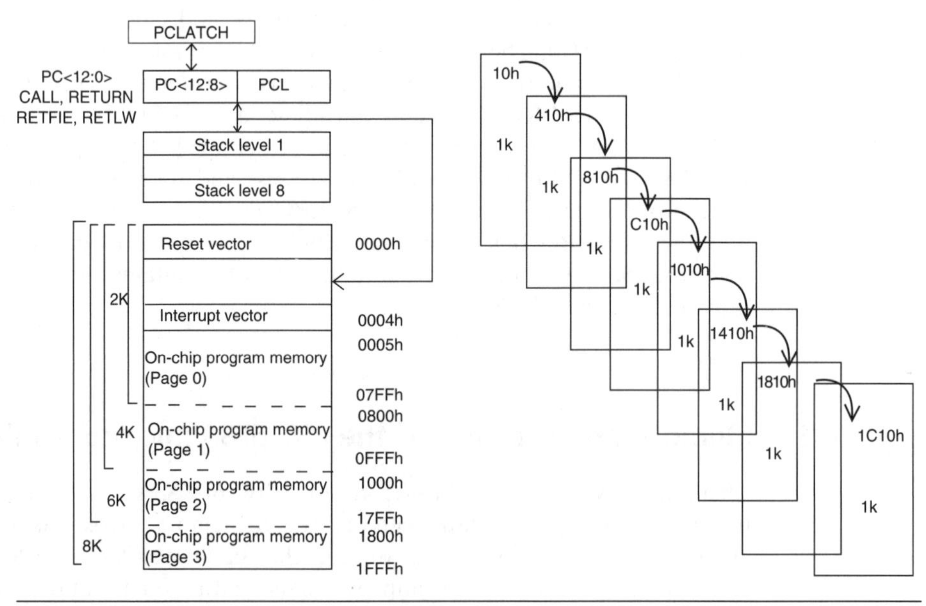

Figure 6.4 PIC16F84 program memory organization.

6.3.3.2 ***Data memory (RAM) organization***

The PIC16F8X data memory is partitioned into two areas:

- 12 special function registers (SFRs);
- 68 general purpose registers (GPRs within the RAM).

Portions of data memory are banked (for both the SFR area and GPR area). Figure 6.5 shows the configuration. The EEPROM is not included here as it is indirectly addressed.

The *special function registers* (SFRs) are used by the CPU and peripheral functions to control the device operation (these registers are SRAM) and occupy the first 12 file addresses corresponding to 80h (128) up to 8Bh (139), (i.e. 128, 129, 138, 139). Alternatively, we can obtain the number as $(139 - 128 + 1) = 12$.

File address			File address
00h	Indirect address[1]	Indirect address[1]	80h
01h	TMR0	OPTION	81h
02h	PCL	PCL	82h
03h	STATUS	STATUS	83h
04h	FSR	FSR	84h
05h	PORTA	TRISA	85h
06h	PORTB	TRISB	86h
07h			87h
08h	EEDATA	EECON1	88h
09h	EEADR	EECON2	89h
0Ah	PCLATH	PCLATH	8Ah
0Bh	INTCON	INTCON	8Bh
0Ch			8Ch
	68 General	Addresses map	
	Purpose registers	Back to Bank 0	
	(SRAM)		
4Fh			CFh
50h			D0h
7Fh			FFh

[1]Not a physical register

Figure 6.5 The PIC16F84 data memory organization.

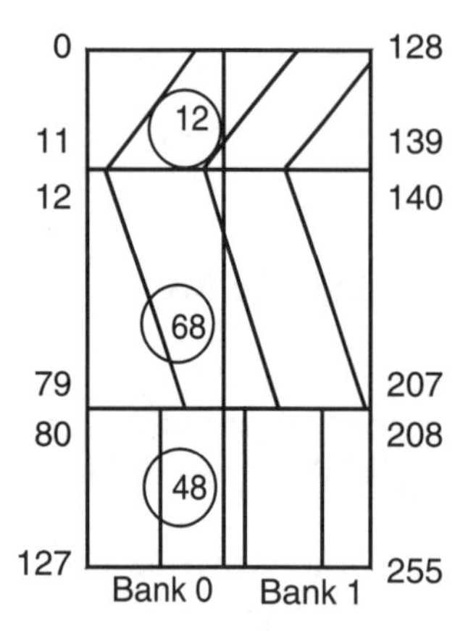

Figure 6.6 Summary of PIC16F84 data memory organization.

The *general-purpose registers* (GPRs) commence from 8Ch (140) up to CFh (207), so that there are a total of $(207 - 140 + 1) = 68$.

The unimplemented data memory locations commence from D0h (208) up to FFh (255), so that there are a total of $(255 - 208 + 1) = 48$.

There are two banks: banks 0 and 1. For bank 0, the 12 SFRs are from 00h (0) to 0Bh (11) or $11 - 0 + 1 = 12$ addresses. The 68 GPRs are from 0Ch (12) to 4Fh (79) or $(79 - 12 + 1) = 68$, the unimplemented are from 50h (80) to 7Fh (127) or $(127 - 80 + 1) = 48$.

6.3.3.3 *Bank 0 and bank 1 mapping and selection*

The GPR addresses in bank 1 are mapped to addresses in bank 0. For example, address location 4Fh and CFh will access the same GPR. Figure 6.6 summarizes the discussion on data memory organization.

6.3.4 Special features of the CPU

What sets a microcontroller apart from other processors are special circuits to deal with the needs of real-time mechatronics applications. The PIC16F8X has a host of such features intended to maximize system reliability, minimize cost through elimination of external components, and provide power saving operating modes and offer code protection. These features are:

- OSC selection
- Reset

- Interrupts
- Watchdog timer (WDT)
- SLEEP
- Code protection
- ID locations
- In-circuit serial programming

Some of these, as well as other special features are now discussed.

6.3.4.1 *Working (W) register or accumulator*

When a program is compiled and downloaded to a PIC, it is stored as a set of binary machine code instructions in the flash program memory. These instructions are sequentially fetched from memory, placed in the instruction register, and executed. Each instruction corresponds to a low-level function implemented with logic circuits on the chip. For example, one instruction might load a number stored in RAM or EEPROM into the working register (which is also called the W register or accumulator); the next instruction might command the ALU to add a different number to the value in this register; and the next instruction might return this summed value to memory. Since an instruction is executed every four clock cycles, the PIC16F84 can do calculations, read input values, store and retrieve information from memory, and perform other functions very quickly. With a clock speed of 4 MHz an instruction is executed every microsecond and 1 million instructions can be executed every second. The microcontroller is referred to as 8 bit, because the data bus is 8 bits wide, and all data processing and storage and retrieval occur, using bytes.

6.3.4.2 *File registers*

The RAM, in addition to providing space for storing data, maintains a set of special purpose byte-wide locations called *file registers*. The bits in these registers are used to control the function and indicate the status of the microcontroller. Several of these registers are described later.

6.3.4.3 *Watchdog timer*

A useful special purpose timer, called a *watchdog timer*, is included on PIC microcontrollers. This is a countdown timer that, when activated, needs to be continually reset by the running program. If the program fails to reset the watchdog timer before it counts down to 0, the PIC will automatically reset itself. In a critical application, you might use this feature to have the microcontroller reset if the software gets caught in an unintentional endless loop.

6.3.4.4 ***Ports***

The PIC16F84 is packaged on an 18-pin DIP IC that has the pin schematic (pin-out) shown in Figure 6.7. The figure also shows the minimum set of external components recommended for the PIC to function properly. Table 6.1 lists the pin identifiers in natural groupings, along with their descriptions. The five pins RA0 through RA4 are digital I/O pins collectively referred to as PORTA, and the eight pins RB0 through RB7 are digital I/O pins collectively referred to as PORTB. In total, there are 13 I/O lines, called *bi-directional* lines because each can be individually configured in software as an input or output. PORTA and PORTB are special purpose file registers on the PIC that provide the interface to the I/O

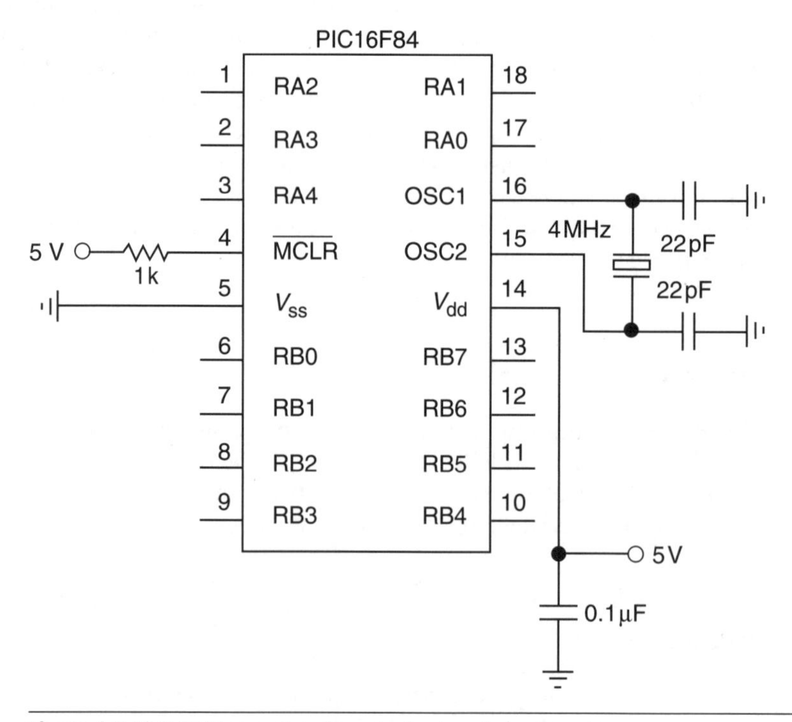

Figure 6.7 PIC16F84 pin-out and required external components.

Table 6.1 PIC16F84 pin identifier and description

Pin identifier	*Description*
RA [0–4]	5 bits of bi-directional I/O (PORTA)
RA [0–7]	8 bits of bi-directional I/O (PORTA)
V_{ss}, V_{dd}	Power supply ground reference
OSC1, OSC2	Oscillator crystal inputs
$\overline{\text{MCLR}}$	Master clear (active low)

pins. Although all PIC registers contain 8 bits, only the five least significant bits (LSBs) of PORTA are used.

6.3.4.5 *Interrupt*

An important feature of the PIC, available with most microcontrollers, is its ability to process interrupts. An interrupt occurs when a specially designated input changes state. When this happens, normal program execution is suspended while a special interrupt handling portion of the program is executed. This is discussed in a later section. On the PIC16F84, pins RB0 and RB4 through RB7 can be configured as interrupt inputs.

6.3.4.6 *Power*

Power and ground are connected to the PIC through pins V_{dd} and V_{ss}. The dd and ss subscripts refer to the drain and source notation used for MOS transistors, since a PIC is a CMOS device. The voltage levels (e.g. $V_{dd} = 5\,V$ and $V_{ss} = 0\,V$) can be provided using a d.c. power supply or batteries (e.g. four AA batteries in series or a 9 V battery connected through a voltage regulator).

6.3.4.7 *Reset*

The master clear pin (MCLR) is active low and provides a reset feature. Grounding this pin causes the PIC to reset and restart the program stored in EEPROM. This pin must be held high during normal program execution. This is accomplished with the pull-up resistor shown in Figure 6.7. If this pin is left unconnected (floating), the chip might spontaneously reset itself. To provide a manual reset feature to a PIC design, you can add a normally open (NO) pushbutton switch as shown in Figure 6.8. Closing the switch grounds the pin and causes the PIC to reset.

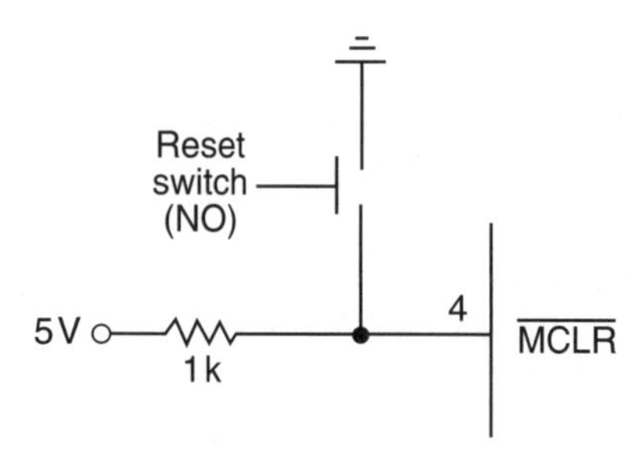

Figure 6.8 Reset switch circuit.

6.3.4.8 *Clock*

The PIC clock frequency can be controlled using different methods, including an external *RC* circuit, an external clock source, or a clock crystal. Figure 6.7 shows the use of a clock crystal to provide an accurate and stable clock frequency at relatively low cost. Connecting a 4 MHz crystal across the OSC1 and OSC2 pins with the 22 pF capacitors grounded as shown, sets the clock frequency.

6.4 Programming a PIC using assembly language

To use a microcontroller in mechatronics system design, software must be written, tested, and stored in the microcontroller ROM. Usually, the software is written and compiled using a personal computer (PC) and then downloaded to the microcontroller ROM as machine code. If the program is written in assembly language, the PC must have software called a *cross-assembler* that generates machine code for the microcontroller. An assembler is software that generates machine code for the microprocessor in the PC, whereas a cross-assembler generates machine code for a different microprocessor, in this case the microcontroller.

6.4.1 Software development tools

Various software development tools can assist in testing and debugging assembly language programs written for a microcontroller.

6.4.1.1 *Simulator*

This is software that runs on a PC and allows the microcontroller code to be simulated (run) on the PC. Most programming errors can be identified and corrected during simulation.

6.4.1.2 *Emulator*

This is a hardware that connects a PC to the microcontroller in a prototype mechatronics system. It usually consists of a printed circuit board connected to the mechatronics system through ribbon cables. The emulator can be used to load and run a program on the actual microcontroller attached to the mechatronics system hardware (containing sensors, actuators, and control circuits). The emulator allows the PC to monitor and control the operation of the microcontroller while it is embedded in the mechatronics system.

6.4.1.3 *Instruction set*

The assembly language used to program a PIC16F84 consists of 35 commands that control all functions of the PIC. This set of commands is called the *instruction set* for the microcontroller. Every microcontroller brand and family has its own specific instruction set that provides access to the resources available on the chip. The complete instruction set and brief command descriptions for the PIC16F84 are listed in Table 6.2. Each command consists of a name called the *mnemonic* and, where appropriate, a list of operands. Values must be provided for each of these operands. The letters f, d, b, and k correspond, respectively; to a file register address (a valid RAM address), a result destination (0: W register, 1: file register), a bit number (0 through 7), and a literal constant (a number between 0 and 255). Note that many of the commands refer to the working register W. As discussed earlier, this is a special CPU register used to temporarily store values (e.g. from memory) for calculations or comparisons. At first, the mnemonics and descriptions in the table may seem cryptic, but after you compare functionality with the terminology and naming conventions, it becomes much more understandable. Example 6.1 introduces a few of the statements. Example 6.2 illustrates how to write a complete assembly language program. For more information (e.g. detailed descriptions and examples of each assembly statement), refer to the PIC16F8X data sheet available on Microchip's website (www.microchip.com).

EXAMPLE 6.1

Table 6.3 gives more detailed descriptions and examples of a few of the assembly language instructions to help you better understand the terminology and the naming conventions.

EXAMPLE 6.2

The purpose of this example is to write an assembly language program that will turn on an LED when the user presses a push-button switch. When the switch is released, the LED is to turn off. After the switch is pressed and released a specified number of times, a second LED is to turn on and stay lit.

Table 6.2 PIC16F84 instruction set

Mnemonic operands	*Description*
ADDWF f, d	Add W and f
ANDWF f, d	AND W with f
CLRF f	Clear f
CLRW	Clear W
COMF f, d	Complement f
DECF f, d	Decrement f
DECFSZ f, d	Decrement f, Skip if 0
INCF f, d	Increment f
INCFSZ f, d	Increment f, skip if 0
IORWF f, d	Inclusive OR W with f
MOVWF f	Move W to f
MOVF f, d	Move f
NOP	No operation
RLF f, d	Rotate f left 1 bit
RRF f, d	Rotate f right 1 bit
SUBWF f, d	Subtract W from f
SWAPF f, d	Swap nibbles in f
XORWF f, d	Exclusive OR W with f
BIT-ORIENTED FILE	**REGISTER OPERATIONS**
BCF f, b	Bit clear f
BSF f, b	Bit set f
BTFSC f, b	Bit test f, skip if clear
BTFSS f, b	Bit test f, skip if set
LITERAL AND CONTROL	**OPERATIONS**
ADDLW k	Add literal and W
ANDLW k	AND literal with W
CALL k	Call subroutine
CLRWDT	Clear watchdog timer
GOTO k	Go to address
IORLW k	Inclusive OR literal with W
MOVLW k	Move literal to W
OPTION k	Load OPTION register
RETLW k	Return with literal in W
SLEEP	Go into standby mode
SUBLW k	Subtract W from literal
TRIS f	Load TRIS register
XORLW k	Exclusive OR literal with W

The hardware required for this example is shown in Figure 6.9.

The push-button switch is assumed to be bounce free, implying that when it is pressed and then released, a single pulse is produced (the signal goes high when it is pressed and goes low when it is released).

Table 6.3 Some assembly language instructions

Instruction	*Read as*	*Function*	*For example*
BCFf, b	bit clear f	clears bit *b* in file register *f* to 0, where the bits are numbered from 0 (LSB) to 7 (MSB)	*BCF PORTB, 1* makes bit 1 in PORTB go low (where PORTB is a constant containing the address of the PORTB file register). If PORTB contained the hexadecimal (hex) value FF (binary 11111111) originally, the final value would be hex FC (binary 11111101). If PORTB contained the hex value A8 (binary 10101000) originally, the value would remain unchanged.
MOVLWk	move literal to W	stores the literal constant *k* in the accumulator (the W register)	*MOVLW* 0*x*48 would store the hex value A8 in the W register. In assembly language, hexadecimal constants are identified with the 0*x* prefix.
RLFf, d	rotate f left	shifts the bits in file register *f* to the left 1 bit, and store the result in *f* if *d* is 1 or in the accumulator (the W register) if *d* is 0. The value of the LSB will become 0, and the original value of the MSB is lost.	If the current value in PORTB is hex IF (binary 00011111), then *RLF PORTB. 1* would change the value to hex 3E (binary 00111110).
SWAPFf, d	swap nibbles in f	exchanges the upper and lower nibbles (a nibble is 4 bits or half a byte) of file register *f* and store the result in *f* if *d* is 1 or in the accumulator (the W register) if *d* is 0	if the memory location at address hex 10 contains the value hex AB, then *SWAPF* 0*x*10, 0 would store the value hex BA in the W register. *SWAPF* 0*x*10, *1* would change the value at address hex 10 from hex AB to hex BA

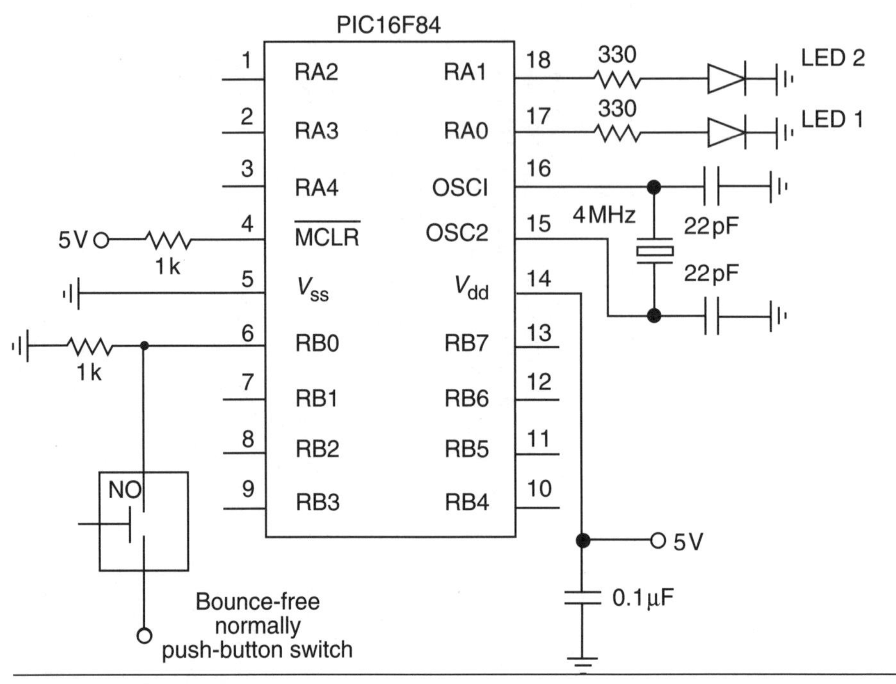

Figure 6.9 LED switch circuit.

Assembly language code that will accomplish the desired task follows the text below. The line numbers are included solely to aid identification during the flowing description and do not form part of the code. A remark or comment can be inserted anywhere in a program by preceding it with a semicolon (;). Comments are used to clarify the associated code. The assembler ignores comments when generating the hex machine code. The first four active lines (*list* ... *target*, lines 9–13) are assembler directives that designate the processor and define constants that can be used in the remaining code. Defining constants (with the *equ* directive, lines 12 and 13) at the beginning of the program is a good idea because the names, rather than hex numbers, are easier to read and understand in the code and because the numbers can be conveniently located and edited later. Assembly language constants such as addresses and values are written in hexadecimal, denoted with a *0x* prefix.

The next two lines of code, starting with *movlw* (line 15), move the literal constant *target* into the W register and then from the W register into the *count* address location in memory (line 17). The target value (0x05) will be decremented until it reaches 0x00. The next section of code (lines 20 to 24) initializes the special function registers PORTA and TRISA to allow output to pins RA0 and RA1, which drive the LEDs. These registers are located in different banks of memory, hence the need for the *bsf* and *bcf* statements (lines 20, 22, and 24 in the program. All capitalized words in the program are constant addresses or values predefined in the processor-dependent include file (p16f84.inc, line 10).

The function of the TRISA register is discussed later; but by clearing the bits in the register, the PORTA pins are configured as outputs. The main loop (lines 27 to 43) uses the *btfss* (bit test in file register; skip the next instruction if the bit is set) and *btfsc* (bit test in file register; skip the next instruction if the bit is clear) statements to test the state of the signal on pin RB0. The tests are done continually within loops created by the *goto* statements (lines 29, 35, and 40). The words *begin* and *wait* are statement labels used as targets for the *goto* loops. When the switch is pressed, the state goes high and the statement *btfss* skips the *goto begin* instruction; then LED1 turns on. When the switch is released, pin RB0 goes low and the statement *btfsc* skips the *goto wait* instruction; then LED1 turns off.

After the switch is released and LED1 turns off, the statement *decfsz* (decrement file register; skip the next instruction if the count is 0) executes (line 39). The *decfsz* decrements the *count* value by 1. If the *count* value is not yet 0, *goto begin* executes and control shifts back to the label *begin*. This resumes execution at the beginning of the main loop, waiting for the next switch press. However, when the *count* value reaches 0, *decfsz* skips the *goto begin* statement and LED2 is turned on. The last *goto begin* statement (line 40) causes the program to again jump back to the beginning of the main loop.

```
1    ; bcount.asm (program file name)
2    ; Program to turn on an LED every time a push-button switch is pressed
     and turn on
3    ; a second LED once it has been pressed a specified number of times
4    ; I/O:
5    ; RB0: bounce-free push-button switch (1:pressed, 0:not pressed)
6    ; RA0: count LED (first LED)
7    ; RA1: target LED (second LED)
8    ; Define the processor being used
9    list p=16f84
10   include <p16F84.inc>
11   ; Define the count variable location and the initial countdown value
12   count equ 0x0c ; address of countdown variable
13   target equ 0x05 ; number of presses required
14   ; Initialize the counter to the target number of presses
15   movlw target ; move the count-down value into the
16   ; W register
17   movwf count ; move the W register into the count memory
18   ; location
19   ; Initialize PORTA for output and make sure the LEDs are off
```

```
20  bcf STATUS, RP0 ; select bank 0
21  clrf PORTA ; initialize all pin values to 0
22  bsf STATUS, RP0 ; select bank 1
23  clrf TRISA ; designate all PORTA pins as outputs
24  bcf STATUS, RP0 ; select bank 0
25  ; Main program loop
26  ; Wait for the push-button switch to be pressed
27  begin
28  btfss PORTB, 0
29  goto begin
30  ; Turn on the count LED1
31  bsf PORTA, 0
32  ; Wait for the push-button switch to be released
33  wait
34  btfsc PORTB, 0
35  goto wait
36  ; Turn off the count LED1
37  bcf PORTA, 0
38  ; Decrement the press counter and check for 0
39  decfsz count, 1
40  goto begin ; continue if count-down is still > 0
41  ; Turn on the target LED2
42  bsf PORTA, 1
43  goto begin ; return to the beginning of the main loop
44  end ; end of instructions
```

Learning to program in assembly language can be very difficult at first and may result in errors that are difficult to debug. Fortunately, high-level language compilers are available that allow a PIC to be programmed at a more user-friendly level. The particular programming language we discuss in the remainder of the section is CC5X (C Compiler for the PIC microcontrollers). We will later introduce the CCS compiler due to its superiority over CC5X.

6.5 Programming a PIC using C

PIC programs can be written in a form of C called CC5X. The CC5X compiler can compile these programs, producing their assembly language equivalents, and this assembly code can then be converted to hexadecimal machine code (hex code) that can be downloaded directly to the PIC flash EEPROM through a programming device attached to a PC. Once loaded, the program begins to execute when power is applied to the PIC if the necessary additional components, such as those shown in Figure 6.9, are connected properly. CC5X programming is merely introduced and we do not intend to cover all of its aspects. Rather, we present an introduction to some of the basic programming principles, provide a brief summary of the statements, and then provide some examples. A compiler manual is available on the Internet and it is a necessary supplement to this section if the user needs to solve problems requiring more functionality than the examples we present here.

6.5.1 Initializing ports

Irrespective of the higher level programming language used, the I/O status of the PORTA and PORTB bits are configured in two special registers called TRISA and TRISB. The prefix *TRIS* is used to indicate that the tri-state gate control is set whether or not a particular pin provides an input or an output. The input and output circuits for PORTA and PORTB on the PIC16F84 are discussed in a later section on interfacing. When a TRIS register bit is set high (1), the corresponding PORT bit is considered an input, and when the TRIS bit is low (0), the corresponding PORT bit is considered an output. For example, TRISB = %01110000 would designate pins RB4, RB5, and RB6 as inputs and the other PORTB pins as outputs. At power-up, all TRIS register bits are set to 1 (i.e. TRISA and TRISB are both set to $FF or %11111111), so all pins in PORTA and PORTB are treated as inputs by default. If required as outputs, they have to be explicitly redefined in the program's initialization statements.

6.5.2 Programming the PIC using CC5X

The CC5X compiler is a C language compiler available online for PIC microcontrollers. The compiler was designed to generate optimized code. An optimizer within the compiler squeezes the codes to a minimum. CC5X can be selected as a tool in MPLAB (available from Microchip), which offers an integrated environment including editor and tool support. The CC5X files have to be installed in the root directory of the MPLAB program. This chapter concentrates on the basic aspects of the programming by delivering important terms and expressions. Further details can be obtained from the currently available CC5X manual or online.

The first set of codes defines the PICmicro device. This is most commonly done in C by

```
#include "16F84.h"
```

This statement will make reference to the header file of the 16F84 device for register address and other information. It must be noted that the header file must be contained within the same folder as the program codes.

One of the first tasks in the *main* function (main is the primary function in C) is to initialize and define the ports. This is done by the following codes.

```
PORTA = 0b.0000.0000;    //Initialize all Port A bits (first 5 bits)
TRISA = 0b.0000.0000;    //Sets Port A bits as outputs (0 - output)

PORTB = 0b.0000.0000;    //Initialize all Port B bits (all 8 bits)
TRISB = 0b.1111.1111;    //Sets Port B bits as inputs (1 - input)
```

The first statement initializes the pins. If a nibble high is required initially, these bits can be rewritten as 1s. The first bit written is the MSB. The TRIS statement decides whether the bit is to behave as an input or as an output. When a bit is given TRIS = 1, it is an input, TRIS = 0 refers to output. In both the statements, the last letter determines the port. PORTA and TRISA refer to the port A.

Thereafter, in between the definition of the PIC device and the main function, the pins have to be assigned. CC5X allows each pin to be named. This makes the program logical to read and easy to program. The following statement assigns bit 2 of port b the name 'motor'. That means, in the source codes, *Motor* will always refer to Port B bit 2.

```
bit Motor @ PORTB.2;
```

Once these are done, the actual programming begins. The programming uses most of the features of C.

In order to send outputs, the pin names (assignments) have to be referred to. If Port B bit 2 is to be turned high, the code will read,

```
Motor = 1;
```

The following code will turn the pin low.

```
Motor = 0;
```

The rest of the operations such as generation of pulses and duty cycles follow this. Implementing an output response is the only way of testing inputs. Hence, there can be conditional statements to check if the bit is high or low. The following code will check a pin assigned the name *Trigger*. If *Trigger* is high, motor will turn on.

```
If (Trigger==1)
{
  motor = 1;
}
```

C programs are compiled by CC5X and converted into assembly. These then get transferred into hex files. Hex files are downloaded to PIC. It therefore becomes necessary to appreciate the codes of CC5X and link them up with the difficult-to-program assembly codes. Table 6.4 compares the codes and describes their functions. Application is similar to Borland C++ codes. Some

Table 6.4 CC5X and assembly code comparison

Assembly		*CC5X code*	*Description*
NOP		nop();	no operation
MOWF	F	f=w	move w to f
CLRW		w=0	clear w
CLRF	f	f=0	clear f
SUBWF	f,d	d=f−w	subtract w from f
DECF	f,d	d=f−1	decrement f
IORWF	f,d	d=f|w	inclusive OR w and f
ANDWF	f,d	d=f&w	AND w and f
XORWF	f,d	d=f^w	exclusive OR w and f
ADDWF	f,d	d=f+w	add w and f
MOVWF	f,d	d=f	move f
COMF	f,d	f=f^255	Complement f
INCF	f,d	d=f+1	increment f
DECFSZ	f,d	d=f−1	skip of zero
RRF	f,d		rotate right through carry bit
SWAPF	f,d		rotate left through carry bit
INCFSZ	f,d		increment f, skip if 0
BCF	f,b	f.b=0	bit clear f
BSF	f,b	f.b=1	bit set f
BTFSC	f,b		bit test f, skip if clear
BTFSS	f,b		bit test f, skip if set
OPTION		option=w	load option register
SLEEP			go into standby mode
CLRWDT		WDT=0	clear watchdog timer
TRIS	f		tri-state port f
RETLW	k		return, put literal in W
CALL	k		call subroutine
GOTO	k		go to address
MOVLW	k	w=k	move literal to w
IORLW	k	w=w|k	inclusive OR literal and w
ANDLW	k	w=w&k	AND literal and w
XORLW	K	w=w^k	Exclusive OR literal and W
ADDLW	K	w=k+w	Add literal to W
SUBLW	K	w=k−w	Subtract W from literal

descriptions have been picked up from the Borland C++ compiler. A statement summary is given in Table 6.5.

6.5.2.1 *Programming a robot*

To illustrate the fundamentals of CC5X, we discuss an interesting example of a robot built in-house for an international competition. We reproduce part of the CC5X code, which drives the PIC microcontroller, since all that is important is

Table 6.5 Summary of the CC5X compiler statements

Statement	*Syntax*	*Description*
AUTO	[auto] <data-definition>;	Defines a local variable as having a local lifetime.
BREAK	break;	Passes control to the first statement following the innermost enclosing brace.
CASE	case<constant expression> <statement>; [break;]	Case statement in conjunction with switches to determine which statements to evaluate.
CHAR	char <variable name>	Char to define a character data type. Variables of type char are 1 byte in length.
CONST	const <variable name>	Use the const modifier to make a variable value un-modifiable.
CONTINUE	continue;	Passes control to the end of the innermost enclosing brace, allowing the loop to skip intervening statements and re-evaluate the loop condition immediately.
DEFAULT	default: <statement>;	Use the default statement in switch statement blocks.
DOUBLE	[long] double <identifier>	Use the double type specifier to define an identifier to be a floating point data type.
ENUM	enum [<type tag >] {constant_name> [=value>} [var_list]	Use the enum keyword to define a set of constants of type int, called an enumeration data type.
EXTERN	extern <data definition>;	Use the extern modifier to indicate that the actual storage and initial value of a variable, or body of a function, is defined in a separate source code module.
DO	do statement while (condition);	Executes the specified statement until the value of the specified condition becomes FALSE.
ELSE	else <new line> final section	Else is used as an alternative condition for which all previous tests have proved false.
FLOAT	float <identifier>	Float type specifier to define an identifier to be a floating-point data type.
FOR	For ([initialization]; [condition]; [expression]) statement	Executes the specified statement as long as the condition is TRUE.
GOTO	goto <identifier>;	Use the goto statement to transfer control to the location of a local label specified by <identifer>.
IF	if (condition) statement;	Implements a conditional statement.
INT	[signed/unsigned] int <identifier>;	Use the int type specifier to define an integer data type.
LONG	long [int] <identifier>;	It doubles the number of bytes available to store the integer value.
RETURN	return [expression];	A module, by default, returns TRUE if successfully run.
SHORT	short int <variable>;	Use the short type modifier when you want a variable smaller than an int.
SIGNED	signed <type> <variable>	Use the signed type modifier when the variable value can be either positive or negative.
SIZEOF	sizeof unary-expression	Sizeof gives the size of the pointer; when applied to structures and unions, sizeof gives the total number of bytes, including any padding.

STATIC	Static <data definition>;	Use the static storage class specifier with a local variable to preserve the last value between successive calls to that function.
STRUCT	Struct [<struct type name>]	Use a struct to group variables into a single record.
SWITCH	Switch (switch_expresn){	Chooses one of the several alternatives.
TYPEDEF	Typedef <type definition> <identifier>;	Use the typedef key word to assign the symbol name <identifier> to the data type definition <type definition>.
UNION	union [<union type name>]{	Use unions to define variables that share storage space.
UNSIGNED	unsigned<type><variable>	Use the unsigned type modifier when variables values will always be positive.
VOID	void identifier	Void is a special type indicating the absence of any value.
WHILE	While (<condition>) <statement>	Use the while keyword to conditionally iterate a statement.
DEFINE	#define macro_indent <token_sequence>	The #define directive defines a macro.
ELIF	<#elif constant_exp-2 newline section-2>	Conditional operator.
IFDEF	#ifdef identifier	The #ifdef and ifndef conditional directives allow one test whether an identifier is currently defined or not.
IFNDEF	#ifndef identifier	The #ifdef and ifndef conditional directives allow one test whether an identifier is currently defined or not.
INCLUDE	#include <file name>	Pulls other script files into the source code.
ENDIF	#endif	Conditional operator.
ERROR	Virtual void Error (Twindosw owner);	Error is an abstract function called by Valid when it detects that the user has entered invalid information.
PRAGMA	#pragma directive_name	With #pragma, Borland C++ can define the directives it wants without interfering with other compilers that support #pragma.
UNDEF	#undef macro_identifer	#undef detaches any previous token sequence from the macro identifier; the macro definition has been forgotten, and the macro identifier is undefined.

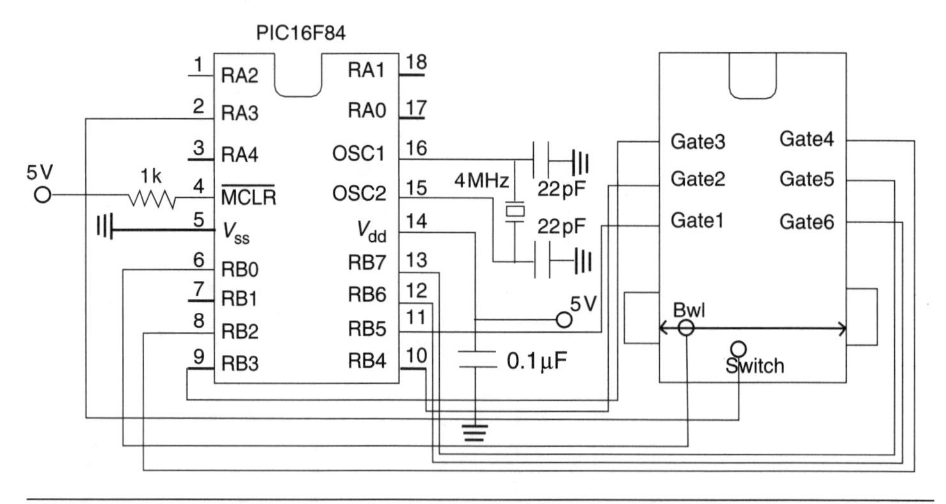

Figure 6.10 Circuit schematic for the robot (relay.c) example.

the understanding of how a higher-level language can be used rather than the traditional assembly language. The code for this program, called *relay.c,* follows. The hardware required is shown in Figure 6.10.

```
#include "16F84.H"
/*
Program: PACSEA, Left start zone
Copyright: University of the South Pacific; V.1]
Requirements, Battery voltage=12.84V
Name of Robot: PACSEA
*/
//pin configuration
bit bwl @ PORTB.0; //back wheel
//bit fwl @ PORTB.1; //forward wheel
bit gt4 @ PORTB.2; //gate 4
bit gt3 @ PORTB.3; //gate 3
bit gt2 @ PORTB.4; //gate 2
bit gt1 @ PORTB.5; //gate 1
bit gt6 @ PORTB.6; //gate 6
bit gt5 @ PORTB.7; //gate 5
bit gt7 @ PORTA.0; //gate 7
bit sens @ PORTA.2;//sensor
bit sw @ PORTA.3; //switch
/*
Rules:
Gate 1 must always open before gate 2
Gate 6 must always open before gate 5
*/
//variable declaration
bit sensed;
```

```
int nump;//number of pulses
int tubes;//number of tubes detected by sensor
uns16 temp;
uns16 count;
//function declarations
//function declaration
void delayvar(uns16 time, int var); //var,125:millisec, 62: half
millisec, 31: quarter millisec,
void fwd(uns16);//move forward
void opGt(int );
/*
1 rev~1 sec
r=8in~20.3 cm
1 sec - -> 63.78 cm
*/
void main (void)
{
 //initialize the ports
 PORTB=0b.0000.0000;
 TRISB=0b.0000.0000; //All b pins shouldbe output's
 PORTA=0b.0000.0000;
 TRISA=0b.0000.1100; //bits 0&1=output, 2&3=input
  sw=0; //initialize the switch.
 //Wait for switch to be on
 while(sw!=1)
 {
  nop();

 }
 //switch is on. start executing program
 fwd(8600);//approximately 3.9 m
 opGt(6);

 delayvar(2800,125); //3 second for ball to fall
 fwd(2800);//aprox 1.5 m
 opGt(5);
 opGt(2);
 delayvar(3000,125);// 3 second for ball to fall
 fwd(2860);//approx 1.5 m
 opGt(4);
 delayvar(1000,125); //3 second for ball to fall
 //fwd(8600);
 //do nothing else
 while(1)
 {
       bwl=0;
       nop();
       sleep();
 }
}
```

As can be seen, the program structure is similar to ANSI C, except for some specific features peculiar to CC5X.

Table 6.6 Selected mathematical operators and functions

Math operator or function	*Description*
A + B	Add A and B
A − B	Subtract B from A
A . B	Multiply A and B
A / B	Divide A by B
A<< n	Shift A n bits to the left
A >> n	Shift A n bits to the right
COS A	Return the cosine of A
A MAX B	Return the maximum of A and B
A MIN B	Return the minimum of A and B
SIN A	Return the sine of A
SQR A	Return the square root of A
A & B	Return the bitwise AND of A and B
$A \mid B$	Return the bitwise OR of A and B
A^B	Return the bitwise Exclusive OR of A and B
~A	Return the bitwise NOT of A

6.5.3 Examples of CC5X programming

In this section, we list a number of practical problems to which the CC5X compiler could be easily applied. The source codes are given.

6.5.3.1 *Example 1: A LED*

Design a PIC-based circuit and write a CC5X code for switching two LEDs alternately.

The hardware required has already been shown in Figure 6.9. Pins RA0 and RA1 are used as output to source current to an LED each through a current limiting resistor.

Normally, constants and results of calculations are assumed to be unsigned (i.e. zero or positive), but certain functions, such as *sin* and *cos,* use a different byte format, where the MSB is used to represent the sign of the number. In this case, the byte can take on values between −127 and 126. Some of the fundamental expressions using *mathematical operators* and functions available in CC5X are listed in Table 6.6; logical operators are shown in Table 6.7. Other operators and functions, and more detail and examples, can be found in the CC5X compiler manual.

The keywords *and, or, xor*, and *not* can also be used in conjunction with parentheses to create general Boolean expressions for use in logical comparisons. The CC5X source code for this example is given as follows:

```
#include "16F84.H"        //Header of the PIC used
bit switch @ PORTB.0;
```

Table 6.7 CC5X Pro logical comparison operators

Operator	*Description*
= or ==	equal
<> or !=	not equal
<	less than
>	greater than
<=	less than or equal to
>=	greater than or equal to

```
bit LED1 @ PORTA.1;
bit LED2 @ PORTA.2;
void main()
{
PORTA=0b.0000.0000;     //Initialize all Port A bits (Low Nibble)
TRISA=0b.0000.0000;       //Sets Port A bits as outputs (0 - output)
PORTB=0b.0000.0000;     //Initialize all Port B bits (Low Nibble)
TRISB=0b.1111.1111;       //Sets Port B bits as inputs (1 - input)
int counter=0;
while(1)
        {
                if (switch==1)
        {
                LED1=1;
                LED2=0;
                counter++;
        }
                if(counter==10)
        {
                LED2=1;
        }
                else
                LED1=0;
                LED2=0;
        }
}
```

6.5.3.2 *Example 2: A security system*

Design a PIC-based circuit and write a CC5X code for a home security system that will trigger when an intruder enters a house or when a door or window is opened.

The hardware required for PIC implementation is shown in the Figure 6.11.

The door and window sensors are assumed to be normally open (NO) switches that are closed when the door and window are closed. They are wired

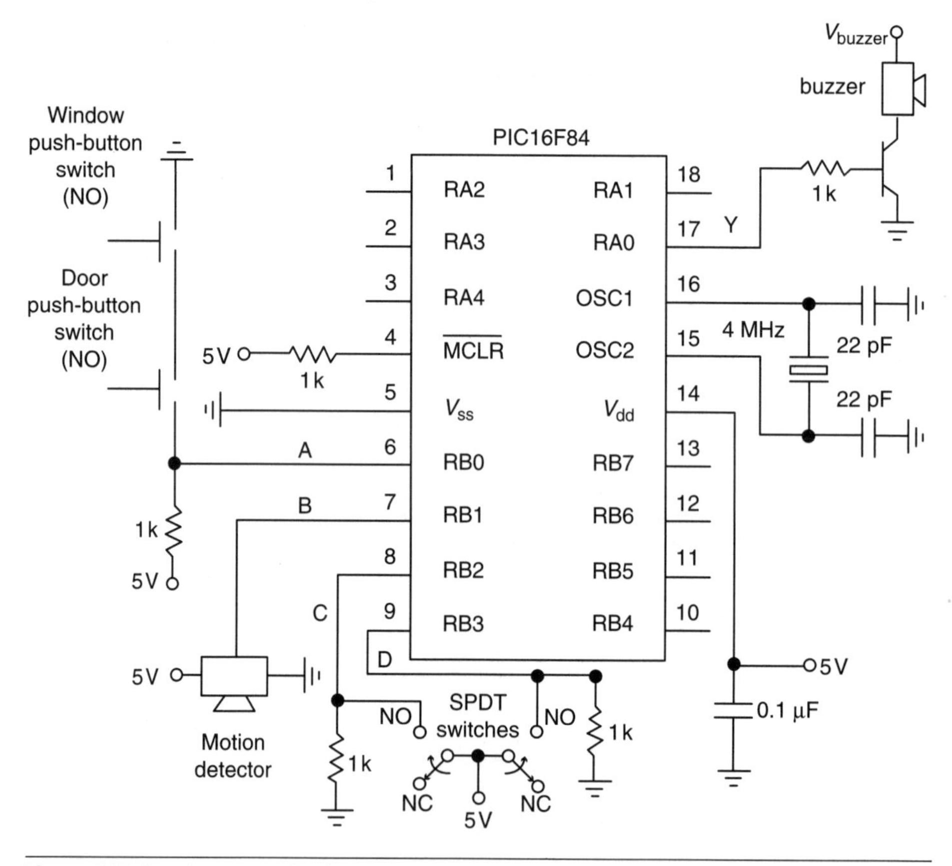

Figure 6.11 Security system circuit.

in series and connected to 5 V through a pull-up resistor; therefore, if either switch is open, then signal A will be high. Both the door and window must be closed for signal A to be low. This is called a *wired-AND* configuration since it is a hardwired solution providing the functionality of an AND gate.

The motion detector produces a high on line B when it detects motion. Single-pole, double-throw (SPDT) switches are used to set a 2-bit code C D.

In the figure, the switches are both in the normally closed (NC) position; therefore, code C D is 0 0. The alarm buzzer sounds when signal Y goes high, forward biasing the transistor. When Y is high, the 1 kΩ base resistor limits the output current to approximately 5 mA, which is well within the output current specification for a PORTA pin.

The CC5X source code for this example is given as follows:

```
#include "16F84.H"              //Header of the PIC used
bit DOOR_WINDOW @ PORTB.0;      //assign bit 0 of Port B to DOOR_WINDOW
```

```
bit MOTION @ PORTB.1;
bit C @ PORTB.2;
bit D @ PORTB.3;
bit ALARM @ PORTA.0;
void main()
{
PORTA = 0b.0000.0000;   //Initialize all Port A bits (Low Nibble)
TRISA = 0b.0000.0000;   //Sets Port A bits as outputs (0 - output)
PORTB = 0b.0000.0000;   //Initialize all Port B bits (Low Nibble)
TRISB = 0b.1111.1111;   //Sets Port B bits as inputs (1 - input)
        while (1)       //Keeps alarm always running
        {
                if(C==0 && D==1 && DOOR_WINDOW==1) //Operating State 1
        {
                ALARM=1;
        }
                if(C==1 && D==0)
        {
                if (DOOR_WINDOW==1 || MOTION==1)    // Operating State 2
        {
                ALARM=1;
        }
        }
        }
}
```

6.5.3.3 *Example 3: A seven-segment digital display*

Design a PIC-based circuit 7-segement LED display system for displaying a decimal digit.

Sometimes it is necessary to display a decimal digit using a seven segment LED display (see Figure 6.12). The display could represent some calculated or counted value (e.g. the number of times a switch is pressed). One approach is to drive the seven LED segments directly from seven output pins of a PIC. This would

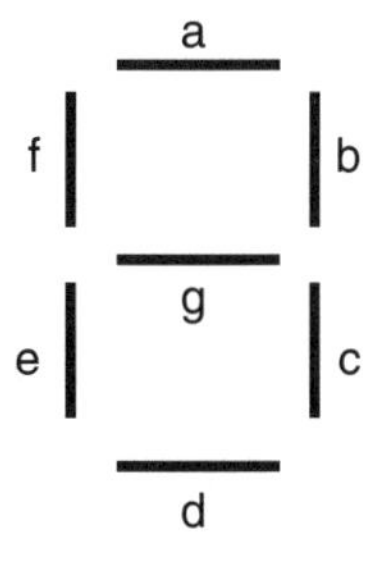

Figure 6.12 A seven-segment display.

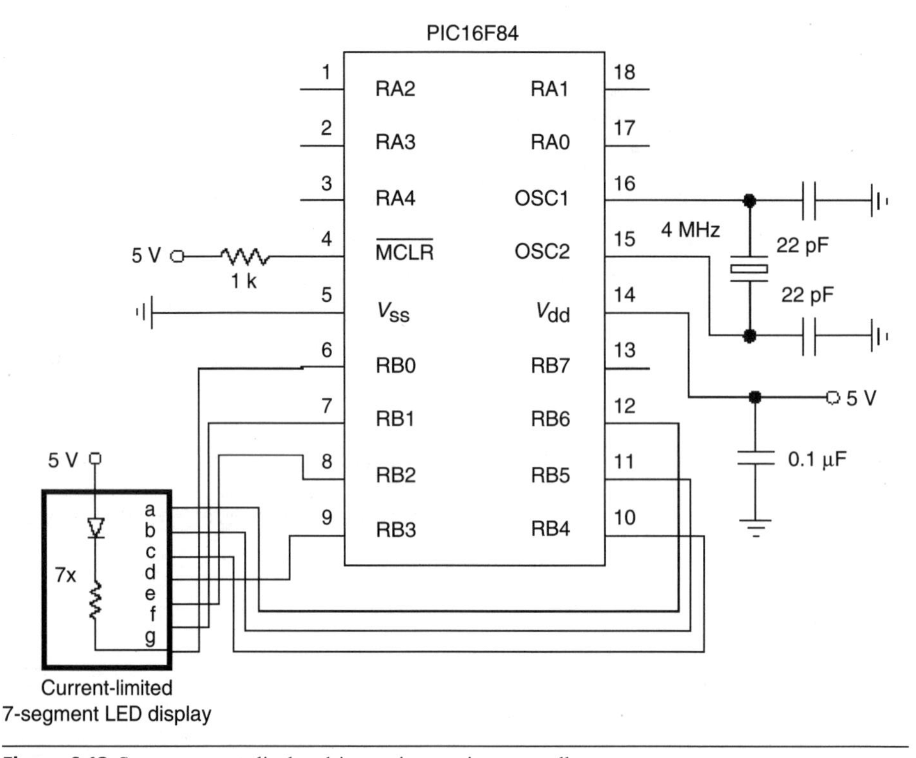

Figure 6.13 Seven-segment display driver using a microcontroller.

involve decoding in software to determine which segments need to be on or off to display the digit properly. If we connect the LED segments to the PORTB pins (Figure 6.13), where the segments are connected to 5 V through a set of current-limiting resistors, then the following initialization code must appear at the top of your program:

```
' Declare variables
number var BYTE ' digit to be displayed (value assumed to be from 0
' to 9)
pins var BYTE[10] ' an array of 10 bytes used to store the 7-segment
' display codes for each digit
' Initialize I/O pins
TRISB = %00000000 ' designate all PORTB pins as outputs (although, pin 7 is
' not used)
' Segment codes for each digit where a 0 implies the segment is on and a 1
implies
' it is off, because the PIC sinks current from the LED display
' %gfedcba display
pins[0] = %1000000 ' 0
pins[1] = %1111001 ' 1
```

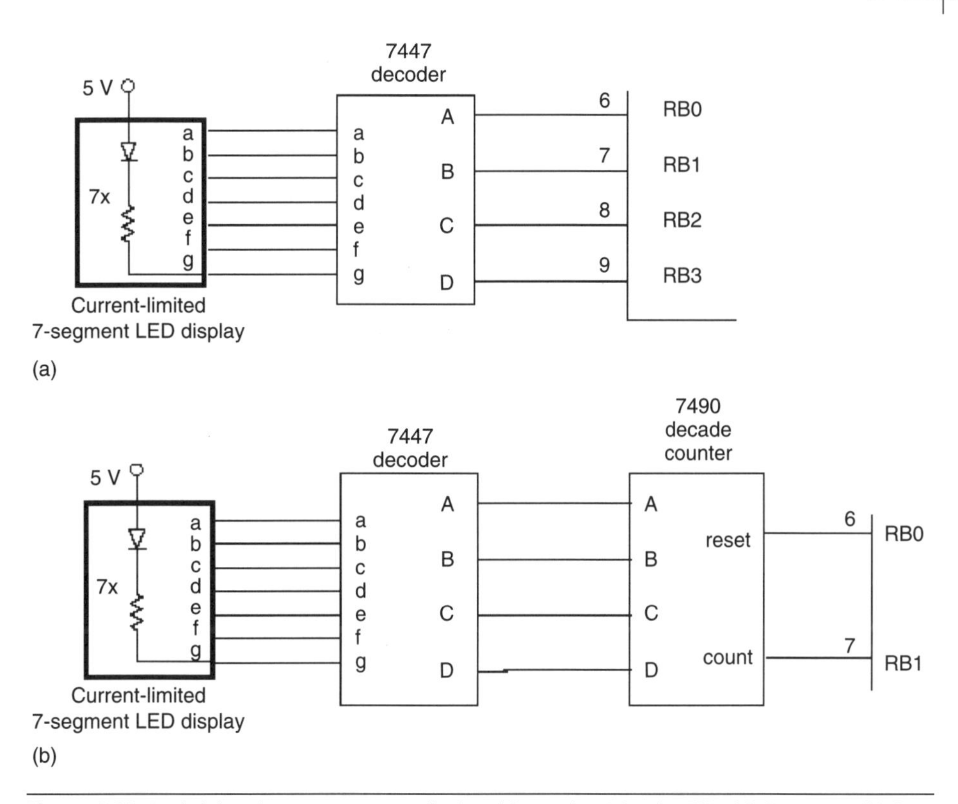

Figure 6.14 Optimizing the seven-segment display driver using 74-series ICs: (a) 4 outputs; (b) 2 outputs.

```
pins[2] = %0100100 ' 2
pins[3] = %0110000 ' 3
pins[4] = %0011001 ' 4
pins[5] = %0010010 ' 5
pins[6] = %0000011 ' 6
pins[7] = %1111000 ' 7
pins[8] = %0000000 ' 8
pins[9] = %0011000 ' 9
```

The solution just presented above requires seven output pins. Since the PIC16F84 has a total of only 13 I/O pins, this could limit the addition of other I/O functions in your design. An alternative design that requires fewer output pins uses a seven-segment decoder IC. Here, only four I/O pins are required as shown in Figure 6.14(a).

If using four I/O pins still constrains your design, another alternative is to use a decade-counter IC with reset and count inputs. The reset input is assumed to have positive logic, so when the line goes high, the counter is reset to 0. The count input is edge triggered, and in this example it does not matter if it is positive or negative edge triggered. Only two PIC I/O pins are required as shown in Figure 6.14(b).

The CC5X source code for this example is given as follows:

```
#include "16F84.H"        //Header of the PIC used

void display(int);
void delay_variable(uns16 time, int var);

void main()
{
PORTA = 0b.0000.0000;     //Initialize all Port A bits
TRISA = 0b.0000.0000;     //Sets Port A bits as outputs (0 - output)
PORTB = 0b.0000.0000;     //Initialize all Port B bits (Low Nibble)
TRISB = 0b.0000.0000;     //Sets Port B bits as inputs (1 - input)
int counter=0;
while(1)
        {
                while (counter<10)
                {
                display(counter);
                delay_variable(200, 125);
                counter++;
                } //end inner while
                counter=0;
        } // end outer while
} //end main

//+++++++++++++

void display(int x)
{
switch(x)
        {
        case 0:
        PORTB=0b.1011.1111; // Port B& is always high for Millennium board
        break;

        case 1:
        PORTB = 0b.1000.0110;
        break;

        case 2:
        PORTB = 0b.1101.1011;
        break;

        case 3:
        PORTB = 0b.1100.1111;
        break;

        case 4:
        PORTB = 0b.1110.0110;
        break;

        case 5:
        PORTB = 0b.1110.1101;
```

```
        break;

        case 6:
        PORTB = 0b.1111.1100;
        break;

        case 7:
        PORTB=0b.1000.0111;
        break;

        case 8:
        PORTB=0b.1111.1111;
        break;

        case 9:
        PORTB=0b.1110.0111;
        break;

        } // end switch
} // end display

//++++++++++++

void delay_variable(uns16 time, int var)
{
int counter=0;
        OPTION=2;
        do
{
TMR0=0;
clrwdt();
while (TMR0<var);
} // end do
        while (--time>0);
} // end delay
```

6.5.3.4 *Example 4: Controlling a stepper motor*

The code below shows part of a program used to read a sensor input to a PIC microcontroller.

When the sensor goes high, the motor rotates 90 degrees clockwise and after a delay of 1 second the motor rotates 90 degrees anticlockwise.

```
#include "C:\GCO\PIC_Codes\head.h"

int16 i;
float a=1.8;
void base_motorc();
void base_motorac();

void main() {

/*
```

```
    // Use these depending on the applications
    setup_adc_ports(NO_ANALOGS);
    setup_adc(ADC_CLOCK_DIV_2);
    setup_psp(PSP_DISABLED);
    setup_spi(FALSE);
    setup_counters(RTCC_INTERNAL,WDT_18MS);
    setup_timer_1(T1_DISABLED);
    setup_timer_2(T2_DISABLED,0,1); */
    while (input(PIN_A0))          //sensor goes high (input)
    {
      for (i=1;i<=90/a;i++)        //rotate motor 90deg. Clockwise
        {
          base_motorc();
        }
      delay_ms(1000);              //delay 1 second
      for (i=1;i<=90/a;i++)        //rotate motor 90deg. Anticlockwise
        {
          base_motorac();
        }
      delay_ms(1000);              //delay 1 second
    }
}
void base_motorc()
{
    output_high(PIN_C0);           //fullstep/preset
    output_high(PIN_C2);           //direction(clockwise)
    output_high(PIN_C1);           //clock (freq. 500Hz)
    delay_ms(1);
    output_low(PIN_C1);
    delay_ms(1);
}
void base_motorac()
{
    output_high(PIN_C0);           //fullstep/preset
    output_low(PIN_C2);            //direction(anticlockwise)
    output_high(PIN_C1);           //clock (freq. 500Hz)
    delay_ms(1);
    output_low(PIN_C1);
    delay_ms(1);
}
```

6.6 Interfacing common PIC peripherals: the PIC millennium board

The PIC millennium board has at least four practical peripherals for interfacing:

- 6 light-emitting diodes, LEDs (J9);

1	2	3	F
4	5	6	E
7	8	9	D
A	0	B	C

Figure 6.15 16-button numeric keypad.

- 7-segment display (U2, U3, U4);
- 16-button keyboard (U1);
- liquid crystal display, LCD (LCD1).

In this section we describe how to interface a PIC with two common peripheral devices. The first is a 16-button numeric *keypad* that can be used to input numeric data. The second is an LCD that can be used to output messages and numeric information.

6.6.1 The numeric keypad

Figure 6.15 shows the four-row, four-column 16-button keypad. Each key is attached to a normally open (NO) pushbutton switch. When a key is pressed, the switch closes. Figure 6.16 shows the electrical schematic of the keypad with a recommended interface to the PIC16F84. A standard keypad has a seven-pin header for connection to a ribbon cable socket. There is one pin for each row and one pin for each column as numbered in Figure 6.16.

The four rows (row 1, row 2, row 3, row 4) are connected to pins RB7 through RB4, which are configured as inputs. Internal pull-up resistance is available as a software option on these pins, so external pull-up resistors are not required. The four columns (col 1, col 2, col 3, col 4) are connected to pins RB0 through RB3, which are configured as outputs.

The circuit is based on flow of current across a continuous path. RB0:RB2 are outputs. They send current to the keypad. This is done sequentially. When a pin is high, the rows, which are Pins RB7:RB4 are checked. For every high column this is done continuously and the numbers are found.

6.6.2 An LCD display

The other common peripheral device we want to highlight is a standard Hitachi 44780-based liquid crystal display (LCD). Applications of LCDs include

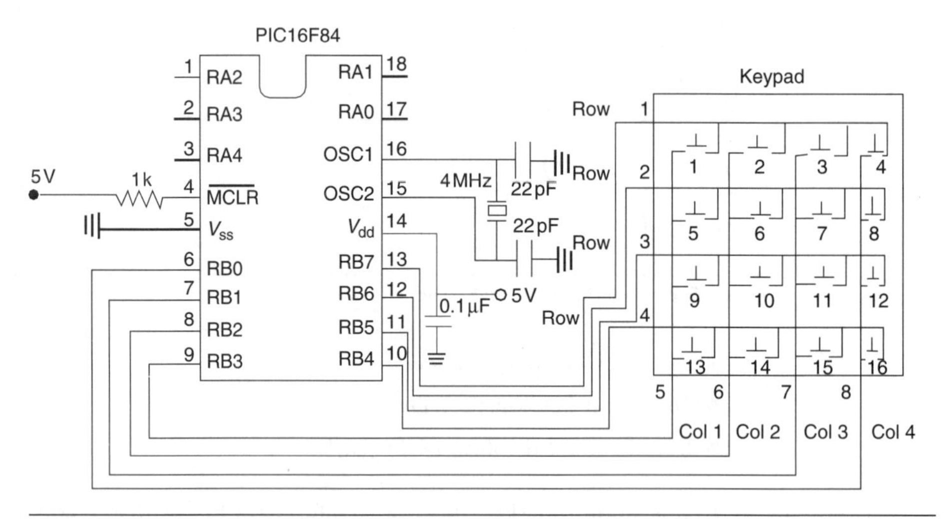

Figure 6.16 PIC/numeric keypad circuit.

displaying messages or information to the user (e.g. a home thermostat display, a microwave oven display, or a digital clock) and displaying a hierarchical input menu for changing settings and making selections (e.g. a fax machine display).

LCDs come in different shapes and sizes that can support different numbers of rows of text and different numbers of characters per row. The standard choices for the number of characters and rows are 8×2, 16×1, 16×2, 16×4, 20×2, 24×2, 40×2, and 40×4. The commonly used 20×2 LCD is illustrated connected to a microcontroller in Figure 6.17. For an LCD display with 80 characters or less (all but the 40×4), the display is controlled via 14 pins. The names and descriptions of these pins are listed in Table 6.8. LCD displays with more than 80 characters use a 16-pin header with different pin assignments. A 14-pin LCD can be controlled via four or eight data lines. It is recommended that you use four lines to minimize the number of I/O pins required. Figure 6.17 shows the recommended connections to the PIC using a 4-line data bus. Commands and data are sent to the display via lines DB4 through DB7, and lines DB0 through DB3 are not used. PIC pin RA4 is connected to 5 V through a pull-up resistor since it is an open drain output (see details in Section 6.8). The variable resistor connected to LCD V_{ee} is used to adjust the contrast between the foreground and background shades of the display. The LCD RS, R/W, and E lines are controlled automatically by the CC5X code when communicating with the display.

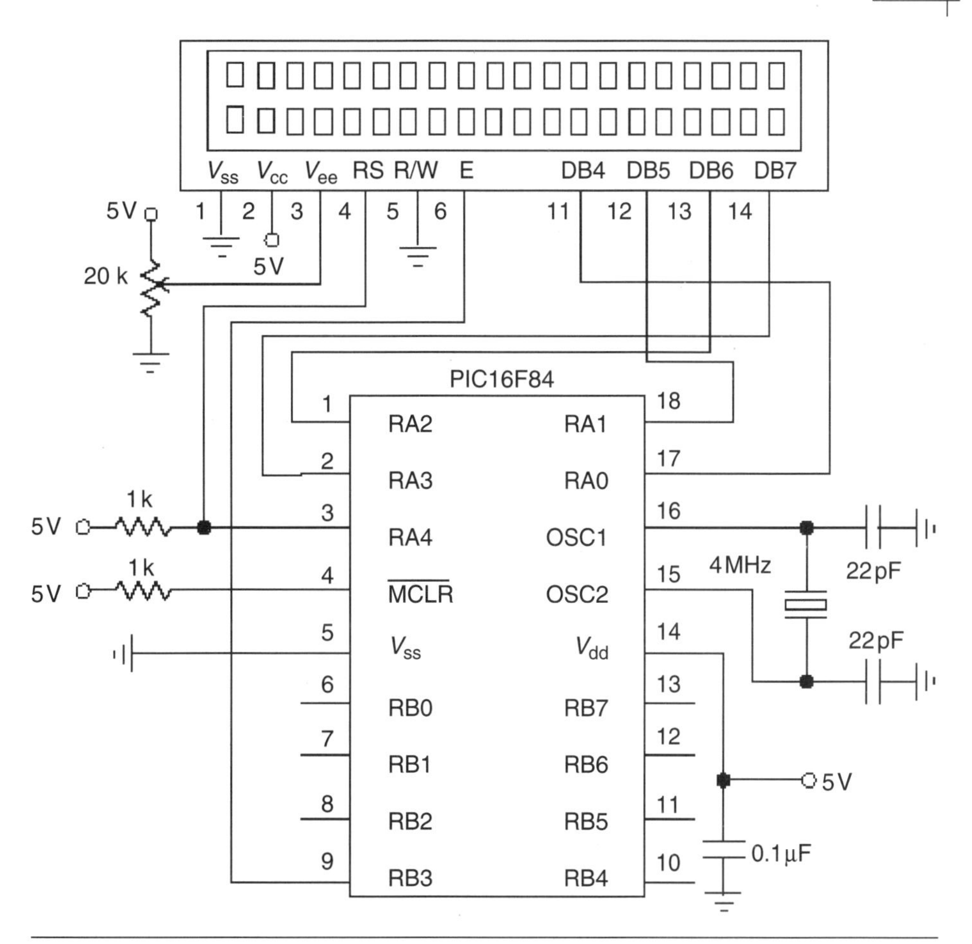

Figure 6.17 PIC/LCD circuit.

Table 6.8 LCD pin description

Pin	*Symbol*	*Description*
1	V_{ss}	Ground reference
2	V_{cc}	Power supply (5 V)
3	V_{ee}	Contrast adjustment voltage
4	RS	Register select (0: instruction input; 1: data input)
5	R/W	Read/write status (0: write to LCD; 1: read from LCD RAM)
6	E	Enable signal
7–14	DB0–DB7	Data bus lines

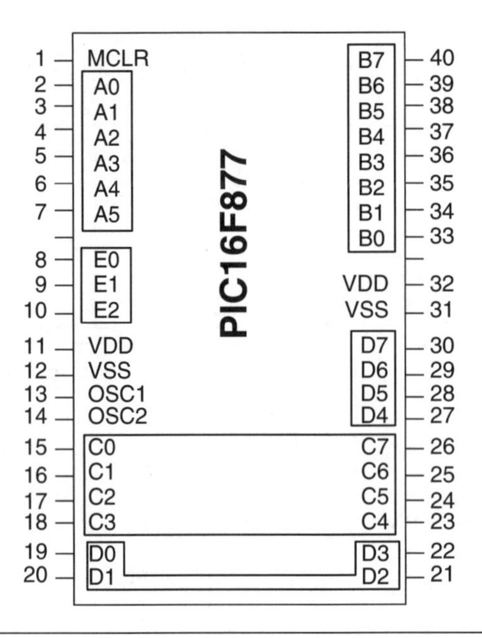

Figure 6.18 The PIC16F877 microcontroller.

6.7 The PIC16F877 microcontroller

The PIC microcontroller, PIC16F877 is a 40-pin device with five ports giving a total of 33 I/O pins as shown in Figure 6.18. It has 198 k RAM data memory, 128 k EEPROM data memory, and 4 k flash program memory. Some pins are reserved for specific functions: pins C1 and C2 can be configured as a timer or a pulse width modulator, pins C3 and C4 can be configured with I^2C used to communicate with a compass, while pins C6 and C7 can be configured as a USART (universal synchronous/asynchronous transmitter/receiver) clock. Moreover, C5 can also be configured with the SPI data out (SPI mode). Table 6.9 gives a description of the PIC16F877 pins.

6.8 Interfacing to the PIC

In this section, we discuss interfacing the PIC to a variety of input and output devices. Each pin of the I/O ports may be configured in software as an input or an output. In addition, the port pins may be multiplexed with other functions to use additional features of the PIC. In this section, we examine the electronic schematics of the different I/O ports of the PIC16F84. The ports are different combinations of TTL and CMOS devices and have voltage and current limitations

Table 6.9 PIC16F877 pin description

Pin name	*Description*
OSC1/CLKIN	Oscillator crystal input/external clock source input.
OSC2/CLKOUT	Oscillator crystal output.
MCLR	Master Clear (Reset) input or programming voltage input. This pin is an active low RESET to the device.
	PORT A is a bi-directional input/output port
RA0	RA0 can also be analog input0.
RA1	RA1 can also be analog input1.
RA2	RA2 can also be analog input2 or negative analog reference voltage.
RA3	RA3 can also be analog input3 or positive analog reference voltage.
RA4	RA4 can also be the clock input to the Timer0 timer/counter.
RA5	RA5 can also be analog input4 or the slave select for the synchronous serial port.
	PORT B is a bi-directional input/output port. PORT B can be software programmed for internal weak pull-up on all inputs.
RB0	RB0 can also be the external interrupt pin.
RB1	
RB2	
RB3	RB3 can also be the low voltage programming input.
RB4	Interrupt on charge pin.
RB5	Interrupt on charge pin.
RB6	Interrupt on charge pin or In-Circuit Debugger pin.
RB7	Interrupt on charge pin or In-Circuit Debugger pin.
	PORT C is a bi-directional input/output port.
RC0	RC0 can also be the Timer1 oscillator output or a Timer1 clock input.
RC1	RC1 can also be the Timer1 oscillator input or a Capture2input/Compare output/PWM2 output.
RC2	RC2 can also be the Capture1 input/Compare1 output/PWM1 output.
RC3	RC3 can also be the synchronous serial clock input/output for both SPI and I^{2}C modes.
RC4	RC4 can also be the SPI data in SPI mode.
RC5	RC5 can also be the SPI data out (SPI mode).
RC6	RC6 can also be the USART clock.
RC7	RC7 can also be the USART data.
RD0	
RD1	
RD2	PORT D is a bi-directional I/O or parallel slave port
RD3	when interfacing to a microprocessor bus.
RD4	
RD5	
RD6	
RD7	
	PORTE is a bi-directional I/O port.
RE0	RE0 can also be read control for the parallel slave port, or analog input 5.
RE1	RE1 can also be write control for the parallel slave port, or analog input 6.
RE2	RE2 can also be select control for the parallel slave port, or analog input 7.
V_{ss}	Ground reference for logic and I/O pins.
V_{DD}	Positive supply for logic and I/O pins.

that must be considered when interfacing other devices to the PIC. You should first refer to Section 6.8.2 to review details of TTL and CMOS equivalent output circuits and open drain outputs. We begin by looking at the architecture and function of each of the ports.

The input and output pins of the PIC are the features that the user is most interested in. The PIC16F84 has 13 I/O pins, five of which are in Port A and eight in Port B. The pins can individually be set to be an input or output by controlling the status of the tri-state gate more commonly referred as TRIS. TRIS can be set for every pin in each of the ports. When TRIS is high, a defined pin is set as an input. When TRIS is low, a defined pin is set as an output.

The operation of the pins in each of the ports differs by virtue of its design. Port B pins, unlike those in port A, provide weak pull-ups for control switches and sensors. The pins in port A also differ from each other. Pin RA4 has a Schmitt trigger input buffer, pins RA3–RA0 have a TTL input buffer. Pins RB7–RB4 differ from pins RB3–RB0 since these have an interrupt feature. Pins RB3–RB0 however, have Schmitt trigger buffers. Such differences must be studied before assigning peripherals to each of the pins. Some pins are suited for one operation while the others for another. For example, an encoder giving pulsed waveform must be read by a PIC. The ideal pins to choose for reading the waveform are the ones that have a weak pull-up with a Schmitt trigger (RB3–RB0). Weak pull-ups will allow the signal to be easily read in while the Schmitt trigger will filter out noise in the waveform.

6.8.1 Port A pins

PORTA is a 5-bit-wide latch with the pins denoted by RA0 through RA4. The block diagram for pins RA0 through RA3 is shown in Figure 6.19 and the block diagram for pin RA4 is shown in Figure 6.20. The five LSBs of the TRISA register configure the 5-bit-wide latch for input or output. Setting a TRISA bit high causes the corresponding PORTA pin to function as an input, and the CMOS output driver is in high impedance mode, essentially removing it from the circuit. Clearing a TRISA bit low causes the corresponding PORTA pin to serve as an output, and the data on the data latch appears on the pin. Reading the PORTA register accesses the pin values. RA4 is slightly different in that it has a Schmitt trigger input buffer that triggers sharply on the edge of a slowly changing input. Also, the output configuration of RA4 is open drain, and external components (e.g. a pull-up resistor to power) are required to complete the output circuit.

The circuit is a digital one like all the components of the PIC and thus all data can be represented with 1s and 0s. D-type flip-flops are used (Table 6.10 shows the truth table). The outputs Q and $\overline{Q}$ are then transferred to an AND and OR gate. The n- and p-channel MOSFETS then receive this data and work

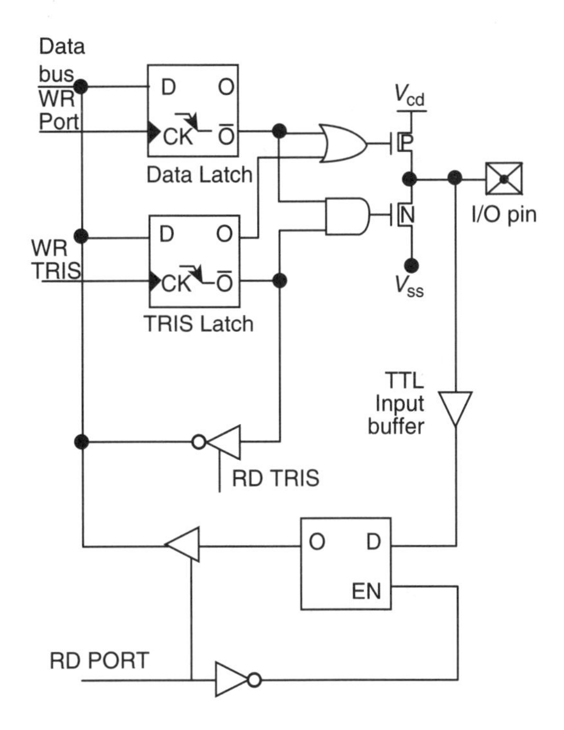

Figure 6.19 PIC16F84 block diagram for pins RA0–RA3. (Courtesy of Microchip Technology Inc.)

accordingly to receive or transmit the appropriate signal through the selected pin. Table 6.11 shows the truth table of the procedure. When the p- and n-channel MOSFETS are open, there is no route for the current to flow in apart from the path that leads to the TTL input buffer. From that, through another D-type flip-flop, it enters the data bus. For an output, when the p-channel MOSFET closes, there is a path for V_{DD} to the output pin. Hence, when the data is 1, the output is 1. The output becomes 0 when data is 0.

As indicated above, pin RA4 has a Schmitt trigger in its input section and together with a single n-channel MOSFET to switch between input and output states. Table 6.12 summarizes the effects of the data and TRIS inputs on pin RA4.

Contrary to the behavior of most other pins, RA4 sees the data also going low. In the first instance, both data and TRIS are high. This causes $\overline{Q1}$ and $\overline{Q2}$ to be low, hence the output of the AND gate to be low. Due to its characteristics, the n-channel MOSFET remains open and the current therefore flows to the Schmitt trigger. In the second instance, both data and TRIS turn low. This causes $\overline{Q1}$ and $\overline{Q2}$ to be high and thus the output of the AND gate to be high. The n-channel

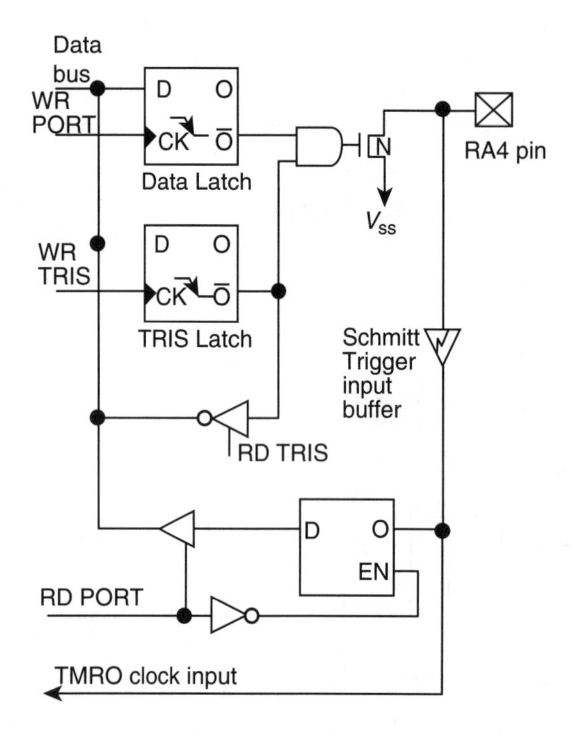

Figure 6.20 PIC16F84 block diagram for pin RA4. (Courtesy of Microchip Technology Inc.)

Table 6.10 Truth table for D-type flip-flop

D	*CLK*	*Q*
0	↑	0
1	↑	1

Table 6.11 Truth table for pins RA3–RA0

D	*TRIS*	*Q*	*Data* $\overline{Q}$	*AND*	*OR*	*P*	*N*	*I/O, state*
1	0	0	0	0	0	1	0	O,1
1	1	1	0	1	0	0	0	I,1
0	0	0	1	1	1	0	1	O,0
0	1	0	1	0	1	0	0	I,1

Table 6.12 Truth table for pin RA4

Data	*TRIS*	$\overline{Q1}$	$\overline{Q2}$	*AND*	*N*	*RA4*
1	1	0	0	0	open	Input
0	0	1	1	1	closed	Output

MOSFET is forced to close, opening the path to ground. The pin links the peripheral to ground, thus reaffirming its open drain output status.

6.8.2 Port B pins

PORTB is also bi-directional but is 8-bits wide. Its data direction register is denoted by TRISB. Figure 6.21 shows the schematic for pins RB4 through RB7 and Figure 6.22 shows pins RB0 through RB3. A high on any bit of the TRISB register sets the tri-state gate to the high impedance mode, which disables the output driver. A low on any bit of the TRISB register places the contents of the data latch on the selected output pin. Furthermore, all of the PORTB pins have weak pull-up FETs. A single control bit called RBPU (active low register B pull-up) controls these FETs. When this bit is cleared, the FET acts like a weak pull-up resistor. This pull-up is automatically disabled when the port pin is configured as an output. RBPU can be set in software through the OPTION_REG special purpose register.

The port B pins are different in two aspects. Firstly, they have weak pull-ups, and secondly, they have an RBPU register. An inverted tri-state gate is also an important aspect of the port since it determines the status of the respective pins. Table 6.13 shows the truth table for pins RB4–RB7.

As in most cases, D is taken to be always high and CLK2 corresponds to TRIS. When TRIS is 1, the output Q2 is high. Q1 will be high since data and PORT select are high. Q2 goes to the inverting tri-state gate and becomes $\overline{Q2}$. A high Q2 therefore becomes a low Q2, allowing no current to flow through. The output of this gate is 0. The setup is that of an input. However, if RBPU is set to 0 in the OPTION register, then the output of the NAND gate is 0, and thus the p-channel MOSFET is closed, enabling the weak pull-up.

In the alternative case, where CLK2 of TRIS is 0, Q1 will still provide a high signal. However, Q2 will be low. The inverting tri-state gate receives a high signal allowing current to pass through the I/O pin; confirming an output. Since Q2 is now low, the output of the NAND gate will be 1, which means that the weak pull-up is not in effect since the p-channel MOSFET is open. These inputs can also be used to set flag bits in RBIF.

The pins RB3–RB0 are able to process interrupts as well. This is because of a bypass route available in the circuitry from the TTL input buffer. Interrupts lead

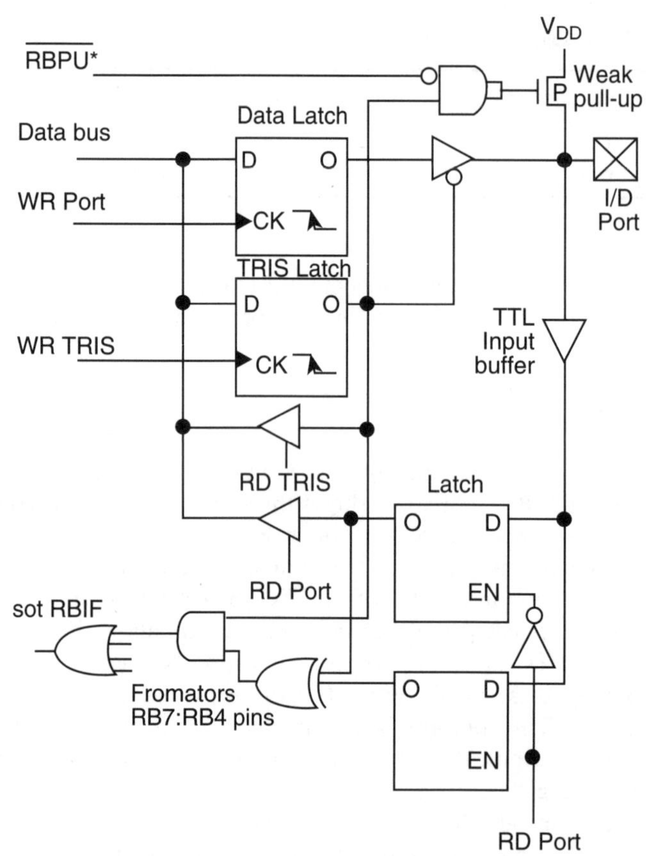

Figure 6.21 PIC16F84 block diagram for pins RB4–RB6. (Courtesy of Microchip Technology Inc.)

to the RBIF registers set the rest of the port B pins. They are reset once the interrupt is processed. Apart from this, the behavior of the rest of the pins is the same as RB7–RB4.

6.8.3 Data input to the PIC

Figure 6.23 illustrates how to properly connect different types of components and digital families of devices as inputs to the PIC. All I/O pins of the PIC that are configured as inputs interface through a TTL input buffer (pins RA0 through RA3 and pins RB0 through RB7) or a Schmitt trigger input buffer (RA4).

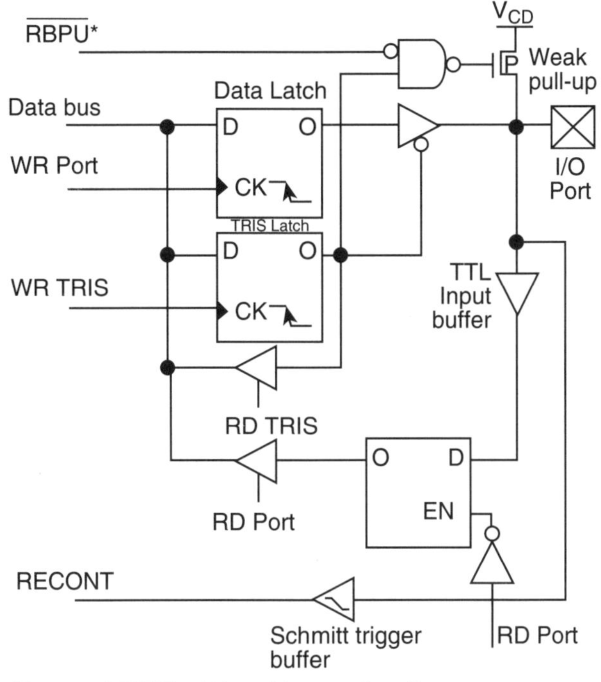

Note 1: TRISB = '1' enables weak pull-up
(RBPU* = '0' in the OPTION_REG register)
2: I/O pins have diode protection to V_{CD} and V_{SS}.

Figure 6.22 PIC16F84 block diagram for pins RB0–RB3. (Courtesy of Microchip Technology Inc.)

Table 6.13 Truth table for pins RB7–RB4

Data	*CLK2 (TRIS)*	*Q1*	*Q2*	*Inverter*	*I/O pin*
1	1	1	1	0	Input
1	0	1	0	1	Output

The Schmitt trigger enhances noise immunity for a slowly changing input signal. Since an input pin is TTL buffered in the PIC, connecting a TTL gate to the PIC can be done directly unless it is has an open collector output. In this case, an external pull-up resistor is required on the PIC output pin. Since the output of a 5 V powered CMOS device swings nearly from 0 to 5 V, the device will drive a PIC input directly. The weak pull-up option on pins RB0 through RB7 is useful when using mechanical switch or keypad inputs. The pull-up FET maintains a 5 V input until the switch is closed, bringing the input low. Although a TTL input

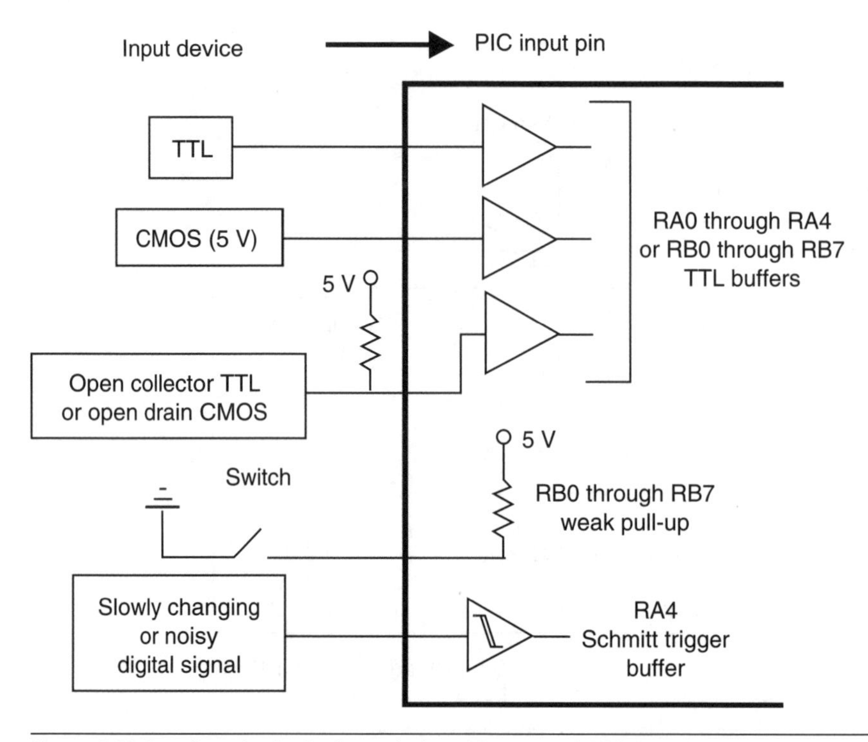

Figure 6.23 Interface circuits for input devices. (Courtesy of Alciatore, D.G. and Histand, M.B. (2002)).

usually floats high if it is open, the FET pull-up option is useful, since it simplifies the interface to external devices (e.g. keypad input). Finally, one must be aware of the current specifications of the PIC input and output pins. For the PIC16F84, there is a 25 mA sink maximum per pin with an 80 mA maximum for the entire PORTA and a 150 mA maximum for PORTB.

6.8.4 Data output from the PIC

Figure 6.24 illustrates how to properly connect different types of components and digital families of devices to outputs from the PIC. Pins RA0 through RA3 have full CMOS output drivers, and RA4 has an open drain output. RB0 through RB7 are TTL buffered output drivers. A 20 mA maximum current is sourced per pin with a 50 mA maximum current sourced by the entire PORTA and a 100 mA maximum for PORTB. CMOS outputs can drive single CMOS or TTL devices directly. TTL outputs can drive single TTL devices directly but require a pull-up resistor to provide an adequate high-level voltage to a CMOS device. To drive multiple TTL or CMOS devices, a buffer can be used to provide adequate current for the fan-out. Because pin RA4 is an open drain output, external power

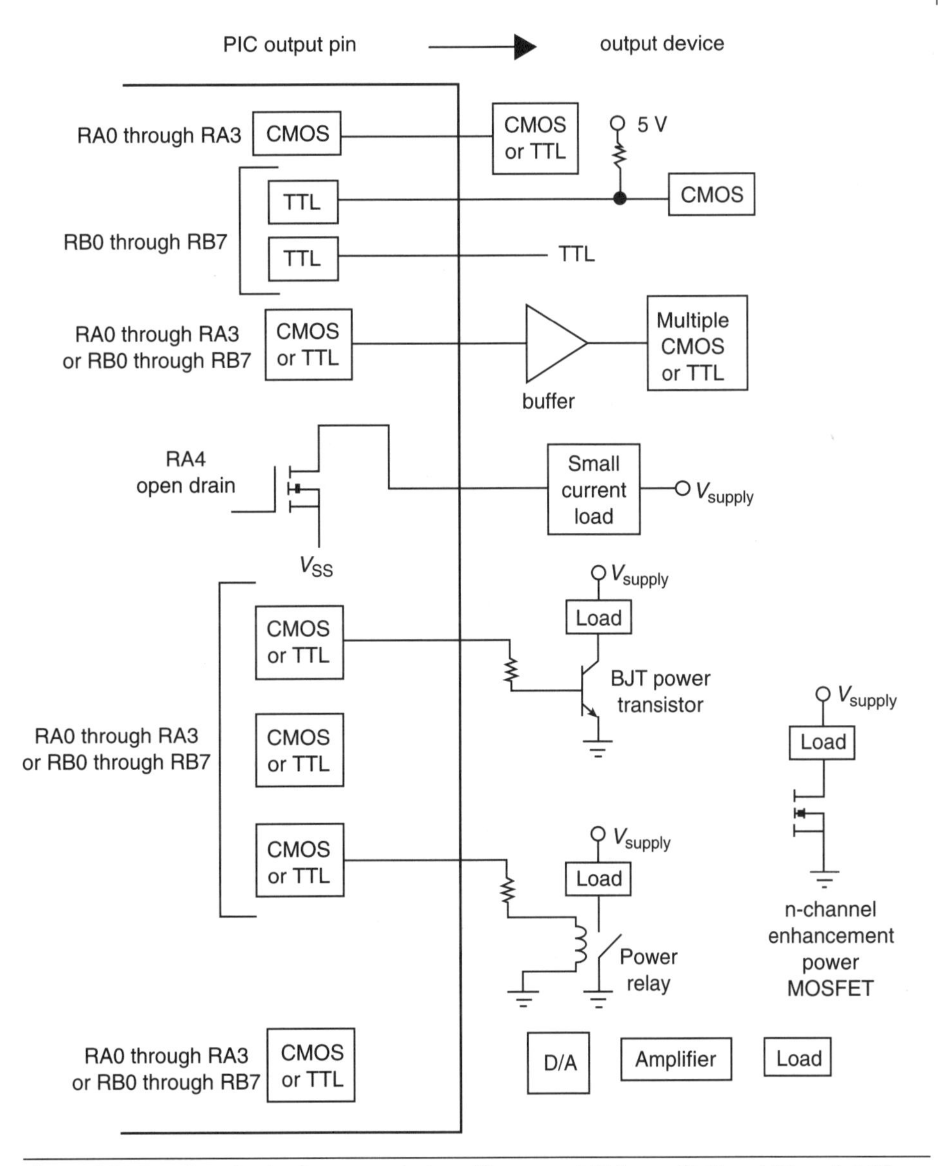

Figure 6.24 Interface circuits for output devices. (Courtesy of Alciatore, D.G. and Histand, M.B. (2002)).

is required. When connecting transistors, power transistors, thyristors, triacs, and SCRs, current requirements must be considered for a proper interface. If the PIC contains a DAC, it can be used with an amplifier to drive an analog load directly. Otherwise, as shown in the figure, an external D/A IC can be used with the digital I/O ports.

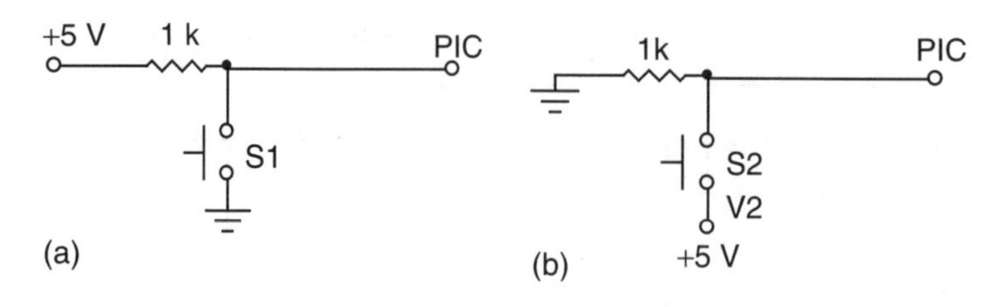

Figure 6.25 Switch connections to the PIC16F84.

6.8.5 Connecting the PIC to peripherals

Having described the internal architecture of PIC16F84, we consider connecting a PIC to peripherals (actuators). First, the PIC has to be powered up using a standard circuit. Complexities can be added but in basic operations, it is rarely required. Figure 6.7 shows the power up and oscillator connections for PIC16F84. A 5 V supply is considered ideal, but the PIC16F84 has an operating range of 2–6 V. The 0.1 μF capacitor ensures that the supply is maintained at 5 V by smoothing any surges. If a regulated 5 V supply is not available, a normal supply can be used with a linear voltage regulator. The master clear pin is also connected to 5 V. A 1 kΩ resistor controls the flow of current through the system. Often, switches are required. Figure 6.25 shows two of the various ways that switches can be configured. Arrays of switches can also be connected. Figure 6.25(a) shows a setup in which the pin of the PIC reads HIGH unless the switch is pressed. Figure 6.25(b) shows an alternative configuration in which the pin of the PIC reads low unless the switch is pressed.

D.c. motors are often used in mechatronics applications. Often, designers have difficulty with the limited source/drain current of the PIC. Motors in particular, require a lot of current. One of the best and proven methods of driving motors is using a Darlington driver IC with a relay. 5 V relays are available, but with the current limitations of the PIC, it will not trigger without the driver. If very high current is used, diodes must be used to prevent the electromotive force from de-energizing the relay and harming the PIC (Figure 6.26).

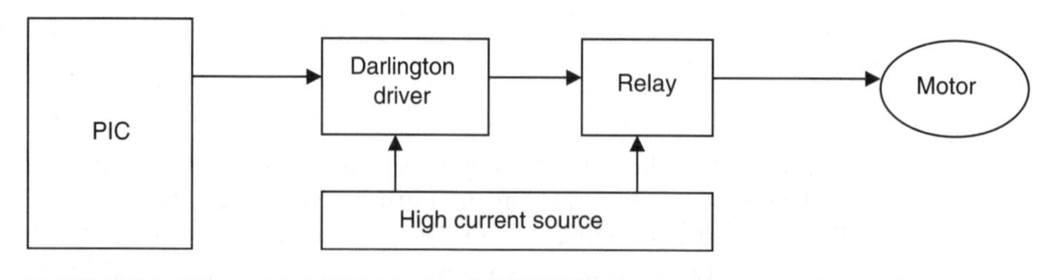

Figure 6.26 Block diagram for connecting high current peripherals.

Even more can be done with PIC16F84. However, the rest of it is basic electronics. The interfacing is the same as it is for a parallel port or for the Z80 microprocessor boards.

6.9 Communicating with the PIC during programming

In using the PIC Starter Plus programmer, there is the tendency for the user to regularly handle the PIC microcontroller chip each time changes are made in the C program. Normally, after programming on a PC, the MPLAB C compiler is used to generate the corresponding assembly code which is then transmitted to the PIC mounted on the PIC Starter Plus programmer. Thereafter, the PIC is manually removed and placed on the hardware application circuit board. An electrostatic hand band should be worn to avoid failure due to possible charges transferred from the human user and the PIC. This material handling takes place every time the PIC is reprogrammed, and hence the probability of the PIC failing becomes high.

Tera Term is free software available for downloading from the Internet (www.ayera.com/teraterm). It uses a serial file transfer protocol for communicating with other computers or any hardware.

Problems

Q6.1 Design a PIC-based system for graphically displaying the value of a potentiometer.

Q6.2 Design a relay PIC-based system to control a d.c. motor for a robotic motion.

Q6.3 Write a PIC code to rotate the d.c. motor Q6.2.

Further reading

[1] Alciatore, D.G. and Histand, M.B. (2002) *Introduction to Mechatronics and Measurement Systems* (2nd. ed.), McGraw-Hill.

[2] Data. B. (2000) *C Compiler for the PICmicro devices, V3.1*, Trondheim, Norway. (See B. Knudsen Data, Trondheim, 1992–2000).

[3] Horowitz, P. and Hill, W. (1989) *The Art of Electronics* (2nd. ed.), New York: Cambridge University Press.
[4] Microchip Technology Inc. (1998) *PIC16F8X Data Sheet*, www.microchip.com.
[5] Microchip Technology Inc. (2000) *MPLAB User's Guide,* Chandler, AZ.
[6] Mitchell, R.J. (1995) *Microprocessor Systems: An Introduction*, London: MacMillan Press Ltd.
[7] Peatman, J. (1988) *Design with Microcontrollers,* New York: McGraw-Hill.
[8] Predko, M. (2002) *Programming and Customizing PICmicro® Microcontrollers* (2nd ed.), McGraw-Hill.
[9] Stiffler, A.K. (1992) *Design with Microprocessors for Mechanical Engineers,* McGraw-Hill.

CHAPTER 7

Data acquisition

Chapter objectives

When you have finished this chapter you should be able to:

- understand fundamentals of data acquisition systems;
- understand sampling, aliasing, and quantization;
- understand digital-to-analog conversion hardware;
- understand analog-to-digital conversion hardware.

7.1 Introduction

The purpose of most electronic systems is to measure or control some physical quantity, hence a system will need to acquire data from the environment, process this data and record it. Acting as a control system it will also have to interact with the environment.

Data acquisition can be divided into the steps shown in Figure 7.1. The flow of information in a typical data acquisition system can be described as follows:

1. The input transducers measure some property of the environment.
2. The output from the transducers is conditioned (amplified, filtered, etc.).
3. The conditioned analog signal is digitized using an analog-to-digital converter (ADC).
4. The digital information is acquired, processed and recorded by the computer.

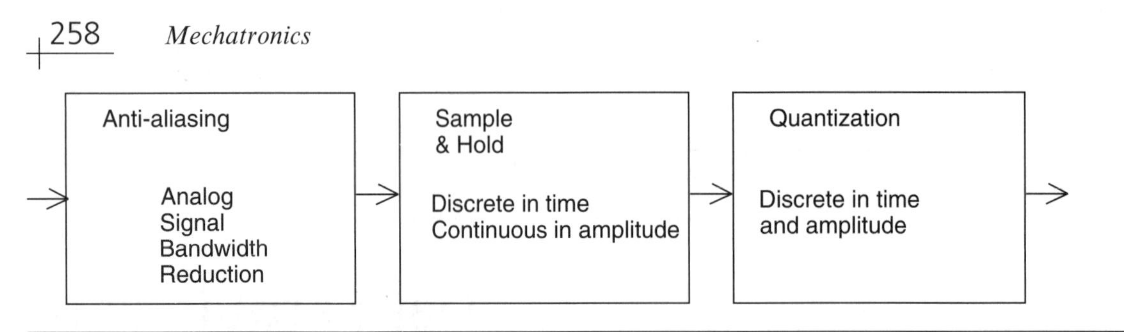

Figure 7.1 Data acquisition.

5. The computer may then modify the environment by outputting control signals. The digital control signals are converted to analog signals using a digital-to-analog converter (DAC).
6. The analog signals are conditioned (e.g. amplified and filtered) appropriately for an output transducer.
7. The output transducer interacts with the environment.

7.2 Sampling and aliasing

In many common situations in engineering, a function $f(t)$ is sampled. When a function is evaluated by numerical procedures, it is always necessary to sample the function in some manner, because digital computers cannot deal with analog, continuous functions except by sampling them. Often the function is not even explicitly defined, but only known as a series of values recorded on tape, by a data logger or a computer. Examples of where sampled time domain data is used are areas where computers and microcontrollers are used to automate processes and react to input data from the processes. It is also used in engineering applications, which include simple and complex vibration analysis of machinery, as well as measurements of other variables such as boiler pressures, temperatures, flow rates and turbine speeds and many other machine parameters.

7.2.1 Sampling

If Δ is the time interval between consecutive samples, then the sampled time data can be represented as:

$$h_n = h(\Delta n) \quad n = 0, \pm 1, \pm 2, \ldots \tag{7A}$$

Consider an analogue signal $x(t)$ that can be viewed as a continuous function of time, as shown in Figure 7.2(a). We can represent this signal as a discrete time

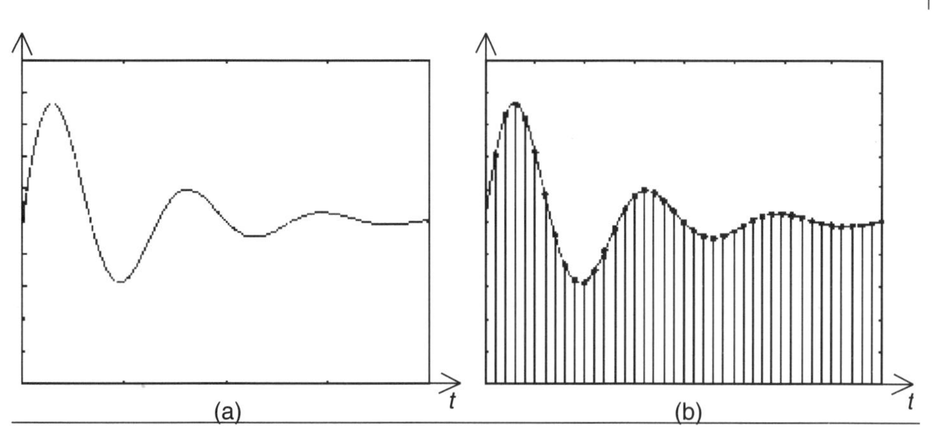

Figure 7.2 Digitizing a waveform: (a) analog signal; and (b) sampled equivalent.

signal by using values of $x(t)$ at intervals of nT_s to form $x[n]$ as shown. In this case, we are mapping points from the function $x(t)$ at regular intervals of time, T_s, called the sampling period.

It is usual to specify a sampling rate or frequency f_s rather than the sampling period. The frequency is given by $f_s = 1/T_s$, where f_s is in Hertz. If the sampling rate is high enough, then the signal $x(t)$ can be constructed from $x[n]$ by simply joining the points by small linear portions, thus approximating to the analog signal. The discrete samples are digitized for processing by a computer or similar device (Figure 7.2(b)).

7.2.1.1 *The sampling theorem*

The sampling theorem, or more correctly Shannon's sampling theorem, states that we need to sample a signal at a rate at least twice the maximum frequency component in order to retain all frequency components in the signal. This is expressed as

$$f_s > 2f_{max}, \tag{7.1}$$

where f_s is the *sampling rate* (frequency), f_{max} is the highest frequency in the input signal, and the minimum required rate ($2f_{max}$) is called the *Nyquist frequency*.

The time interval between the digital samples is

$$\Delta t = \frac{1}{f_s}. \tag{7.2}$$

7.2.2 Aliasing

One would expect that if the signal has significant variation then T_s must be small enough to provide an accurate approximation of the signal $x(t)$. Significant signal variation usually implies that high frequency components are present in the signal. It could therefore be inferred that the higher the frequency of the components present in the signal, the higher the sampling rate should be. If the sampling rate is not high enough to sample the signal correctly then a phenomenon called aliasing occurs as shown in Figure 7.3. In other words, we do not correctly obtain the frequency in the original signal if a signal is sampled at less than twice its maximum frequency component.

The term aliasing refers to the distortion that occurs when a continuous time signal has frequencies larger than half the sampling rate. The process of aliasing describes the phenomenon in which components of the signal at high frequencies are mistaken for components at lower frequencies. The frequency domain view of sampling is that, when a continuous time signal is sampled, its spectrum will show the aliasing effect if aliasing occurs because regions of the frequency domain will be shifted by an amount equal to the sampling frequency.

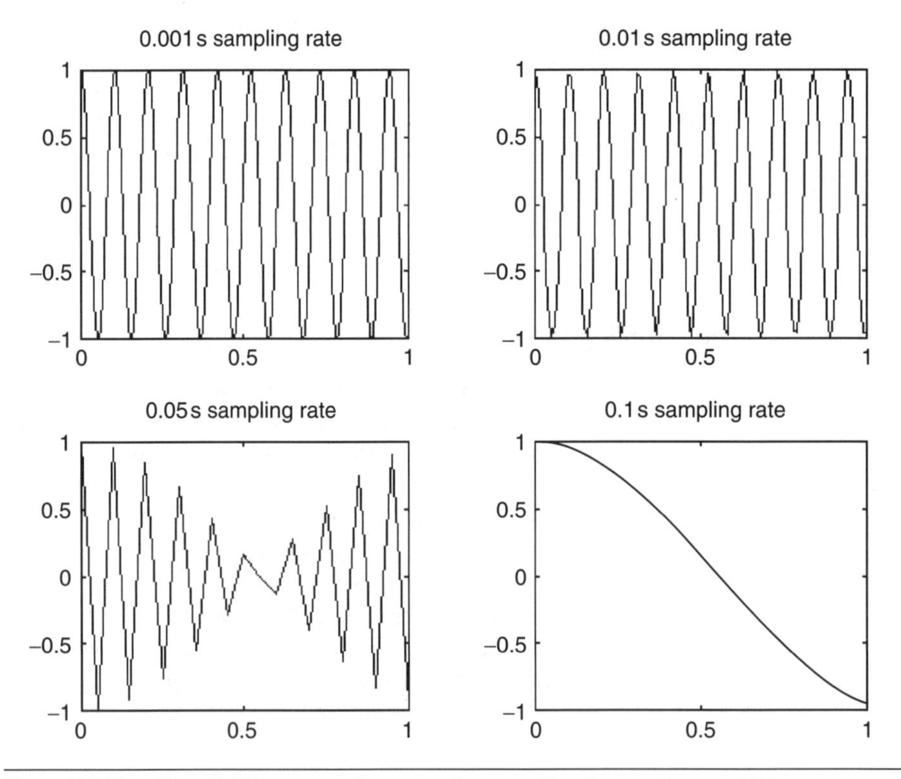

Figure 7.3 The effect of different sampling rates when sampling the waveform cos(60t).

7.2.2.1 *Anti-aliasing and the Nyquist sampling theorem*

The Nyquist sampling theorem states that to avoid aliasing occurring in the sampling of a signal, the sampling rate should be greater than or equal to twice the highest frequency present in the signal. This is referred to as the Nyquist sampling rate.

Many scientific observation instruments critically sample data to permit high throughput scanning and minimal data transfer. Minimal sampling refers to sampling at a rate that corresponds to twice the highest frequency content in the source signal. The frequency that corresponds to the sampling rate, 1/2, is known as the Nyquist frequency ($1/T$). Instruments that minimally sample a signal may pose one of the most difficult problems associated with the activity of signal enhancement. Resolution enhancement increases the high frequency content from observed data, since we are narrowing the narrowest features in the observation. Potential problems exist when we increase the resolution of observed data. The required sampling interval shortens to maintain the critical sampling criteria. If the data is sampled at the critical sampling rate then that sampling rate is less than that necessary to properly sample the enhanced data, then we must be concerned with our effects on the observed data. It is important to understand what effect aliasing has on our observations. In some instruments, data is sampled at a rate less than that necessary to properly characterize data at the Nyquist frequency, that is, the data is under sampled. It has been the general consensus that a mild under sampling of the observable does not produce significant fictitious information to appear in the under-sampled observation.

The Nyquist criterion dictates that all signals must be bandlimited to less than half the sampling rate of the sampling system. Many signals already have a limited spectrum, so this is not a problem. However, for broad spectrum signals, an analog low-pass filter must be placed before the data acquisition system. The minimum attenuation of this filter at the aliasing frequency should be at least:

$$A_{\min} = 20 \log\left(\sqrt{3} + 2^{B}\right), \quad (7.2A)$$

where B is the number of bits of the ADC. This formula is derived from the fact that there is a minimum noise level inherent in the sampling process and there is no need to attenuate the sensor signal more than this noise level.

7.2.2.2 *Problems with the anti-aliasing filter*

- **Time response:** In designing an anti-aliasing filter, there is a temptation to have it's attenuation roll-off extremely quickly. The way to achieve this is to increase the order of the filter. A so-called brick-wall filter (one with infinitely high order), however, causes a sinc function time response that decays proportionally to $1/t$. What this means is that an extremely high order filter that eliminates all signals above the cut-off frequency will cause signals that change rapidly to ring on for a long time. This has a very undesirable effect.

- **Phase distortion/time delay:** Most analog filters have a non-linear phase response. This a problem since non-linear phase causes an unequal time (group) delay as a function of frequency. The higher frequency signals will arrive later than low frequency signals. This can especially be a problem when multiple sensor outputs are compared such as when using a microphone array.
- **Amplitude distortion:** By definition, the filter will modify the frequency structure of the sensor signal which is usually not desired.

The solutions to these problem include:

- Increase the sampling rate of the ADC. This allows the anti-aliasing filter to have a higher cut-off frequency and still eliminate aliasing. This enables the following:

    1. the filter roll off can be more shallow, allowing a better time response;
    2. the frequency response of the filter does not attenuate the lower sensor frequencies of interest;
    3. phase distortion is strongest around the cut-off frequency of the filter so if this is pushed higher, it will not affect the sensor frequencies.

- Use linear phase filters. This, of course, will reduce the phase distortion problems.

7.3 Quantization theory

Analog to digital conversion is a two-step process, which changes a sampled analog voltage into digital form. These processes are quantization and coding. *Quantization* is the transformation of a continuous analog input into a set of data represented by discrete output states. *Coding* is the assignment of a digital code word or number to each output state as shown in Figure 7.4.

The number of possible states N is equal to the number of bit combinations:

$$N = 2^n, \tag{7.3}$$

where n is the number of bits.

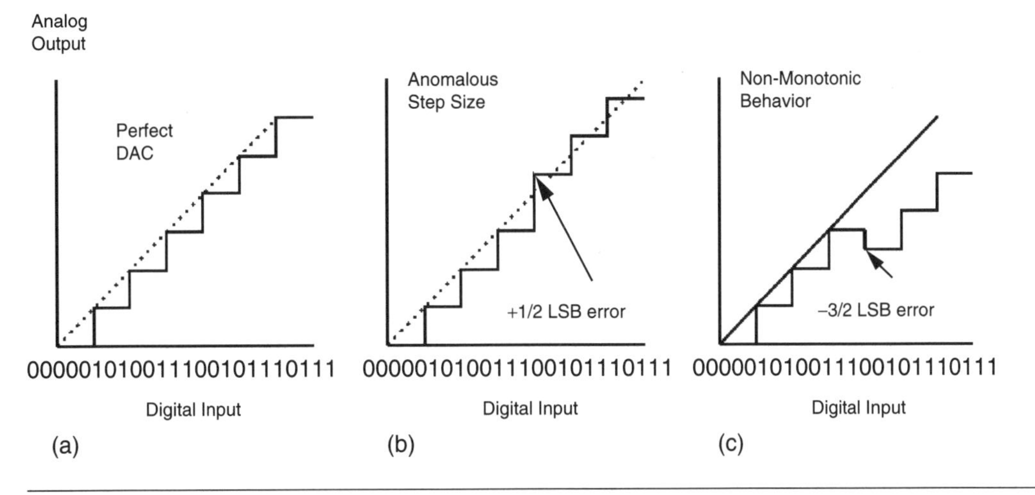

Figure 7.4 Quantization.

The *analog quantization size* (or *resolution*) Q is defined as the full scale range of the ADC divided by the number of output states:

$$Q = \frac{(V_{\max} - V_{\min})}{2^n - 1}. \tag{7.4}$$

The range of an ADC is given as

$$R = (V_{\max} - V_{\min}). \tag{7.5}$$

Knowing the resolution, Q, and range, R, of an ADC, we can determine the number of bits required as follows:

$$Q = \frac{(V_{\max} - V_{\min})}{2^n - 1} = \frac{R}{2^n - 1} \tag{7.6}$$

$$2^n = 1 + \frac{R}{Q} \tag{7.7}$$

$$n = \frac{\log\left(1 + \frac{R}{Q}\right)}{\log 2} \tag{7.8}$$

EXAMPLE 7.1

Determine the smallest step size (resolution) of a 4-bit ADC, which has a maximum output voltage of 12 V.

Solution

$$Q = \frac{(12 - 0)}{2^4 - 1} = \frac{12}{15} = 0.8\,\text{V} \tag{7.8A}$$

EXAMPLE 7.2

Determine the smallest step size (resolution) of an 8-bit ADC in the range −5 V to +5 V.

Solution

$$Q = \frac{(5 - (-5))}{2^8 - 1} = \frac{10}{255} = 0.04\,\text{V} \tag{7.8B}$$

EXAMPLE 7.3

An ADC has a range of 5 V and a resolution of 5 mV. Determine the number of bits required.

Solution

$$n = \frac{\log\left(1 + \frac{5}{0.005}\right)}{\log 2} = 9.97 = 10\,\text{bits} \tag{7.8C}$$

7.4 Digital-to-analog conversion hardware

The process of converting a number held in a digital register to an analog voltage or current is accomplished with a digital-to-analog converter (DAC). The DAC is a useful interface between a computer and an output transducer.

7.4.1 Binary-weighted ladder DAC

Current summing and IC devices are used to build DACs. DACs are nothing more than operational amplifiers whose gains can be programmed digitally. An inverting op amp has a selection of fixed resistances which scales a varying input voltage.

DACs fix the input voltage but switch a series of input resistors to vary the gain and voltage output level.

DACs are normally switched current devices designed to drive the current-summing junction of an operational amplifier. In Figure 7.5 all the input voltages are equal to the reference voltage V_R. Each input resistor is twice as large as the one preceding it. Thus the construction is called a binary-weighted ladder. Each current is proportional to the value of a bit position in a binary number.

The output voltage from each input, is one-half the value produced by the preceding input:

$$\text{1-bit only} \quad V_{out} = -\frac{V_R}{2} \tag{7.9}$$

$$\text{2-bit only} \quad V_{out} = -\frac{V_R}{4} \tag{7.10}$$

$$\text{3-bit only} \quad V_{out} = -\frac{V_R}{8} \tag{7.11}$$

$$\vdots \qquad \vdots$$

$$N\text{-bit only} \quad V_{out} = -\frac{V_R}{2^N} \tag{7.12}$$

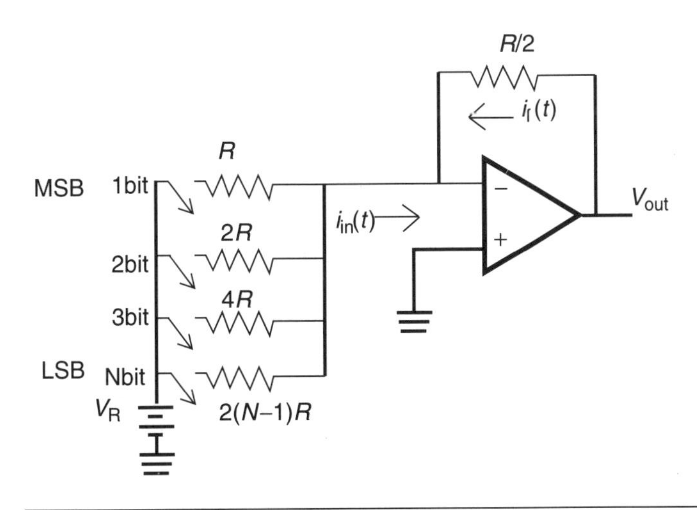

Figure 7.5 The binary-weighted ladder DAC.

The LSB, or bit N, produces the minimum step size, called the resolution, Q:

$$Q = \frac{V_R}{2^N} \tag{7.13}$$

The full-scale output occurs when all bits are closed:

$$V_{out} = V_R\left(\frac{1}{2} + \frac{1}{4} + \frac{1}{8} + \cdots + \frac{1}{2^N}\right) = -V_R\left(1 - \frac{1}{2^N}\right). \tag{7.14}$$

As the number of bits increases, the full scale output approaches the reference voltage but is never equal to it. Generalizing for an arbitrary binary input, we have

$$V_{out} = -V_R \sum_{i=1}^{N} \frac{b_i}{2^i}, \tag{7.15}$$

where b_1 = MSB and b_N = LSB. The design requires a number of different precisely defined resistor values. We can improve the circuit by replacing it with a circuit that requires fewer distinct resistor values.

EXAMPLE 7.4

Determine the output of the DAC for an input of 10011001.

Solution

$$V_{out} = V_R\left(\frac{1}{2} + \frac{0}{4} + \frac{0}{8} + \frac{1}{16} + \frac{1}{32} + \frac{0}{64} + \frac{0}{128} + \frac{1}{256}\right) = -\frac{153}{256} V_R \tag{7.15A}$$

7.4.1.1 *Limitations of binary-weighted ladder DAC*

The limitations of binary-weighted ladder DAC are as follows:

- resistance values require precision trimming (difficult to achieve);
- increasing the resolution leads to the R_N becoming very large;
- their practical realization has stray capacitance (pF), which combined with very large resistances can cause undesirable RC time constants or slow conversion times.

7.4.2 Resistor ladder DAC

The simplest type of DAC converter and perhaps the most popular single package DAC is the resistor ladder network connected to an inverting summer op amp circuit. It requires two precision resistance values (R and $2R$) and resolves the problems of the binary-weighted ladder. Each digital input bit in the circuit controls a switch between ground and the inverting input of the op amp as shown in Figure 7.6.

If the binary number is 0001, only the b_0 switch is connected to the op amp, and the other bit switches are grounded.

The resistance between node V_0 and ground is R since it is the parallel combination of two $2R$s. Therefore, V_0 is the result of voltage division of V_1 across two series resistors of equal value R: $V_0 = V_1/2$. Similarly, we can obtain $V_1 = V_2/2$ and $V_2 = V_3/2$.

Consequently, $V_0 = V_3/8 = V_s/8$.

Since the gain of the inverting amplifier is

$$-\frac{R}{2R} = -\frac{1}{2}, \tag{7.15B}$$

the analog output voltage corresponding to the binary input 0001 is $V_{out0} = -V_s/16$.

Similarly, for the inputs 0010, 0100, and 1000, $V_{out1} = -V_s/8$, $V_{out2} = -V_s/4$, and $V_{out3} = -V_s/2$ respectively.

The output for any combination of bits is $V_{out} = b_3 V_{out3} + b_2 V_{out2} + b_1 V_{out1} + b_0 V_{out0}$.

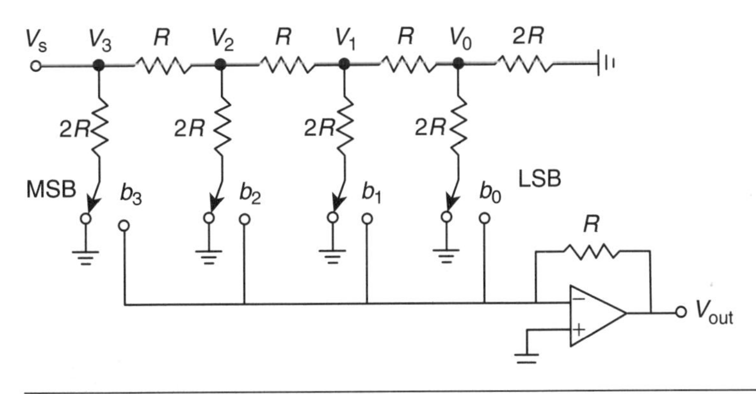

Figure 7.6 The resistor ladder DAC.

7.4.3 DAC limitations

The output of a DAC can only assume discrete values. The relationship between the input binary number and the analog output of a perfect DAC is shown in Figure 7.4. The figure shows output signals from DACs showing (a) the ideal result, and (b) a differential non-linearity or (c) non-monotonic behavior, both caused by imperfectly matched resistors.

Common DAC limitations are an anomalous step size between adjacent binary numbers, non-monotonic behavior, or a zero output.

7.5 Analog-to-digital conversion hardware

The purpose of the analog to digital converter is to digitize the input signal from the sample and hold circuit to 2^B discrete levels, where B is the number of bits of the ADC. The input voltage can range from 0 V to V_{ref} (or $-V_{ref}$ to $+V_{ref}$ for a bipolar ADC). What this means is that the voltage reference of the ADC is used to set the range of conversion of the ADC. For a monopolar ADC, a 0 V input will cause the converter to output all zeros. If the input to the ADC is equal to or larger than V_{ref} then the converter will output all ones. For inputs between these two voltage levels, the ADC will output binary numbers corresponding to the signal level.

There are three widely used ADC technologies today: (1) flash or parallel converter, (2) successive-approximation converters and (3) dual slope converters. These are discussed in the following sections.

To properly convert an analog voltage signal for computer processing the following elements must be properly selected and used in this sequence (Figure 7.7):

1. Buffer amplifier (chosen to provide a signal in a range close to but not exceeding the full input voltage range of the ADC).

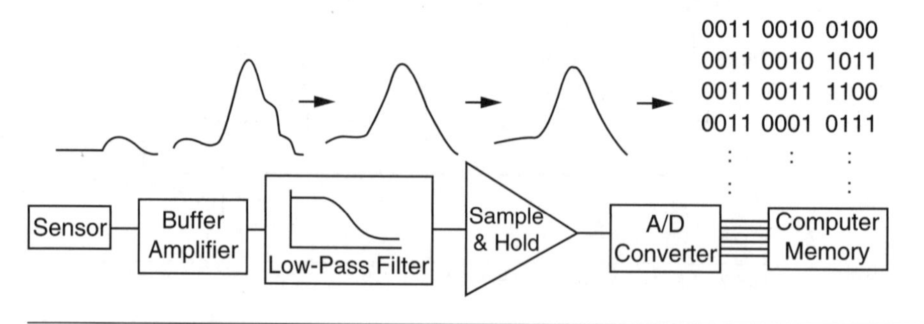

Figure 7.7 ADC block diagram.

2. Low-pass filter (to remove undesirable high-frequency components in the signal that could produce aliasing).
3. Sample and hold amplifier (to maintain a fixed input value during the short conversion time).
4. Analog to digital converter (ADC)
5. Computer

Other chapters discuss the buffer amplifier and the low-pass filter. Here we look at the sample and hold amplifier.

7.5.1 The sample and hold circuit

The sample and hold circuit takes a 'snapshot' of a signal level at a particular time in readiness for digitization. A signal value (sample) is stored in a capacitor which is prevented from discharging using FETs.

An illustrative sample and hold circuit made from discrete components is shown in Figure 7.8. A dual op amp is ideal for this purpose, because their input bias currents are practically zero. The FET is used to isolate the capacitor from the source used to charge it. A 'hold' is initiated by connecting HOLD to −12 V, and leaving it disconnected for a 'sample'. Note the back-to-back signal diodes at the output of the first op amp (any diodes can be used). The purpose of these diodes is to 'catch' the output when the feedback loop is broken when the FET turns OFF. The resistor R isolates this action from the output feedback loop. If these diodes are not present, the op amp saturates at the supply rail and the FET will not turn off. It is important to test the circuit, and to make sure it *holds* properly, and observe the droop of the output.

The choice of hold capacitor is important. The leakage of an electrolytic and the transient behavior of ceramics rule them out completely in this application.

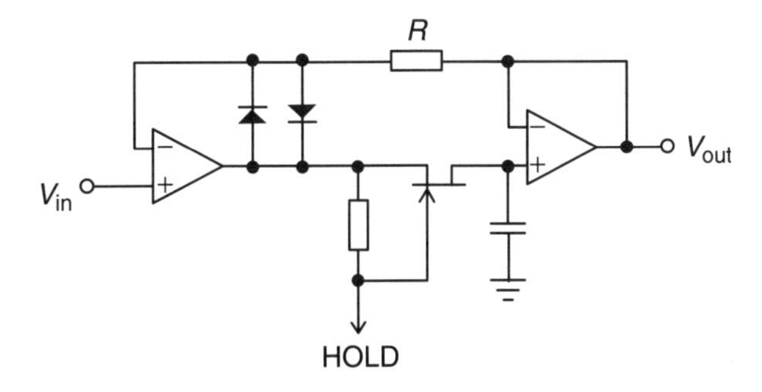

Figure 7.8 Sample and hold circuit.

The best choice is probably polypropylene, and after that polystyrene or Mylar. Polycarbonate is the least preferred. The greatest problem (after leakage, which should be practically zero) is dielectric hysteresis in which the voltage changes on charge and discharge are not the same. There is also dielectric absorption, where there is a 'memory' of past states. A capacitor freshly discharged may acquire a small voltage as time goes by. All of these phenomena are the result of the complexity of dielectric structure and behavior.

Some properties of sample-and-hold circuits are important in critical, dynamic applications. The hold step is the change in output voltage when the circuit is switched OFF, the result of various capacitive effects. The settling time is the time required after the HOLD command for the output to stabilize. The aperture time is the time after the HOLD command at which changes in the input have no effect. The acquisition time is the time at which the output settles after a change at the input. Finally, the dynamic sampling error is the difference between the voltage held and the instantaneous input voltage at the instant of the *hold* command.

Everything necessary for a sample-and-hold except the hold capacitor can be put on chip, so monolithic sample and hold circuits are available and very easy to use. The sample/hold command is given through a digital logic level, so these circuits interface directly with logic.

7.5.1.1 *Limitations of a sample and hold*

- Finite aperture time: The sample and hold takes a period of time to capture a sample of the sensor signal. This is called the aperture time. Since the signal will vary during this time, the sampled signal can be inaccurate.
- Signal feedthrough: When the sample and hold is not connected to the signal, the value being held should remain constant. Unfortunately, some signal does bleed through the switch to the capacitor, causing the voltage being held to change slightly.
- Signal droop: The voltage being held on the capacitor starts to slowly decrease over time if the signal is not sampled often enough.
- The main solution to these problems is to have a small aperture time relative to the sampling period. This means that if the designer uses a high sampling rate, the aperture time of the sample and hold must be quite small.

7.5.2 Noise problem

Because the ADC outputs only 2^B levels there is inherently noise in the quantized output signal. The ratio of the signal to this quantization noise is called SQNR. The SQNR (in dB) is approximately equal to six times the number of bits of the ADC: $20\log(\text{SQNR}) = 6 \times \text{bits}$. Therefore, for a 16-bit ADC this means that the

SQNR is approximately equal to 96 dB. There are, of course, other sources of noise that corrupt the output of the ADC. These include noise from the sensor, from the signal conditioning circuitry, and from the surrounding digital circuitry.

The key to reducing the effects of the noise is to maximize the input signal level. What this means is that the designer should increase the gain of the signal conditioning circuitry until the maximum sensor output is equal to the V_{ref} of the ADC. It is also possible to reduce V_{ref} down to the maximum level of the sensor. The problem with this is that the noise will corrupt the small signals. A good rule of thumb is to keep V_{ref} at least as large as the maximum digital signal, usually 5 V.

7.5.3 The parallel-encoding (flash) ADC

The parallel-encoding or flash ADC design provides the fastest operation at the expense of high component count and high cost (Figure 7.9). For a 2-bit converter, the relationship between the code bit G_i and the binary bits B_i is:

$$B_0 = G_0 \cdot \overline{G}_1 + G_2 \text{ and } B_1 = G_1 \tag{7.16}$$

Table 7.1 shows the comparator output codes and corresponding binary outputs for each state, when an input voltage range of 0 V to 4 V is assumed.

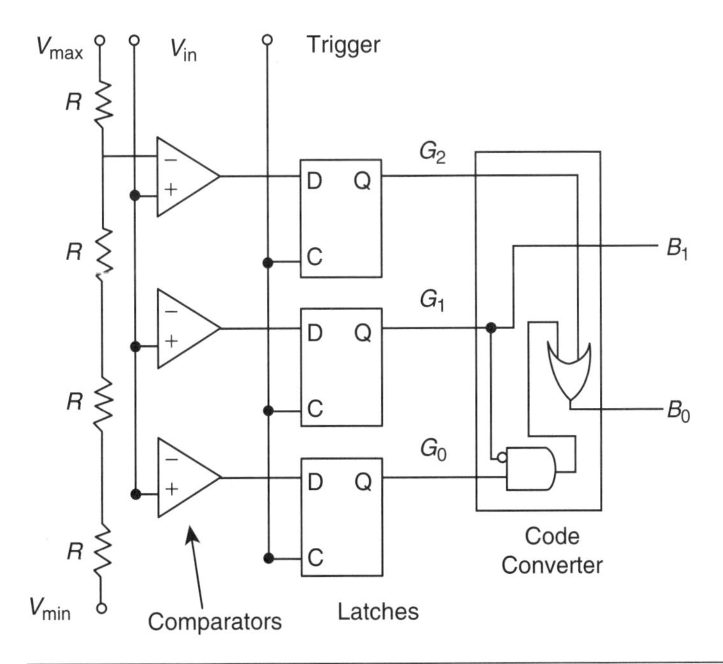

Figure 7.9 The flash ADC.

Table 7.1 Output of a 2-bit flash converter

State	Code ($G_2G_1G_0$)	Binary (B_1B_0)	Voltage range
0	000	00	0–1
1	001	01	1–2
2	011	10	2–3
3	111	11	3–4

7.5.4 The successive-approximation ADC

The successive-approximation ADC is the most commonly used design (Figure 7.10). This design requires only a single comparator and will be only as good as the DAC used in the circuit. The analog output of a high-speed DAC is compared against the analog input signal. The digital result of the comparison is used to control the contents of a digital buffer that both drives the DAC and provides the digital output word. The successive-approximation ADC uses fast control logic, which requires only n comparisons for an n-bit binary result.

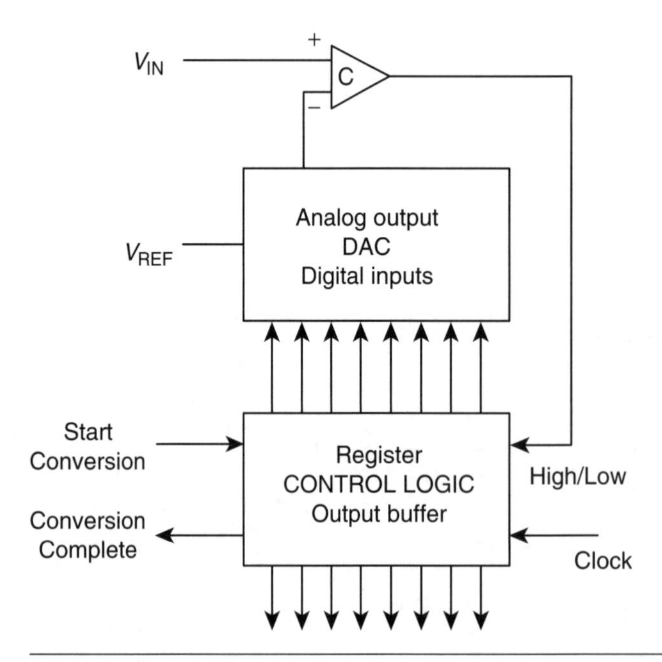

Figure 7.10 Block diagram of an 8-bit successive-approximation ADC.

7.5.4.1 ***The successive-approximation procedure***

The procedure involved in a successive-approximation ADC (see Figure 7.11(a)) is as follows:

- When the start signal is applied, the sample and hold (S&H) latches the analog input.
- The control unit begins an iterative process where the digital value is approximated, converted to an analog value with the DAC, and compared to the analog input with the comparator.

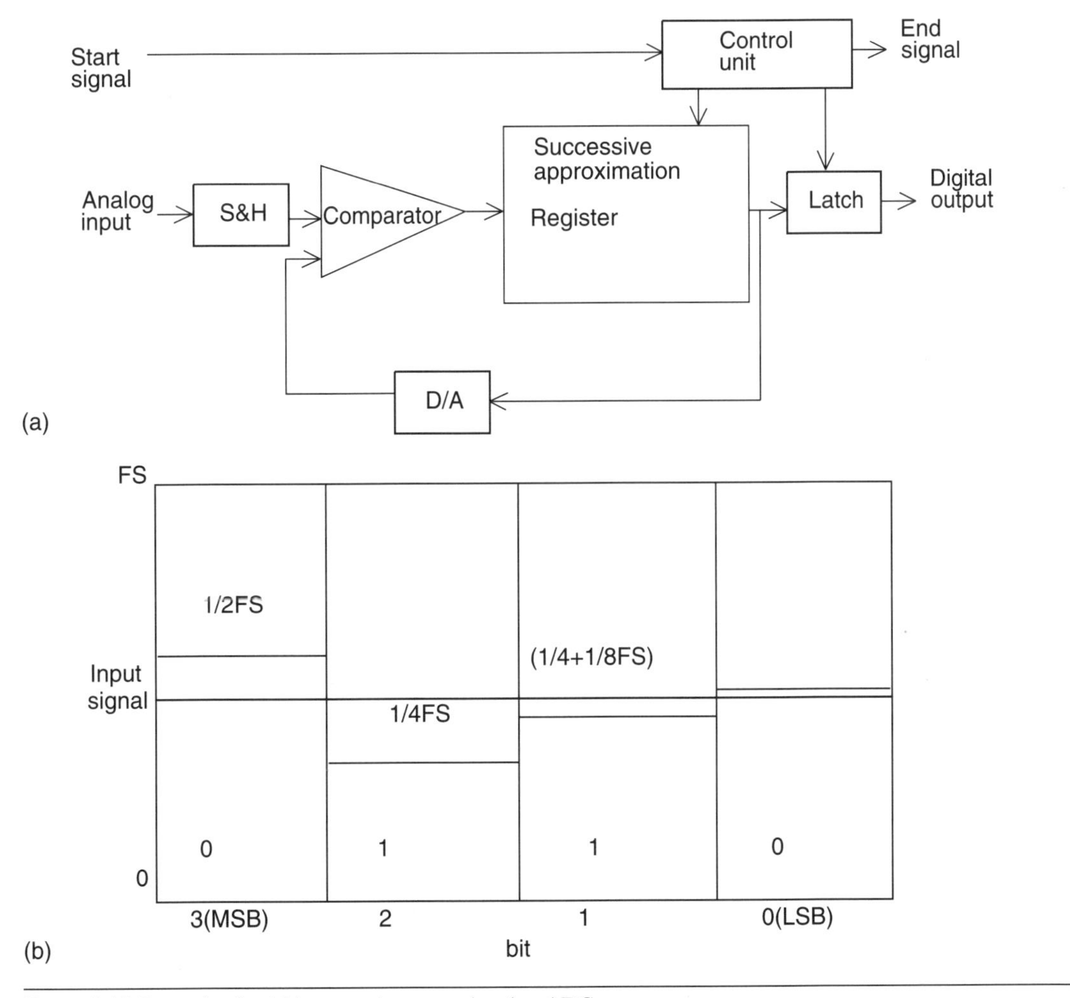

Figure 7.11 Example of a 4-bit successive-approximation ADC.

- When the DAC output equals the analog input, the control unit sets the end-signal and the correct digital output is available at the input.
- The input is compared to standard values in a decreasing binary sequence defined by 1/2, 1/4, 1/8, ..., $1/n$ of the full scale (FS) value of the ADC.

We illustrate the successive approximation method using an example shown in Figure 7.11(b), in which the MSB is 1/2 FS, which in this case is greater than the signal; therefore, the bit is turned off. The second bit is 1/4 FS and is less than the signal, so it is turned on. The third bit gives $1/4 + 1/8$ of FS, which, is still less than the input signal, so the third bit is turned on. The fourth provides $1/4 + 1/8 + 1/16$ of FS and is greater than the signal, so the fourth bit is turned off and the conversion complete.

We note that an n-bit ADC has a conversion time of $n\Delta T$, where ΔT is the cycle time for the DAC and control unit.

A *unipolar* output is either positive or negative, but not both. A *bipolar* output ranges over negative and positive values. We note that a single ADC can digitize several analog signals, using an analog *multiplexer*, which simply switches between several analog inputs. Important parameters in selecting an ADC are the input voltage range, output resolution, and conversion time.

7.5.5 The dual-slope ADC

The limitations associated with the DAC in a successive-approximation ADC can be avoided by using the analog method of charging a capacitor with a constant current; the time required to charge the capacitor from zero to the voltage of the input signal becomes the digital output. When charged by a constant current the voltage on a capacitor is a linear function of time and this characteristic can be used to connect the analog input voltage to the time as determined by a digital counter. Although there are several versions of the integrating method: single slope, dual slope, and multiple, we discuss the dual-slope ADC, which is shown in Figure 7.12 since it is the most common.

An unknown voltage is applied to the input where an analog switch connects it to an integrator. The integrator drives a comparator. Its output goes *high* as soon as the integrator output is more than several millivolts. When the comparator output is *high*, an AND gate passes clock pulses to a binary counter. The binary counter counts pulses until the counter overflows. This time period T_1 is fixed by the clock frequency f_c and counter size M, and is independent of the unknown voltage. However, the integrator output voltage is proportional to the unknown voltage:

$$V_1 = \left(\frac{T_1}{RC}\right) V_{\text{in}} = \frac{M V_{\text{in}}}{f_c RC} \tag{7.17}$$

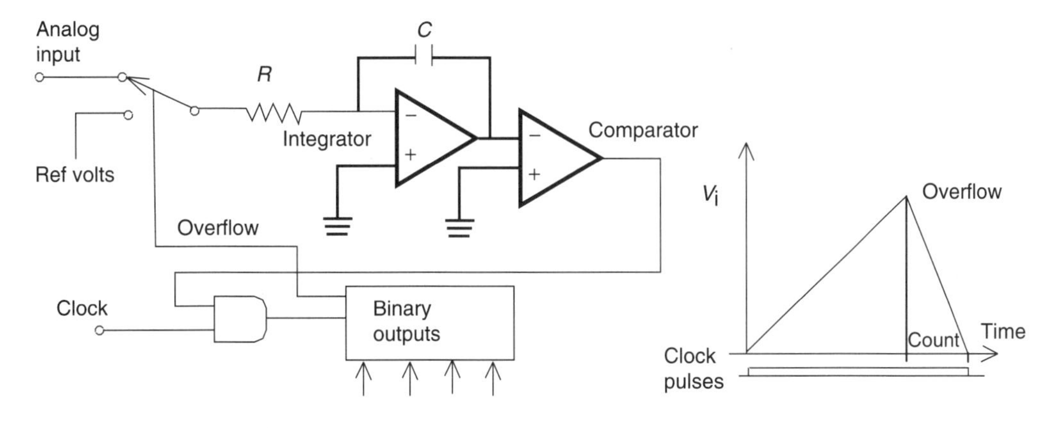

Figure 7.12 Dual-slope ADC.

When the counter overflows, it resets to zero and sends a signal to the analog switch, which disconnects the unknown voltage and connects a reference voltage. The polarity of the reference voltage is opposite that of the unknown voltage. The integrator voltage decreases at a rate proportional to the reference voltage. In the meantime the counter has started counting again from zero at the instant the reference voltage is applied to the integrator. When the integrator output reaches zero, the comparator goes *low*, bringing the AND gate LOW. Clock pulses no longer pass through, and the counter stops counting at m.

$$V_1 = \left(\frac{T_2}{RC}\right) V_{\text{ref}} = \frac{mV_{\text{ref}}}{f_c RC}. \tag{7.18}$$

The base of the triangle, V_i (Figure 7.12(b)) is common to both, hence, the count is

$$m = \left(\frac{M}{V_{\text{ref}}}\right) V_{\text{in}}. \tag{7.19}$$

The accuracy of the dual slope method depends only upon V_{ref}. The integrator action of the converter averages out random negative and positive contributions over the sampling period T_1, and hence it is excellent in rejecting noise.

Problems

Quantization

Q7.1 A data acquisition system uses a DAC with a range of ±15 V and a resolution of 0.02 V. Determine the number of bits the DAC must have.

Q7.2 A data acquisition system uses a DAC with a range of ±10 V and a resolution of 0.05 V. Determine the number of bits the DAC must have.

Q7.3 A data acquisition system uses a DAC with a range of −10 V to +15 V and a resolution of 0.005 V. Determine the number of bits the DAC must have.

Q7.4 A DAC is used to deliver velocity commands to the motor of a drilling machine whose maximum velocity is to be 2000 revolutions per minute, and the minimum non-zero velocity is to be 1 revolution per minute. Determine:
(a) the number of bits required in the DAC; (b) the resolution required.

Q7.5 The analog input voltage to a particular ADC is assumed to be the full-scale value of 12 V.

(a) What is the resolution of the output if this is a 4-bit device?

(b) What is the resolution of the output if this is an 8-bit device?

(c) Generalize the relationship between the number of bits and the resolution of an ADC.

Q7.6 The voltage range of feedback signal from a mechatronic process is −3 V to +12 V, and a resolution of 0.05% of the voltage range is required. Determine the number of bits required for the DAC.

Q7.7 What is the minimum number of bits required to digitize an analog signal with a resolution of: (a) 2.5%; (b) 5%; (c) 10%?

Q7.8 The output voltage of an aeroplane altimeter is to be sampled using an ADC. The sensor outputs 0 V at 0 m altitude and outputs 10 V at 50 km altitude. The altimeter datasheet shows that the allowable error in sensing (±½LSB) is 5 m. Determine the minimum number of bits required for the ADC.

Q7.9 An 8-bit ADC with a 0 V to 10 V range is used for the purpose of sampling the voltage of an analog sensor. Determine the digital output code that would correspond to the following: (a) 2.5 V; (b) 5 V; (c) 7.5 V.

Digital-to-analog conversion

Q7.10 The unsigned decimal number 20_{10} is input to a 4-bit DAC. Given that $R_F = R_o/31$, logic 0 corresponds to 0 V and logic 1 corresponds to 5 V.

Determine:

(a) the output of the DAC;

(b) the maximum voltage that can be output from the DAC;

(c) the resolution over the range of 0 to 5 V;

(d) the number of bits required in the DAC if an improved resolution of 30 mV is desired.

Q7.11 Solve Q7.10 for an unsigned decimal number 250_{10}, 8-bit DAC, $R_F = R_o/255$, logic 0 and logic 1 corresponding to 0 V and 12 V, respectively, and improved resolution of 5 mV is desired.

Q7.12 Solve Q7.10 for an unsigned decimal number 100_{10}, 8-bit DAC, $R_F = R_o/255$, logic 0 and logic 1 corresponding to 0 V and 5 V respectively, and improved resolution of 1 mV is desired.

Q7.13 Solve Q7.10 for an unsigned decimal number 1500_{10}, 12-bit DAC, $R_F = R_o/4095$, logic 0 and logic 1 corresponding to 0 V and 12 V respectively, and improved resolution of 1 mV is desired.

Analog-to-digital conversion

Q7.14 A 4-bit ADC utilizes the successive-approximation method for converting an input signal of 2 V and an input range of −5 V to +5 V. Determine the digital output.

Q7.15 Repeat Q7.14 for a 5-bit ADC.

Further reading

[1] Alciatore, D.G. and Histand, M.B. (2002) *Introduction to Mechatronics and Measurement Systems* (2nd. ed.), McGraw-Hill.

[2] Fletcher, W.I. (1980) *An Engineering Approach to Digital Design*, Englewood Cliffs, NJ: Prentice-Hall.

[3] Givone, D.D. (2002) *Digital Principles and Design* (1st. ed.), McGraw-Hill.

[4] Horowitz, P. and Hill, W. (1989) *The Art of Electronics* (2nd. ed.), New York: Cambridge University Press.

[5] Rashid, M.H. (1996) *Power Electronics: Circuits, Devices, and Applications*, Prentice Hall.
[6] Rizzoni, G. (2003) *Principles and Applications of Electrical Engineering* (4th. ed.), McGraw-Hill.
[7] Stiffler, A.K. (1992) *Design with Microprocessors for Mechanical Engineers*, McGraw-Hill.

CHAPTER 8

Sensors

Chapter objectives

When you have finished this chapter you should be able to:

- understand the fundamentals of distance sensors;
- understand the fundamentals of movement sensors;
- understand the fundamentals of proximity sensors;
- understand the fundamentals of stress/strain/force sensors;
- understand the fundamentals of temperature sensors.

8.1 Introduction

In virtually every engineering application there is the need to measure some physical quantities, such as displacements, speeds, forces, pressures, temperatures, stresses, flows, and so on. These measurements are performed using physical devices called sensors, which are capable of converting a physical quantity to a more readily manipulated electrical quantity.

The key issues in the selection of sensors are: (a) the field of view and range; (b) accuracy; (c) repeatability and resolution; (d) responsiveness in the target-domain; (e) power consumption; (f) hardware reliability; (g) size; and (h) interpretation reliability.

Often the active element of a sensor is referred to as a transducer. Most sensors, therefore, convert the change of a physical quantity (e.g. pressure, temperature) to a corresponding and usually proportional change in an electrical quantity (e.g. voltage or current). Often the direct output from a sensor needs additional manipulation before the electrical output is available to the user.

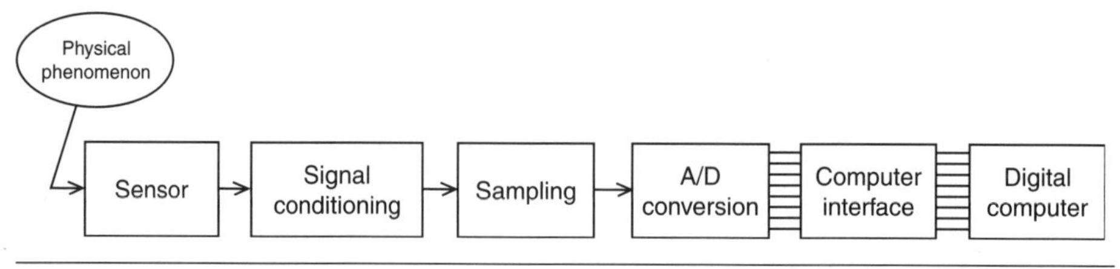

Figure 8.1 Measurement system.

For example, the output of a strain gage used for measuring stress resistance in an element must first be converted into an appropriate voltage using a suitable electrical circuit such as a Wheatstone bridge and then amplified from the millivolt level to appropriate voltage level. This process is known as *signal conditioning*, which we have already covered. Often, the conditioned sensor signal output is transformed into a digital form for display on the computer or other display unit. The apparatus for manipulating the sensor output into a digital form for display is referred to as a measuring instrument (see Figure 8.1 for a typical computer-based measuring system). In this chapter, we classify sensors into distance, movement, proximity, stress/strain/force, and temperature. There are many commercially available sensors but we have picked on the ones that are frequently used in mechatronics applications.

8.2 Distance sensors

Distance (or displacement) sensors form the basis of many different types of measurement system. In this section methods of non-contact distance measurement will receive most attention since they are most widely suited to the inspection of manufacturing parts in industry. Light and sound form the basis of most distance measurement apparatus, with ultrasonic distance sensing being popular due to its low cost and ease of installation.

8.2.1 The potentiometer

Potentiometers are variable resistance devices. A change in the linear or angular displacement of a potentiometer varies the effective length of its conductor, and therefore the resistance of the device. This change in resistance can be related to the displacement through a change in output voltage. Potentiometers have a tendency for non-linearity, and care must be taken when a high degree of accuracy is required.

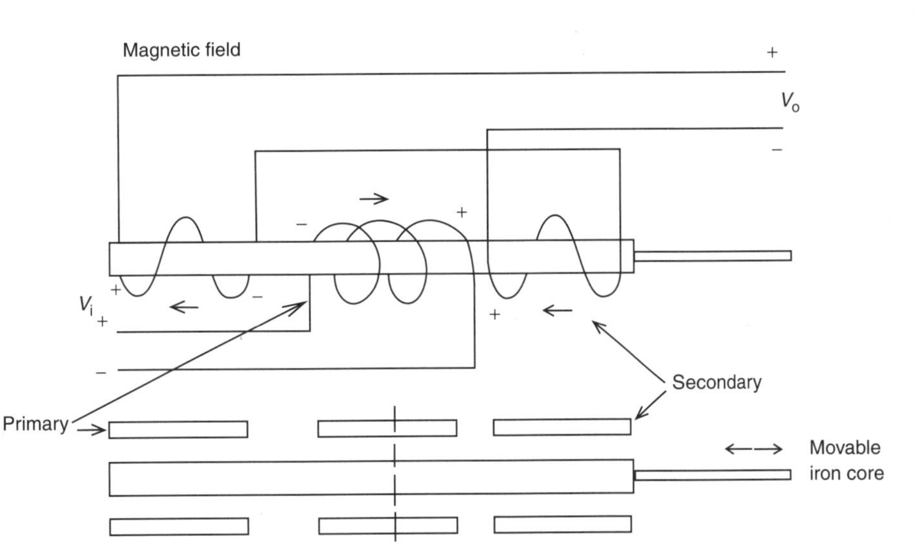

Figure 8.2 Linear variable differential transformer (LVDT) schematic.

8.2.2 The linear variable differential transformer (LVDT)

The linear variable differential transformer is a mechanical displacement transducer. It gives an a.c. voltage output proportional to the distance of the transformer core to the windings. The LVDT is a mutual-inductance device with three coils and a core (see Figure 8.2). An external a.c. power source energizes the central coil and the two-phase opposite end coils are used as pickup coils. The output amplitude and phase are dependent on the relative position of the two pickup coils and the power coil. Theoretically there is a null or zero position between the two end coils, although in practice this is difficult to obtain perfectly.

A typical representation of core displacement to output voltage is shown in Figure 8.3. The output voltage on either side of the null position is approximately proportional to the core displacement. The phase shift that occurs on passing through the null position can be sensed by a phase sensitive demodulator and used to detect the side that the output voltage is from.

The sensitivity of an LVDT can be determined by the following equation:

$$\text{Sensitivity} = (\text{output} \times \text{input})/(\text{excitation voltage} \times \text{displacement})$$

The typical range for LVDT sensitivity is $0.4 - 2.0\,\text{mV/V} \times 10^{-3}\,\text{cm}$. LVDTs are typically used in force, displacement and pressure measurement. They offer the advantages of being relatively insensitive to temperature change, and provide high outputs without intermediate amplification. The appreciable mass of the core is a disadvantage in the area of dynamic measurements.

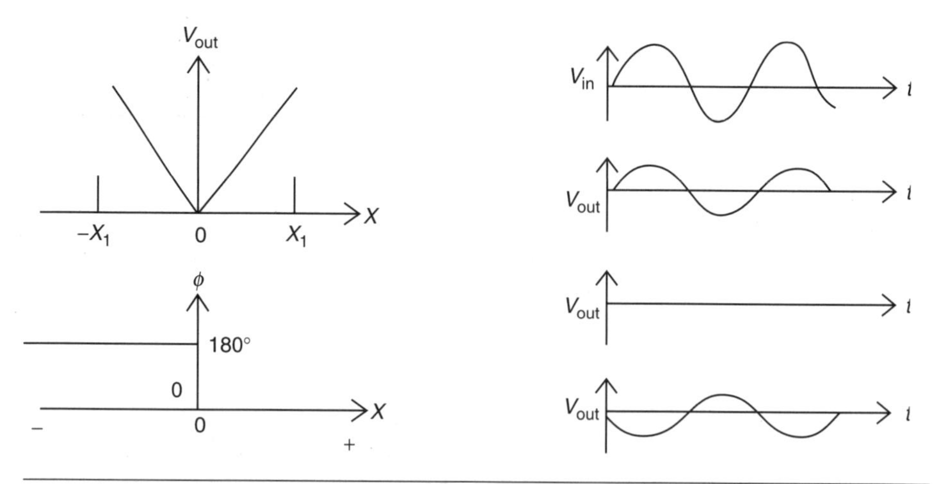

Figure 8.3 Output from a differential transformer.

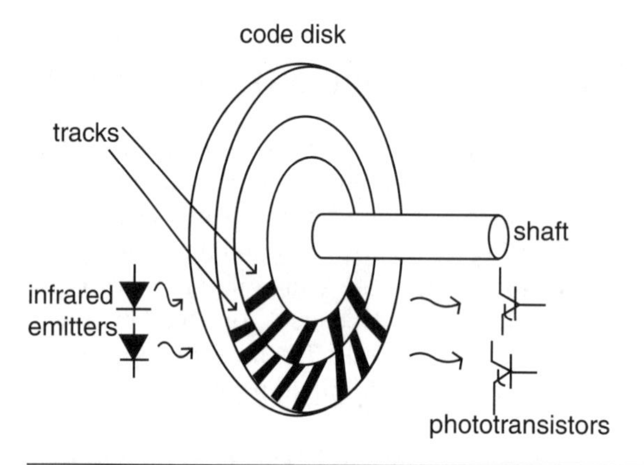

Figure 8.4 The rotary optical encoder.

8.2.3 The digital optical encoder

A digital optical encoder is a device that converts motion into a sequence of digital pulses. By counting a single bit or by decoding a set of bits, the pulses can be converted to relative or absolute position measurements. Encoders have both linear and rotary configurations, but the most common type is rotary, shown in Figure 8.4. Rotary encoders are manufactured in two basic forms: the absolute encoder where a unique digital word corresponds to each rotational position of the shaft, and the incremental encoder, which produces digital pulses as the

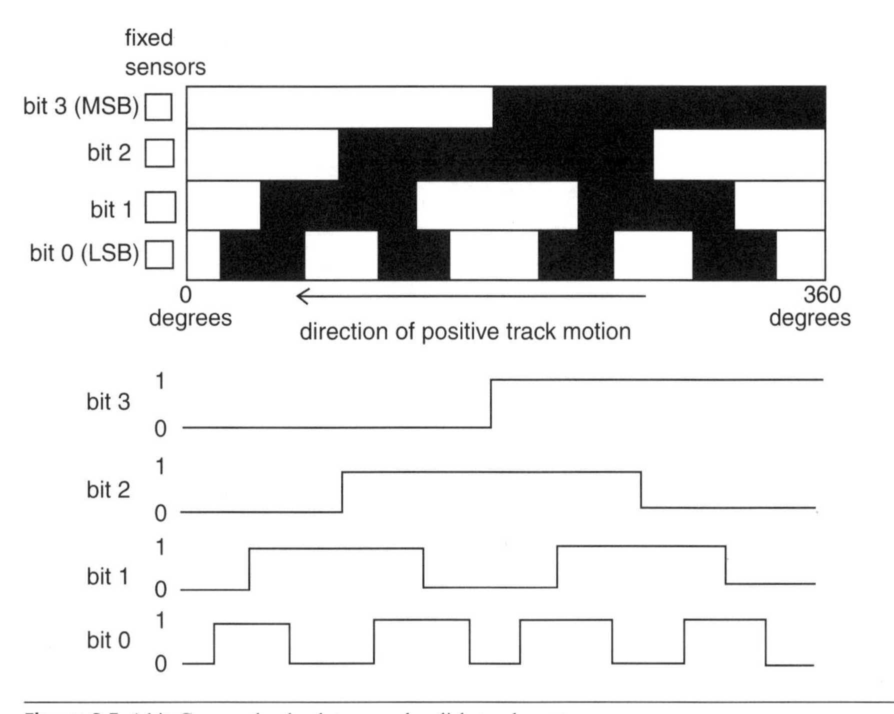

Figure 8.5 4-bit Gray code absolute encoder disk track pattern.

shaft rotates, allowing measurement of relative position of shaft. Most rotary encoders are composed of a glass or plastic code disk with a photographically deposited radial pattern organized in tracks. As radial lines in each track interrupt the beam between a photo-emitter detector pair, digital pulses are produced.

8.2.3.1 *The absolute encoder*

The optical disk of the absolute encoder is designed to produce a digital word that distinguishes N distinct positions of the shaft. For example, if there are eight tracks, the encoder is capable of producing 256 distinct positions or an angular resolution of 1.406 ($=360/256$) degrees. The most common types of numerical encoding used in the absolute encoder are the Gray and binary codes. To illustrate the action of an absolute encoder, the Gray code and natural binary code disk track patterns for a simple 4-track (4-bit) encoder are shown in Figures 8.5 and 8.6. The output bit codes for both coding schemes are listed in Table 8.1.

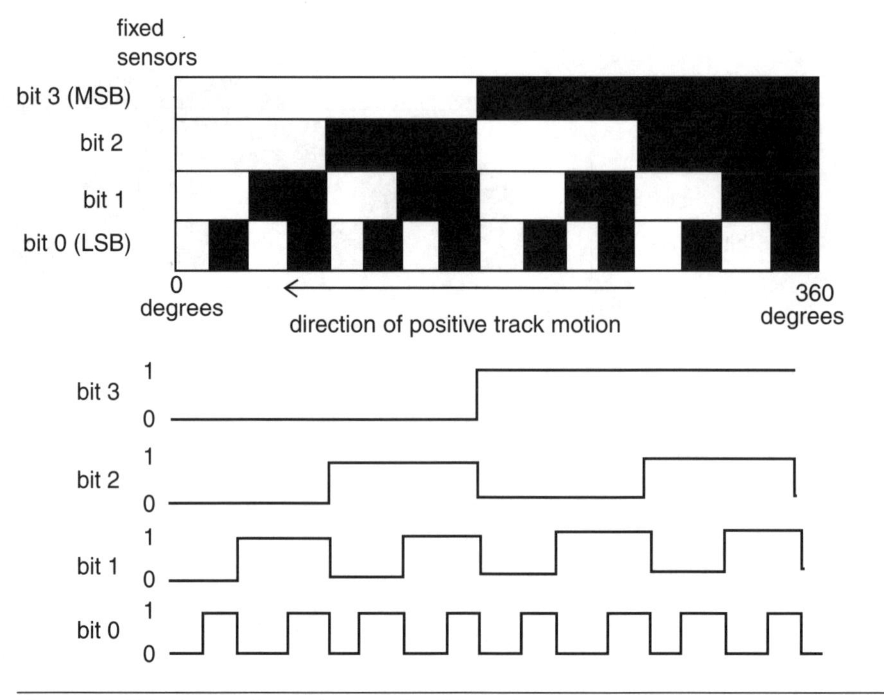

Figure 8.6 4-bit binary code absolute encoder disk track pattern.

Table 8.1 4-bit Gray and natural binary codes

Decimal code	*Rotation range (deg.)*	*Binary code*	*Gray code*
0	0–22.5	0000	0000
1	22.5–45	0001	0001
2	45–67.5	0010	0011
3	67.5–90	0011	0010
4	90–112.5	0100	0110
5	112.5–135	0101	0111
6	135–157.5	0110	0101
7	157.5–180	0111	0100
8	180–202.5	1000	1100
9	202.5–225	1001	1101
10	225–247.5	1010	1111
11	247.5–270	1011	1110
12	270–292.5	1100	1010
13	292.5–315	1101	1011
14	315–337.5	1110	1001
15	337.5–360	1111	1000

The Gray code is designed so that only one track (one bit) will change state for each count transition, unlike the binary code where multiple tracks (bits) change at certain count transitions. This effect can be seen clearly in Table 8.1. For the Gray code, the uncertainty during a transition is only one count, unlike with the binary code, where the uncertainty could be multiple counts.

Since the Gray code provides data with the least uncertainty but the natural binary code is the preferred choice for direct interface to computers and other digital devices, a circuit to convert from Gray to binary code is desirable. The Boolean expressions that relate the binary bits to the Gray code bits are as follows:

$$\begin{aligned} B_3 &= G_3 \\ B_2 &= B_3 \oplus G_2 \\ B_1 &= B_2 \oplus G_1 \\ B_0 &= B_1 \oplus G_0 \end{aligned} \tag{8.1}$$

Figure 8.7 shows a simple circuit that utilizes exclusive-OR gates to perform this function. For a Gray code to binary code conversion of any number of bits N, the most significant bits (MSB) of the binary and Gray code are always

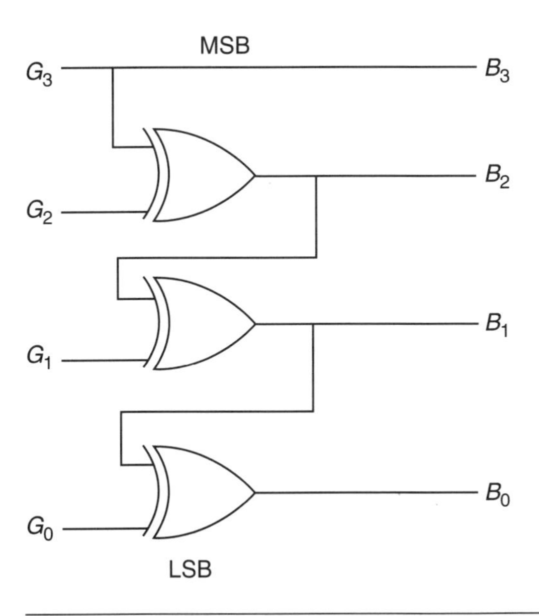

Figure 8.7 Gray code to binary code conversion.

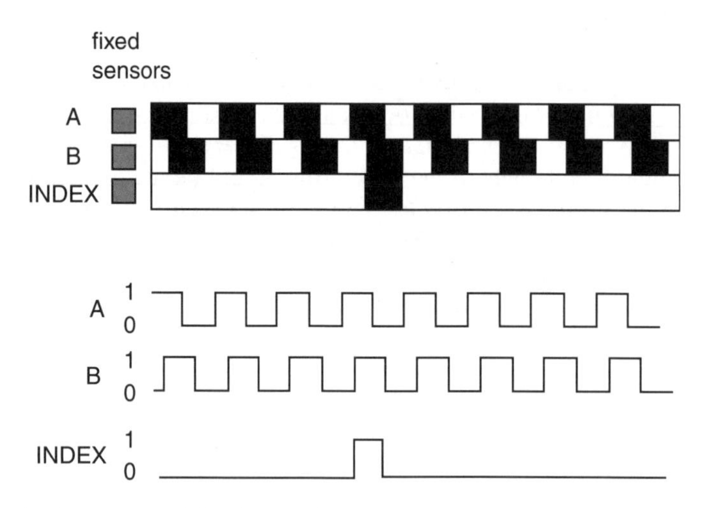

Figure 8.8 Incremental encoder disk pattern.

identical, and for each other bit, the binary bit is the exclusive-OR combination of adjacent Gray code bits.

8.2.3.2 *The incremental encoder*

The incremental encoder, sometimes called a relative encoder, is simpler in design than the absolute encoder. It consists of two tracks and two sensors whose outputs are called channels *A* and *B*. As the shaft rotates, pulse trains occur on these channels at a frequency proportional to the shaft speed, and the phase relationship between the signals yields the direction of rotation. The code disk pattern and output signals *A* and *B* are illustrated linearly in Figure 8.8. By counting the number of pulses, and knowing the resolution of the disk the angular motion can be measured. The *A* and *B* channels are used to determine the direction of rotation by assessing which channel *leads* the other. The signals from the two channels are a 1/4 cycle out of phase with each other and are known as quadrature signals. Often a third output channel, called INDEX, yields one pulse per revolution, which is useful in counting full revolutions. It is also useful as a reference to define a home base or zero position.

Figure 8.8 illustrates two separate tracks for the *A* and *B* channels, but a more common configuration uses a single track with the *A* and *B* sensors offset a 1/4 cycle on the track to yield the same signal pattern. A single-track code disk is simpler and cheaper to manufacture.

The quadrature signals *A* and *B* can be decoded to yield the direction of rotation as shown in Figure 8.9. Decoding transitions of *A* and *B* by using

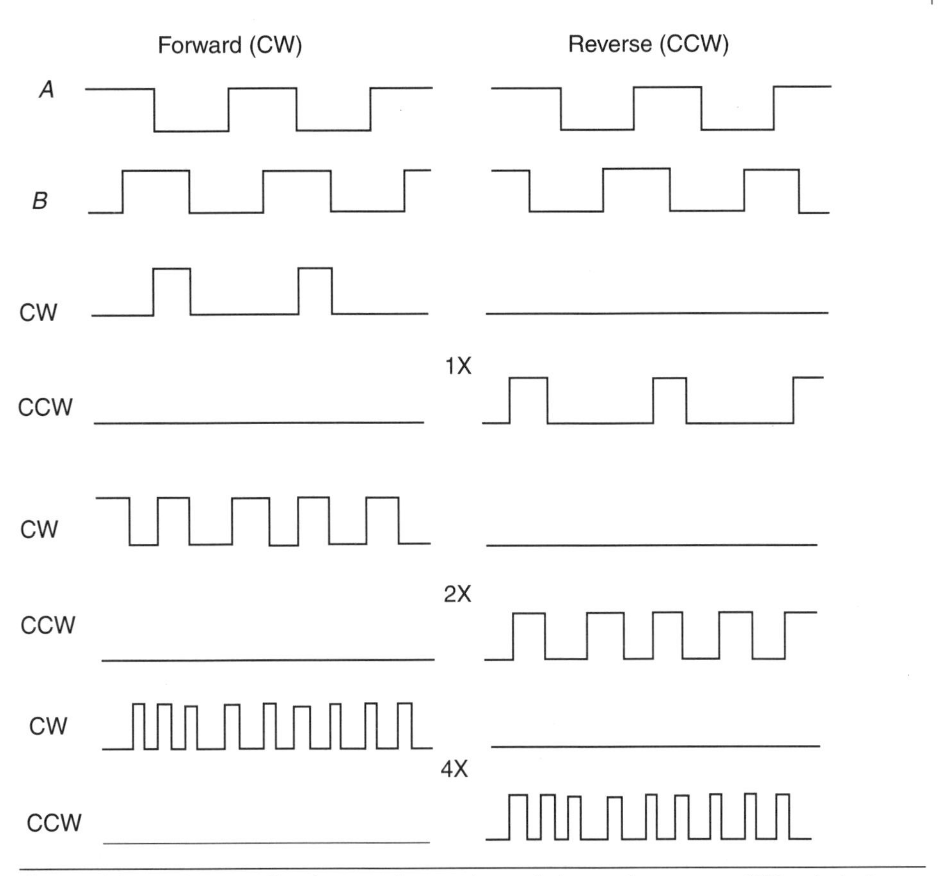

Figure 8.9 Quadrature direction sensing and resolution enhancement (CW = clockwise, CCW = counter-clockwise).

sequential logic circuits in different ways can provide three different resolutions of the output pulses: 1×, 2×, 4×, 1× resolution only provides a single pulse for each cycle in one of the signals A or B, 4× resolution provides a pulse at every edge transition in the two signals A and B providing four times the 1× resolution. The level of one signal determines the direction of rotation (clockwise or counter-clockwise) during an edge transition of the second signal. For example, in the 1× mode, $A = \downarrow$ with $B = 1$ implies a clockwise pulse, and $B = \downarrow$ with $A = 1$ implies a counter-clockwise pulse. If there is only a single output channel A or B, it would be impossible to determine the direction of rotation. Furthermore, shaft jitter around an edge transition in the single signal would result in erroneous pulses.

Figure 8.10 shows the 1× quadrature circuit, where the input to the clocks of the D-type flip-flops is tied to the data input of the other flip-flop and there is

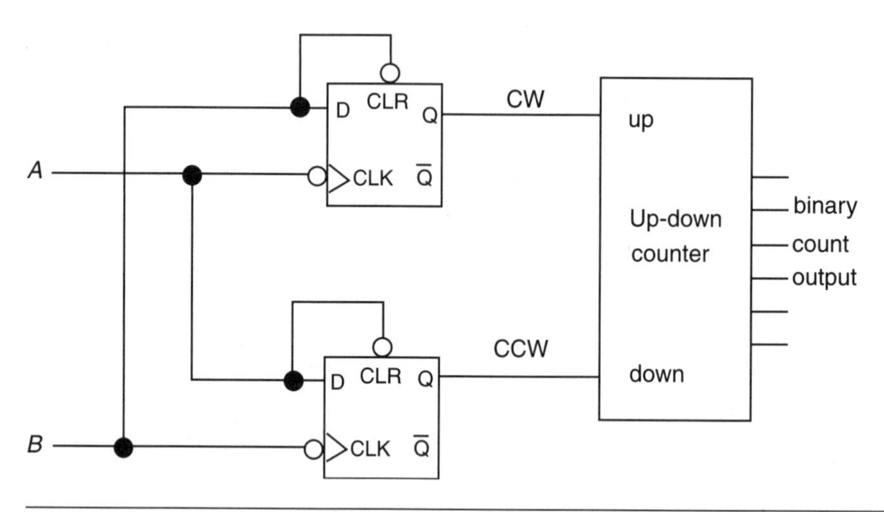

Figure 8.10 The 1× quadrature circuit.

a feedback from the clear (CLR) to the data line. The D-type flip-flops decode the direction of the shaft as clockwise (CW) or counter-clockwise (CCW) which are then input to the up-down counter.

8.3 Movement sensors

8.3.1 Velocity sensors

The measurement of motion has been heavily applied to security systems. Transducers used in the field of motion detection will be described in the following sections.

8.3.1.1 *The Doppler effect*

The Doppler effect is a phenomenon that has been used in conjunction with a number of different technologies to measure motion. The Doppler effect states that if a wave source and corresponding receiver are moving relative to each other, the frequency observed by the receiver will be greater than or smaller than the actual source frequency. This is also known as the Doppler frequency shift. When the wave source and the receiver are moving towards each other the observed frequency will be greater than the actual frequency. If the source and receiver are moving away from each other, the observed frequency will be less than the actual frequency.

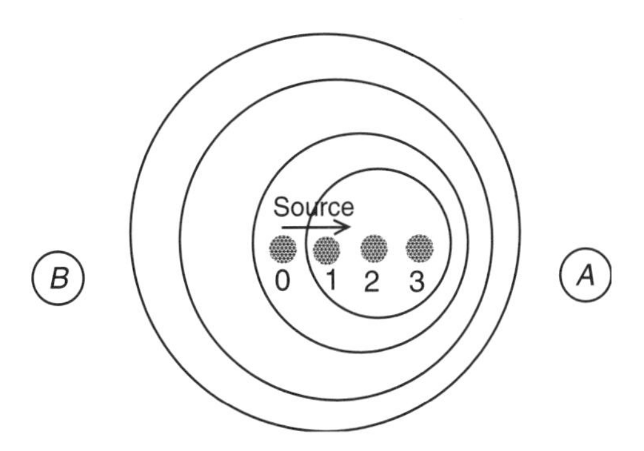

Figure 8.11 The Doppler effect.

If the target is moving either towards or away from the sensor then the Doppler effect becomes relevant. If the source of sound is moving then there is a compression of wavefronts in front of the source and an elongation behind it as shown in Figure 8.11. The effect of this is an apparent increase in the frequency of sound for a listener at point *A* and a proportionate decrease for the listener at point *B*. The apparent frequency may be computed as follows:

$$f' = \frac{V_s}{V_s - V_r} \times f, \tag{8.2}$$

where V_S is the velocity of sound, V_r is the relative velocity of the source/listener (positive when moving together), f is the frequency of source, and f' is the apparent frequency of the source.

Perhaps the best-known applications of the Doppler effect in engineering have been in conjunction with ultrasound. Electromagnetic waves or ultrasonic waves are reflected or emitted from a moving object. The resulting Doppler shift is detected with demodulators and filtering techniques.

The Doppler effect has also been used in conjunction with radioactive isotopes to detect motion in the 10 nm range. A radioactive isotope, which is a gamma-ray source, is attached to the object of interest. A gamma-ray absorber is used to measure the Doppler shift of the gamma rays. Velocity information can be extracted from this information.

8.3.2 Acceleration sensors

8.3.2.1 *Spring-mass accelerometers*

Accelerometers are used to measure acceleration. The majority of these sensors are, by definition, force sensors, being based on Newton's second law of motion: force = mass × acceleration.

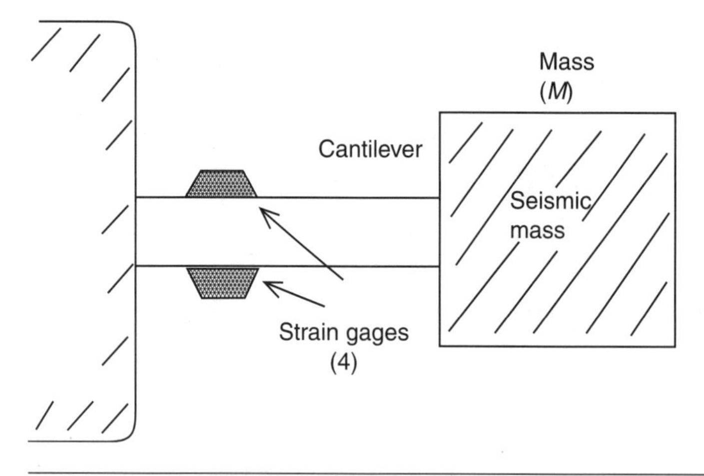

Figure 8.12 Cantilever beam spring-mass system.

The basic sensor comprises a mass that is attached to one surface of the force sensing transducer. The mass is often referred to as the seismic mass. A spring-mass system is devised such that an acceleration of the mass exerts a force on the spring. A force transducer senses the force and acceleration can be calculated from the above expression. A spring-mass system using a cantilever beam and strain gages is depicted in Figure 8.12. Systems using strain gages or capacitative force transducers are popular. Strain gage-based systems have the advantage of relative ease of use during field calibration. Commercial systems are generally capable of measuring 0.5 g to 200 g. Viscous fluids and permanent magnets are incorporated as dampers in spring-mass systems.

When the object (and the accelerometer) accelerates, the mass exerts a force, which the transducer translates into a voltage output. The mass governs the specification of the physical size; large masses are normally used for highly sensitive accelerometers, whereas small masses are utilized for those that will be subject to very high accelerations.

The design rule is that the accelerometer mass should be not more than 10 percent of the dynamic mass of the vibrating part on which it is mounted.

8.3.2.2 *Piezoelectric accelerometers*

In 1880, Pierre and Jacques Curie discovered the piezoelectric effect. Certain materials were observed to generate a voltage when subjected to a mechanical strain, or undergo a change in physical dimensions under an applied voltage. In response to a mechanical strain the piezoelement generates a charge, which is temporarily stored in the inherent capacitance of the piezoelement. With time

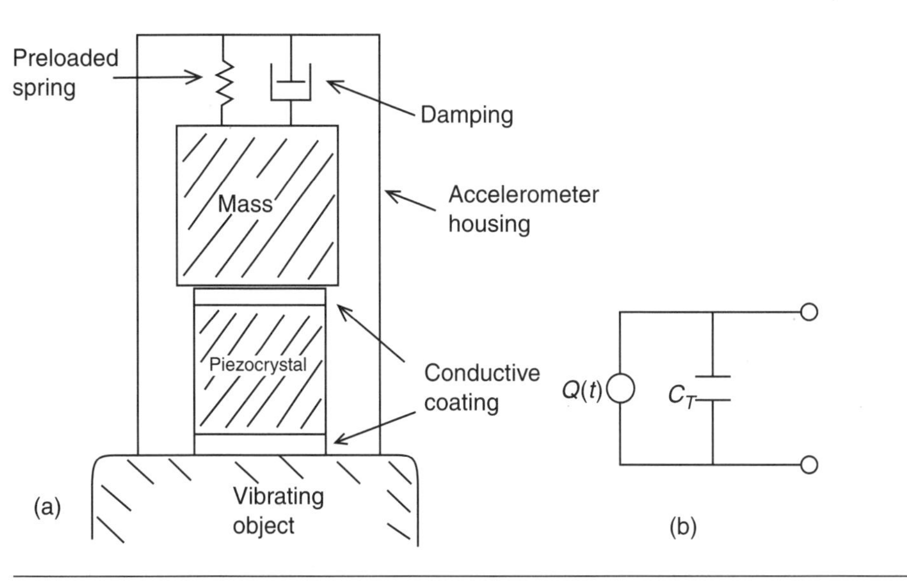

Figure 8.13 Piezoelectric accelerometer and an equivalent circuit.

the charge dissipates due to leakage. This makes piezoelements most useful for dynamic measurements, such as acceleration. The different modes of deformation for piezoelectric sensors and an equivalent circuit are shown in Figure 8.13.

Quartz, lead zirconate titanate (PZT), potassium sodium tartarate (Rochelle salt) and ordinary sugar all exhibit the piezoelectric effect. Piezoelectric materials that are usable in transducer design must have the following properties: stability, high output, insensitivity to temperature and humidity changes and the ability to be worked into a desired configuration or shape. As no material is optimal for all the above properties, transducer design requires certain compromises.

The relationship between an applied force and the surface charge generated is: $Q = DF$, where Q is the surface charge, D is the piezoelectric constant and F is the applied force.

8.3.2.3 *Piezoresistive transducers*

In piezoresistive transducers, a single silicon beam with the seismic mass at one end and rigidly supported at the other, has two piezoresistive strain elements: one diffused into the upper side of the silicon beam and the other into the lower. The piezoresistive strain elements are solid-state silicon resistors, which change their resistance when they are subjected to a mechanical stress.

Interfacing piezoresistive accelerometers are very straightforward to interface since they have a high level output and low output impedance and are more immune to noise sources than their base strain and thermal transient counterparts.

8.3.2.4 *Variable capacitance*

Variable capacitance accelerometers comprise a pair of parallel plates the capacitance of which varies as one plate is moved towards the other by some external force. These devices can be fabricated from micro-machined single crystals of silicon. In a capacitance transducer, the seismic mass is attached to a metallic plate that is held between two plates of a capacitor as shown in Figure 8.14. The metallic plate is attached to a spring.

8.4 Proximity sensors

Proximity sensors are used to detect the presence of an object (or obstacle). They are in much demand in robotic and automated machinery applications.

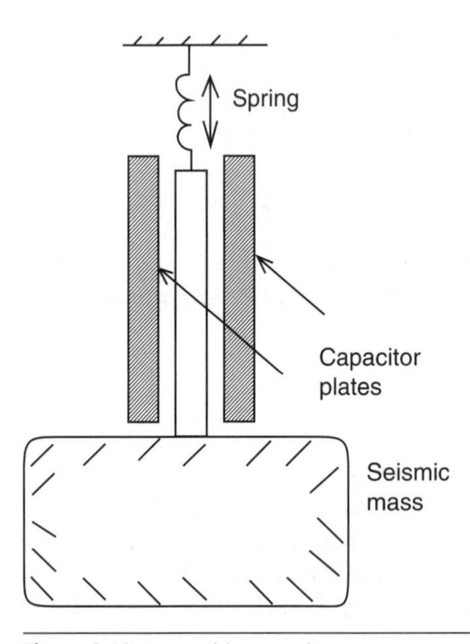

Figure 8.14 A capacitive transducer.

8.4.1 Inductive proximity sensors

Inductive proximity sensors are used to detect the presence of ferrous and non-ferrous metals and are capable of switching at high speed (as high as 2 kHz) with high repeatability, resolution (as much as 2 μm), and with sensing at distances up to 20 mm.

There are basically two types of inductive proximity sensors. The first relies on inducing a *magnetic field* within the object to be sensed, which is known as the target, while the second induces *eddy currents* to flow within the target. The eddy current-based approach is used most often when a current flows through a wire, a magnetic field, which encircles the wire, is created. When an a.c. current is used, the magnetic field can change direction as the flow of current reverses. In the same way that an electric current causes a magnetic field to be created, the converse is also true, that is, a moving magnetic field will create a moving electric field. This inter-relationship between electric and magnetic fields is the basis on which proximity sensors are designed.

An inductive proximity sensor comprises a coil sited at the end of the sensor with the axis of the coil being in line with the sensor body. The current is flowing around the coil and so an alternating magnetic field is created that emerges from the end of the sensor body as shown in Figure 8.15. If a piece of conducting material is placed near the end of the sensor then the lines of magnetic field induce circular eddy currents to flow in the material at right angles to the

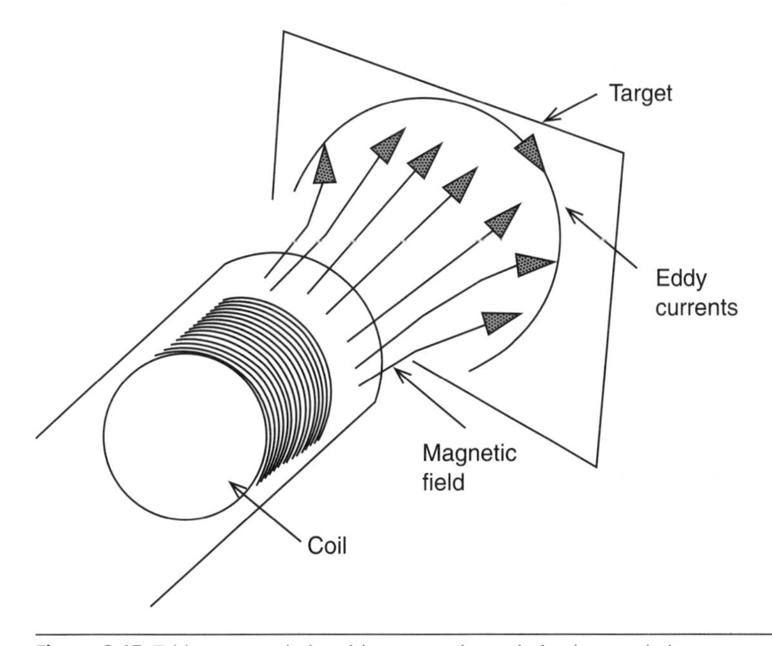

Figure 8.15 Eddy currents induced in a target by an inductive proximity sensor.

magnetic field. These eddy currents consume energy due to the resistance of the material and also produce an opposing magnetic field, which the sensor radiates and absorbs. This has the effect of loading the oscillating circuit in the sensor, and if the target is brought close enough to the sensor it will impose such a load that it will cease to oscillate altogether. A circuit, which monitors the amplitude of the oscillator will then detect that this amplitude is reduced and will accordingly trigger the output switch in the sensor circuit.

Inductance is a measure that relates electrical flux to current. Inductive reactance is a measure of the inductive effect and can be expressed as:

$$X = 2\pi fL, \tag{8.3}$$

where X is the inductive reactance in ohms, f is the frequency of the applied voltage in hertz and L is the inductance in henrys.

The inductance of a circuit is influenced by a number of factors, including: the number of turns in a coil, the coil size, and the permeability of the flux path. As a result of a mechanical displacement the permeability of the flux path is altered and a resulting change in the inductance of the system occurs. Inductance is monitored through the resonant frequency of the inductance coils to an applied voltage. As inductance changes, the resonant frequency of the coils changes. Electronic circuits that convert frequency to voltage are used to gain a voltage output to inductive transducers.

An important aspect to consider in this kind of sensor is the skin effect, a phenomenon in which high frequencies current flows near the surface of a conductor with very little current flowing within the body of the conductor. Tables 8.2, 8.3, and 8.4, show values for skin depth, inductive sensing for some materials, and target size, respectively.

8.4.2 Capacitive proximity sensors

Capacitance proximity sensors are similar to inductive proximity sensors except that they sense the presence of almost any material. Their advantages

Table 8.2 Skin depth

	Frequency (kHz)		
Skin depth (mm)	*100*	*500*	*10000*
Copper	0.2	0.093	0.066
Aluminum	0.25	0.115	0.082
Mild steel	0.5	0.225	0.159

Table 8.3 Inductive sensing

Material	Factor
Mild steel	1.0
Cast iron	1.1
Aluminum foil	0.9
Stainless steel	0.7
Brass	0.4
Aluminum	0.35
Copper	0.3

Table 8.4 Target size

Area (% standard)	Sensing distance
75%	95%
50%	90%
25%	85%

include: (a) adjustable sensitivity, and (b) detection over longer ranges than inductive proximity sensors.

The capacitance for a parallel plate capacitor is given by:

$$C = \frac{8.85 \times 10^{-12} \varepsilon A}{d}, \tag{8.4}$$

where ε is the relative permittivity, A is the area of the plates in square centimeters, and d is the separation between the two plates in centimeters. If the two parallel plates move relative to each other, the distance and area of the plates affecting the capacitance is altered, and a corresponding change in the capacitance occurs. The relationship between capacitance and mechanical displacement for a given system can be determined. Capacitive transducer systems have been used to measure displacements in the micrometer range.

8.4.3 Photoelectric proximity sensors

Photoelectric proximity sensors comprise an infrared light-emitting diode (LED) source and a light sensitive switch known as a detector. There are different configurations of photoelectric proximity sensors: (a) the diffuse reflector, (b) the retro reflector, (c) the through beam, (d) the fixed focus type, and

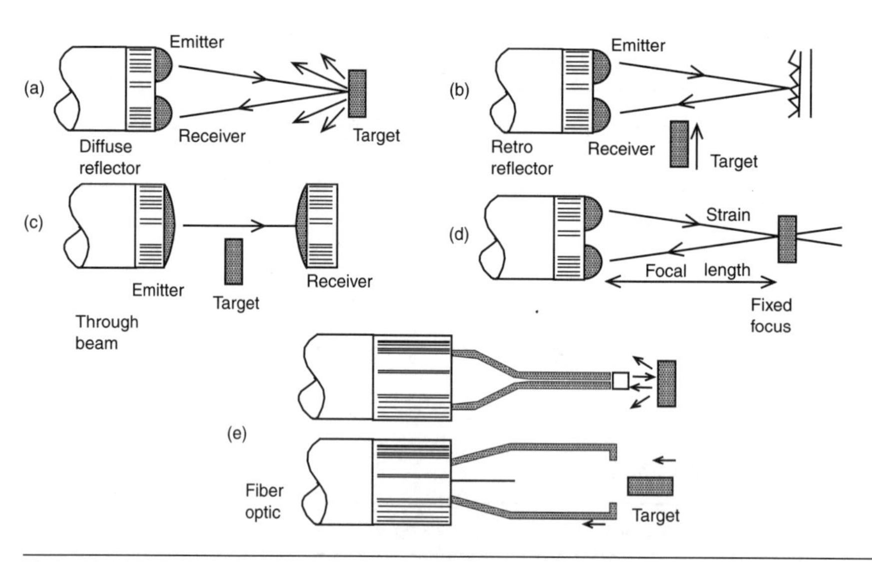

Figure 8.16 Photoelectric proximity sensor configurations: (a) diffuse reflector; (b) retro reflector; (c) through beam, (d) fixed focus, and (e) fiber optics.

(e) fiber optics. These are shown in Figure 8.16 and are briefly discussed below.

- **Diffuse reflector:** These sensors operate by an emitter sending out a pulsed beam of infrared light, which is scattered (or diffused) by the target.
- **Retro reflector:** These sensors have a special reflector, which is used to reflect the light from the emitter directly towards the detector. The target is detected by its blocking the returning beam.
- **Through beam:** As with retro reflective sensors, a through-beam target obstructs the light beam, however the detector is placed in line with the emitter and the system does not operate on reflected light.
- **Fixed focus:** These are similar in operation to diffuse reflective sensors as they rely on light being diffused from the target; however fixed focus types only switch when the object is at one set point.
- **Fiber optics:** These are particularly useful for checking the alignment of various parts of an assembly process.

Figure 8.17 shows the front of a robot base with proximity sensors clearly visible.

8.4.4 Sensors in robotics

Proximity sensors are used to sense the presence of an object close to a mechatronics device. In robotic applications, sensors are needed for obstacle detection,

Figure 8.17 A robotic base showing sensor positions. (Courtesy of The Open Automation Project.)

line tracing and direction monitoring. For obstacle detection, the sensors need to see far and only a logic response is required. Line tracing is normally required to distinguish between a white surface and a black one in order to provide guidance by the demarcation. For direction monitoring, the obvious sensor to use is a compass, which echoes the bearing of the mobile robot in real time.

8.5 Electrical strain and stress measurement

Electrical strain gages are divided into the following classes: resistance, capacitance, photoelectric, and semiconductor. For a good strain gage, some of the most important features are:

- small size and mass;
- ease of production over a range of sizes;
- robustness;
- good stability, repeatability and linearity over large strain range;
- good sensitivity;

- freedom from (or ability to compensate for) temperature effects and other environmental conditions;
- suitability for static and dynamic measurements and remote recording;
- low cost.

8.5.1 Resistance strain gages

The use of strain gages is based on the fact that the resistance of a conductor changes when the conductor is subjected to strain. The resistance of an electrically conductive material changes with dimensional changes which take place when the conductor is deformed elastically. When such a material is stretched, the conductors become longer and narrower, which causes an increase in resistance.

For many engineering materials the strain is a small number. It is often multiplied by a million and expressed in microstrain units or parts per million. Strain gages can be calibrated to measure force or pressure, and operate by detecting the variation in electrical resistance, piezoresistance, capacitance, inductance, piezoelectricity or photoelectricity. Simple resistance strain gages are generally the most popular as they have the advantage of small size and mass, and low cost. Inductive and capacitative strain gages are generally more rugged, and able to maintain calibration over an extended period of time. These are more likely to be used for permanent installations and specialized applications. Semiconductor and piezoresistive strain gages are sensitive to strain but tend to be non-linear and sensitive to temperature change.

There are several types of resistance strain gages: wire, foil and the Wheatstone bridge.

8.5.1.1 *The wire gages*

Wire gages can be divided into two types: *flat wound* and *wrap around*. In flat wound gages, the filament wire is zigzagged between two pieces of paper. With wrap around gages, the wire is wrapped around a paper support. The advantage of this is the possibility of smaller grid dimensions. The disadvantage is that they experience higher levels of creep.

8.5.1.2 *The foil gages*

Foil gages (Figure 8.18) are made from very thin metal strips (2–10 micrometers thick). They are essentially a printed circuit, and therefore require the best

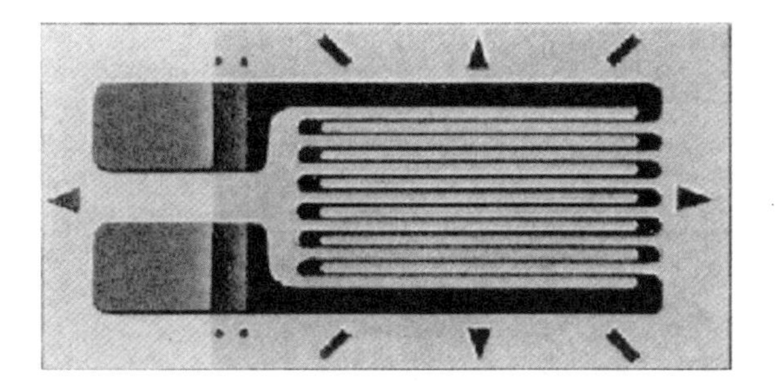

Figure 8.18 A foil resistance strain gage.

manufacturing techniques and careful handling to ensure good quality measurements. It is possible to mass-produce foil gages, whereas wire gages must still be largely manufactured by hand.

The electrical resistance of a cylindrical conductor is given by:

$$R = \frac{\rho L}{A},$$

where L is the length of the cylindrical conductor, ρ is the resistivity of the conductor material, and A is cross-sectional area of the conductor.

Taking the natural logarithm, gives

$$\ln R = \ln \rho + \ln L - \ln A. \tag{8.5}$$

Differentiating gives the following expression for the change in resistance:

$$\frac{\mathrm{d}R}{R} = \frac{\mathrm{d}\rho}{\rho} + \frac{\mathrm{d}L}{L} - \frac{\mathrm{d}A}{A}. \tag{8.6}$$

Adapting this for a foil strain gage, the cross-sectional area (width × height) of the conductor is

$$A = wh. \tag{8.7}$$

Differentiating gives the following expression for the change in area:

$$\frac{\mathrm{d}A}{A} = \frac{w \times \mathrm{d}h + h \times \mathrm{d}w}{w \times h} = \frac{\mathrm{d}h}{h} + \frac{\mathrm{d}w}{w}. \tag{8.8}$$

From the definition of Poisson's ratio

$$\begin{aligned} \frac{\mathrm{d}h}{h} &= -\nu \frac{\mathrm{d}L}{L} \\ \frac{\mathrm{d}w}{w} &= -\nu \frac{\mathrm{d}L}{L} \end{aligned} \tag{8.9}$$

$$\frac{dA}{A} = -2v\frac{dL}{L} = -2v\varepsilon_{axial} \tag{8.10}$$

$$\frac{dR/R}{\varepsilon_{axial}} = (1+2v) + \frac{d\rho/\rho}{\varepsilon_{axial}}. \tag{8.11}$$

The resistivity also changes as a result of the stresses within the material of the wire, but these variations are only slight in normal conductors and so the vast majority of the change results from the deformation. In semi-conductive materials, this situation is reversed such that the change in resistivity prevails.

The relationship between strain and resistance variation is almost linear, and the constant of proportionality is known as the *sensitivity factor* (or the K factor). The *strain sensitivity* of the gage is called the *gage factor*, and is given the symbol F. Typical values for K (and F) lie between 2 and 4, and depend on the material used. We can write:

$$\frac{dR/R}{\varepsilon_{axial}} = F. \tag{8.12}$$

8.5.1.3 ***The Wheatstone bridge***

The operation of the Wheatstone bridge is adequately covered in many other books. There are two modes of operation of the Wheatstone bridge when measuring resistance changes: (a) static balanced bridge circuit, and (b) dynamic deflection operation.

8.5.1.3.1 Static balanced bridge circuit

In this mode (Figure 8.19), R_2 and R_3 are precision resistors, and R_1 is the strain gage resistance (to be measured). A precision variable resistor (potentiometer) R_4 is adjusted until the voltage between nodes A and B is zero. Then

$$i_1R_1 = i_2R_2. \tag{8.13}$$

A high impedance voltmeter between A and B is assumed not to draw any current, hence:

$$i_1 = i_4 = \frac{V_{ex}}{(R_1 + R_4)} \tag{8.14}$$

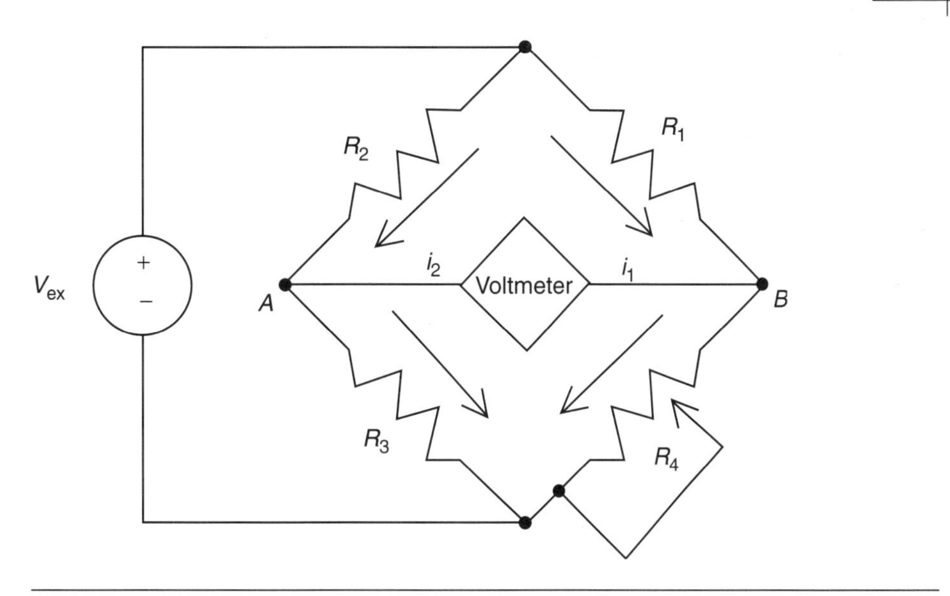

Figure 8.19 The static balanced Wheatstone bridge.

and

$$i_2 = i_3 = \frac{V_{ex}}{(R_2 + R_3)}. \tag{8.15}$$

Consequently, we have

$$\frac{V_{ex} \times R_1}{(R_1 + R_4)} = \frac{V_{ex} \times R_2}{(R_2 + R_3)}. \tag{8.16}$$

Clearing the terms, leads to

$$\frac{R_1}{R_4} = \frac{R_2}{R_3}, \tag{8.17}$$

and finally,

$$R_1 = \frac{R_2 R_4}{R_3}. \tag{8.18}$$

Knowing the values of the other resistors, it is now possible to calculate the resistance of R_1 (the strain gage resistance).

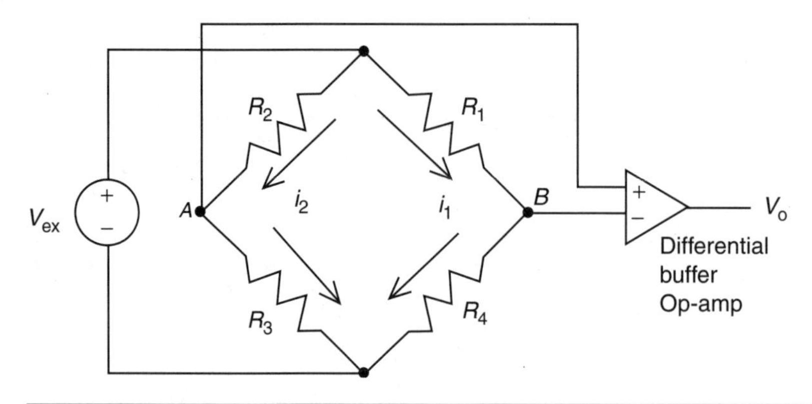

Figure 8.20 The Wheatstone bridge dynamic deflection mode.

8.5.1.3.2 Dynamic deflection operation

Again in this mode, R_2 and R_3 are precision resistors, and R_1 is the strain gage resistance for which the change is to be measured (see Figure 8.20). A precision variable resistor (potentiometer) R_4 is adjusted until the voltage between nodes A and B is zero. When the mechanical component is loaded, changes will occur in the resistance R_1 resulting in change in an output voltage from the differential buffer amplifier, expressed as

$$V_0 = i_1 R_1 - i_2 R_2 = i_1 R_4 + i_2 R_3. \tag{8.19}$$

Moreover, the excitation voltage is

$$V_{ex} = i_1(R_1 + R_4) = i_2(R_2 + R_3). \tag{8.20}$$

Solving for the output voltage yields the expression

$$V_0 = \left(\frac{R_1}{R_1 + R_4} - \frac{R_2}{R_2 + R_3}\right) V_{ex}. \tag{8.21}$$

When the bridge is balanced the amplifier voltage is zero. As mentioned, when the mechanical component is loaded, changes will occur in the resistance R_1 (ΔR_1) resulting in changes in the output voltage (ΔV_o). Consequently, we have

$$\frac{\Delta V_0}{V_{ex}} = \left(\frac{R_1 + \Delta R_1}{R_1 + \Delta R_1 + R_4} - \frac{R_2}{R_2 + R_3}\right). \tag{8.22}$$

Let us define $\alpha = \Delta V_o / V_{ex}$ and $\beta = R_2/(R_2 + R_3)$, then using these to simplify Equation 8.22 yields

$$\alpha = \left(\frac{R_1 + \Delta R_1}{R_1 + \Delta R_1 + R_4} - \beta \right) \tag{8.23}$$

or

$$\alpha = \left(\frac{1 + \Delta R_1/R_1}{1 + \Delta R_1/R_1 + R_4/R_1} - \beta \right). \tag{8.24}$$

This leads to

$$(\alpha - \beta)(1 + \Delta R_1) + (\alpha + \beta)\frac{R_4}{R_1} = \left(1 + \frac{\Delta R_1}{R_1} \right) \tag{8.25}$$

$$(1 - \alpha - \beta)\left(1 + \frac{\Delta R_1}{R_1} \right) = (\alpha + \beta)\frac{R_4}{R_1} \tag{8.26}$$

$$\left(1 + \frac{\Delta R_1}{R_1} \right) = \frac{(\alpha + \beta)}{(1 - \alpha - \beta)} \frac{R_4}{R_1} \tag{8.27}$$

$$\therefore \; \frac{\Delta R_1}{R_1} = \frac{(\alpha + \beta)}{(1 - \alpha - \beta)} \frac{R_4}{R_1} - 1. \tag{8.28}$$

8.5.1.4 ***Strain gage load cell***

Commercial load cells (see Figure 8.21) provide convenient, prepackaged, calibrated and ready to use strain gage systems. Various types of commercial strain gage-based load cells designed for different loading ranges and conditions are available. Resistance gages are commonly the transducers used in the design of load cells, although piezo-type load cells are also used. Piezoelectric transducer based load cells generally permit a wide loading range, good frequency response and high resolution. However, the inherent properties of the piezoelectric transducer limit it to dynamic loads; long-term static measurements are not practical.

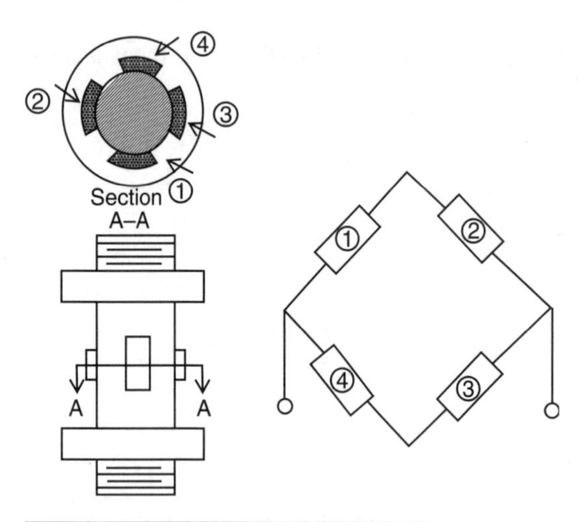

Figure 8.21 Tension-compression load cell.

8.5.2 Capacitance strain gages

Capacitance devices, which depend on geometric features, can be used to measure strain. The capacitance, C, of a simple parallel plate capacitor is given by $C = \varepsilon_0 A/t$, where A is the plate area, ε_0 is the dielectric constant, and t is the separation between plates. The capacitance can be varied by changing the plate area and the gap. The electrical properties of the materials used to form the capacitor are relatively unimportant, so capacitance strain gage materials can be chosen to meet the mechanical requirements. This allows the gages to be more rugged, providing a significant advantage over resistance strain gages.

8.5.3 Photoelectric strain gages

An extensometer (an apparatus with mechanical levers attached to the specimen) is used to amplify the movement of a specimen. A beam of light is passed through a variable slit, actuated by the extensometer, and directed to a photoelectric cell. As the gap changes, the amount of light reaching the cell varies, causing a change in current generated by the cell.

8.5.4 Semiconductor strain gages

Semiconductor strain gages are widely used today, and these differ in many aspects from the metallic wire and foil strain gages. Most importantly, they produce much greater sensitivity (10 to 50 times). This was at one time thought to herald the downfall of metallic gages, but semiconductor gages are very limited as a general-purpose gage, and so there is a place for both types in modern strain measurement.

In piezoelectric materials, such as crystalline quartz, a change in the electronic charge across the faces of the crystal occurs when the material is mechanically stressed. The piezoresistive effect is defined as the change in resistance of a material due to an applied stress and this term is used commonly in connection with semiconducting materials. The resistivity of a semiconductor is inversely proportional to the product of the electronic charge, the number of charge carriers, and their average mobility. The effect of applied stress is to change both the number and average mobility of the charge carriers. By choosing the correct crystallographic orientation and dope type, both positive and negative gage factors may be obtained. Silicon is now almost universally used for the manufacture of semiconductor strain gages.

8.6 Force measurement

Force measurement can be divided into direct and indirect comparison methods. Direct comparison is generally based on some form of a beam balance. Indirect comparison makes use of a calibrated transducer system. Force transducers have been applied to muscle testing and evaluation, as well as research systems to simulate tactile sensation in the skin.

8.6.1 Optoelectric force sensors

LEDs and photodiode light sensor packages have been used as force transducers. Such sensors often work on the principle of light occlusion in response to an applied force. A force sensing element or material is placed in the path of the LED and photodiode. The amount of light that passes through to the photodiode is dependent on the applied force.

Fiber optics has also been used in conjunction with LEDs and photodiodes. Pressure to the fiber causes a distortion and leakage of the light and a resulting attenuation. The variation in light transmission can be related to the applied force.

8.7 Time of flight sensors

Time of flight sensors emit a signal or pulse which is transmitted through a medium. The transit time through the medium is measured electronically and the distance traveled can be deduced. An applied force causes a deformation of the medium and alters the measured transit time. Sensors using ultrasound pulses through a silicone rubber medium have been developed.

8.8 Binary force sensors

Binary force sensors using discrete switching elements have been developed. The number of short-circuiting switch elements corresponds to the application of force. Increasing force causes successive switch closures. Elastomers and metals have been used as sensing elements to force switch closure.

8.9 Temperature measurement

Temperature measurement instrumentation can be divided into contact and non-contact measurement systems. Most instrumentation systems involve contact measurement. Contact temperature measurement involves heat conduction from body tissue to the sensor. A portion of the body's thermal radiation is absorbed by the sensor and converted to a useful signal. This introduces a degree of unavoidable error as heat is being removed from the tissue to warm up the sensor. However, this error can be minimized through proper sensor design.

Non-contact temperature measurement systems based on the principles of radiation thermometry have been developed quite recently for medical use. The technology for radiation thermometry has been widely used in the metallurgical industry previous to its application for medical purposes. Non-contact temperature measurement offers the advantages of no heat absorption from the body or tissue being monitored, and temperature measurement can be made from a distance.

The different types of thermometers that are commonly used are:

- Mechanical
- Electrical
- Semiconductor devices and thermal sensors

8.9.1 The liquid expansion thermometer

The volumetric expansion of liquids and solids can be used for temperature measurement. Mercury- and alcohol-filled thermometers work under this principle.

8.9.2 The bimetallic strip thermometer

Two different metals with different coefficients of thermal expansion are bonded together. As a change in temperature occurs the unequal expansion to the two metals will cause the bimetal strip to curl. If one end of the metal strip is fixed then the other end will be displaced in response to temperature changes. Bimetal strips can be fabricated into coils, spirals and disks.

8.9.3 The gas thermometer

A gas filled thermometer operating under the principles of the ideal gas law ($PV = nRT$) can be used for temperature measurement. If pressure is held constant the change in gas volume can be used to indicate temperature.

8.9.4 The resistance temperature detector (RTD)

The resistance of metal wires changes with temperature. Change in resistance can be expressed by the following equations:

$$R_t = R(1 + AT + BT^2 + CT^3) \tag{8.29}$$

$$R_t = R_0(1 + \alpha(T - T_0)), \tag{8.30}$$

where A, B, C and α are constants based on material properties of the RTD, T is temperature and R_t is the resistance, R_0 is the reference resistance, and T_0 is a reference temperature.

Platinum, nickel and copper wires are used in RTD devices. Platinum has a linear curve in the range of $-200°C$ to $+850°C$, and is accurate to $0.01°C$. Several sources of error are possible for RTD devices: they include poor wire contact, thermoelectric electromotive forces and self heating of the wire from current flow through the wire. RTDs can be designed using wires and thin or thick films. Very small probes down to the micrometer scale can be built.

8.9.5 The thermocouple

An electromotive force exists at the junction formed between two dissimilar metals. This phenomenon is known as the Seebeck effect, named after its discoverer. The Seebeck effect is actually the combined result of two other phenomena named after their discoverers: Thomson and Peltier. Thomson observed the existence of an electromotive force due to the contact of two

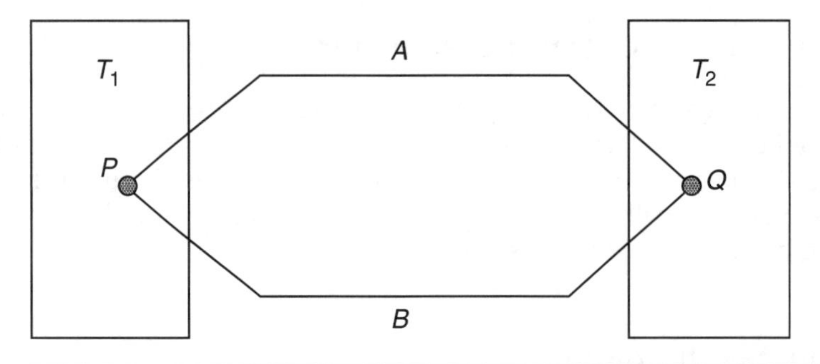

Figure 8.22 Typical thermocouple circuit.

dissimilar metals and the junction temperature. Peltier discovered that temperature gradients existing along the conductors in a circuit generated an electromotive force. The Thomson effect is normally much smaller in magnitude than the Peltier effect and can be minimized and disregarded through correct thermocouple design.

A typical thermocouple circuit is shown in Figure 8.22. Two conductors and two junctions are required. One of the junctions is kept at a known or reference temperature, typically known as the cold or reference junction. The other junction is known as the hot or measuring junction. The relationships between temperature and voltage are determined empirically and can be expressed as:

$$V = \alpha(T_1 - T_2) + \gamma(T_1^2 - T_2^2), \tag{8.31}$$

where α and γ are constants for each of the two materials.

The sensitivity of the thermocouple system, also known as the thermoelectric power, is obtained by taking the derivative of the voltage with respect to T_1:

$$S = \alpha + 2\gamma T_1 \tag{8.32}$$

8.9.6 Semiconductor devices and integrated circuit thermal sensors

Semiconductor junction diodes have thermal sensing properties, which can be expressed by the following equations:

$$I = I_0(e^{qV/kT} - 1) \tag{8.33}$$

$$I_0 = T^{3/2}e^{-qVg/kT}, \tag{8.34}$$

where k is Boltzmann's constant, I is current, T is temperature, V is voltage and q is a constant.

8.9.6.1 *The diode thermometer*

Diodes, transistors and integrated circuits made of silicon or germanium can be used as thermal sensors over the range of – 40°C to +150°C. Output signals can be in the form of either current or voltage. Integrated circuit thermal sensors have the advantage of not needing any additional interfaces and can be connected directly to signal processing circuits.

8.9.6.2 *The thermistor*

Thermistors are temperature sensitive resistors made of semiconductor materials. Oxides of manganese, nickel and cobalt have been used in the fabrication of thermistors. Thermistors can have a negative temperature coefficient (NTC) or positive temperature coefficient (PTC). NTC thermistors are generally made from metallic oxides, while PTC thermistors are formed from barium and strontium titanate mixtures.

The relationship of resistance to temperature is of the form:

$$R_t = R_0 e^{\beta/t}, \tag{8.35}$$

where R_t is the resistance of the thermistor, R_0 is a reference resistance of the thermistor, β is a constant based on material properties, and t is the temperature.

The relationship between thermistor resistance and temperature is exponential and therefore non-linear. Various techniques for linearization of a portion (over a specific temperature range) of the thermistor output are employed. However, the sensitivity of the device may be sacrificed as a result of linearization. Thermistors can be configured with operational amplifiers to give current or voltage outputs. The physical size of thermistors start in the sub-millimeter range. Temperature response times range from milliseconds.

8.10 Pressure measurement

All force transducers can be adapted to measure pressure if an area can be defined and held constant (pressure = force/area). Pressure transducer design often uses a secondary force or displacement transducer to measure elastic deformation or displacement of a mechanical element such as a diaphragm, bellows or optical fiber. Some pressure transducer designs are:

- Gravitational: liquid columns, pistons, weights and loose diaphragms;
- Direct acting elastic: elastic diaphragms, bellows and loaded tubes;
- Indirect-acting elastic: piston with elastic restraining member.

8.10.1 Pressure gradient flow transducers

Bernoulli's equation for the flow of incompressible fluids between two points can be written as:

$$(P_1 - P_2)/\rho = (V_2^2 - V_1^2)/2g_c + (Z_2 - Z_1)g/g_c, \tag{8.36}$$

where P is pressure, ρ is the density, V is the linear velocity, g_c is a dimensional constant, Z is the elevation and g is the acceleration due to gravity. The equation assumes there is no mechanical work and no heat transfer between points 1 and 2. If the effects of change in elevation are eliminated, it can be seen that a change in pressure is proportional to the velocity change. This provides the basis for many flow measuring devices which relate fluid pressure change in a vessel to velocity.

Figure 8.23 shows an arrangement for measuring flow rate using the differential pressure at two points in a fluid stream. In steady state flow conditions the relationship between pressure drop and fluid flow rate can be described by

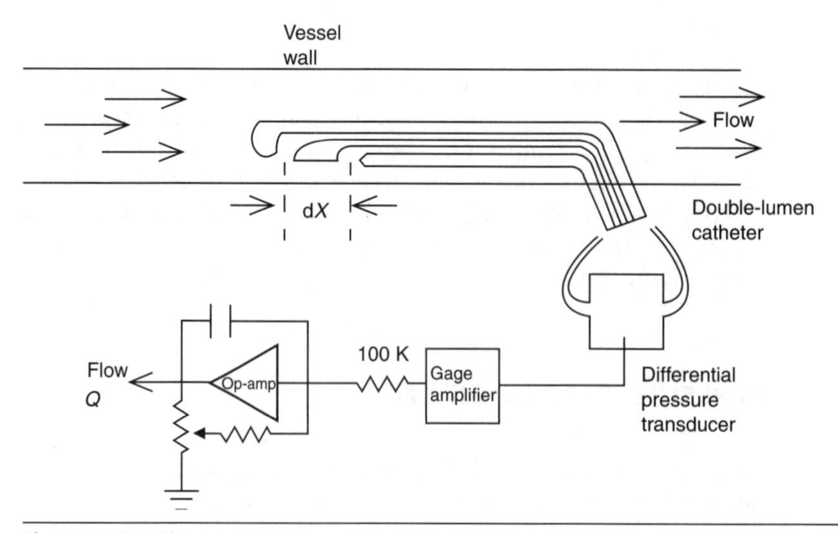

Figure 8.23 Differential pressure flow rate measuring system.

the following equation:

$$\frac{\Delta P}{\Delta X} = \frac{-12.8\mu Q}{g\pi a^4}, \qquad (8.37)$$

where ΔP is the differential pressure (cm H_2O), ΔX is the distance between the two sensing tubes in the stream of the flow (cm), μ is the viscosity (poise), $g = 980\,cm\,s^{-2}$, a is the inner diameter of the sensing tube or lumen, and Q is the flow rate ($cm^3\,s^{-1}$].

The main disadvantage of the differential pressure method in biological systems is that the values of ΔX and ΔP tend to be small and are prone to background noise and interference. In addition the introduction of sensing tubes in a flow may appreciably change the characteristics being monitored.

Problems

Q8.1 A new experimental strain gage is mounted on a 6 mm diameter steel bar in the axial direction. The gage has a measured resistance of 360 Ω. When the bar is loaded with 2 kN in tension, the gage resistance increases by 0.03 Ω. Determine the gage factor of the transducer. Take the modulus of elasticity E, of steel to be 209 GPa.

Q8.2 A 20 mm diameter steel bar is loaded in tension with an axial load of 200 kN. The strain gage resistance, which is mounted in the bar in the axial direction, has a resistance of 200 Ω and gage factor of 1.85. Determine the change in resistance of the strain gage.

Q8.3 The strain gage in Q8.2 is placed in one branch of a Wheatstone bridge with the other three branches having the same resistance value (each 150 Ω). Determine for the strained state, the output voltage of the bridge, V_o. (Thermocouple data is given in Table 8.5.)

Table 8.5 Thermocouple data

Polynomial order	*Type J*
c_0	−0.048 868 3
c_1	$1.987\,31 \times 10^4$
c_2	$-2.186\,15 \times 10^5$
c_3	$1.156\,92 \times 10^7$
c_4	$-2.649\,18 \times 10^8$
c_5	$2.018\,44 \times 10^9$

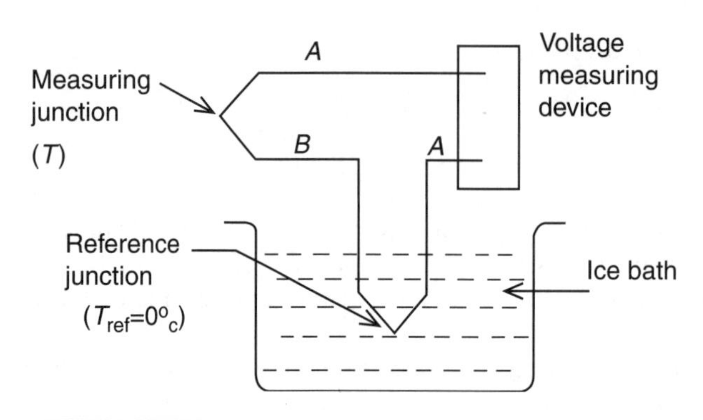

Figure 8.24 Thermocouple circuit for Q8.4.

Q8.4 In a J-type thermocouple used in a standard two-junction thermocouple configuration with 0°C reference temperature (see Figure 8.24), determine the voltage that would result for an output temperature of 300°C. (Thermocouple data is given in Table 8.5.)

Q8.5 If a J-type thermocouple is used in a standard two-junction thermocouple configuration with 50°C reference temperature, determine the temperature that would correspond to a measured voltage of 40 mV. (Thermocouple data is given in Table 8.5.)

Further reading

[1] Beckwith, T.G., Buck, N.L. and Marangoni, R.D. (1982) *Mechanical Measurements* (3rd. ed.), Reading, MA: Addison-Wesley.
[2] Doeblin, E. (1990) *Measurement Systems Applications and Design* (4th. ed.), New York: McGraw-Hill.
[3] Figliola, R. and Beasley, D. (1995) *Theory and Design of Mechanical Measurements* (2nd. ed.), New York: John Wiley.
[4] Loughlin, C. (1993) *Sensors for Industrial Inspection*, Dordrecht, The Netherlands: Kluwer Academic.
[5] Welkowitz, W., Deutsch, S. and Akay, M. (1992) *Biomedical Instruments, Theory and Design* (2nd. Ed.), San Diego, CA: Academic Press.

Internet resources

- Open Automation Project robot base: http://oap.sourceforge.net/prototype .php

- http://mechatronics.mech.northwestern.edu/mechatronics/design_ref/sensors/encoders.html
- http://www.dur.ac.uk/~des0rhs/gages.html
- http://bits.me.berkeley.edu/beam/sg_2a.html
- http://uhaweb.hartford.edu/biomed/gateway/Electrodes.html

CHAPTER 9

Electrical actuator systems

Chapter objectives

When you have finished this chapter you should be able to:

- distinguish between the roles of solenoids and relays;
- distinguish between shunt and separately excited d.c. motors;
- control d.c. motor speed by adding an armature resistance;
- control d.c. motor speed control by adjusting armature voltage;
- control d.c. motor speed control by pulse width modulation;
- understand how a stepper motor works;
- understand the hardware for stepper motor speed control;
- choose motors based on some practical guidelines.

9.1 Introduction

So far, we have concentrated on electronic components, sensors and related signal-processing methods which on their own, cannot produce mechanical actions or motion. While a *sensor* is a device that can convert mechanical energy to electrical energy, a *transducer* (in this case, it is often called an actuator) is a device that can convert electrical energy to mechanical energy. Actuators are the devices which are used to produce motion or action, such as linear motion or angular motion. Some of the important actuators used in mechatronic systems are solenoids, electric motors, hydraulic pumps, and hydraulic cylinders and pneumatic cylinders. These actuators are instrumental in moving physical objects in mechatronic systems. Before discussing these actuators in more detail, it is appropriate to briefly cover the basic principles of moving-iron transducers.

9.2 Moving-iron transducers

One important class of electro-magneto-mechanical transducer is that of moving-iron transducers, which form the basis for electromagnets, solenoids, and relays. The simplest example of a moving-iron transducer is the electromagnet shown in Figure 9.1, in which the U-shaped element is fixed and the bar is moveable.

Table 9.1 summarizes the analogies that exist between electric and magnetic circuits.

In order for a mass to be displaced, some work needs to be done. This work corresponds to a change in the energy stored in the electromagnetic field, which causes the mass to be displaced. Referring to Figure 9.1, let the magnetic force acting on the bar be designated as f_e and let the displacement equal x, and let the net work due to the electromagnetic field be W_m. The force acting to pull the bar toward the electromagnet structure will be opposite relative to f_e. It can be easily shown that

$$f = -f_e = -\frac{dW_m}{dx}. \tag{9.1}$$

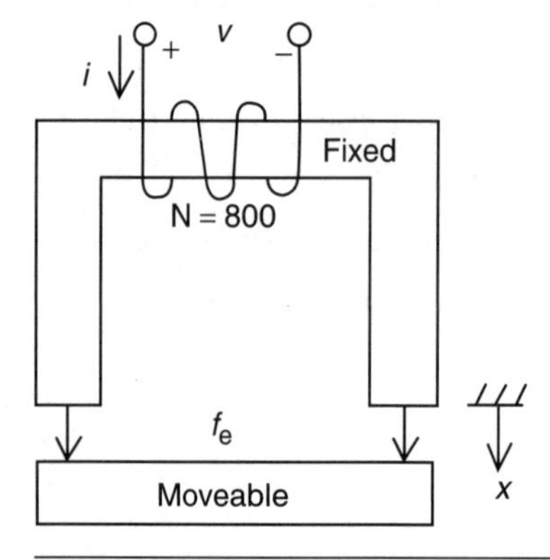

Figure 9.1 The electromagnet.

Table 9.1 Electric and magnetic analogies

Electrical quantity	*Magnetic quantity*
Electrical field intensity E (V/m)	Magnetic field intensity H (A-turns m^{-1})
Voltage v (V)	Magneto-motive force F (A-turns)
Current i (A)	Magnetic flux φ (Wb)
Current density J (A/m^2)	Magnetic flux density B (Wb m^{-2})
Resistance R (Ω)	Reluctance $\Re = l/\mu A$ (A-turns Wb^{-1})
Conductivity σ (1/Ω-m)	Permeability μ (Wb A-m^{-1})

The energy stored in a magnetic structure is given by

$$W_m = \frac{\phi \times F}{2}. \tag{9.2}$$

The flux and the magneto-motive force F (mmf) are related by the expression

$$\phi = \frac{N \times i}{\Re} = \frac{F}{\Re}. \tag{9.3}$$

The stored energy can be related to the reluctance of the structure according to the following equation

$$W_m = \frac{\phi^2 \times \Re(x)}{2}. \tag{9.4}$$

Then using Equations 9.1 and 9.4, the magnetic force acting on moving iron is:

$$f = -\frac{dW_m}{dx} = -\frac{\phi^2}{2} \times \frac{d\Re(x)}{dx}. \tag{9.5}$$

Equation 9.5 is the basis for the operation of solenoids and relays.

9.3 Solenoids

One of the more common practical applications of the moving-iron transducer discussed in this section is the solenoid shown in Figure 9.2, in which the movable part is connected to springs to return it. Solenoids find application in a variety of electrically controlled valves.

9.4 Relays

Another electromechanical device that finds common industrial practical applications is the relay (Figure 9.3). The relay is essentially an electromechanical switch that permits the opening and closing of electrical contacts by means of an electromagnetic structure similar to the moving-iron transducer. The relay works as follows. When the push button is pressed, an electric current flows through the coil and generates a field in the magnetic structure. The resulting force draws the moveable part toward the fixed part, thereby causing an electric contact to be

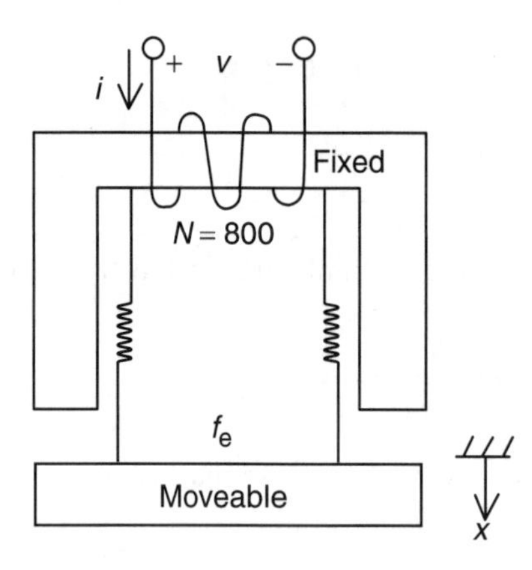

Figure 9.2 The solenoid.

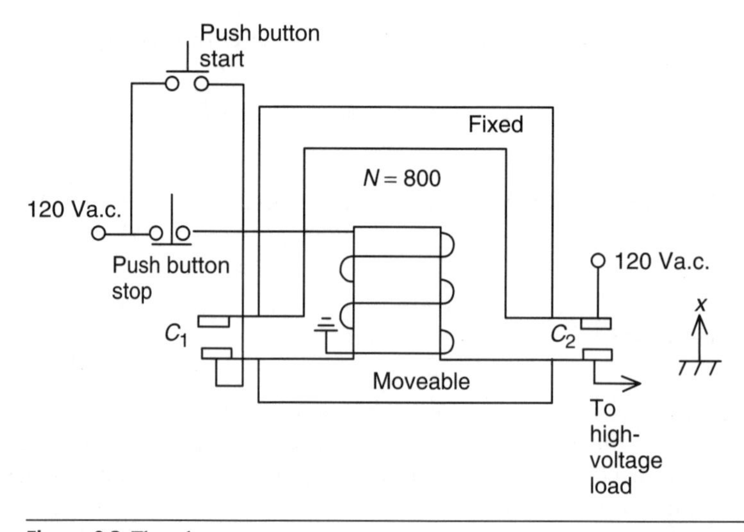

Figure 9.3 The relay.

made. The advantage of a relay is that a small current can be used to control the opening and closing of a circuit that carries large currents. Industrial applications include power switches and electrochemical control elements.

9.5 Electric motors

Electric motors are prominent actuators that find applications in virtually all electromechanical systems. When trying to physically move things with microcontrollers, there are basically three kinds of motor that are most useful: d.c. motors, servomotors, and stepper motors.

Most motors work on the electrical principle of induction. When an electric current flows through a wire, it generates a magnetic field around the wire. Conversely, by placing a charged coil of wire in an existing magnetic field (say, between two magnets), the coil will be attracted to one magnet and repelled by the other, or vice versa, depending on the current flow. When current flows through a wire, the higher the current, the greater the magnetic field generated; therefore, the greater the attraction or repulsion. The coil is mounted on a spinning shaft in the middle of the motor. As one magnet alternately attracts the coil and the other repulses the coil, it spins from one magnet to the other resulting in circular motion. These concepts are captured in Figures 9.4 and 9.5.

All inductive loads (like motors, electromagnets, and solenoids) work on this same principle; when a magnetic field is induced by current through a wire, it is used to attract or repulse a magnetic body. By spinning a wire in an existing magnetic field, the field induces a current in the wire. When a motor spinning is

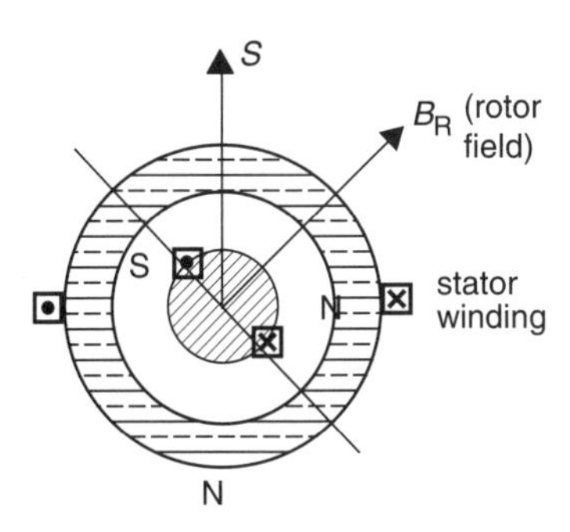

Figure 9.4 A rotating electric motor.

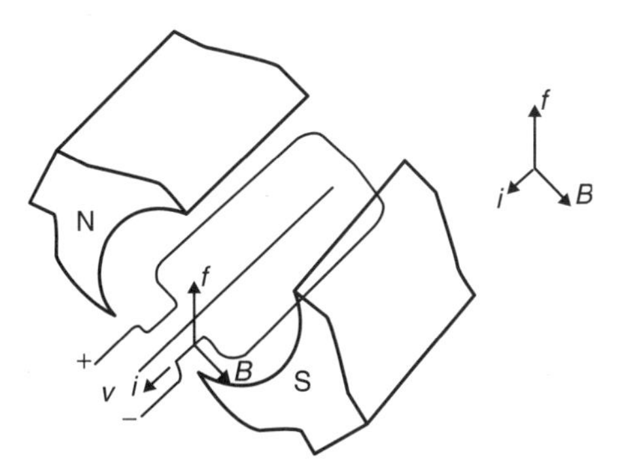

Figure 9.5 Stator and rotor electric fields and the forces acting on a rotating machine.

turned off, the fact that the motor's coil is spinning in a magnetic field will generate a current in the wire for a brief amount of time. This current comes back in the reverse direction of the current flow generated to run the motor. It is called blowback, or back voltage, and it can cause damage to the motor's electronics. A diode in line with the motor gives protection.

9.6 Direct current motors

The direct current (d.c.) motor is the simplest of the motors discussed here. It works on exactly the principle outlined in the previous section. A d.c motor has two terminals, and when d.c. flows in one terminal (the other being grounded), the motor spins in one direction. When current flows in the other terminal (and the first is grounded), the motor spins in the opposite direction. That is, by switching the polarity of the terminals, the direction of the motor is reversed. The motor's speed is controlled by varying the current supplied. Specific techniques for doing these tasks are discussed later.

Many appliances and power tools used at home, such as washers, circular saws and blenders, are d.c. machines. They are very popular because:

- they are usually very fast, spinning at several thousand revolutions per minute (rpm);
- they are simple to operate.
- their starting torque is large, which is the main reason for using them in several traction applications;
- in a special form, they can be used with either an a.c. or d.c. supply.

Electric motors have a variety of speed–torque characteristics during steady-state and transient operation. For a given drive application, in the past, engineers often selected motors with characteristics matching the needed operation. Because of advances in power electronic devices and circuits, such stringent restrictions no longer exist. The characteristics of most motors can now be altered to match the desired performance when external power converters are used and advanced control strategies are employed.

There are three main types of electric motors: d.c., induction, and synchronous. Although there are several other types of motor, such as the brushless, inductance, linear, and stepper, they all share common features with these main types of electric motor. For example, the brushless machine can be considered to be a special form of a synchronous machine switched to imitate a d.c. motor. The linear induction motor is also considered a special form of the induction motor.

9.6.1 Fundamentals of d.c. motors

Figure 9.6 shows the main components of the d.c. machine: field circuit, armature circuit, commutator, and brushes. The field is normally an electric magnet fed by a d.c. power source. In small machines, the field is often a permanent magnet.

The armature circuit is composed of the windings, commutator, and brushes. The windings and the commutator are mounted on the rotor shaft.

The brushes are mounted on the stator and are stationary, but in contact with the rotating commutator segments. The rotor windings are composed of several coils; each has two terminals connected to the commutator segments on opposite sides. The commutator segments are electrically isolated from one another. The segments are exposed, and the brushes touch two opposing segments. The brushes allow the commutator segments to be connected to an external d.c. source.

Figure 9.7 illustrates the operation of a typical d.c. machine. The stator field produces flux, φ, from the north pole to the south pole. The brushes touch the

Figure 9.6 The main components of a d.c. machine. (Source: www.csd.ijs.si/applications/doptimel/Euromicro2001.pdf).

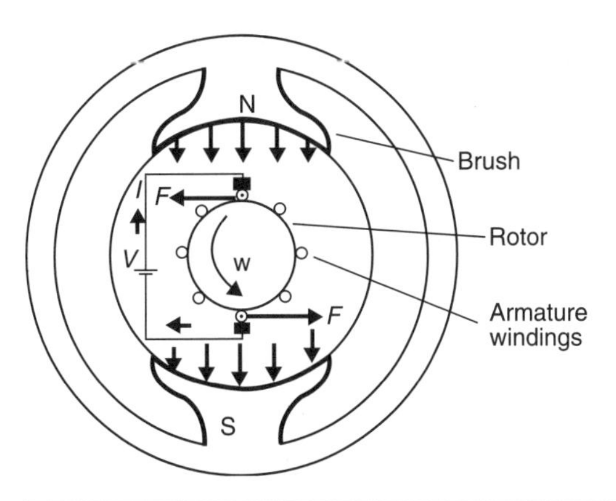

Figure 9.7 Operation of a typical d.c. machine.

terminals of the rotor coil under the pole. When the brushes are connected to an external d.c. source of potential, V, a current, I, enters the terminal of the rotor coil under the north pole and exits from the terminal under the south pole. The presence of the stator flux and rotor current produces a force, F, on the coil known as the Lorentz force. The direction of F is shown in the figure. This force produces torque that rotates the armature counterclockwise. The coil that carries the current moves away from the brush and is disconnected from the external source. The next coil moves under the brush and carries the current I. This produces a continuous force F and continuous rotation. We note that the function of the commutator and brushes is to switch the coils mechanically.

The rotation of the machine is dependent on the magneto-motive force, MMF, of the field circuit, which is described by $\text{MMF} = NI$, where N is the number of turns and I is the field current. The desired MMF can be achieved by the design of the field windings. There are basically two types of field windings: the first has a large number of turns and low current, and the second type has a small N and high current. Both types achieve the desired range of MMF. Actually, any two different windings can produce identical amounts of MMF if their current ratio is inversely proportional to their turns ratio. The first type of winding can handle higher voltage than the second type. Moreover, the cross section of the wire is smaller for the first type since it carries a smaller current.

Direct current motors can be classified into four groups based on the arrangement of their field windings. Motors in each group exhibit distinct speed–torque characteristics and are controlled by different means. These four groups are:

- **Separately excited machines:** The field winding is composed of a large number of turns with small cross section wire. This type of field winding is designed to withstand the rated voltage of the motor. The field and armature circuits are excited by separate sources.
- **Shunt machine:** The field circuit is the same as that for separately excited machines, but the field winding is connected in parallel with the armature circuit. A common source is used for the field and armature windings.
- **Series machines:** The field winding is composed of a small number of turns with a large cross section wire. This type is designed to carry large currents and is connected in series with the armature winding.
- **Compound machines:** This type uses the shunt and series windings.

9.6.2 Separately excited motors

The equivalent circuit of a separately excited motor is shown in Figure 9.8. The motor consists of two circuits: the field and the armature. The field circuit is mounted on the stator of the motor and is energized by a separate d.c. voltage

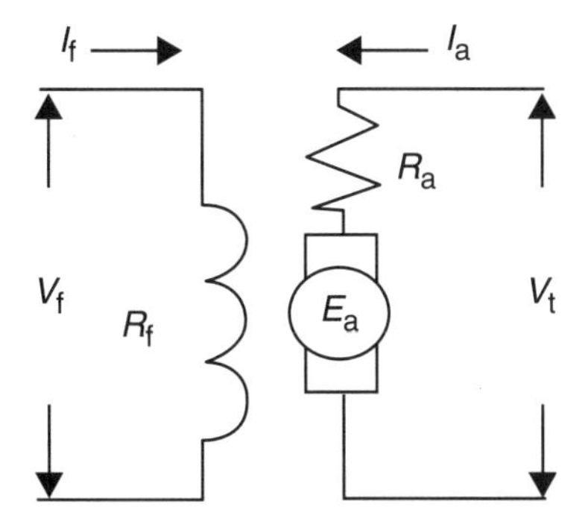

Figure 9.8 Separately excited motor operation.

source, V_f. The field has a resistance R_f and a high inductance L_f. The field inductance has no impact in the steady-state analysis, since the source is a d.c. type. The field current I_f is represented by

$$I_f = \frac{V_f}{R_f}. \tag{9.6}$$

For small motors (up to a few hundred watts), the field circuit is a permanent magnet. In such cases, the flux of the field is constant and cannot be adjusted.

The armature circuit, mounted on the rotor, is composed of a rotor winding and commutator segments. An external voltage source, V_t, is connected across the armature to provide the electric energy needed to drive the load. The source is connected to the armature circuit via the commutator segments and brushes. The direction of the current in the armature winding is dependent on the location of the winding with respect to the field poles.

Relative to the field circuit, the armature carries a much higher current. Therefore, the wire cross section of the armature winding is much larger than that for the field circuit. The armature resistance, R_a, is, therefore, much smaller than the field resistance R_a, being in the range of a few ohms (smaller for larger wattage motors). The field resistance is a hundred times larger than the armature resistance. The field current is usually about 1–10 percent of the rated armature current. The field voltage is usually in the same order of magnitude as the armature voltage.

The back electromagnetic force, E_a, (shown in Figure 9.8) is equal to the voltage of the source minus the voltage drop due to the armature resistance. The armature current I_a can then be expressed by

$$I_a = \frac{V_t - E_a}{R_a}. \tag{9.7}$$

The product $I_a \times E_a$ represents the developed power, P_d. In mechanical representation, the developed power is also equal to the developed torque multiplied by the angular speed.

$$P_d = E_a I_d = T_d \omega, \tag{9.8}$$

where ω is the angular speed in radians per second.

P_d is equal to the output power consumed by the mechanical load plus rotational losses (frictional and windage). Similarly, the developed torque, T_d, is equal to the load torque plus the rotational torque.

Using Faraday's law and the Lorentz force expressions, the relationships that govern the electromechanical motion are

$$e = Blv \tag{9.8A}$$

$$F = Bli, \tag{9.8B}$$

where B is the flux density, l is the length of a conductor carrying the armature current, v is the speed of the conductor relative to the speed of the field, and i is the conductor current. F and e are the force and the induced voltage on the conductor, respectively. If we generalize these equations by including all conductors, using the torque expression instead of the force F ($T_d \sim F$), and using the angular speed instead of v ($\omega \sim v$), we can rewrite E_a and T_d as

$$E_a = K\phi\omega \tag{9.9}$$

$$T_d = K\phi I_a, \tag{9.10}$$

where φ is the flux, which is almost proportional to I_f for separately excited motors. The constant K is dependent on design parameters such as the number of poles, number of conductors, and number of parallel paths.

The speed–torque equation can be obtained by first substituting I_a of Equation 9.7 into Equation 9.10.

$$T_d = K\phi \frac{(V_t - E_a)}{R_a}, \tag{9.11}$$

then, by substituting E_a of Equation 9.9 into Equation 9.11, we get

$$T_d = K\phi \frac{(V_t - K\phi\omega)}{R_a} \tag{9.12}$$

$$\omega = \frac{V_t}{K\phi} - \frac{R_a}{(K\phi)^2} T_d. \tag{9.13}$$

The speed–current equation can be obtained if $T_d/K\varphi$ of Equation 9.13 is replaced by I_a.

$$\omega = \frac{V_t}{K\phi} - \frac{R_a I_a}{K\phi}. \tag{9.14}$$

If we ignore the rotational losses, the developed torque, T_d, is equal to the shaft torque, and the no-load armature current is equal to zero. Hence, the no-load speed can be calculated from Equation 9.13 or 9.14 by setting the armature current and load torque equal to zero.

$$\omega_0 = \frac{V_t}{K\phi} \tag{9.15}$$

In reality, the mass of the drive system and the rotational losses are the base load of the motor. The no-load speed, ω_0, is therefore slightly smaller than the value computed in Equation 9.15. Nevertheless, Equation 9.15 is an acceptable approximation.

In the steady state, the T_d is equal to the load torque T_m. At a given value of T_m, the speed of the motor drops by an amount $\Delta\omega$ that is equal to the second term on the right side of Equation 9.13.

$$\Delta\omega = \frac{R_a}{(K\phi)^2} T_d. \tag{9.16}$$

The speed of the motor can then be expressed by using the no-load and speed drop.

$$\omega = \omega_0 - \Delta\omega. \tag{9.17}$$

Figures 9.9 and 9.10, show the speed–torque and speed–current characteristics when the field and armature voltages are kept constant.

For large motors (greater than 0.75 kW), the armature resistance, R_a, is very small, because the armature carries higher currents, and the cross section of the

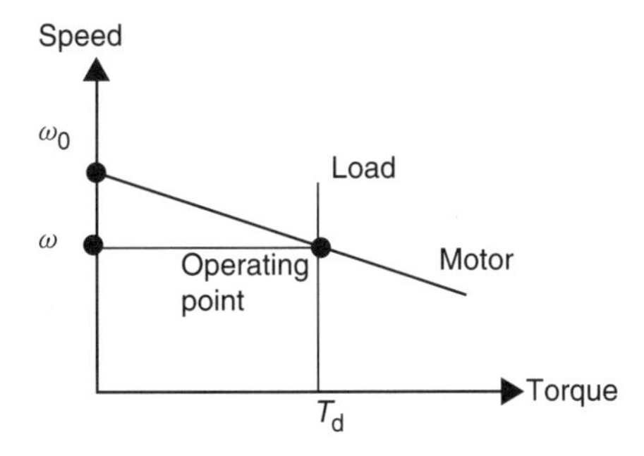

Figure 9.9 Speed–torque characteristic of a d.c. separately excited motor.

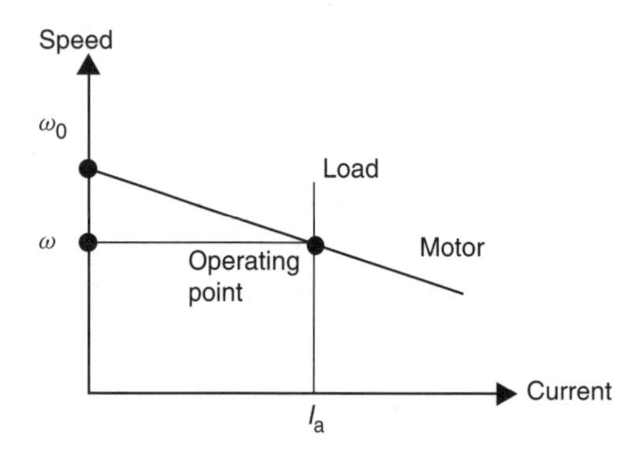

Figure 9.10 Speed–current characteristic of a d.c. separately excited motor.

wire must then be larger than normal. For these motors, the speed drop, $\Delta\omega$, is small, and the motors can be considered constant-speed machines.

The developed torque at starting, T_{st}, and the starting armature current, I_{st}, can be calculated from Equations 9.13 and 9.14 by setting the motor speed to zero.

$$T_{st} = K\phi \frac{V_t}{R_a} \tag{9.18}$$

$$I_{st} = \frac{V_t}{R_a} \tag{9.19}$$

Equations 9.18 and 9.19 provide important information about the starting behavior of the d.c. separately excited motor. As stated earlier, R_a is usually small. Hence, the starting torque of the motor is very large when the source voltage is equal to the rated value. This is an advantageous feature, and is highly desirable when motors start under heavy loading conditions. A problem, however, will arise from the fact that the starting current is also very large, as seen in Equation 9.19. Large starting currents might have a damaging effect on the motor windings. Excessive currents flowing inside a winding will result in large losses due to the winding resistance. These losses, when accumulated over a period of time, may result in excessive heat that could melt the insulations of the winding, causing an eventual short circuit. This is illustrated by the next example.

EXAMPLE 9.1

A d.c. separately excited motor has the following specification:

$$K\varphi = 4.0\,\text{V s (volt second)}$$
$$V_t = 12.0\,\text{V}$$
$$R_a = 2.00\,\Omega$$
$$I_a = 5.0\,\text{A (armature current at full load)}$$

Determine:

(a) the rated torque;
(b) the starting torque;
(c) the starting current at full voltage;
(d) the starting speed; and
(e) the speed at the rated torque condition.

Solution

(a) Rated torque, $T_d = K\varphi I_a = 4 \times 5 = 20\,\text{Nm}$
(b) Starting torque, $T_{st} = K\varphi(V_t/R_a) = (4 \times 12)/2 = 24\,\text{Nm}$
(c) Starting current, $I_{st} = V_t/R_a = 12/2 = 6\,\text{A}$
(d) Starting speed, $\omega_0 = V_t/K\varphi = 12/4 = 3\,\text{rad s}^{-1}$
(e) Speed at rated torque, $\omega = (V_t - R_a I_a)/K\varphi = 0.5\,\text{rad s}^{-1}$

One important parameter missing in this example is the inductance of the armature winding. This inductance reduces the value of the current during transient conditions such as starting or braking. Nevertheless, the starting current under full voltage conditions is excessively large, and methods must be implemented to bring this current to a lower and safer value.

By examining Equation 9.19, the starting current can be reduced by lowering the terminal voltage or inserting a resistance in the armature circuit. Let us assume that the starting current must be limited to five times the rated value. This can be achieved by reducing the terminal voltage at starting to

$$V_{st} = I_{st} R_a = 5 \times 6 \times 2 = 60.0\,\text{V}. \qquad (9.19\text{A})$$

Figure 9.11 illustrates the effect of reducing the terminal voltage during starting. When the voltage is reduced from V_{t1} to V_{t2} the slope of the speed–current characteristic remains unchanged, whereas the no-load speed is reduced. Note that the starting current I_{st2} is less than I_{st1}.

Another method to reduce the starting current is to add a resistance R to the armature circuit.

$$\begin{aligned} R + R_a &= \frac{V_t}{I_{st}} \\ R &= \frac{V_t}{I_{st}} - R_a. \end{aligned} \qquad (9.19\text{B})$$

Figure 9.12 illustrates the effect of reducing the starting current by adding a resistance to the armature circuit. The resistance increases the slope of the speed–current characteristic but keeps the no-load speed unchanged.

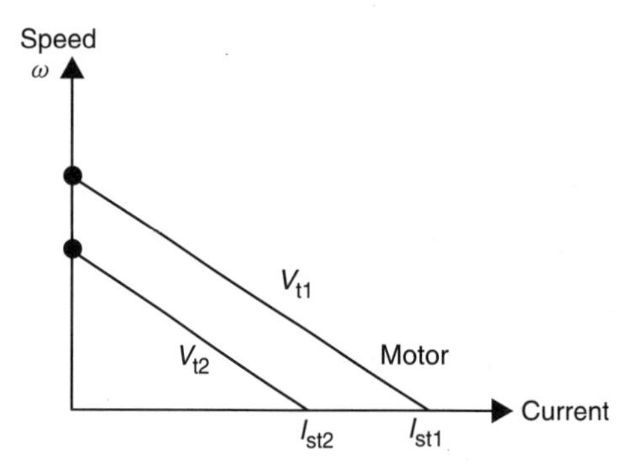

Figure 9.11 Effect of reducing source voltage on starting.

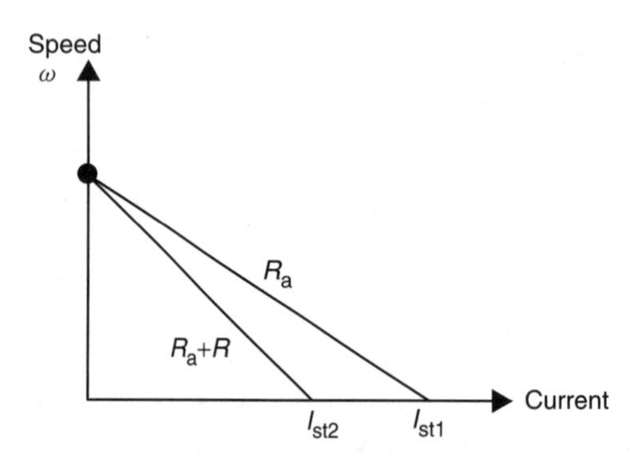

Figure 9.12 Effect of inserting a resistance in the armature circuit on starting.

9.6.3 Shunt motors

A shunt motor has its field winding connected across the same voltage source used for the armature circuit, as shown in Figure 9.13. The source current, I, is equal to the sum of the armature current, I_a, and the field current, I_f. The shunt motor exhibits characteristics identical to those of the separately excited motor.

EXAMPLE 9.2

See website for downloadable MATLAB code to solve this problem

A d.c. shunt motor has the following specification:

$$K\varphi = 4.0\,\text{V s}$$

$$V_t = 12.0\,\text{V}$$

$$R_f = 150\,\Omega$$

$$R_a = 2\,\Omega$$

$$I_a = 5.0\,\text{A}$$

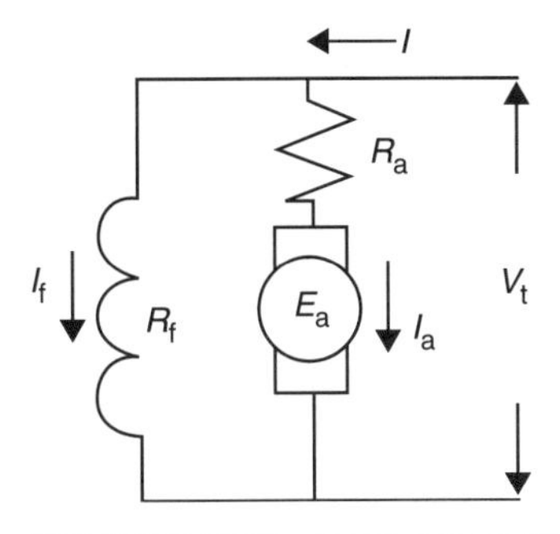

Figure 9.13 Shunt motor operation.

Determine:

(a) the rated torque;
(b) the starting torque;
(c) the starting current at full voltage;
(d) the starting speed; and
(e) the speed at the rated torque.

Solution

Equivalent resistance, $R_e = (2 \times 150)/(2 + 150) = 1.9737\ \Omega$

(a) Rated torque, $T_d = K\varphi I_a = 20\,\text{Nm}$
(b) Starting torque, $T_{st} = K\varphi(V_t/R_e) = 24.32\,\text{Nm}$
(c) Starting current, $I_{st} = V_t/R_e = 6.08\,\text{A}$
(d) Starting speed, $\omega_0 = V_t/K\varphi = 12/4 = 3\,\text{rad s}^{-1}$
(e) Speed at rated torque, $\omega = (V_t - R_e I_a)/K\varphi = 0.5329\,\text{rad s}^{-1}$

9.6.4 Series motors

The field winding of a series motor is connected in series with the armature circuit, as shown in Figure 9.14. There are several distinct differences between the field winding of a series machine and that of a shunt machine; among them are

- The series field winding is composed of a smaller number of turns than the shunt field winding.

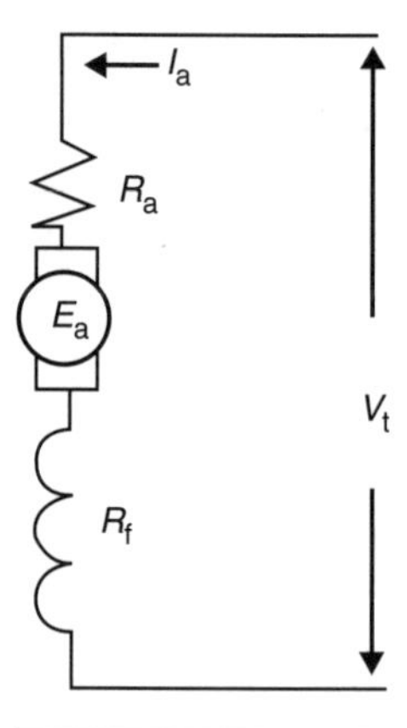

Figure 9.14 Series motor operation.

- The current of the series winding is equal to the armature current, whereas the current of the shunt field is equal to the supply voltage divided by the field resistance. Hence, the series field winding carries a much larger current than the shunt field winding.
- The field current of the shunt machine is constant regardless of loading conditions (armature current). The series machine, on the other hand, has a field current varying with the loading of the motor: the heavier the load, the stronger the field. At light or no-load conditions, the field of the series motor is very small.

When analyzing series machines, one should keep in mind the effect of flux saturation due to high field currents. A flux saturation curve is shown in Figure 9.15. The field coil is wound around the metal core of the stator. The current of the field winding produces the flux inside the core. When the current increases, the flux increases in a linear proportion unless the core is saturated.

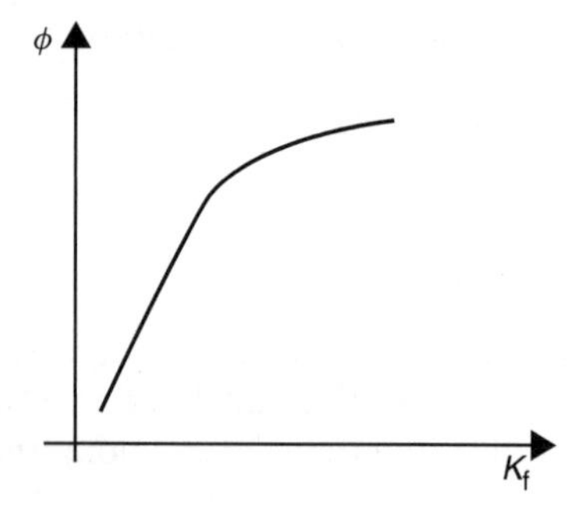

Figure 9.15 Flux saturation curve.

At saturation, the flux tends to increase at a progressively diminishing rate when the field current increases.

The series motor has the same basic equations used for shunt motors: Equations 9.9 and 9.10. The armature current is calculated by using the loop equation of the armature circuit.

$$I_a = \frac{(V_t - E_a)}{R_a + R_f} \tag{9.20}$$

Note that R_f is present in Equation 9.20. The machine torque can be calculated in a similar manner as used in Equation 9.12.

$$T_d = K\phi\frac{(V_t - E_a)}{R_a + R_f}, \tag{9.21}$$

leading to

$$T_d = K\phi\frac{(V_t - K\phi\omega)}{R_a + R_f}. \tag{9.22}$$

9.6.5 Control of d.c. motors

Direct current motors have several intrinsic properties, such as the ease by which they can be controlled, their ability to deliver high starting torque, and their near-linear performance. Direct current motors are widely used in applications such as actuation, manipulation, and traction, but they do have drawbacks that may restrict their use in some applications. For example, they are relatively high-maintenance machines due to their commutator mechanisms, and they are large and expensive compared to other motors, such as the induction. They may not be suitable for high-speed applications due to the presence of the commutator and brushes. Also, because of the electrical discharging between the commutator segments and brushes, d.c. machines cannot be used in clean or explosive environments unless they are encapsulated. Nevertheless, d.c. motors still hold a large share of the ASD (adjustable speed drive) market. Newer designs of d.c. motors have emerged that eliminate the mechanical commutator. The brushless motor, for example, is a d.c. motor that has the armature mounted on the stator and the field in the rotor. Like the conventional d.c. motor, the brushless motor switches the armature windings based on motor position. The switching, however, is done electronically, thereby eliminating the mechanical switching of the conventional d.c. motor.

9.6.6 Speed control of shunt or separately excited motors

As shown, the speed–torque characteristics of a d.c. separately excited (or shunt) motor can be expressed by the formula

$$\omega = \frac{V_t}{K\phi} - \frac{R_a}{(K\phi)^2} T_d = \omega_0 - \Delta\omega \tag{9.23}$$

or

$$\omega = \frac{V_t}{K\phi} - \frac{R_a}{K\phi} I_a = \omega_0 - \Delta\omega, \tag{9.24}$$

where ω_0 is the no-load speed and $\Delta\omega$ is the speed drop. The no-load speed is computed when the torque and current are equal to zero. The speed drop is a function of the load torque. The load torque and rotational torques (such as friction) determine the magnitude of the motor's developed torque at steady state. For a given torque, the motor speed is a function of the following three quantities:

- **Resistance in armature circuit:** When a resistance is inserted in the armature circuit, the speed drop, $\Delta\omega$, increases and the motor speed decreases.
- **Terminal voltage (armature voltage):** Reducing the armature voltage, V_t, of the motor reduces the motor speed.
- **Field flux (or field voltage):** Reducing the field voltage reduces the flux, φ, and the motor speed increases.

As explained earlier, electric motors cannot be operated with voltages higher than the rated value. Therefore, we cannot control the motor speed by increasing the armature or field voltages beyond the rated values. Only voltage reduction can be implemented. Hence, the second method of speed control (armature voltage) is only suitable for speed reduction, whereas the third method (field voltage) is suitable for speed increase. For a full range of speed control, more than one of the three methods must be employed.

9.6.7 Controlling speed by adding resistance

Figure 9.16 shows a d.c. motor with resistance added in the armature circuit. Figure 9.17 shows the corresponding speed–torque characteristics. Let us assume that the load torque is unidirectional and constant. A good example of this type of torque is an elevator. Also assume that the field and armature voltages are constant. At point 1, no external resistance is in the armature circuit. If a

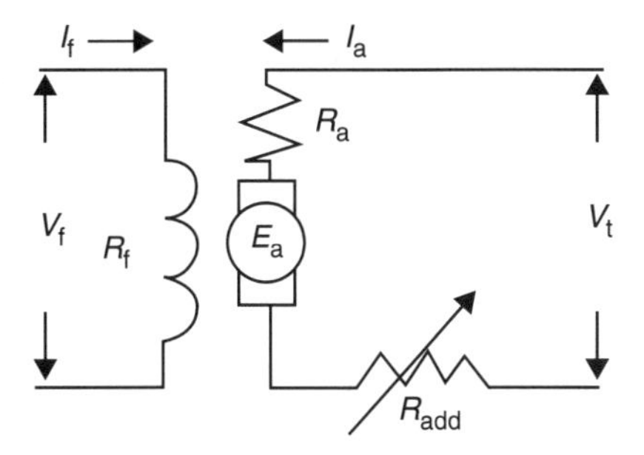

Figure 9.16 Motor speed change by adding an armature resistance.

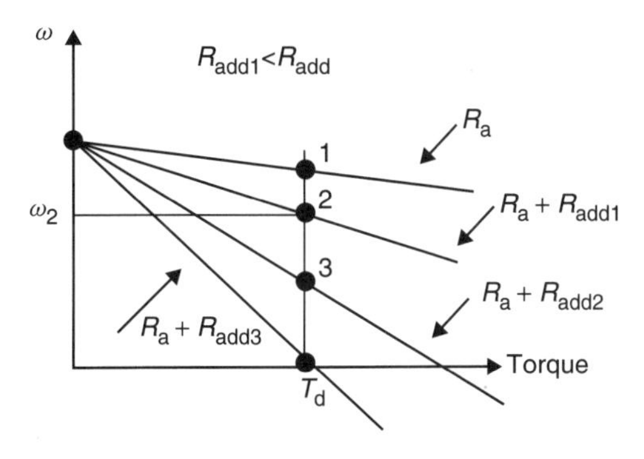

Figure 9.17 Effect of adding an armature resistance.

resistance, R_{add1}, is added to the armature circuit, the motor operates at point 2, where the motor speed ω_2 is

$$\omega_2 = \frac{V_t}{K\phi} - \frac{(R_a + R_{add_1})}{(K\phi)^2} T_d = \omega_0 - \Delta\omega_2 \tag{9.25}$$

or

$$\omega_2 = \frac{V_t}{K\phi} - \frac{(R_a + R_{add_1})}{K\phi} I_a = \omega_0 - \Delta\omega_2 \tag{9.26}$$

We note that the no-load speed, ω_0, is unchanged regardless of the value of resistance in the armature circuit. The second term of the speed equation is the speed drop, $\Delta\omega$, which increases in magnitude when R_{add} increases. Consequently, the motor speed is reduced.

If the added resistance keeps increasing, the motor speed decreases until the system operates at point 4, where the speed of the motor is zero. The operation of the drive system at point 4 is known as *holding*. It is quite common to operate the motor under electrical holding conditions in applications such as robotics and actuation. An electrical drive system under holding may jiggle unless a feedback control circuit is used to stabilize the system.

When the motor is operating under a holding condition, the speed drop, $\Delta\omega_4$, is equal in magnitude to the no-load speed ω_0. Consequently, we have the following condition:

$$\omega_4 = \omega_0 - \Delta\omega_4 = \frac{V_t}{K\phi} - \frac{(R_a + R_{add_3})}{(K\phi)^2} T_d = 0. \tag{9.27}$$

The resistance R_{add3} in this case is

$$R_{add_3} = \frac{K\phi V_t}{T_d} - R_a = 0 \tag{9.28}$$

or

$$R_{add_3} = \frac{V_t}{I_d} - R_a. \tag{9.29}$$

We keep in mind that operating a d.c. motor for a period of time with a resistance inserted in the armature circuit is a very inefficient method. The use of resistance is acceptable only when the heat produced by the resistance is utilized as a by-product or when the resistance is used for a very short period of time.

EXAMPLE 9.3

See website for downloadable MATLAB code to solve this problem

A 50 V, d.c. shunt motor drives a constant-torque load at a speed of 1200 rpm. The armature and field resistances are 2 Ω and 100 Ω, respectively. The motor draws a line current of 15 A at the given load.

(a) Calculate the resistance that should be added to the armature circuit to reduce the speed by 25 percent.

(b) Assume the rotational losses to be 80 W. Calculate the efficiency of the motor without and with the added resistance.

(c) Calculate the resistance that must be added to the armature circuit to open the motor at the holding condition.

Solution

(a) Let us use Figure 9.17 to help solve this problem. Assume that operating point 1 represents the motor without any added resistance, and point 2 is *f*, the operating point at 25 percent speed reduction. Since the motor is a shunt machine, the line current is equal to the armature current plus the field current.

Shunt current:

$$I_f = \frac{V_t}{R_f} = \frac{50}{100} = 0.5\,\text{A} \tag{9.29A}$$

Armature current:

$$I_a = I - I_f = 15 - 0.5 = 14.5\,\text{A}$$

Power input:

$$P_{in} = V_t \times I = 50 \times 15 = 750\,\text{W}$$

Resistance to be added to achieve reduction:

$$\begin{aligned} R_{add} &= \left[V_t - \frac{(V_t - I_a R_a)}{K} \right] / I_a - R_a \\ &= \left[50 - \frac{(50 - 14.5 \times 2)}{(1/0.75)} \right] / 14.5 - 2 = 0.3621\,\Omega \end{aligned} \tag{9.29B}$$

(b) Losses before adding armature resistance:

$$\text{Loss 1} = I_f^2 R_f + I_a^2 R_a + \text{losses} = 0.5^2 \times 100 + 14.5^2 \times 2 + 80 = 525.5\,\text{W} \tag{9.29C}$$

Losses after adding armature resistance:

$$\begin{aligned} \text{Loss 2} &= I_f^2 R_f + I_a^2 (R_a + R_{add}) + \text{losses} \\ &= 0.5^2 \times 100 + 14.5^2 \times (2 + 0.3621) + 80 = 601.625\,\text{W} \end{aligned} \tag{9.29D}$$

Efficiency before adding armature resistance:

$$\begin{aligned} \eta_1 &= [(P_{in} - \text{Loss1})/P_{in}] \times 100 \\ &= [(750 - 525.5)/750] \times 100 = 29.93\% \end{aligned} \tag{9.29E}$$

Efficiency after adding armature resistance

$$\eta_2 = [(P_{in} - \text{Loss2})/P_{in}] \times 100$$
$$= [(750 - 601.625)/750] \times 100 = 19.8733\% \quad (9.29F)$$

(c) To calculate the resistance to be added to the armature for the holding operation, set the motor speed equal to zero.

$$\omega_2 = \frac{V_t}{K\phi} - \frac{(R_a + R_{add_1})}{K\phi} = 0 \quad (9.29G)$$

leading to

$$R_{add} = \frac{V_t}{I_a} - R_a = \frac{50}{14.5} - 2 = 1.4483\,\Omega. \quad (9.29H)$$

9.6.8 Controlling speed by adjusting armature voltage

A common method of controlling speed is to adjust the armature voltage: a method which is highly efficient and stable and is simple to implement (Figure 9.18). The only controlled variable is the armature voltage of the motor, which is depicted as an adjustable-voltage source in Equation 9.23: when the armature voltage is reduced, the no-load speed, ω_0, is also reduced. Moreover, for the same value of load torque and field flux, the armature voltage does not affect the speed drop $\Delta\omega$. The slope of the speed-torque characteristic is $R_a/(K\varphi)^2$, which is independent of the armature voltage. The characteristics are parallel lines as shown in Figure 9.19. We note that we are assuming the field voltage changes when the armature voltage varies.

$$\omega_i = \frac{(V_t)_i}{K\phi} - \frac{R_a}{(K\phi)^2} T_d = \omega_{0i} - \Delta\omega \quad (9.30)$$

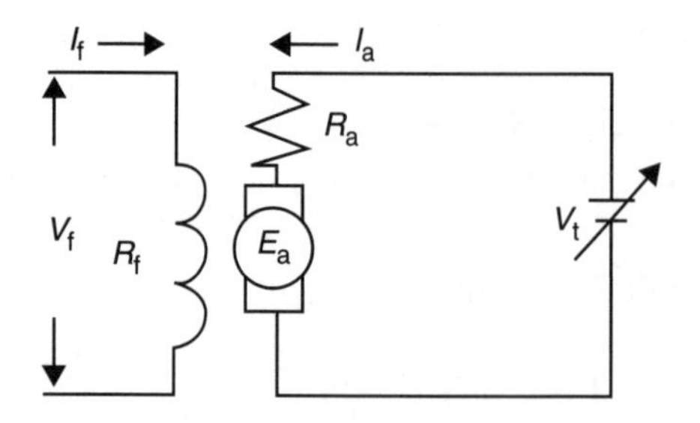

Figure 9.18 Motor speed change by varying the armature voltage.

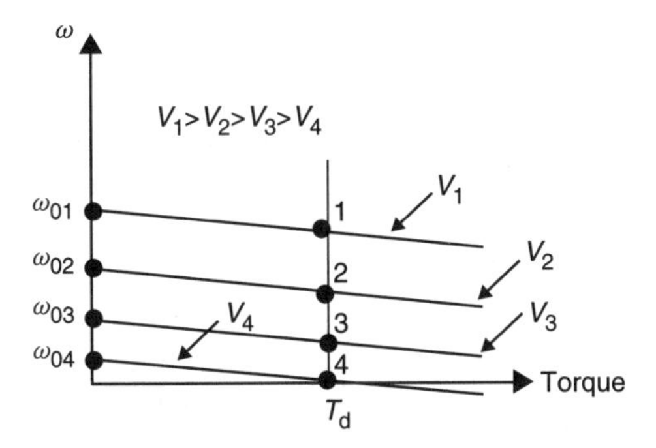

Figure 9.19 Motor characteristics when armature voltage changes.

We can now reach the following conclusions:

$$\frac{\omega_{0\mathrm{i}}}{\omega_{0\mathrm{i}+1}} = \frac{(V_\mathrm{t})_\mathrm{i}}{(V_\mathrm{t})_{\mathrm{i}+1}} \tag{9.31}$$

$$\frac{\omega_{0\mathrm{i}}}{\omega_{0\mathrm{i}+1}} = \frac{\phi_{\mathrm{i}+1}}{\phi_\mathrm{i}} \tag{9.32}$$

$$\frac{\Delta\omega_{0\mathrm{i}}}{\Delta\omega_{0\mathrm{i}+1}} = \left(\frac{\phi_{\mathrm{i}+1}}{\phi_\mathrm{i}}\right)^2 \tag{9.33}$$

Equations 9.31 to 9.33 are the basis of speed control via voltage adjustment.

Electric holding can be done if the armature voltage is reduced until $\Delta\omega$ is equal to ω_0. This operating point is shown in Figure 9.19 at an armature voltage equal to V_4.

$$\omega_4 = \omega_0 - \Delta\omega_4 = \frac{V_4}{K\phi} - \frac{R_\mathrm{a}}{(K\phi)^2} T_\mathrm{d} = 0 \tag{9.34}$$

or

$$V_4 = \frac{R_\mathrm{a}}{K\phi} T_\mathrm{d}. \tag{9.35}$$

EXAMPLE 9.4

See website for downloadable MATLAB code to solve this problem

The initial conditions of a d.c. motor are:

$$K\varphi = 3.0\,\mathrm{V\,s}$$
$$V_1 = 24.0\,\mathrm{V}$$
$$T_\mathrm{d} = 18\,\mathrm{Nm}$$
$$R_\mathrm{a} = 2\,\Omega$$

For a 30 percent increase in the armature voltage, calculate the percentage change in the no-load speed. Assume that the load torque is unchanged.

Solution

Equations 9.31 to 9.33 are the basis of speed control via voltage adjustment; we apply these in solving this problem.

Voltage in state 1: $V_{1t} = 24.0$ V

Speed at state 1:

$$\omega_1 = \frac{V_{1t}}{K\phi} - \frac{R_a}{(K\phi)^2} T_d = \frac{24}{3} - \frac{2}{3^3} \times 18 = 4 \text{ rad s}^{-1} \quad (9.35A)$$

Voltage in state 2: $V_{2t} = 1.3 \times 24 = 31.2$ V

Speed at state 2:

$$\omega_2 = \frac{V_{2t}}{K\phi} - \frac{R_a}{(K\phi)^2} T_d = \frac{31.2}{3} - \frac{2}{3^3} \times 18 = 6.4 \text{ rad s}^{-1} \quad (9.35B)$$

Ratio of final speed to initial speed:

$$\frac{\omega_2}{\omega_1} = \frac{6.4}{4} = 1.6 \quad (9.35C)$$

The analysis shows that when the voltage is increased by 30 percent, the speed increases by 60 percent.

9.6.9 Controlling speed by adjusting field voltage

Equations 9.23 and 9.24 show the dependency of motor speed on the field flux. The no-load speed is inversely proportional to the flux, and the slope of Equation 9.23 is inversely proportional to the square of the flux. Therefore, the speed is more sensitive to flux variations than to variations in the armature voltage.

Figure 9.20 shows a setup for controlling speed by adjusting the field flux. If we reduce the field voltage, the field current and consequently the flux are reduced. Figure 9.21 shows a set of speed–torque characteristics for three values of field voltages. When the field flux is reduced, the no-load speed, ω_0, is increased in inverse proportion to the flux, and the speed drop $\Delta\omega$ also increased. The characteristics show that because of the change in speed-drops, the lines are not parallel. Unless the motor is excessively loaded the motor speed increases when the field is reduced. When motor speed is controlled by adjusting the field current, the following should be kept in mind:

- The field voltage must not exceed the absolute maximum rating.
- Large reductions in field current may result in excessive speed since d.c. motors are relatively sensitive to variations in field voltage.

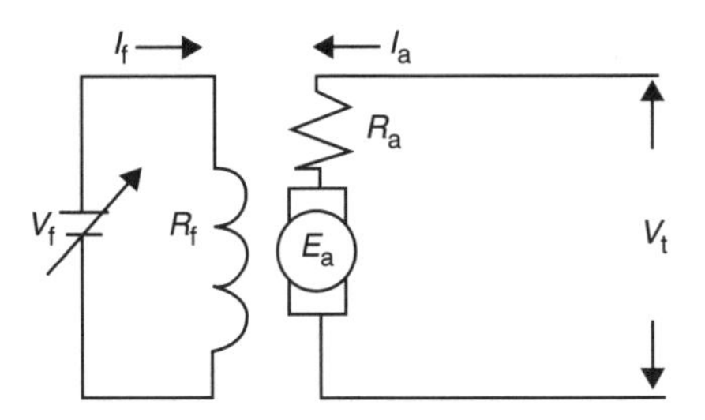

Figure 9.20 Motor speed change by varying the field voltage.

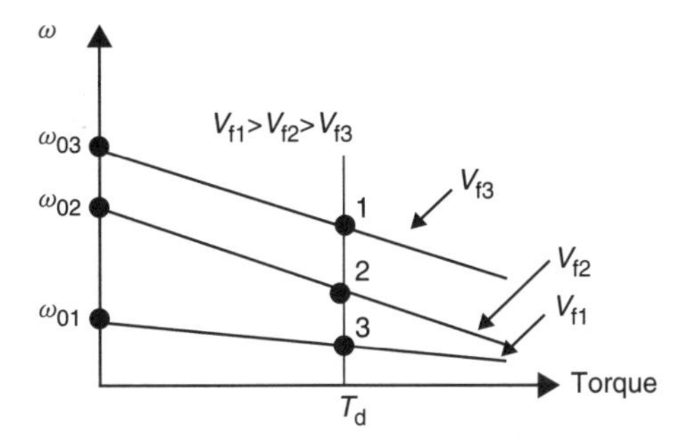

Figure 9.21 Effect of field voltage on motor speed.

- Reducing the field results in an increase in the armature current (assuming that the load torque is unchanged) because the armature current is inversely proportional to the field flux ($I_a = T_d/K\varphi$).

The last two considerations mean that field voltage control should be undertaken with special care to prevent mechanical and electrical damage to the motor. Furthermore, the field current should not be interrupted while the motor is running. If an interruption occurs, the residual magnetism will maintain a small amount of flux in the air gap. Consequently, the motor current will be excessively large, and the motor will accelerate to unsafe speeds. Although the system may have overcurrent breakers, special care should be given to this type of control to avoid an unpleasant experience!

9.7 Dynamic model and control of d.c. motors

The mechanical load on a motor consists of the inertia and the constant torque due to friction or gravity. Consequently, the total torque is given as

$$T = K_T I = J\frac{d\omega}{dt} + T_f. \tag{9.35D}$$

Taking the Laplace transformation of both sides of the equation yields

$$K_T I(s) = Js\omega(s) + T_f(s) \tag{9.35E}$$

from which

$$I(s) = \frac{Js\omega(s) + T_f(s)}{K_T}. \tag{9.35F}$$

But

$$V = L\frac{dI}{dt} + RI + E \tag{9.35G}$$

In Laplace form,

$$V(s) = LsI + RI(s) + E(s) \tag{9.35H}$$

We can now write

$$V(s) = Ls\left(\frac{Js\omega(s) + T_f(s)}{K_T}\right) + R\left(\frac{Js\omega(s) + T_f(s)}{K_T}\right) + K_E\omega(s). \tag{9.35I}$$

We rewrite this as

$$V(s) = \left(\frac{LJs^2}{K_T} + \frac{RJs}{K_T} + K_E\right)\omega(s) + \frac{LT_f(s)s}{K_T} + \frac{RT_f(s)}{K_T} \tag{9.35J}$$

$$V(s) = K_E\left(\frac{LJs^2}{K_T K_E} + \frac{RJs}{K_T K_E} + 1\right)\omega(s) + \frac{R}{K_T}\left(\frac{Ls}{R} + 1\right)T_f(s). \tag{9.35K}$$

Dividing through by K_E

$$\frac{V(s)}{K_E} = \left(\frac{LJs^2}{K_T K_E} + \frac{RJs}{K_T K_E} + 1\right)\omega(s) + \frac{R}{K_T K_E}\left(\frac{Ls}{R} + 1\right)T_f(s). \tag{9.35L}$$

Let us define

$$K_m = \frac{K_T}{\sqrt{R}}; \quad or\ K_m{}^2 = \frac{K_T{}^2}{R} \text{ and let } K_T = K_E$$

$$\frac{V(s)}{K_E} = \left(\frac{LJs^2}{K_m{}^2} + \frac{RJs}{RK_m{}^2} + 1\right)\omega(s) + \frac{R}{K_T{}^2}\left(\frac{Ls}{R} + 1\right)T_f(s) \tag{9.35M}$$

$$\text{and } \tau_e = \frac{L}{R}; \quad \tau_m = \frac{J}{K_m{}^2},$$

then we have the dynamic model to be

$$\frac{V(s)}{K_E} = \left(\tau_e\tau_m s^2 + \tau_m s + 1\right)\omega(s) + \frac{1}{K_m{}^2}(\tau_e s + 1)T_f(s). \qquad (9.35N)$$

For most d.c. motors, $\tau_e = 0$, and hence

$$\frac{V(s)}{K_E} = (\tau_m s + 1)\omega(s) + \frac{1}{K_m{}^2}T_f(s). \qquad (9.35P)$$

Rearranging, leads to

$$(\tau_m s + 1)\,\omega(s) = \frac{V(s)}{K_E} - \frac{T_f(s)}{K_m{}^2}. \qquad (9.35Q)$$

9.7.1 Open-loop control of permanent magnet motors

There are two methods using amplifiers for open-loop control of permanent magnet (PM) motors: (a) the linear transistor, and (b) switching transistors. A block diagram is given in Figure 9.22. A simple H-bridge circuit using pulsed-width modulated (PWM) controlled voltages consists of four switching transistors as shown in Figure 9.23. Motor direction is controlled by which input receives the PWM voltage. In the forward direction mode, $Q1$ and $Q4$ are ON and current flow is left-to-right. In the reverse direction mode, $Q2$ and $Q3$ are ON and current flow is right-to-left.

9.7.2 Closed-loop control of permanent magnet motors

A closed-loop speed control for d.c. motors could be achieved using a tachometer as shown in Figure 9.24. Some manufacturers supply an amplifier box unit that inputs analog voltages and outputs a pulse width modulated signal for switching

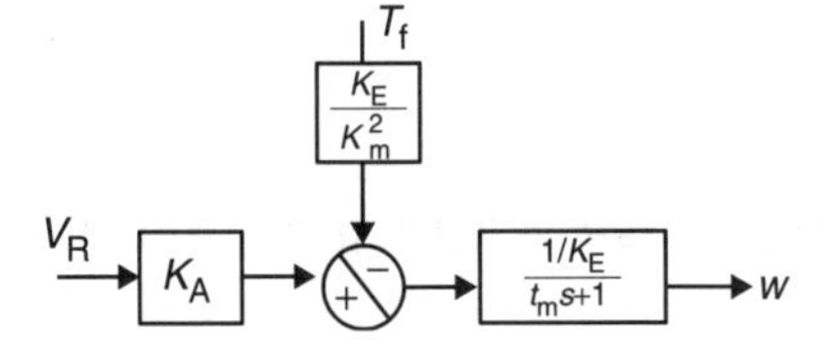

Figure 9.22 Open-loop speed control for d.c. motors.

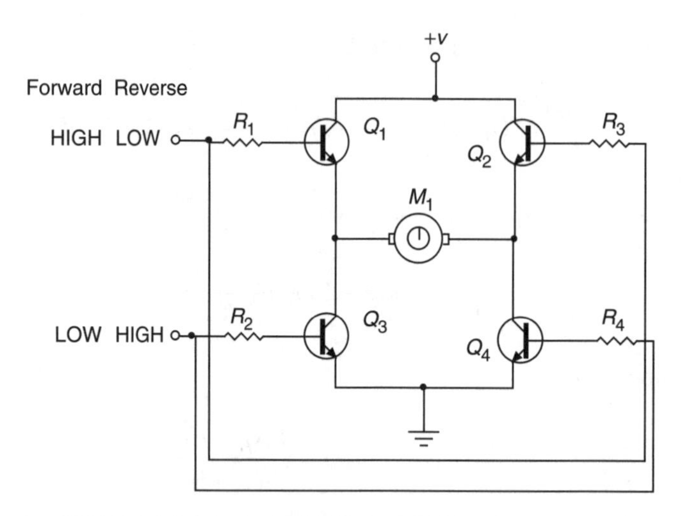

Figure 9.23 Open-loop motor speed control using H-bridge circuit with PWM. (Adapted from Stiffler, 1992.)

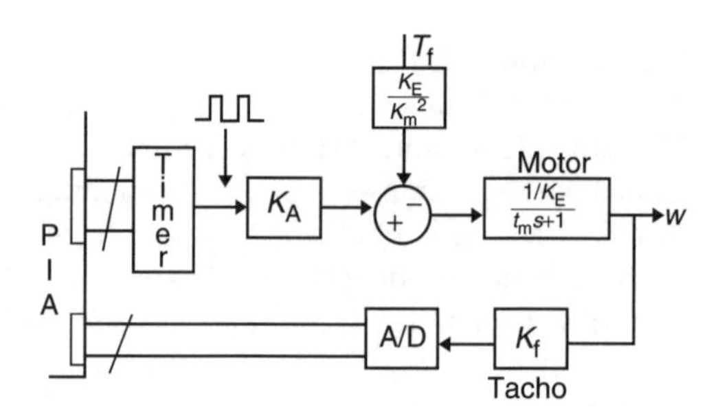

Figure 9.24 Closed-loop motor speed control using PWM. (Adapted from Stiffler, 1992.)

transistors. The tachometer voltages are converted to digital signals with an ADC. The microcontroller then compares the actual motor speed with the desired speed, which is located in the memory. Encoders are very easily designed in-house and installed for feedback control of d.c. motor; buying off-the-shelve encoders can be quite expensive.

9.7.3 Motor speed control using pulse width modulator (PWM)

This section presents two designs for motor speed control using pulse width modulation.

9.7.3.1 *Simplified motor speed control using PWM*

A simplified motor speed control using PWM is shown in Figure 9.25. Three signals are used for the motor speed control switching circuit: PWM, DIR (direction) and EN (Enable).

The truth table in Table 9.2 shows that in clockwise direction (DIR = 1), when the PWM is high, MOSFETs M_2 and M_3 are open and current flows through the motor. When the PWM is low only MOSFET M_3 is open (conducting) and no current flows through the motor. Similarly, in the counterclockwise direction (DIR = 0), when the PWM is high, MOSFETs M_1 and M_4 are open and current flows through the motor. When the PWM is low only M_1 is open and no current flows through the motor.

From Table 9.2,

$$\begin{aligned} A[1] &= EN \cdot \overline{A}; \\ B[3] &= EN \cdot DIR; \\ C[4] &= PWM \cdot A[1]; \text{ and} \\ D[2] &= PWM \cdot B[1]. \end{aligned} \tag{9.35R}$$

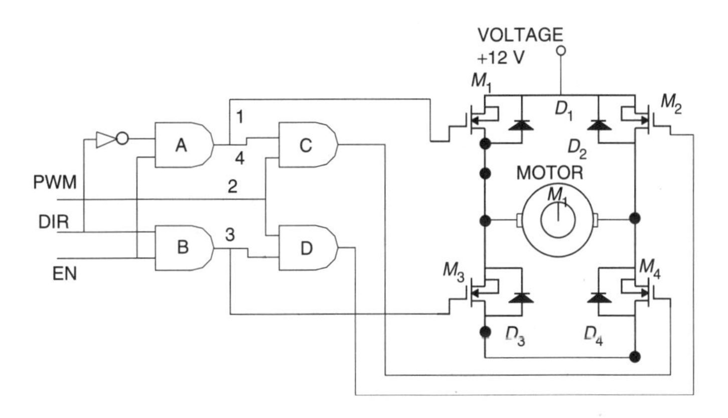

Figure 9.25 Simplified motor speed control using PWM.

Table 9.2 Truth table for simplified motor speed control signals

EN	*DIR*	*PWM*	$\bar{A}$	*A[1]*	*B[3]*	*C[4]*	*B[2]*
1	1	1	0	0	1	0	1
1	1	0	0	0	1	0	0
1	0	1	1	1	0	1	0
1	0	0	1	1	0	0	

9.7.3.2 *Practical motor speed control using PWM*

A more practical motor speed control circuit using PWM, which can be used to control a range of motors (12–24 V) and a range of current ratings, is shown in Figure 9.26.

The following design considerations should be noted before the designing of the H-Bridge driver to control the MOSFETS: (a) this is an n-channel driver, (b) it can switch up to 1 MHz, (c) it has the PWM-mode switching. The HIP4081A IC drives the MOSFETs according to the logic signals it gets from the PIC program. This driver has been used because it makes the control circuit easier to implement. It also has advantages such as preventing the *shoot through* condition (where two MOSFETs on the same side are switched on at the same time), low power consumption and can independently drive four MOSFETs.

9.7.4 Designing for reliability

- External bootstrap capacitors and diodes are required for this H-bridge driver to supply the high instantaneous current needed for turning on the power devices.
- Protection resistors are placed in parallel with the gate channel of the MOSFETs to limit the current going to the gates.
- Flyback diodes are placed across the source and drain of the MOSFETs to protect them from voltage spikes (back e.m.f.) when they switch on and off at a high frequency.

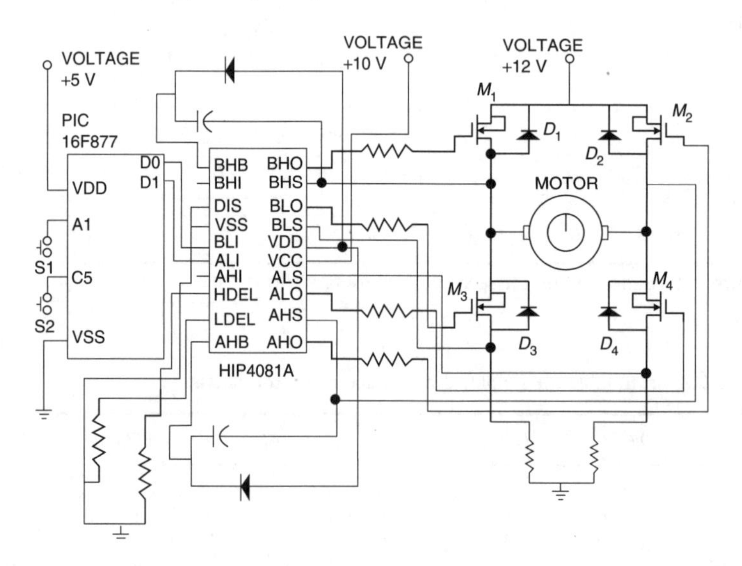

Figure 9.26 Practical motor speed control using PWM.

9.8 The servo motor

Servo motors are a variation on the gear-head motor coupled with a potentiometer to give feedback on the motor's position. The gears of the gearbox on a servo are attached to a potentiometer inside the case. A potentiometer is connected to a capacitor in a resistor–capacitor (*RC*) circuit, and by pulsing this *RC* circuit, the motor is powered to turn. When the motor turns, it changes the resistance of the *RC* circuit, which in turn feeds the motor again. By pulsing the *RC* circuit, you set the motor's position in a range from 0 to 180 degrees.

Servos have three wires to them, unlike most d.c. and gear-head motors, which have two. The first two in a servo are power and ground, and the third is a digital control line. This third line is used to set the position of a servo. Unlike other d.c. motors, you do not have to reverse the polarity of the power connections to reverse direction.

Hobby servos, the kind most often used in small physical computing projects, usually take a pulse of between 1–2 ms every 18–20 ms. This type of servo, rotates from 0 to 180 degrees depending on the pulse width. A pulse of 1 ms will turn the motor to 0 degrees; 2 ms will turn it to 180 degrees. A servo needs to see a pulse every 18–20 ms even when it is not turning, to keep it in its current position.

9.9 The stepper motor

A traditional motor has a series of coils which are automatically switched on and off by a set of brushes in contact with the commutator. Once power is applied, the motor runs at a speed proportional to the voltage and the load.

Stepper motors are different to regular d.c. motors in that they don't turn continuously, but move in a series of steps. A stepper motor is a motor controlled by a series of electromagnetic coils. The center shaft has a series of magnets mounted on it, and the coils surrounding the shaft are alternately given current or not, creating magnetic fields which repulse or attract the magnets on the shaft, causing the motor to rotate. A stepper motor has no commutator. Instead, there are five or six wires coming out of the motor; one wire for each coil (usually four) and one or two common ground wires. Power must be applied to one coil after another in the proper sequence in order to get the motor to turn. In order to obtain the maximum torque, two coils are always on at any time.

Each step only turns the shaft a degree or two. This four-step cycle has to be repeated about 50 times for a full revolution (not just once, as shown in the diagrams of Figure 9.27). If all four coils are switched off, the motor will be free to idle. Otherwise it is always locked in its current position. If the load on a stepper motor is too great or if the stepping sequences are being cycled too fast, it will skip a step.

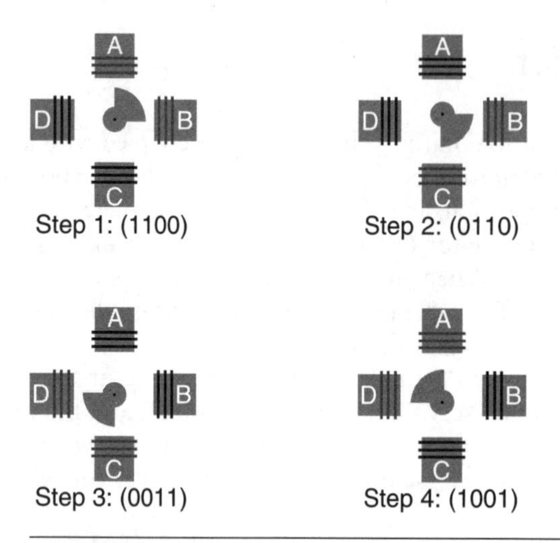

Figure 9.27 Stepper motor movement.

There are other types of stepping sequences that provide smoother motion or lower power consumption. One potential problem is determining which wire is on the stepper motor. The simplest method is to use an ohmmeter to figure out which wire or wires are the common ones. Then plug the remaining ones into the controller randomly. There are only 24 possible permutations and eight of them are correct, so it won't take too long to find one that works.

This design allows for very precise control of the motor: by proper pulsing, it can be turned in very accurate steps of set degree increments (for example, two-degree increments, half-degree increments, etc.).

Stepper motors are used in printers, disk drives, and other devices where precise positioning of the motor is necessary. Steppers usually move much slower than d.c. motors, since there is an upper limit to how fast you can step them (5–600 pulses per second, typically). However, unlike d.c. motors, steppers often provide more torque at lower speeds. They can be very useful for moving a precise distance. Furthermore, stepper motors have very high torque when stopped, since the motor windings are holding the motor in place like a brake.

The characteristics of stepper motors make them quite suitable for variety of applications and continuous current and synchronous motors are now being replaced by stepper motors.

The advantages of stepper motors are:

- speed does not depend on the torque applied on the axis;
- controls are more simple;
- great speed range available.

9.9.1 Stepper motor control

To control a stepper, it is necessary to create a stepper driver that will energize the coils in the right order to make the motor move forward.

The first thing to do is to understand the wiring for a stepper motor. The most common type is a unipolar stepper motor, with six wires and four coils (actually two coils divided by center wires on each coil). To do this, take an ohmmeter to the wires and measure the resistance from one wire to another. The outer wires for each coil will have a definite resistance that is double the resistance between the inner wire and either of the two outer wires, as shown in Figure 9.28.

For example, if the resistance between wires 1 and 2 is $x\,\Omega$, then that between 1 and 3 is $2x\,\Omega$. Remember, two wires that are not connected (e.g. 1 and 4, 5, or 6) have infinite resistance, which should read as an error on your meter. When you apply a voltage across two wires of a coil (e.g. 1 to 3, or 2 to 4), you should find that the motor is very difficult to turn (don't force it as it's bad for the motor).

Like other motors, the stepper requires more power than a microcontroller can deliver, so a separate power supply is required. Ideally the supply rating will be available from the manufacturer, but if not, get a variable d.c. power supply, apply the minimum voltage (hopefully 1 V or so), apply a voltage across two wires of a coil (e.g. 1 to 3 or 4 to 6) and slowly raise the voltage until the motor is difficult to turn. It is possible to damage a motor this way hence it is dangerous to increase the supply voltage too high. Typical voltages for a stepper might be 5 V, 9 V, 12 V, or 24 V. Higher than 24 V is less common and above that it is best not to guess the voltage.

To control the stepper, apply the voltage to each of the coils in a specific sequence. These phasing sequences differ for different types of steppers, but for a 4-phase unipolar stepper like the one described above, the phasing would be as shown in Table 9.3.

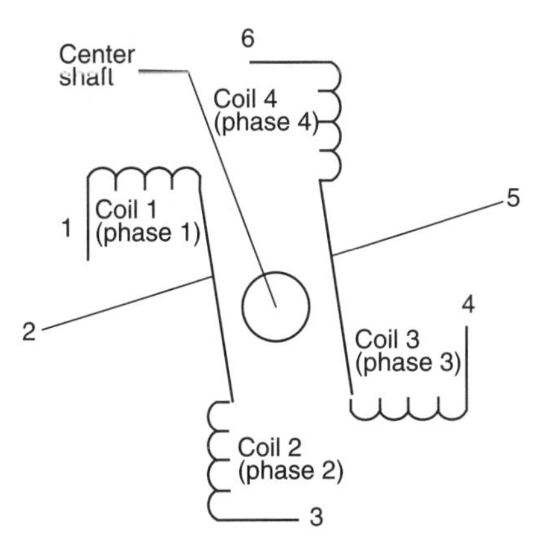

Figure 9.28 Stepper motor wiring.

Table 9.3 Stepper motor phasing sequence

Step	*Wire 1*	*Wire 3*	*Wire 4*	*Wire 6*
1	high	low	high	low
2	low	high	high	low
3	low	high	low	high
4	high	low	low	high

Note: wires 2 and 5 are wired to the supply voltage.

Typically, one would drive the stepper by connecting the four phase wires to a good power transistor or MOSFET, and the two common wires to the supply voltage, as shown in Figure 9.29.

In this diagram, the transistors are Darlington transistors.

Once you have the motor stepping in one direction, stepping in the other direction is simply a matter of doing the steps in reverse order. Knowing the position is a matter of knowing how many degrees per step, and counting the steps and multiplying by that many degrees. So for example, if you have a 2-degree stepper, and it has turned 180 steps, this is 2 × 180 degrees, or 360 degrees, or one full revolution.

Complete control of a stepper motor comprises:

- the power circuit where transistors manage the energy supply to the motor coils;

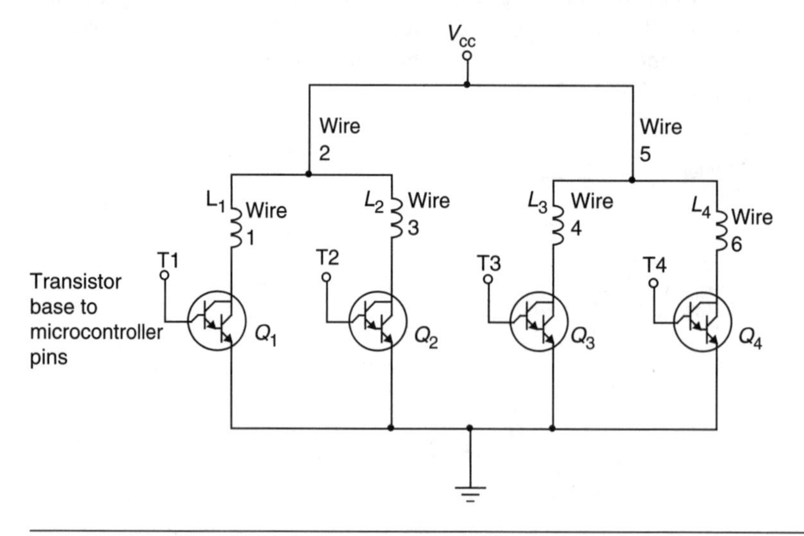

Figure 9.29 Stepper motor drive connections.

- the controller that switches the transistors so that the motor turns to the position requested by the user;
- the oscillator that gives the speed of the motor.

9.10 Motor selection

So far we have reviewed several qualitative features about each motor type as well as motor specifications. There are no hard and fast rules to selecting the best motor. There are always several workable configurations. Constraints can often eliminate several designs. For instance, lack of space can limit the motor diameter or the positioning resolution can rule out stepper motors.

It is the responsibility of the engineer to make the motor drive system work both electrically and mechanically. The engineer should consider the motor-to-load interface before considering the electrical drive-to-motor interface. In a typical motion control application the requirement will be to overcome some load frictional force and move a mass through a certain distance in a specified time. Therefore, the designer should take account of the following requirements (Stiffler, 1992): (1) moment of inertia, (2) torque, (3) power, and (4) cost. These requirements are now discussed.

9.10.1 Load inertia

For optimum system performance, the load moment of inertia should be similar to the motor inertia. When gear reducers intervene between the motor and the load, the reflected load inertia is J_L/N^2, where N is the gear ratio. If the motor inertia, J_M, is equal to the reflected load inertia, the fastest load acceleration will be achieved, or, conversely, the torque to obtain a given acceleration will be minimized. Therefore, matched inertias are best for fast positioning. On the other hand, peak power requirements are minimized by selecting the motor inertia so that the reflected load inertia is 2.5 times as large as the motor inertia. The torque will be increased but the maximum speed will be further reduced. A load inertia greater than 2.5 times the motor inertia is less than ideal, but it should not present any problems if the ratio is less than 5. A larger motor inertia implies that the same performance can be achieved at a lower cost by selecting a smaller motor. In general, we present a rule of thumb:

$$2.5J_m \leq J_L/N^2 \leq 5J_m. \quad (9.35S)$$

9.10.2 Torque

For optimum operation, the motor must supply sufficient torque, T_m, to overcome the load friction and to accelerate over a distance, s, in time τ. The torque and acceleration at the motor are given by the following:

$$T_m = \frac{T_L}{N} + J_m \alpha_m$$
$$\alpha_m = N\alpha_L, \tag{9.35T}$$

where the load torque T_L is given as

$$T_L = T_f + J_L \alpha_L. \tag{9.35U}$$

Here we note that the load torque takes into consideration the frictional torque, T_f. Substituting the load torque in the preceding equation, we obtain the motor torque as

$$T_m = N\left(T_f + \alpha_L\left(J_L + N^2 J_m\right)\right). \tag{9.35V}$$

For linear acceleration $s = ½ \alpha t^2$ over distance s in time τ,

$$\alpha_L = \frac{2s}{\tau^2}. \tag{9.35W}$$

For damped ($\xi = 0.7$) second-order response over distance s, the maximum acceleration is given as

$$\hat{\alpha}_L = \omega_N{}^2 s. \tag{9.35X}$$

The designer should make allowances for variations in load and bearing behavior as well as motor production variations, aiming for a 50 percent torque allowance for most industrial applications.

It is recommended that an initial design should be planned without a gear reducer. In many cases direct drive is not possible because load torque requirements far exceed the torque delivered by a motor of reasonable size. Critical needs on space or weight can lead to gear reducers for otherwise perfectly matched motor/load systems. The problem with gear reducers is gear backlash. If gears mesh too tightly, there is severe sliding friction between the teeth which can cause lockup. Thus, the teeth spacing is a tradeoff between reducing the power loss within the gears (loose fit) or improving the position accuracy (tight fit) of the load. Resolution of stepper motors can be enhanced with gear reducers, but their accuracy remains within the bounds of backlash. Direct drive robots give a

Table 9.4 Comparison of motor design parameters

Type	*Cost*	*Size (D″ × L″)*	*Peak torque/stall torque (oz-in)*	*Time constant (ms)*	*Efficiency (%)*
Iron core	1	2 ×	5–10	20	50–75
Cup	2	4 × 5	5–10	2	50–75
Disk	3	3 × 2	10–15	8	50–75
Brushless	1	2 × 1	5–25	20	75–90
Step	1	2 × 2	1–2	—	25–40
Encoder	1				

Source: (Stiffler, 1992).

repeatability or maximum positioning error, which is an order of magnitude better than traditional robots with gear reducers.

Stall torque is an important factor to be considered for each type of motor over its moment of inertia range. Typically available values represent a composite of several commercially available motors at each designated inertia. Each manufacturer limits its line to a few types and sizes. There are several factors which have considerable effect on the stall torque for a given size motor: (a) the strength of the field permanent magnets (PMs) which is reflected in weight, not size, and (b) the air cooling of the motor which allows larger stall currents through the windings.

Motor types are competitive over their inertia range with regard to stall torque. As expected, the ironless core motors do have better torque magnitudes (Table 9.4) but are more expensive. Small mechanical time constants combined with higher torques make them ideal for fast response systems.

Stall current, thus stall torque, is limited by the maximum permissible core temperature. If the design calls for short bursts of acceleration, peak currents (thus peak design torques) can far exceed their stall counterparts as long as the average power remains the same. All PM motors have excellent peak torque to stall torque ratios, with brushless motors having the highest ratio. Although stepper motors have exceptional stall torque characteristics, their peak torque capability is poor, and they are not good candidates for quick load accelerations.

9.10.3 Power

In addition to maximum torque requirements, torque must be delivered over the load speed range. The product of torque and speed is power. Total power, P, is the sum of the power to overcome friction P_c and the power to accelerate the load P_L. The latter usually the dominant component: $P = T_f\omega + J\alpha\omega$.

Table 9.5 Maximum motor speeds

Motor	*Maximum speed range*
Stepper motor	200–400 steps s^{-1}
Stepper motor (L/nR drive)	200–800 steps s^{-1}
Stepper motor (chopper drive)	10 000 steps s^{-1}
Permanent magnet (PM) motor	10 000 rpm
Brushless motor	> 20 000 rpm

Source: (Stiffler, 1992).

Peak power required during acceleration depends upon the velocity profile. If the load is linearly accelerated over distance s in time T, the maximum, power is

$$\hat{P}_{\mathrm{a}} = \frac{4Js^2}{\tau^3}. \tag{9.35Y}$$

If the load undergoes a damped ($\xi = 0.7$) second-order response over distance s, the maximum power is

$$\hat{P}_{\mathrm{a}} = 0.146\, J\omega_{\mathrm{N}}{}^3 s^2. \tag{9.35Z}$$

Comparison of the expressions for acceleration and power shows that to accelerate a load in one half the time will require eight times the power.

As discussed earlier, the torque–speed curve for d.c. PM motors is a linear line from stall torque to no-load speed. Therefore, the maximum power produced by the motor is the curve midpoint or one-fourth the stall torque and maximum speed product. The maximum speed for various motors is shown in Table 9.5. The designer should as a starting point, choose a motor with double the calculated power requirement.

9.10.4 Cost

Among several designs the single most important criteria is cost. Although it may be more prudent to choose the first workable design when only several units are involved, high-volume applications demand careful study of the economic tradeoffs. For, example, a motor with a given inertia size can deliver a wide range of torques, depending upon the magnet strength.

Price can vary by a factor of three or more over this torque range. Yet, by going to a larger motor with a lower strength magnet the same torque can be achieved at little or no increase in cost. Except for ironless motors, the direct motor cost is similar. However, permanent magnet d.c. motors operate closed loop.

For motor speed control, closed-loop design is essential, and hence the use of encoders becomes paramount. Experience shows that depending on the accuracy required, it may be worthwhile designing a simple encoder in-house for use. The cost of off-the-shelf encoders can equal if not exceed the cost of the motor itself. In addition, stepper and brushless motors have electronic expenses greater than those of brush motors.

The designer should carefully decide if motor controllers arc to bc uscd or if it is feasible to write software to control the motor. In some cases there is a greater burden of control placed on the software. Therefore, the designer should decide whether to buy a motor control board or write software to control the speed of the motor. For stepper motors, it may prove extremely difficult to write software since the controller cards are relatively cheap. These are some of the decisions that a designer must make in controlling motor speeds.

Problems

Q9.1 A d.c. separately excited motor has the following specification:

$$K\varphi = 3.0\,\mathrm{V\,s}$$

$$V_t = 12.0\,\mathrm{V}$$

$$R_a = 4.00\,\Omega$$

$$I_a = 5.0\,\mathrm{A}$$

Determine:

(a) the rated torque;

(b) the starting torque;

(c) the starting current at full voltage;

(d) the starting speed; and

(e) the speed at the rated torque condition.

Q9.2 Outline the programming steps to control the motor in Figure 9.25.

Q9.3 What are the disadvantages of using the motor speed control in Figure 9.25?

Q9.4 Outline the programming steps to control the motor in Figure 9.26.

Q9.5 Outline the design of shaft encoder for controlling the motor in Figure 9.26.

Q9.6 Outline the programming steps to control the motor in Figure 9.26 with encoders.

Further reading

[1] Bolton, W. (1993) *Mechanical Science*, Blackwell Scientific Publications.
[2] Bolton, W. (1995) *Mechatronics: Electronic Control Systems in Mechanical Engineering*, Essex: Longman.
[3] Cathey, J.J. (2001) *Electric Machines: Analysis and Design Applying MATLAB*, McGraw-Hill.
[4] El-Sharkawi, M.A. (2000) *Fundamentals of Electric Drives*, Brooks/Cole Publishers.
[5] Fitzgerald, A.E., Kingsley Jr., C. and Umans, S.D. (2003) *Electric Machinery* (6th. ed.), McGraw-Hill.
[6] Norton, R.L. (1992) *Design of Machinery,* McGraw-Hill.
[7] Stiffler, A.K. (1992) *Design with Microprocessors for Mechanical Engineers*, McGraw-Hill.

Internet resources

- http://perso.wanadoo.fr/hoerni/ol/emoteur.html
- http://fargo.itp.tsoa.nyu.edu/~tigoe/pcomp/motors.shtml

CHAPTER 10

Mechanical actuator systems

Chapter objectives

When you have finished this chapter you should be able to:

- realize the usefulness of hydraulic and pneumatic systems as mechatronics elements;
- realize the usefulness of mechanical elements such as mechanisms, gears, cams, clutches/brakes, and flexible mechanical elements in mechatronics applications.

10.1 Hydraulic and pneumatic systems

Hydraulic and pneumatic systems are similar except that while a hydraulic system uses an incompressible fluid as the working medium, a pneumatic system uses air, which is basically compressible. Advantages of using air as the working medium are that it is readily available and no recycling is necessary. It is non-flammable so that leakage does not create a threat to safety. It has negligible change in viscosity, which controls the system's performance. The major advantage of a hydraulic system is the incompressibility of the fluid helps in positive action or motion, and faster response, unlike pneumatic systems where there are longer time delays.

10.1.1 Symbols for hydraulic and pneumatic systems

Symbols are invaluable to designers in representing complex fluid power systems. Some of the standard ANSI (American National Standard Institute) symbols are shown in Figure 10.1.

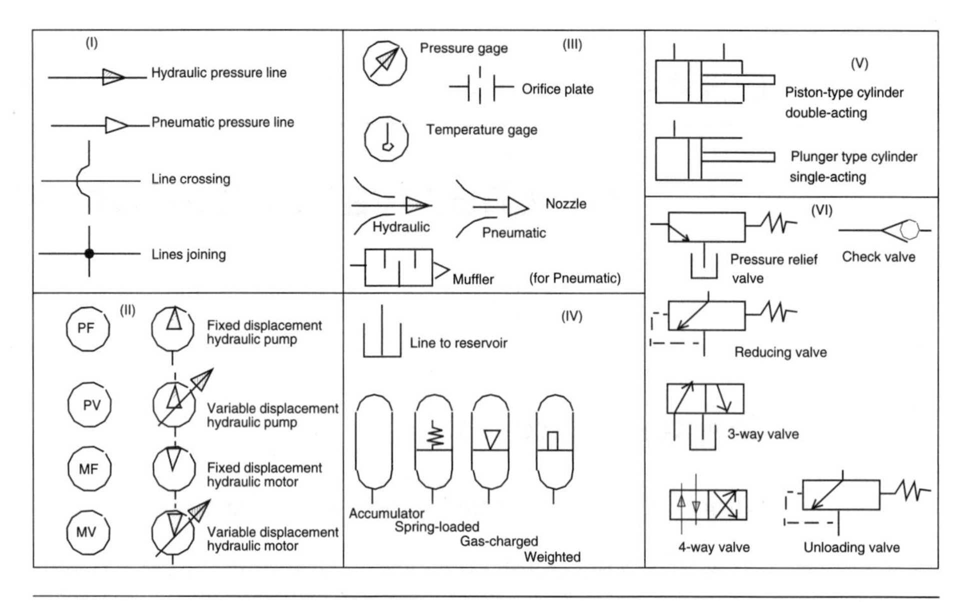

Figure 10.1 ANSI symbols for hydraulic and pneumatic systems.

The major components of hydraulic and pneumatic systems are pumps (compressors for pneumatic systems), valves and receiving units such as motors.

10.1.2 Hydraulic pumps

Pumps are used to supply the high pressure that the mechatronic system requires. Three types are most commonly used: (a) the gear pump; (b) the vane pump; and (c) the piston pump.

10.1.2.1 *The gear pump*

Gear pumps are used in hydraulic systems. For a clockwise rotation of the upper gear as shown in Figure 10.2, fluid is carried between the gear teeth in the same direction from the inlet to the high-pressure discharge side of the pump. The meshing teeth seal the fluid and prevent it from returning to the low-pressure side. This type of pump is cheap but becomes troublesome at high operating speeds and pressures.

10.1.2.2 *The vane pump*

Vane pumps are used in hydraulic systems. When the rotor rotates in a counterclockwise direction as shown in Figure 10.3, a large amount of fluid is carried from the inlet to the outlet. This results from the eccentricity of the

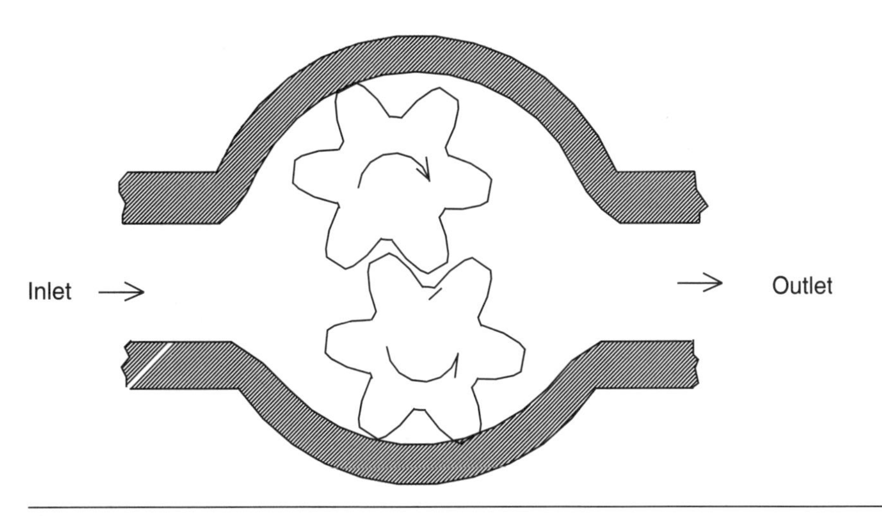

Figure 10.2 Gear pump.

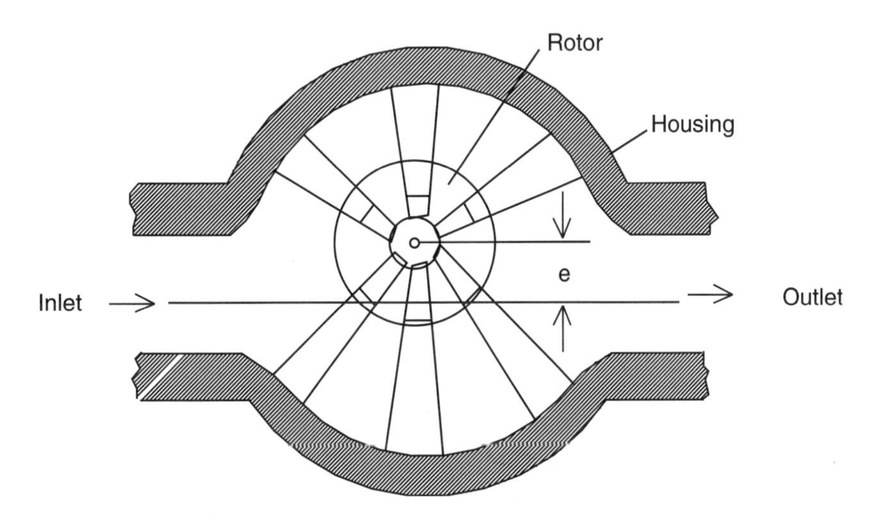

Figure 10.3 Vane pump.

center of the rotor with respect to that of the housing. The net flow of fluid is a function of the eccentricity and when it is varied, the vane pump can then be used as a variable-delivery pump.

10.1.2.3 *The axial piston pump*

The pistons are parallel to and located in the rotor, which is axially driven by the shaft. The swash plate is stationary but inclined at an angle β (see Figure 10.4).

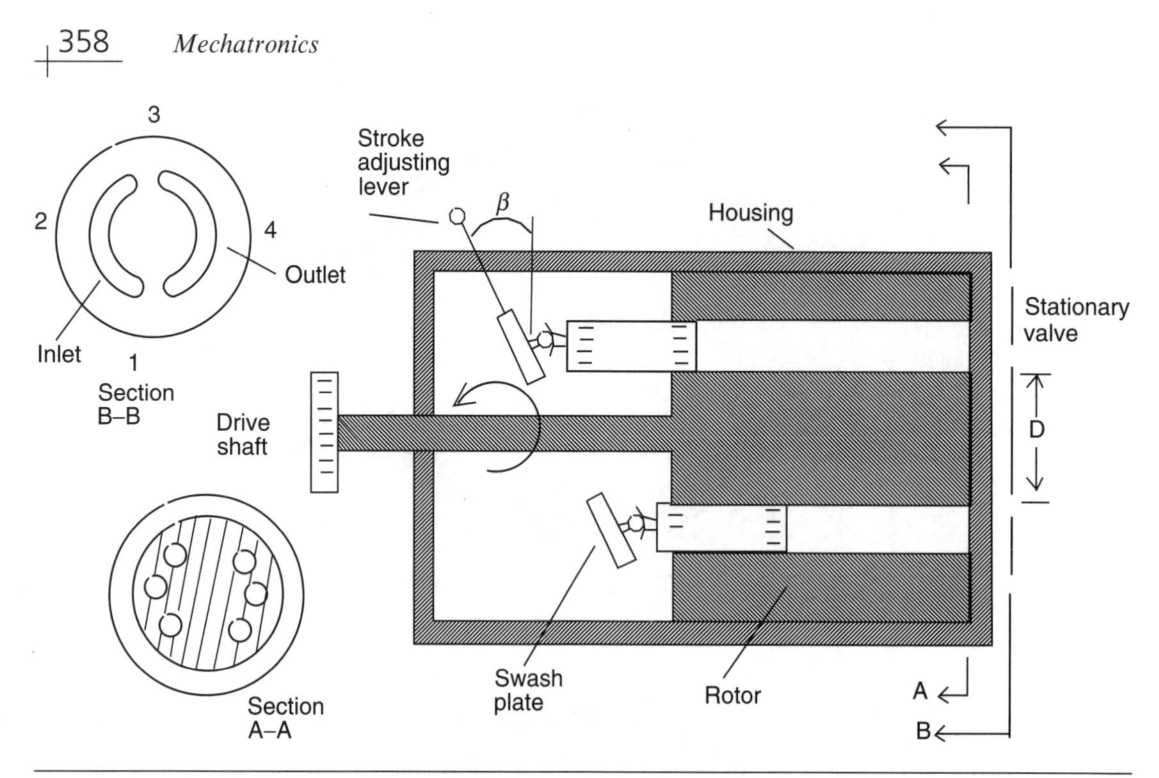

Figure 10.4 Axial piston pump.

The stroke adjusting-lever sets this angle. The axial displacement of each piston is given by

$$x = D \tan \beta \tag{10.1}$$

New settings for β vary this axial displacement. As a piston in the rotor rotates clockwise from positions 1 to 2 to 3, the fluid is admitted as the stroke is increased but discharged to a high pressure side as it rotates from 3 to 4 to 1 as the stroke is decreasing.

10.1.3 Pneumatic compressors

Pneumatic power can be supported with the aid of compressors such as the centrifugal, the axial-flow and the positive-displacement types.

10.1.3.1 *The centrifugal compressor*

In centrifugal compressors, shown in Figure 10.5, air enters the eye or center of the impeller; centrifugal effect throws the air into the volute where it goes to the diffuser.

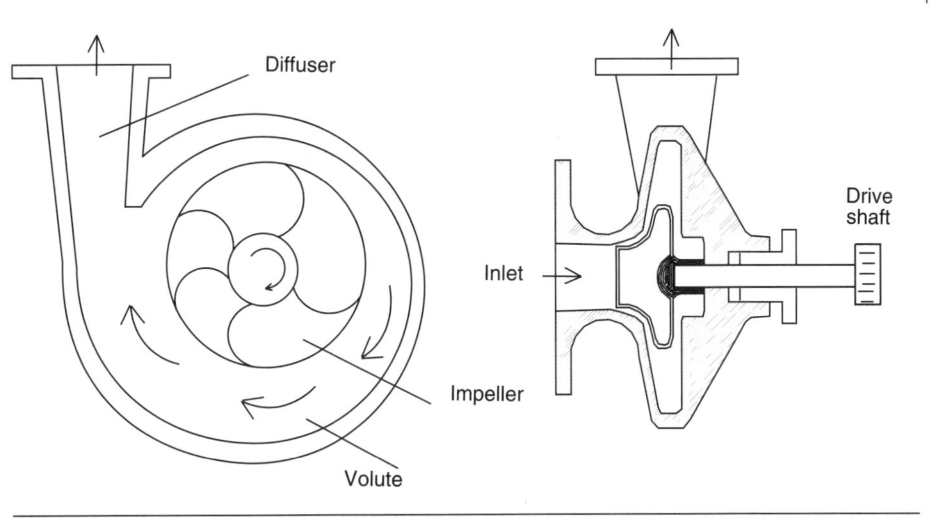

Figure 10.5 Centrifugal compressor (pneumatic).

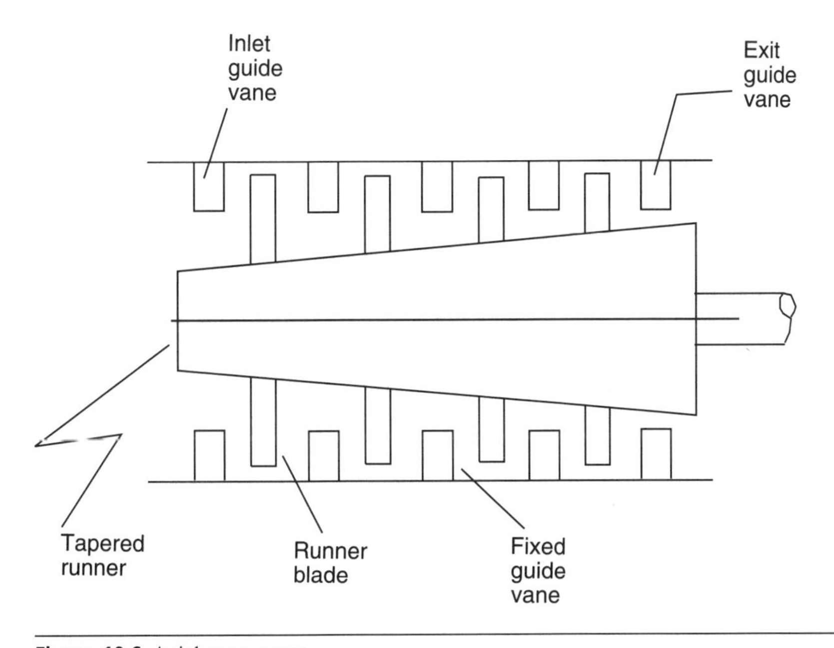

Figure 10.6 Axial compressor.

10.1.3.2 *The axial compressor*

In axial compressors, shown in Figure 10.6, the annular inlet area is much greater than that of the eye of the centrifugal type. Hence it delivers more flow. The blades are attached to the tapered rotor.

10.1.4 Valves

Valves are used to control the direction and amount of flow in a hydraulic or pneumatic system. Several types of commonly used valves are: (a) the relief; (b) the loading; (c) the differential pressure regulating; (d) the three-way; and (e) the four-way.

10.1.4.1 *The relief valve*

In a relief valve (Figure 10.7), the spring exerts force on the plunger when the valve is closed; this force is called the cracking force. When the line pressure, P_1, is high enough to overcome this spring force, the valve opens. This then connects the main line flow to the reservoir, which is at the drain pressure. The valve remains open until P_1 decreases to the value that was required to open it; hence a relief valve limits the maximum obtainable line pressure.

10.1.4.2 *The loading valve*

In a loading valve (Figure 10.8), a combination of an unloading valve and an accumulator is used to maintain constant pressure supply. A check valve is incorporated to prevent reverse flow from the accumulator.

As the accumulator fills, pressure P_1 increases and the plunger rises. Just before the drain port is uncovered, the pressure is acting on both sides of the bottom landing so that the upward force on the valve is P_1A_2. Once the drain port is uncovered, both sides of the top land are under the drain pressure as well as on the topside of the bottom land, so that the upward force on the valve

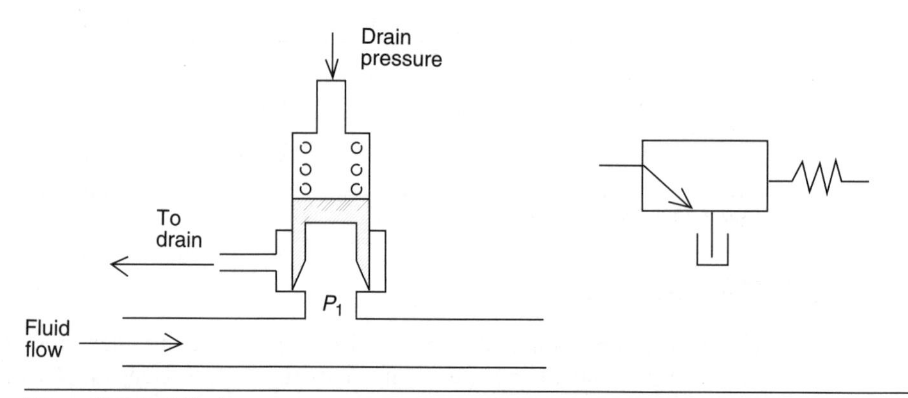

Figure 10.7 Relief valve.

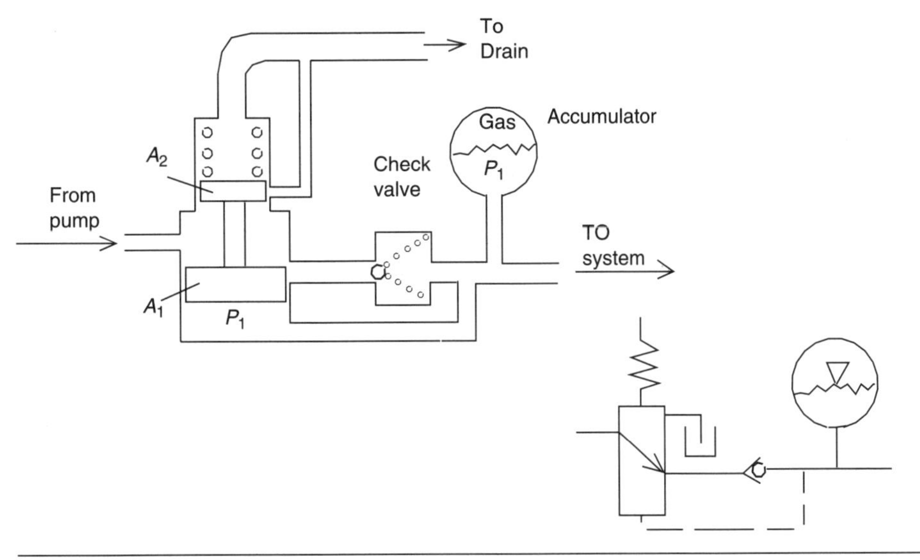

Figure 10.8 Loading valve.

is P_1A_1. Since this is greater than P_1A_2 the pump is immediately unloaded. The pressure in the line is reduced as the fluid in the accumulator is used because the gas expands in the accumulator. Gradually, the spring shuts the drain port and the pressure in the system is maintained.

10.1.4.3 *The differential pressure regulating valve*

A differential pressure regulating valve (Figure 10.9) is used to maintain differential pressure ΔP $(= P_1 - P_2)$ between any two points in the system. When this differential pressure is greater than the spring force, the plunger rises to bypass more flow to drain. This then connects the main line flow to the reservoir, which is at the drain pressure. The flow through the throttle valve is reduced and consequently the pressure ΔP across it also drops.

10.1.4.4 *The three-way valve*

Essentially a three-way valve (Figure 10.10), like the four-way valve, is used to control the direction and amount of flow to a receiving unit. It is often called a control, servo or proportional valve.

The three-way valve has three ports. These are: (a) a high-pressure (supply) port; (b) a cylinder port; and (c) a drain port.

When the valve is moved to the right, the drain port is shut off and the high-pressure line is connected to the cylinder port. The cylinder piston is acted

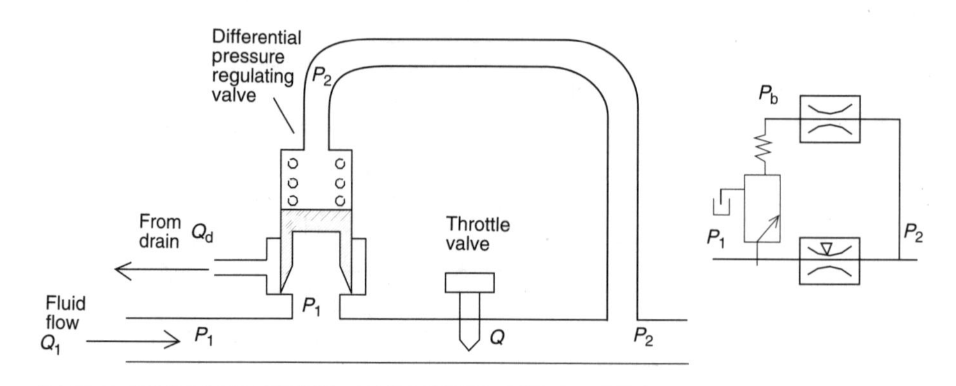

Figure 10.9 Differential pressure regulating valve.

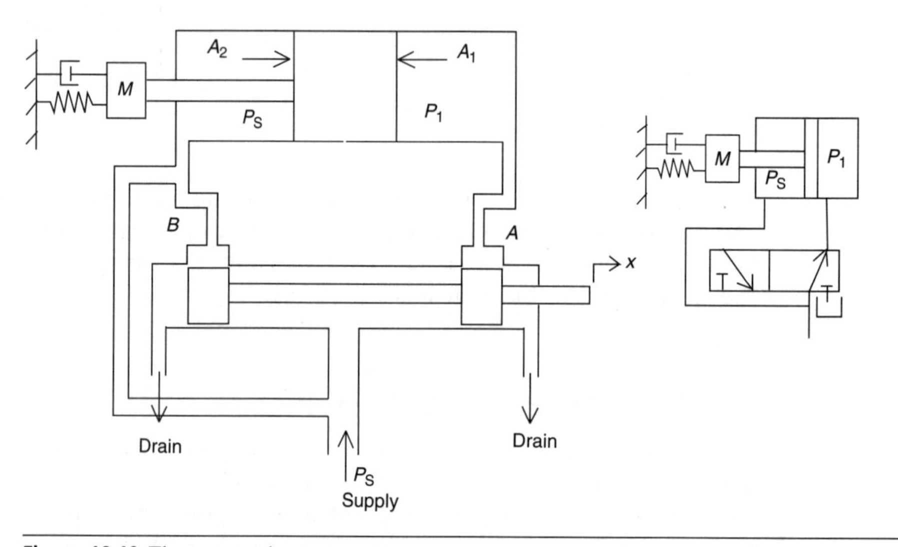

Figure 10.10 Three-way valve.

upon on both sides by the high pressure P_s, but since the area A_1 is greater than the area A_2 (because of the piston rod area), the piston and load move to the left. When the valve is moved to the left of its line-on-line position, the cylinder port is exposed to the drain and the high-pressure on the left of the piston forces it to the right.

The symbolic representation is obtained by considering:

- movement of spool to the right resulting in a high pressure line-cylinder port connection with drain shut off;
- movement of spool to the left resulting in cylinder port connected to drain port.

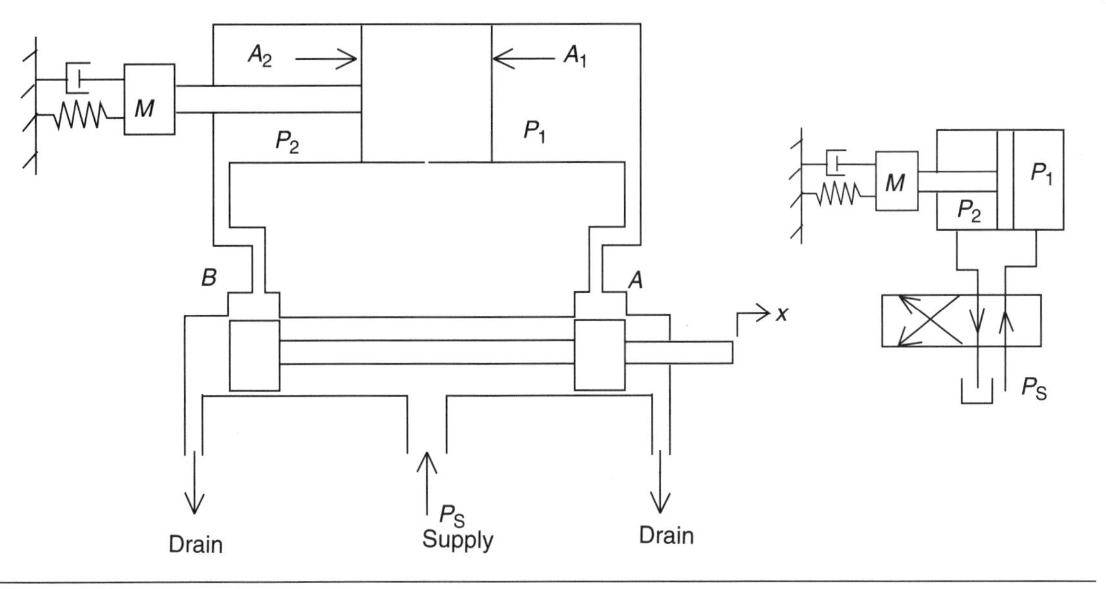

Figure 10.11 Four-way valve.

10.1.4.5 *The four-way valve*

The four-way valve (Figure 10.11) has four ports. These are: (a) a high-pressure (supply) port; (b) and (c) a port each for both ends of the cylinder; and (d) a drain port.

When the valve is moved to the right, port A is connected to the high (supply) pressure and port B is connected to the drain. This results in the piston moving to the left. When the valve is moved to the left the reverse action occurs. This valve is preferred to the three-way valve because it is possible to have the full supply pressure acting on either side of the piston with the other side connected to drain.

10.2 Mechanical elements

Mechanical elements can include the use of linkages, cams, gears, rack-and-pinions, chains, belt drives, etc. For example, the rack-and-pinion can be used to convert rotational motion to linear motion. Parallel shaft gears might be used to reduce a shaft speed. Bevel gears might be used for the transmission of rotary motion through 90 degrees. A toothed belt or chain drive might be used to transform rotary motion in one plane to motion in another. Cams and linkages can be used to obtain motions, which are prescribed to vary in a particular manner.

10.2.1 **Mechanisms**

One definition of a mechanism is the fundamental physical or chemical processes involved in or responsible for an action, reaction or other natural phenomenon. In kinematics, a mechanism consists of rigid bodies connected together by joints, which are used for transmitting, controlling, or constraining relative movement. The term mechanism is applied to the combination of geometrical bodies, which constitute a machine or part of a machine. A mechanism may therefore be defined as a combination of rigid or resistant bodies, formed and connected so that they move with definite relative motions with respect to one another. Mechanisms are devices, which can be considered to be motion converters in that they transform motion from one form to some required form. They might, for example, transform linear motion into rotational motion, or motion in one direction into a motion in a direction at right angles. They might transform a linear reciprocating motion into rotary motion, as in the internal combustion engine where the reciprocating motion of the pistons is converted into the rotation of the crank and hence drive shaft.

10.2.2 **Machines**

A machine is an assemblage of parts that transmit forces, motion and energy in a predetermined manner. A machine is a combination of rigid or resistant bodies, formed and connected so that they move with definite relative motions and transmit force from the source of power to the resistance to be overcome. A machine has two functions: transmitting definite relative motion and transmitting force. These functions require strength and rigidity to transmit the forces. Simple machines include any of various elementary mechanisms having the elements of which all machines are composed. Included in this category are the lever, the wheel and axle, the pulley, the inclined plane, the wedge and the screw.

The similarity between machines and mechanisms is that:

- they are both combinations of rigid bodies;
- the relative motion among the rigid bodies is definite.

The difference between machine and mechanism is that machines transform energy to do work, while mechanisms do not necessarily perform this function. Note that a mechanism is principally concerned with transformation of motion while the term *machine* is used for a system that transmits or modifies the action of a force or torque to do useful work. A machine is thus defined as a system of elements which are arranged to transmit motion and energy from one form to some required form while a mechanism is defined as a system of elements which are arranged to transmit motion and energy from one form to some required form. A mechanism can therefore be thought of as a machine, which is not required to

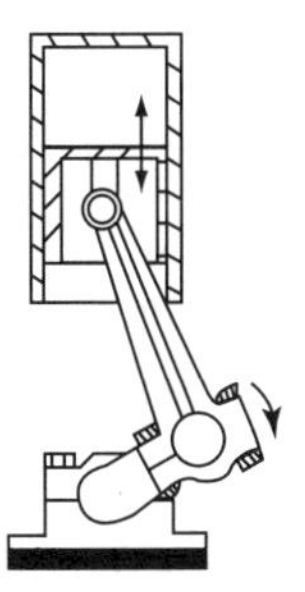

Figure 10.12 Cross-section of a power cylinder (slider-crank mechanism).

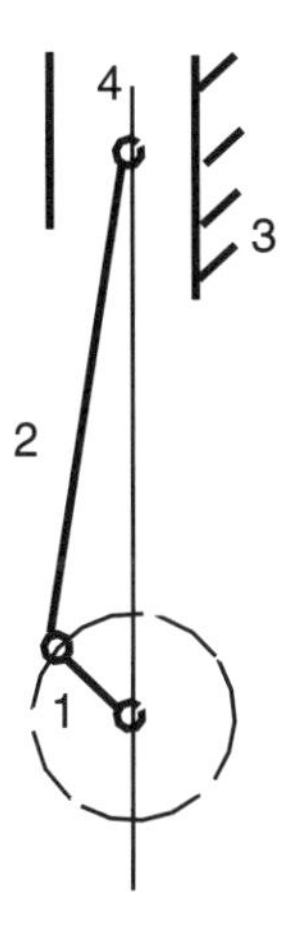

Figure 10.13 The mechanisms of the cylinder-link-crank parts of a diesel engine.

transmit energy but merely to reproduce exactly the motions that take place in an actual machine.

The term machinery generally means machines and mechanisms. Figure 10.12 shows a picture of the main part of a diesel engine. The mechanism of its cylinder-link-crank parts is a slider-crank mechanism, as shown in Figure 10.13.

10.2.3 Types of motion

A rigid body can have a very complex motion, which might seem difficult to describe. However, the motion of any rigid body can be considered to be a combination of translational and rotational motions. By considering the three dimensions of space, a translation motion can be considered to be a movement along one or more of the three axes. A rotation can be defined as a rotation about one or more of the axes.

For example, think of the motion required for you to pick up a pencil from a table. This might involve your hand moving at a particular angle towards the table, a rotation of the hand, and then all the movement associated with opening your fingers and moving them to the required positions to grasp the pencil. However, we can break down all these motions into combinations of translational and rotational motions. Such an analysis is particularly relevant if we are not moving a human hand to pick up the pencil but instructing a robot to carry out the task. Then it really is necessary to break down the motion into combinations of translational and rotational motions. Among the sequence of control signals might be such groupings of signals as those to instruct a joint to rotate by 20 degrees and a link to be extended by 4 mm for translational motion.

10.3 Kinematic chains

The term *kinematics* is used for the study of motion without regard to forces. When we consider just the motions without any consideration of the forces or energy involved then we are carrying out a kinematic analysis of the mechanism. In this case, we can treat the mechanism as being composed of a series of individual links. Each part of a mechanism, which has motion relative to some other part is termed a *link*. A link need not necessarily be a rigid body but it must be a resistant body, which is capable of transmitting the required force with negligible deformation. For this reason is it usually taken as being represented by a rigid body, which has two or more joints, which are points of attachment to other links. Each link is capable of moving relative to its neighboring links. Levers, cranks, connecting rods and pistons, sliders, pulleys, belts and shafts are all examples of links. A sequence of joints and links is known as a *kinematic chain*. For a kinematic chain to transmit motion, one link must be fixed. Movement of one link will then produce predictable relative movements of the others. It is possible to obtain from one kinematic chain a number of different mechanisms by having a different link as the fixed one.

As an illustration of a kinematic chain, consider a motor car engine where the reciprocating motion of a piston is transformed into rotational motion of a crankshaft on bearings mounted in a fixed frame. We can represent this as being four connected links (Figure 10.14). Link 1 is the crankshaft, link 2 the connecting rod, link 3 the fixed frame and link 4 the slider, that is the piston, which moves relative to the fixed frame.

The designs of many machines are based on two kinematic chains, the four-bar chain and the slider-crank chain.

10.3.1 The four-bar chain

In the range of planar mechanisms, the simplest group of lower pair mechanisms is the four-bar linkage. The four-bar chain consists of four links

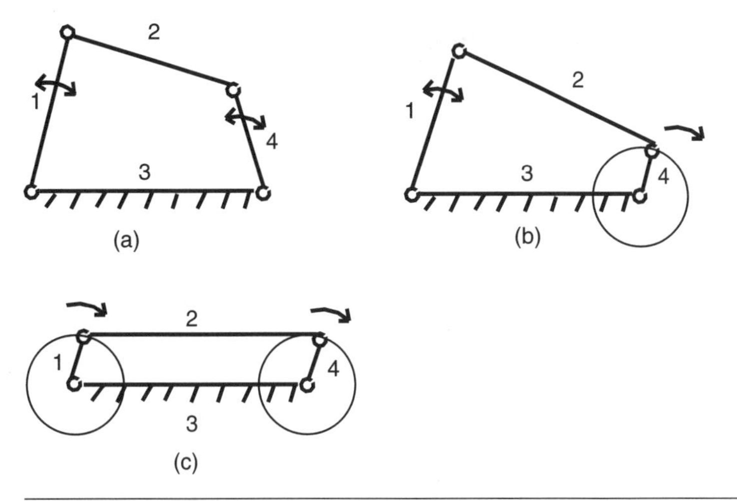

Figure 10.14 Four bar linkage.

connected to give four joints about which turning can occur. Figure 10.14 shows a number of forms of the four-bar chain produced by altering the relative lengths of the links. In Figure 10.14(a), link 3 is fixed and the relative lengths of the links are such that links 1 and 4 can oscillate but not rotate. The result is a *double-lever mechanism*. By shortening link 4 relative to link 1, then link 4 can rotate (Figure 10.14(b)) with link 1 oscillating and the result is termed a *lever-crank mechanism*. With links 1 and 4 the same length, both are able to rotate (Figure 10.14(c)), resulting in the *double-crank mechanism*. By altering which link is fixed, other forms of mechanism can be produced.

The link opposite the frame is called the coupler link, and the links which are hinged to the frame are called side links. A link which is free to rotate through 360 degrees with respect to a second link will be said to revolve relative to the second link (not necessarily a frame). If it is possible for all four bars to become simultaneously aligned, such a state is called a change point.

Some important concepts in link mechanisms are:

- **Crank:** A side link which revolves relative to the frame.
- **Rocker:** Any link which does not revolve.
- **Crank-rocker mechanism:** In a four-bar linkage, where the shorter side link revolves and the other one rocks (i.e. oscillates).
- **Double-crank mechanism:** In a four-bar linkage, where both of the side links revolve.
- **Double-rocker mechanism:** In a four bar linkage, where both of the side links rock.

10.3.2 The slider-crank mechanism

The slider-crank mechanism, which has a well-known application in engines, is a special case of the crank-rocker mechanism. This form of mechanism consists of a crank, a connecting rod and a slider (Figure 10.12). Link 3 is fixed (i.e. there is no relative movement between the centre of rotation of the crank and the housing in which the piston slides). Link 1 is the crank that rotates, link 2 the connecting rod and link 4 the slider which moves relative to the fixed link. When the piston moves backwards and forwards (i.e. link 4 moves backwards and forwards), then the crank is forced to rotate. Hence the mechanism transforms an input of backwards and forwards motion into rotational motion.

10.3.2.1 *Quick-return mechanism*

Figure 10.15 shows another form of the slider-crank mechanism: the *quick-return mechanism*. It consists of a rotating crank, link AB (which rotates round a fixed centre), an oscillating lever CD (which is caused to oscillate about C by the sliding of the block at B along CD as AB rotates, and a link DE which causes E to move backwards and forwards. E might be the ram of a machine and have a cutting tool attached to it. The ram will be at the extremes of its movement when the positions of the crank are AB_1 and AB_2. Thus, as the crank moves counterclockwise from B_1 to B_2, the ram makes a complete stroke, the cutting stroke. When the crank continues its movement from B_2 counterclockwise to B,

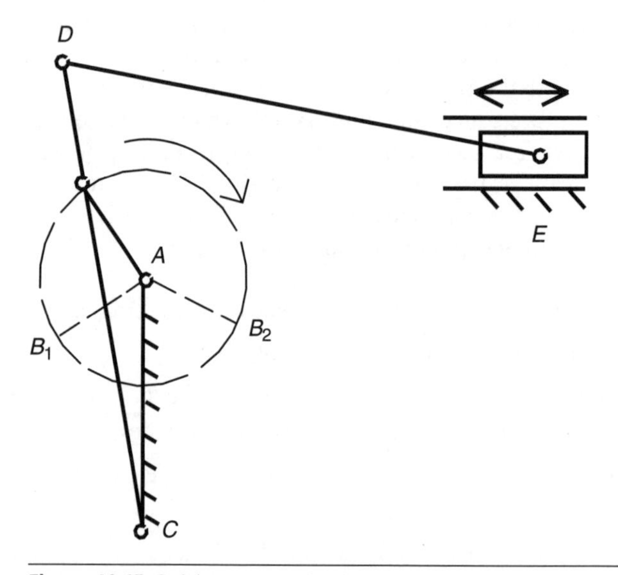

Figure 10.15 Quick-return mechanism.

then the ram again makes a complete stroke in the opposite direction, the return stroke. The angle of crank rotation required for the cutting-stroke is greater than the angle for the return-stroke. With the crank rotating at constant speed, the cutting stroke takes more time than the return stroke; hence, the term, 'quick-return' for the mechanism.

10.4 Cam mechanisms

The transformation of one of the simple motions, such as rotation, into any other motion is often conveniently accomplished by means of a cam mechanism. A cam mechanism usually consists of two moving elements, the cam and the follower, mounted on a fixed frame. Cam devices are versatile, and almost any arbitrarily-specified motion can be obtained. In some instances, they offer the simplest and most compact way to transform motions.

A cam may be defined as a machine element having a curved outline or a curved groove, which, by its oscillating or rotational motion, gives a pre-determined specified motion to another element with which it is in contact, called the follower (Figure 10.16). As the cam rotates so the follower is made to rise, dwell and fall. The lengths of time spent at each of these positions depends on the shape of the cam. The rise section of the cam is the part that drives the follower upwards, its profile determines how quickly the cam follower will

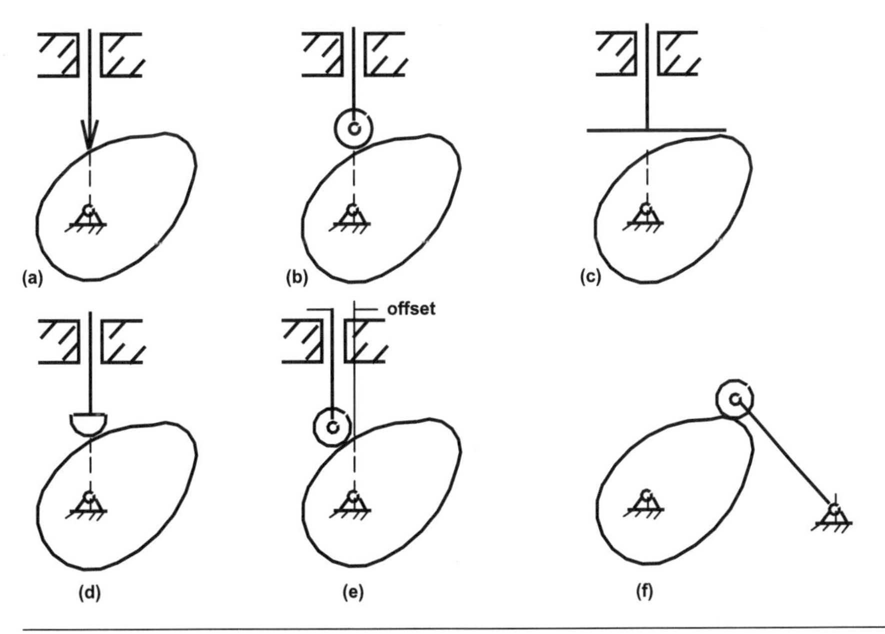

Figure 10.16 Classification of cam mechanisms.

be lifted. The fall section of the cam is the part that lowers the follower, its profile determines how quickly the cam follower will fall. The dwell section of the cam is the part that allows the follower to remain at the same level for a significant period of time. The dwell section of the cam is where it is circular with a radius that does not change. The cam has a very important function in the operation of many classes of machine, especially those of the automatic type, such as printing presses, shoe machinery, textile machinery, gear-cutting machines, and screw machines. In any class of machinery in which automatic control and accurate timing are paramount, the cam is an indispensable part of the mechanism. The possible applications of cams are unlimited, and their shapes occur in great variety. Some of the most common forms will be considered in this chapter.

10.4.1 Classification of cam mechanisms

We can classify cam mechanisms by the modes of input/output motion, the configuration and arrangement of the follower, and the shape of the cam. We can also classify cams by the different types of motion events of the follower and by means of a great variety of the motion characteristics of the cam profile. Figure 10.16 shows a number of examples of different types of cam followers. Roller followers are essentially ball or roller bearings. They have the advantage of lower motion than a sliding contact but can be more expensive. Flat-faced followers are often used because they are cheaper and can be made smaller than roller followers. Such followers are widely used with engine valve cams. While cams can be run dry, they are often used with lubrication and may be immersed in an oil bath.

10.4.1.1 *Modes of input/output motion*

The modes of input/output motion are any of the following:

- Rotating cam – translating follower (Figure 10.16(a–e)).
- Rotating follower (Figure 10.16(f)): The follower arm swings or oscillates in a circular arc with respect to the follower pivot.
- Translating cam – translating follower (Figure 10.17).
- Stationary cam – rotating follower: The follower system revolves with respect to the center line of the vertical shaft.

10.4.1.2 *Follower configuration*

The follower can be configured in any of the following ways:

- Knife-edge follower (Figure 10.16(a)).

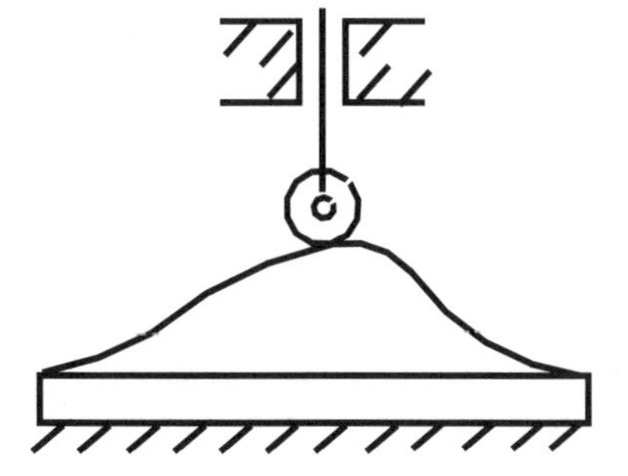

Figure 10.17 Translating cam – translating follower.

- Roller follower (Figure 10.16(b, e and f)).
- Flat-faced follower (Figure 10.16(c)).
- Oblique flat-faced follower.
- Spherical-faced follower (Figure 10.16(d)).

10.4.1.3 *Follower arrangement*

The follower arrangement takes any of the following forms:

- In-line follower: The center line of the follower passes through the center line of the camshaft.
- Offset follower: The center line of the follower does not pass through the center line of the cam shaft. The amount of offset is the distance between these two center lines. The offset causes a reduction of the side thrust present in the roller follower.

10.4.1.4 *Cam shape*

The production of a particular motion of the follower will depend on the shape of the cam and the type of follower used. The eccentric cam (Figure 10.18(a)) is a circular cam with an offset centre of rotation. It produces a follower oscillation, which is simple harmonic motion and is often used with pumps. The heart-shaped cam (Figure 10.18(b)) gives a follower displacement which increases at a constant rate with time before decreasing at a constant rate with time. Hence a uniform speed for the follower is realized. The pear-shaped cam (Figure 10.18(c)) gives a follower motion which is stationary for about half a revolution of the cam and rises and falls symmetrically in each of the remaining quarter revolutions. Such a pear-shaped cam is used for engine valve control. The dwell holds the valve open while the petrol/air mixture passes into the cylinder. The longer the dwell,

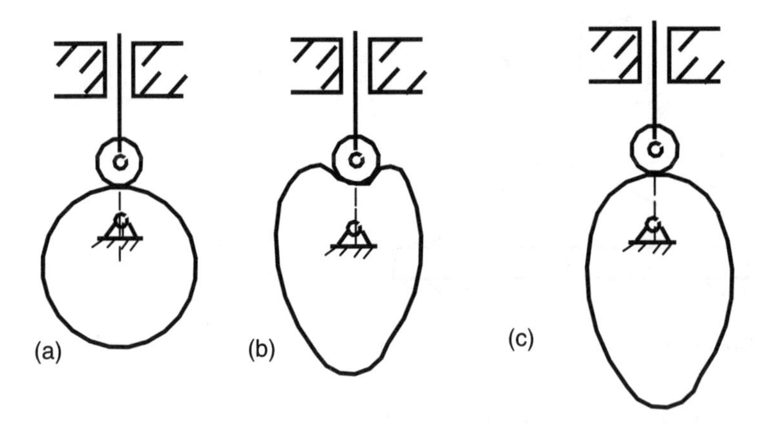

Figure 10.18 Cam shape.

(i.e. the greater the length of the cam surface with a constant radius), the more time is allowed for the cylinder to be completely charged with flammable vapor.

Other shapes of cam include:

- **Plate cam or disk cam:** The follower moves in a plane perpendicular to the axis of rotation of the camshaft. A translating or a swing arm follower must be constrained to maintain contact with the cam profile.
- **Grooved cam or closed cam:** This is a plate cam with the follower riding in a groove in the face of the cam.
- **Cylindrical cam or barrel cam:** The roller follower operates in a groove cut on the periphery of a cylinder. The follower may translate or oscillate. A conical cam results if the cylindrical surface is replaced by a conical one.
- **End cam:** This cam has a rotating portion of a cylinder. The follower translates or oscillates, whereas the cam usually rotates. The end cam is rarely used because of the cost and the difficulty in cutting its contour.

10.4.2 Motion events

When the cam turns through one motion cycle, the follower executes a series of events consisting of rises, dwells and returns. Rise is the motion of the follower away from the cam center, dwell is the motion during which the follower is at rest; and return is the motion of the follower toward the cam center.

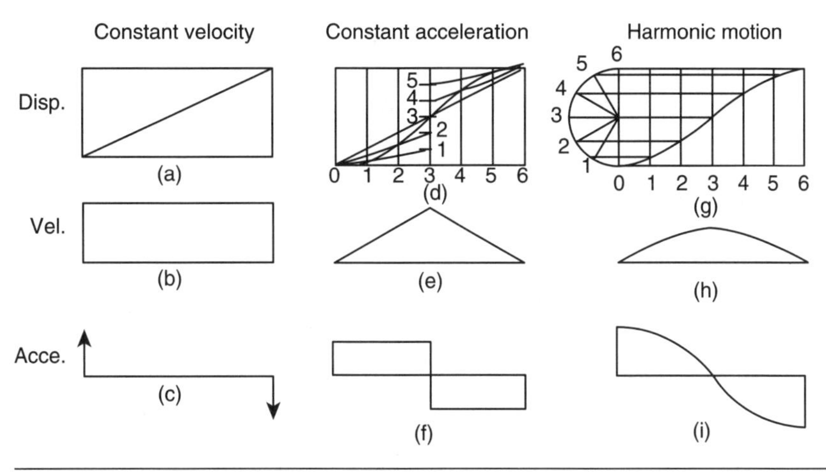

Figure 10.19 Motion events.

There are many follower motions that can be used for the rises and the returns resulting from a number of basic curves. Figure 10.19 shows the type of follower displacement diagrams that can be produced with different shaped cams and either point or knife followers. The radial distance from the axis of rotation of the cam to the point of contact of the cam with the follower gives the displacement of the follower with reference to the axis of rotation of the cam. The figure shows how these radial distances, and hence follower displacements, vary with the angle of rotation of the cams.

10.4.2.1 *Constant velocity motion*

If the motion of the follower describes a straight line (Figure 10.19(a–c)), it would have equal displacements in equal units of time (i.e. uniform velocity from the beginning to the end of the stroke), as shown in Figure 10.19(b). The acceleration, except at the end of the stroke would be zero, as shown in Figure 10.19(c). The diagrams show abrupt changes of velocity, which result in large forces at the beginning and the end of the stroke. These forces are undesirable, especially when the cam rotates at high velocity. The constant velocity motion is therefore only of theoretical interest.

10.4.2.2 *Constant acceleration motion*

Constant acceleration motion is shown in Figure 10.19(d–f). As indicated in Figure 10.19(e), the velocity increases at a uniform rate during the first half of the motion and decreases at a uniform rate during the second half of the motion. The acceleration is constant and positive throughout the first half of the motion, as shown in Figure 10.19(f), and is constant and negative throughout the second half. This type of motion gives the follower the smallest value of maximum acceleration along the path of motion. In high-speed machinery this is particularly important because of the forces that are required to produce the accelerations.

10.4.2.3 *Harmonic motion*

A cam mechanism with a basic motion curve such as Figure 10.19(g) will impart simple harmonic motion to the follower. The velocity diagram in Figure 10.19(h) indicates smooth action. The acceleration, as shown in Figure 10.19(i), is maximum at the initial position, zero at the mid-position, and negative maximum at the final position.

10.5 Gears

A pair of rolling cylinders can transfer rotary motion from one shaft to another (Figure 10.20(a)), however there is a possibility of slip. The transfer of the motion between the two cylinders depends on the frictional *forces* between the two surfaces in contact. Slip can be prevented by the addition of meshing teeth to the two cylinders and the result is then a pair of meshed gear wheels (Figure 10.20(b)). When two gears are in mesh, the larger gear-wheel is often known as the *spur* or *crown wheel*, and the smaller one the *pinion*.

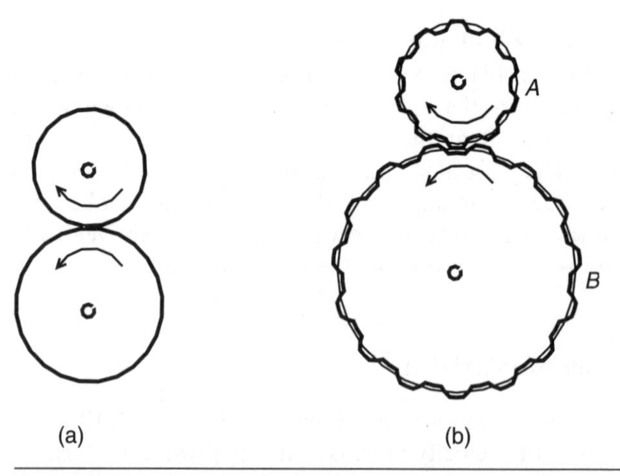

Figure 10.20 Rolling and meshed gears.

Consider two meshed gear wheels A and B (as in Figure 10.20(b)). If there are 20 teeth on wheel A and 40 teeth on wheel B, then wheel A must rotate through two revolutions in the same time as wheel B rotates through one. Thus the angular velocity, ω_A, of wheel A must be twice that of wheel B, that is:

$$\frac{\omega_A}{\omega_B} = \frac{\text{number of teeth on } B}{\text{number of teeth on } A} = \frac{4}{2}. \tag{10.2}$$

Since the number of teeth on a wheel is proportional to its diameter, d, we can write:

$$\frac{\omega_A}{\omega_B} = \frac{\text{number of teeth on } B}{\text{number of teeth on } A} = \frac{d_B}{d_A}. \tag{10.3}$$

Thus for the data we have been considering, wheel B must have twice the diameter of wheel A. The term *gear ratio* is used for the ratio of the angular speeds of a pair of intermeshed gear wheels. Thus the gear ratio for this example is 2.

10.5.1 Spur and helical gears

Gears for use with parallel shafts may have axial teeth with the teeth cut along axial lines parallel to the axis of the shaft. Such gears are then termed *spur gears*. Spur gears are used for the transmission of rotary motion between parallel shafts as shown in Figure 10.21(a).

Alternatively they may have helical teeth with the teeth being cut on a helix and are then termed *helical gears*. Helical gears have the advantage that there is a gradual engagement of any individual tooth and consequently there is a smoother drive and generally prolonged life of the gears. However, the inclination

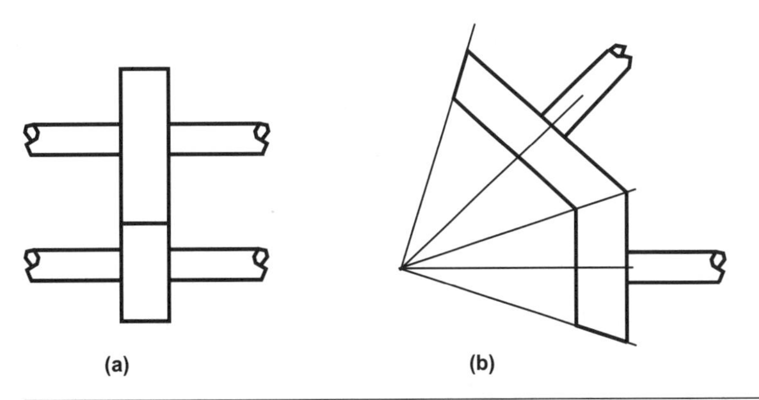

Figure 10.21 Parallel and inclined gear axes.

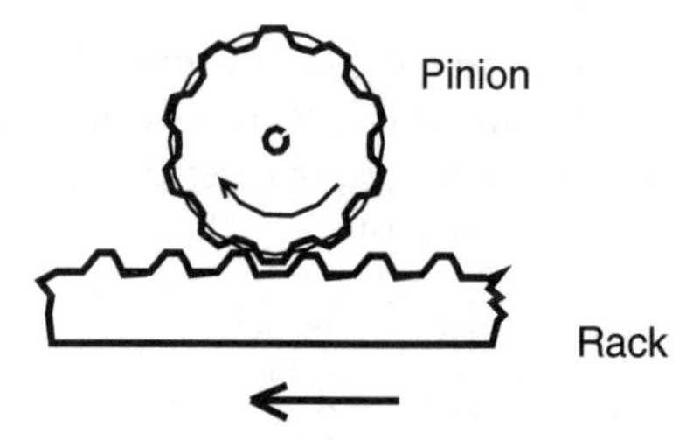

Figure 10.22 The rack and pinion.

of the teeth to the axis of the shaft results in an axial force component on the shaft bearing. This can be overcome by using double helical teeth.

10.5.2 Bevel gears

The term *bevel gear* is used when the lines of the shafts intersect, as illustrated in Figure 10.21(b). Bevel gears are used for the transmission of rotary motion between shafts which have axes inclined to one another.

10.5.3 The rack and pinion

Another form of gear is the *rack and pinion* (Figure 10.22). This transforms either linear motion to rotational motion or rotational motion to linear motion.

10.5.4 Gear trains

Gear trains are mechanisms that are very widely used to transfer and transform rotational motion. They are used when a change in speed or torque of a rotating device is needed. For example, the car gearbox enables the driver to match the speed and torque requirements of the terrain with the engine power available.

10.5.4.1 *Simple gear train*

The term *gear train* is used to describe a series of intermeshed gear wheels. The term *simple gear train* is used for a system where each shaft carries only one gear wheel, as in Figure 10.23(a). For such a gear train, the overall gear ratio is the ratio of the angular velocities at the input and output shafts and is thus ω_A/ω_C.

Consider a simple gear train consisting of wheels A, B and C, as in Figure 10.23(a). A has seven teeth and C has 21 teeth. Then, since the angular

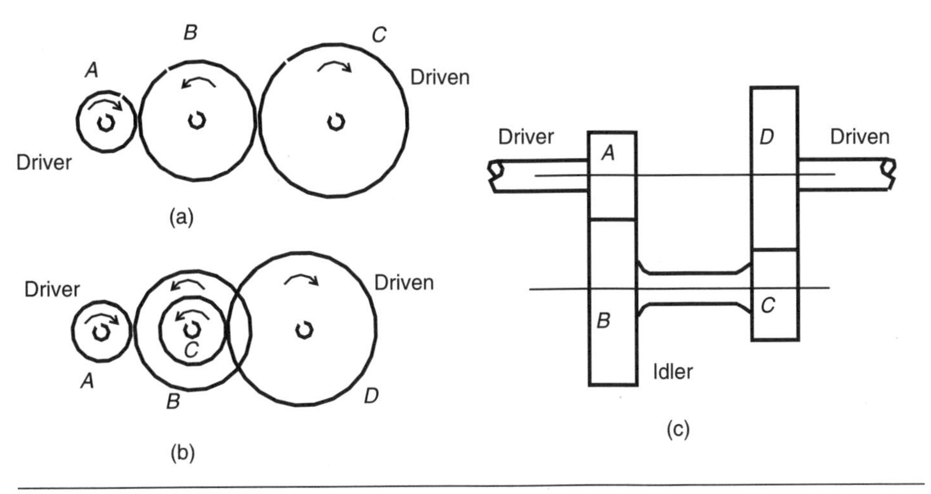

Figure 10.23 Compound gear train.

velocity of a wheel is inversely proportional to the number of teeth on the wheel, the gear ratio is $21/7 = 3$. The effect of wheel B is purely to change the direction of rotation of the output wheel compared with what it would have been with just the two wheels A and C intermeshed. The intermediate wheel, B, is termed the *idler wheel*.

We can rewrite this equation for the overall gear ratio, G, as

$$G = \frac{\omega_A}{\omega_C} = \frac{\omega_A}{\omega_B} \times \frac{\omega_B}{\omega_C}, \tag{10.4}$$

where ω_A/ω_B is the gear ratio for the first pair of gears and ω_B/ω_C is the gear ratio for the second pair of gears.

10.5.4.2 *Compound gear train*

The term *compound gear train* is used to describe a gear train when two wheels are mounted on a common shaft. Figure 10.23(b and c) shows two examples of such a compound gear train. The gear train in Figure 10.23(c) enables the input and output shafts to be in line. An alternative way of achieving this is the epicyclic gear train.

When two gear wheels are mounted on the same shaft they have the same angular velocity. Thus, for both of the compound gear trains in Figure 10.23(b) or (c), ω_B/ω_C. The overall gear ratio, G, is thus

$$G = \frac{\omega_A}{\omega_D} = \frac{\omega_A}{\omega_B} \times \frac{\omega_B}{\omega_C} \times \frac{\omega_C}{\omega_D} = \frac{\omega_A}{\omega_B} \times \frac{\omega_C}{\omega_D}. \tag{10.5}$$

For the arrangement shown in Figure 10.23(c), for the input and output shafts to be in line we must also have

$$r_A + r_B = r_D + r_C. \tag{10.6}$$

Consider a compound gear train of the form shown in Figure 10.23(b), with *A*, the first driver, having 15 teeth, *B* having 30 teeth, *C* having 18 teeth and *D*, the final driven wheel, having 36 teeth. Since the angular velocity of a wheel is inversely proportional to the number of teeth on the wheel, the overall gear ratio is

$$G = \frac{30}{10} \times \frac{18}{9} = 6. \tag{10.6A}$$

Thus, if the input to wheel *A* is an angular velocity of 240 rev min^{-1}, then the output angular velocity of wheel *D* is 240/6 = 40 rev min^{-1}.

A simple gear train of spur, helical or bevel gears is usually limited to an overall gear ratio of about 10. This is because of the need to keep the gear train down to a manageable size if the number of teeth on the pinion is to be kept above a minimum number, which is usually about 10 to 20. Higher gear ratios can, however, be obtained with compound gear trains (or epicyclic gears). This is because the gear ratio is the product of the individual gear ratios of parallel gear sets.

10.5.5 Epicyclic gear trains

In the *epicyclic gear train* one or more wheels is carried on an arm, which can rotate about the main axis of the train. Such wheels are called *planets* and the wheel around which the planets revolve is the *sun*. Figure 10.24 shows such a system with the centers of rotation of the sun wheel, *S*, and the planet wheel, *P*, linked by an arm, *A*. Two inputs are required for such a system. Typically,

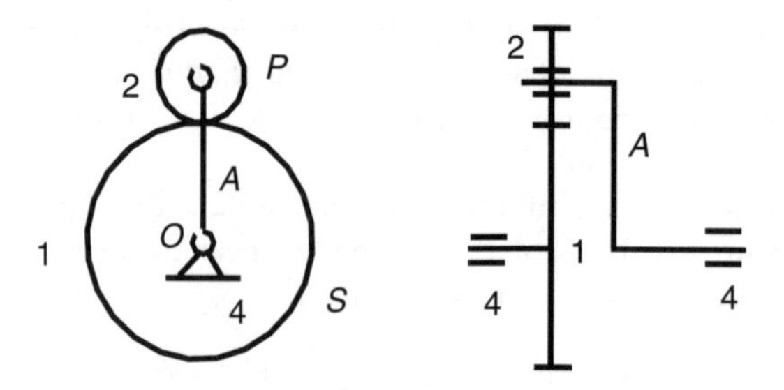

Figure 10.24 Simple epicyclic gear train.

the arm and the sun gear will both be driven and the output taken from the rotation of the planet wheel. In order to determine the amount of this rotation, a technique that can be used is to first imagine the arm to be fixed while S rotates through +1 revolution. This causes P to rotate through $-t_s/t_p$ revolutions, where t_s is the number of teeth on the sun wheel and t_p the number of teeth on the planet wheel. Then we imagine the gears to be locked solid and give a rotation of 1 revolution to all the wheels and the arm about the axis through S. If we had taken S to be fixed with the arm rotating, then the result is just the sum of the above two operations. Thus while the arm rotates through −1 revolution, the planet gear rotates through $-(1 + t_s/t_p)$ revolutions. These results are summarized in Table 10.1.

It is difficult to get a useful output from the orbiting planet as its axis of rotation is moving. A more useful form is shown in Figure 10.25. This has a ring (often termed the annulus) gear, R, added. This has internal teeth and there are three planets, which mesh with it and can rotate about pins through the arms emanating from the centre and the axis of the sun. There are usually three or four planets. We can use the same technique as above to determine the relative motion of the wheels and arm. Consider the arm to be fixed and the ring is

Table 10.1 Analysis of epicyclic gear train in Figure 10.24

	Rotation		
Operation	*Arm*	*S*	*P*
A Fix and rotate *S* by +1 rev.	0	+1	t_s/t_p
B Give all −1 rev.	−1	−1	−1
Adding *A* and *B*	−1	0	$-(1 + t_s/t_p)$

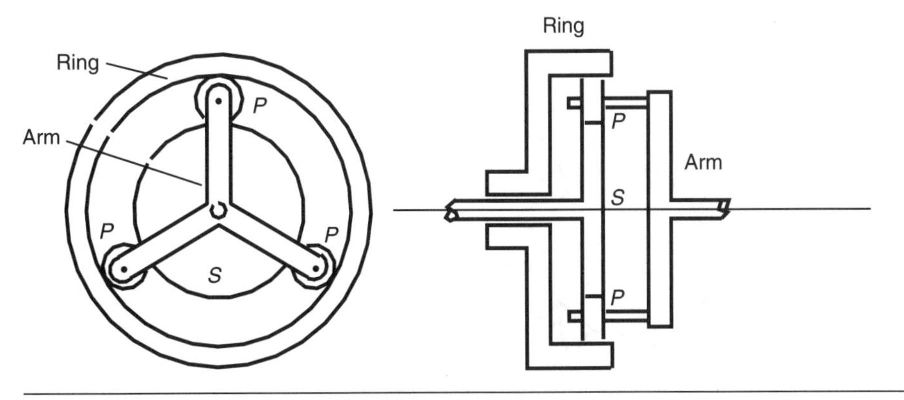

Figure 10.25 Epicyclic gear train with ring output.

Table 10.2 Analysis of epicyclic gear train in Figure 10.25

Operation	Rotation			
	Arm	Ring	S	P
A Fix, arm and rotate arm by +1 rev.	0	+1	$-t_R/t_S$	$+t_R/t_P$
B Give all −1 rev.	−1	−1	−1	−1
Adding *A* and *B*	−1	0	$(-1-t_R/t_S)$	$(-1+t_R/t_P)$

rotated through +1 revolution. This causes the sun to rotate through $-t_R/t_S$ and the planets through $-t_R/t_P$. Then we consider the gears to be locked solid and all given −1 rotation about the axis through the sun. Now if we had taken the ring to be fixed with the arm rotating, then the result is the same as the sum of the above two operations. Thus −1 revolution of the arm, with the ring fixed, results in a revolution for the sun of $(-1-t_R/t_S)$ and for the planets of $(-1+t_R/t_P)$. These results are summarized in Table 10.2.

By fixing different parts of the epicyclic gear, different gear ratios can be obtained. The above discussion related to a fixed ring; we could, however, have had the ring rotate and kept the arm or the sun fixed. Epicyclic gears are the basis of most car automatic gearboxes.

10.6 Ratchet mechanisms

Figure 10.26 shows the basic form of a ratchet mechanism. It consists of a wheel, called a *ratchet,* with saw-shaped teeth, which engage with an arm called a *pawl.* The arm is pivoted and can move back and forth to engage the wheel. The shape of teeth are such that rotation can occur in only one direction.

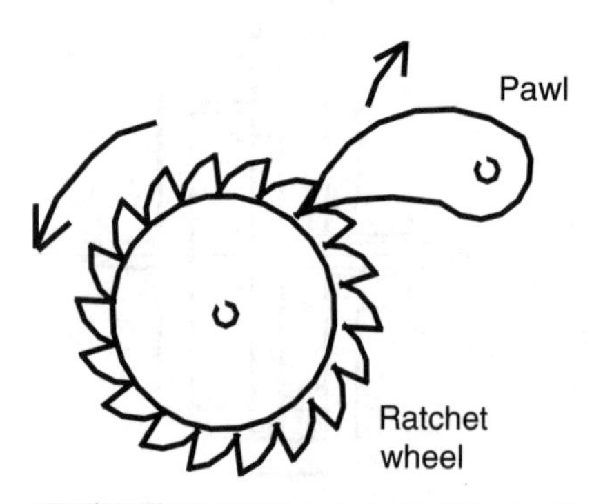

Figure 10.26 Ratchet mechanism.

10.7 Flexible mechanical elements

Flexible mechanical elements, such as belts and chains, are used to usually replace a group of gears, bearings, and shafts or similar power transmission devices. These flexible mechanical transmission devices are employed for power transmission when comparatively long distances are involved. They have the following functions:

- to increase torque by reducing speed;
- to reduce torque by increasing speed;
- to change axis of rotation;
- to convert linear motion into rotary motion; and
- to convert rotary motion into linear motion.

The aim of any design is to reduce the weight and inertia of each machine part, particularly on the links near the output of a device. Moreover, it is preferable to use small actuators that can fit inside the link rather than large ones which would obstruct the user. Without any transmission device, the actuators or motors that drive an output of most machines would have to be bulky and heavy. Therefore, designers can reduce size and weight of each actuator of a machine by cleverly using mechanical transmission devices.

10.7.1 Belt drives

Belt drives are essentially just a pair of rolling cylinders, with the motion of one cylinder being transferred to the other by a belt (Figure 10.27). Belt drives use the motion that develops between the pulleys attached to the shafts and the belt around the arc of contact in order to transmit torque. Since the transfer

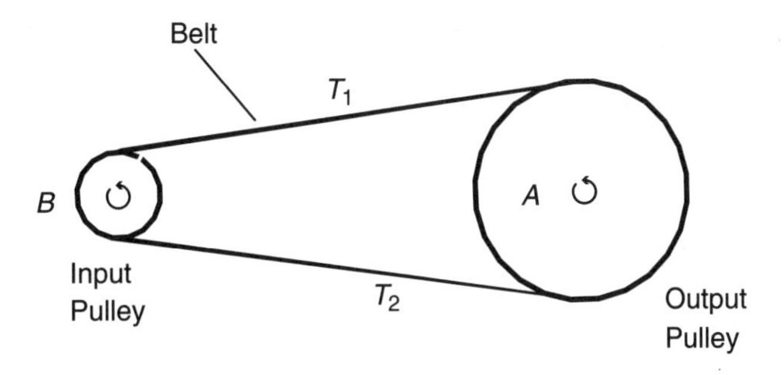

Figure 10.27 Belt drive.

relies on motion forces, slip can occur. As a method of transmitting power between two shafts, belt drives have the advantage that the length of the belt can easily be adjusted to suit a wide range of shaft-to-shaft distances. In this case, the system is automatically protected against overload because slipping occurs if the loading exceeds the maximum tension that can be sustained by frictional forces. If the distances between shafts are large, a belt drive is more suitable than gears, but over small distances gears are to be preferred. Different size pulleys can be used to give a gearing effect. However, the gear ratio is limited to about 3:1 because of the need to maintain an adequate arc of contact between the belt and the pulleys. The transmitted torque is due to the differences in tension that occur in the belt during operation. This difference results in a tight side and a slack side for the belt. If the tension on tight side is T_1, and that on the slack side is T_2, then with pulley A in Figure 10.27 as the driver

$$\text{Torque on } A = (T_1 - T_2)r_A, \qquad (10.7)$$

where r_A is the radius of pulley A.

For the driven pulley B we have

$$\text{Torque on } B = (T_1 - T_2)r_B, \qquad (10.8)$$

where r_B is the radius of pulley B.

Since the power transmitted is the product of the torque and the angular velocity, and since the angular velocity is v/r_A for pulley A and v/r_B for pulley B, where v is the belt speed, then for either pulley we have

$$\text{Power} = (T_1 - T_2)v. \qquad (10.9)$$

10.7.2 Chain drives

Slip can be prevented by the use of chains, which lock into teeth on the rotating cylinders to give the equivalent of a pair of intermeshing gear wheels. A chain drive has the same gear ratio relationship as a simple gear train. The drive mechanism used with a bicycle is an example of a chain drive.

10.8 Friction clutches

A coupling may be used to connect shafts in line so that the one will drive the other. If the connection is to tie permanently, bolted couplings are used. When overloading may have serious consequences a slipping coupling may be used

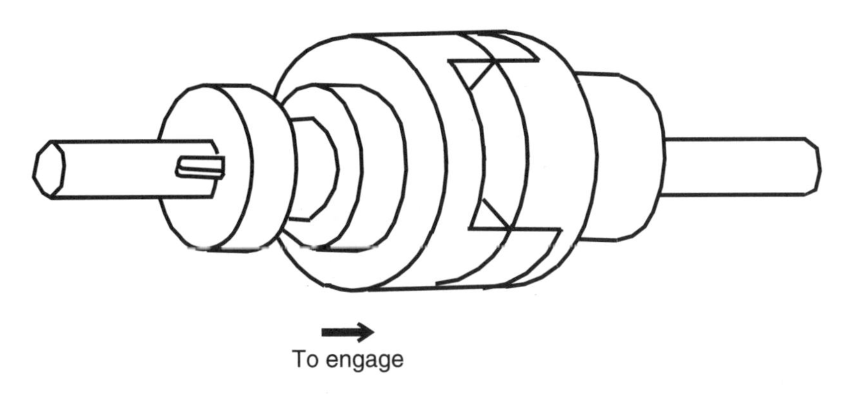

Figure 10.28 Dog clutch.

(the engagement is permanent but slip will take place in an emergency). If frequent disconnection is called for, a coupling that can be readily broken or renewed is required. Such a coupling is called a *clutch*.

10.8.1 The dog clutch

A simple clutch is the dog clutch shown in Figure 10.28. Often, however, engagement has to be gradual and then a *friction* clutch is required. Slipping will take place as engagement proceeds and the speed of the driven shaft builds up until, when the shafts are running at the same speed, slipping ceases. A clutch, therefore, is used to transmit motion from a power source to a driven component and bring the two to the same speed. Once full engagement has been made, the clutch must be capable of transmitting, without slip, the maximum torque that can be applied to it.

10.8.2 The cone clutch

The earliest form of friction clutch was the cone clutch, see Figure 10.29. An external cone is keyed to the driving shaft and against this is pressed an internal cone sliding on a feather on the driven shaft. The movable portion should be on the driven shaft since then it will be at rest when disengaged. The internal cone may be faced with leather or a synthetic material. To ensure a constant engagement force a spiral spring is used to press the faces together, the internal cone being then disengaged against the spring pressure.

Alternatively the driving cone can be tapered in the opposite direction (Figure 10.30) giving what is known as the internal cone clutch. It is very

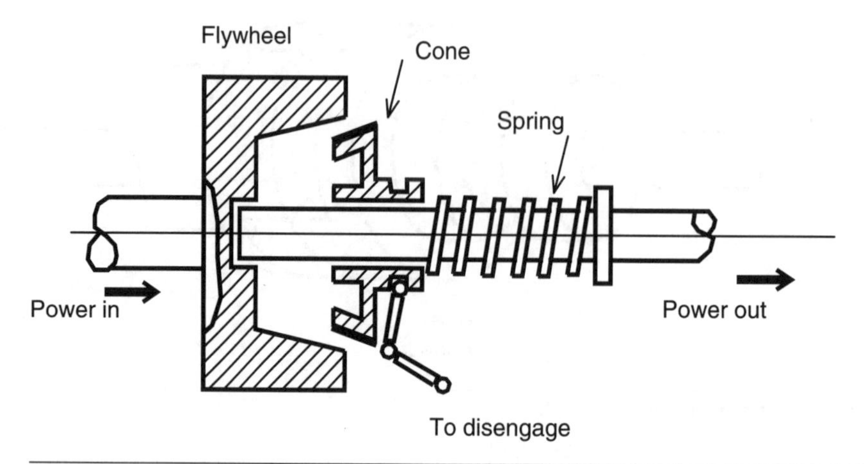

Figure 10.29 External cone clutch.

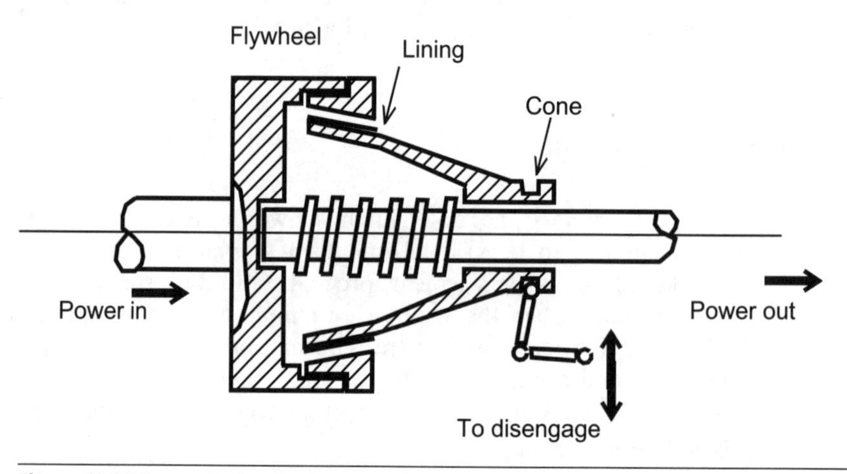

Figure 10.30 Internal cone clutch.

necessary, particularly with fast running shafts, that the cones mate intimately with some provision to enable them to adjust themselves to seat exactly. Wear of the friction lining can be appreciable and relining can be time consuming.

10.8.3 The plate clutch

A more compact and trouble-free type of clutch is the plate type (Figure 10.31) and this is by far the most common type of friction clutch. Early plate

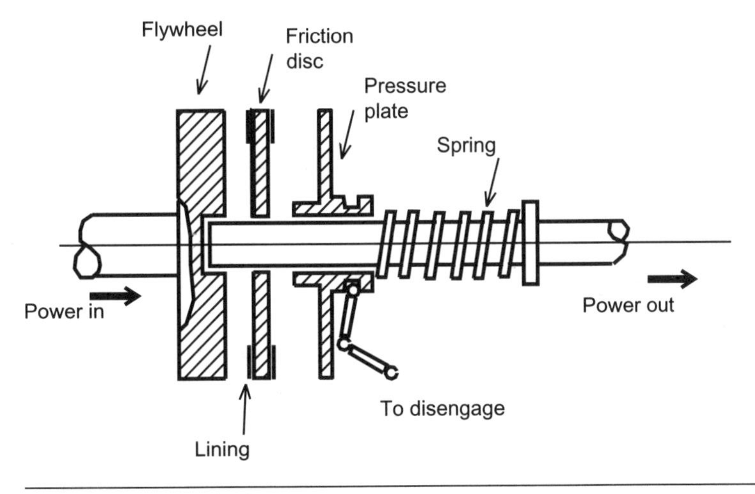

Figure 10.31 Plate clutch.

clutches were of the multi-plate metal-to-metal type and this type is still widely used usually in the oil immersed form. The oil provides a cushioning effect giving smooth engagement and also carries away the heat energy generated, resulting in lower working temperature and prolonged life of the mating parts.

The big and obvious disadvantage is the reduction in the coefficient of friction which has to be counteracted by the use of higher working pressures, but it must be remembered that too high a coefficient of friction will result in a rapid drive take up with shock loading of the parts involved. However, when compactness and price are the main considerations, a high coefficient of friction may have to be tolerated. In dry clutches it is common practice to use fabric linings which, having little inherent mechanical strength, have to be attached to steel plates. In order to transmit the maximum contact it is highly desirable that mating surfaces should be in close contact at all points. Consequently pressure plates are designed to flex sufficiently to adapt themselves to variations in the contacting surfaces and not to distort under high temperature working.

10.8.4 The band clutch

A band clutch, is essentially a belt sliding on a drum and the conditions are such that the tension in the belt is zero for zero torque transmission. The operating mechanism exerts a pull on the free end of the band making it grip the driving or driven member material. The band may be plain or faced with anti-friction material.

10.8.5 The internal expanding clutch

In the internal expanding clutch, the expanding force is generally exerted on the lined shoes by means of right- and left-handed screws, which are rotated by levers operated by a sleeve sliding along the driven shaft.

10.8.6 The centrifugal clutch

Centrifugal clutches, are similar in construction to internal expanding clutches but the operating force is that due to centrifugal action. They are used to enable an electric motor to get started before picking up the load, the weight of the shoes being such that the full load is taken up and slipping ceases when the motor is running at full speed.

10.8.7 Clutch selection

The operating characteristics of different clutch designs, and the requirements of the application, can be used as a guide to the selection of an appropriate clutch type (see Table 10.3).

10.8.8 Clutch facings

10.8.8.1 *Lining materials*

Impregnated woven cotton-based linings are used to obtain high friction, but the maximum operating temperature is limited to that at which cotton begins to char; therefore, asbestos has replaced cotton for applications where greater heat resistance is required.

The fibers are woven to produce a fabric which is impregnated with a resin solution to improve strength and wear rate, and also to produce good friction characteristics.

Zinc or copper wire is often introduced to increase mechanical strength and improve heat conductivity. The zinc acts as a friction modifier, and also helps to conduct heat away.

Asbestos-based molded friction materials consist basically of a cured mix or combination of short asbestos fibers, fillers and bonding resins containing metal particles.

Sintered metals are used for a limited number of friction applications. The metal base is usually bronze, to which is added lead, graphite and iron in

Table 10.3 Guide to selecting clutches.

Type of clutch	*Special characteristics*	*Typical applications*
Cone type	Embodies the mechanical advantage of the wedge, which reduces the axial force required to transmit a given torque. It also has greater facilities for heat dissipation than a plate clutch of similar size and so may be more heavily rated.	In general engineering its use is restricted to more rugged applications such as contractors' plant. Machine tool applications include feed drives, and bar feed, for auto lathes.
Single-plate (disc)	Used where the diameter is not restricted. Springs usually provide the clamping pressure by forcing the spinner plate against the driving plate. Simple construction, and if of the open type ensures no distortion of the spinner plate by overheating.	Wide applications in automobile and other traction drives.
Multi-plate	Main feature is that the power transmitted can be increased by using more plates, thus allowing a reduction in diameter. If working in oil,* it must be enclosed, whereas a dry plate clutch can often have circulating air to carry away the heat generated.	Extensively used in machine tool headstocks, or in any gearbox drive where space is limited between shaft centers.
Expanding ring	Will transmit high torque at low speed. Centrifugal force augments gripping power, so withdrawal force must be adequate. Both cases show positive engagement.	Large excavators. Textile machinery drives. Machine where clutch is located in driving pulley.
Centrifugal	Automatic in operation, the torque without spring control increasing as the square of the speed. An electric motor with a low starting torque can commence engagement without shock, the clutch acting as a safety device against stalling and overload. Shoes are often spring-loaded to prevent engagement until 75% of full speed has been reached.	Wide applications on all types of electric motor drives, generally reducing motor size and cost. Industrial diesel engine drives.

*Working in oil gives a reduction in friction, but this can be counteracted by higher operating pressures. As long as there is an oil film on the plates, the friction and the engagement torque remain low, but as soon as the film breaks the engagement torque rises rapidly and may lead to rapid acceleration. The friction surface pressure should not exceed $1\,MN\,m^{-2}$ with a sliding speed maximum of $20\,m\,s^{-1}$ for steel on steel. With oil immersed clutches having steel and sintered plates, the relationship between static and dynamic coefficient of friction is more favourable. Friction surface pressure and sliding speed may then be up to $3\,MN\,m^{-2}$ and $30\,m\,s^{-1}$.

powder form. The material is suitable for applications where very high temperatures and pressures are encountered. It is rigid and has a high heat conductivity, but gives low and variable friction.

10.8.8.2 *Mating surfaces*

The mating surfaces need to have the following characteristics: (a) requisite strength and low thermal expansion; (b) hardness sufficient to give long wear life and resist abrasion; and (c) heat soak capacity sufficient to prevent heat spotting and crazing.

Close-grained pearlitic grey cast iron meets these requirements; a suitable specification being an iron with the following percentage additions: 3.3 carbon, 2.1 silicon, 1.0 manganese, 0.3 chromium, 0.1 sulphur, 0.2 phosphorus, 4.0 molybdenum, 0.5 copper plus nickel. Hardness should ideally be in the range 200–30 BHN.

10.8.8.3 *Oil-immersed clutches*

Multi-plate clutches are suitable for working in oil. Oil acts as a cushion and energy released by heat is carried away by oil. The main disadvantage is a reduction in friction, but this can be counteracted by higher operating pressures. As long as there is an oil film on the plates, the friction characteristic and engagement torque remain low, but as soon as the film breaks the engagement torque rises rapidly and may lead to rapid acceleration. The friction surface pressure should usually not exceed $1\,MN\,m^{-2}$ with a sliding speed maximum of $20\,m\,s^{-1}$, steel on steel. With oil-immersed clutches having steel and sintered plates the relationship between the static and dynamic coefficient of friction is more favorable. Friction surface pressure and sliding speed may be up to $3\,MN\,m^{-2}$ and $30\,m\,s^{-1}$. The use of facing grooves helps to prevent the formation of an oil film that would lower the coefficient of friction. They also provide space for the oil to be absorbed during clutch engagement.

10.8.8.4 *Clutch facing materials*

The materials for clutch facing are listed in Table 10.4.

10.9 Design of clutches

When a plate clutch is new it is perhaps true to say that the pressure holding the two plates together is uniform. However, as the clutch wears the outer

Table 10.4 Clutch facing materials

Type	*Uses*
Woven	Industrial band, plate and cone clutches, cranes, lifts, excavators, winches and general engineering applications
Millboard	Mainly automotive and light commercial vehicles
Wound tape/yarn	Mainly automotive and light commercial vehicles
Asbestos tape	Agriculture and industrial tractors
Molded	Automotive, commercial vehicles, agriculture and industrial tractors
Sintered	Tractors, heavy vehicles, road rollers, winches, machine tool applications
Cermet	Heavy earth-moving equipment, crawler tractors, sweepers, trenchers and graders
Oil immersed paper	Automotive and agricultural automatic transmissions
Woven	Band linings and segments for automatic transmissions
Molded	Industrial transmissions and agricultural equipment
Sintered	Power shift transmission, presses, heavy-duty general engineering applications
Resin/graphite	Heavy-duty automatic transmissions

portion of the plate, where the velocity is high, will wear more than the inner portion. After initial wearing in it is reasonable to assume that the profile of the clutch face will remain thus it can be considered that the wear rate over the clutch face will be constant. Let us consider these two cases.

10.9.1 Constant pressure

Figure 10.32 shows an elemental area on the two faces of a simple plate clutch. In the radial direction, an element of width, dr, is considered, and analyzed for the axial force, average pressure, and average torque.

$$\text{The element area } dA = (2\pi r\, dr). \qquad (10.10)$$

$$\text{The differential normal force } dF = p dA = p(2\pi r\, dr). \qquad (10.11)$$

$$\text{The differential frictional force } dQ = \mu\, dF = \mu p(2\pi r\, dr). \qquad (10.12)$$

$$\text{The differential frictional torque } dT = r dQ = \mu p\left(2\pi r^2\, dr\right). \qquad (10.13)$$

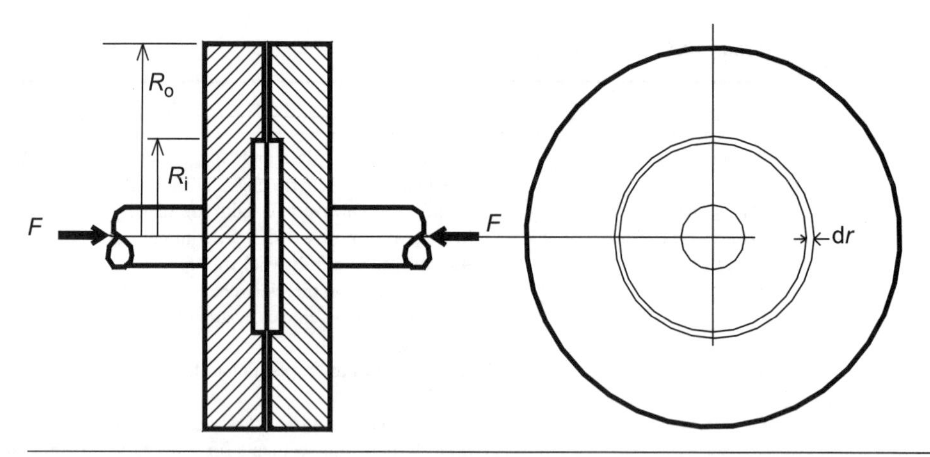

Figure 10.32 Simple plate clutch detail.

The total torque transmitted by the clutch = T.

$$T = 2\pi\mu p \int_{R_1}^{R_0} r^2\, \mathrm{d}r = 2\pi\mu p\left(\frac{R_o^3 - R_i^3}{3}\right). \tag{10.14}$$

$$\text{The axial force } F = p\pi\left(R_o^2 - R_i^2\right), \tag{10.15}$$

$$\text{thus the average pressure } p = \frac{F}{\pi\left(R_o^2 - R_i^2\right)}, \tag{10.16}$$

$$\text{and } T = \mu F\left[\frac{2}{3}\frac{\left(R_O^3 - R_i^3\right)}{\left(R_O^2 - R_i^2\right)}\right]. \tag{10.17}$$

10.9.2 Constant wear

In any time interval the work done per unit area is constant. Uniform wear can be expressed as:

$$C_1 = \frac{\text{friction force} \times \text{velocity}}{\text{area}} = \frac{\mu p(2\pi r\, \mathrm{d}r)\omega r}{(2\pi r\, \mathrm{d}r)} \tag{10.18}$$

$$\text{or } p = \frac{c}{\mu\omega r}, \text{ or as } \mu \text{ and } \omega \text{ are constants, } p = \frac{c}{r}. \tag{10.19}$$

The total torque T is given by

$$T = 2\pi\mu \int_{R_i}^{R_o} pr^2 \mathrm{d}r = 2\pi\mu c\left(\frac{R_o{}^2 - R_i{}^2}{2}\right). \tag{10.20}$$

To find c

$$F = \int_{R_i}^{R_o} p(2\pi r\,\mathrm{d}r) = 2\pi c(R_o - R_i). \tag{10.21}$$

$$\text{Thus } c = \left(\frac{F}{2\pi(R_o - R_i)}\right). \tag{10.22}$$

$$\text{Therefore } T = \mu F\left(\frac{R_o - R_i}{2}\right). \tag{10.23}$$

$$T_{cp} = \mu F R'_{cp} \tag{10.24}$$

$$T_{cw} = \mu F R'_{cw} \tag{10.25}$$

Plotting R'/R_o against R_i/R_o shows that in all cases, $R'_{cp} > R'_{cw}$. Consequently, one should design for constant wear. Thus in designing a plate clutch one should design for constant wear. For a multiple plate disk the torque capacity is the torque transmitted by a single plate clutch times the number of pairs of surfaces transmitting power. Consider the same situation for a cone clutch. Consider a differential element bounded by circles of radii r and $(r + \mathrm{d}r)$. The area of the differential frustum of a cone is

$$\mathrm{d}A = 2\pi r\left(\frac{\mathrm{d}r}{\sin\,\alpha}\right), \tag{10.26}$$

for a uniform pressure the differential torque $\mathrm{d}T$ is

$$\mathrm{d}T = 2\pi r\left(\frac{\mathrm{d}r}{\sin\,\alpha}\right)p\mu r. \tag{10.27}$$

Thus total torque T is

$$T = \frac{2\pi\mu p}{\sin\,\alpha}\left(\frac{R_o{}^3 - R_i{}^3}{3}\right). \tag{10.28}$$

Define normal force, F_n, as that due to pressure applied to the area as if it were stretched out.

$$F_n = p(2\pi R_m b), \tag{10.29}$$

$$\text{where } R_m \text{ is the mean radius} = \left(\frac{R_o + R_i}{2}\right). \tag{10.30}$$

b is width of clutch $= R_2 - R_1$.

To relate the normal force to the axial force consider a differential element with central angle $d\varphi$. The differential area is

$$dA = 2\pi R_m b\left(\frac{d\phi}{2\pi}\right) = R_m b \, d\phi. \tag{10.31}$$

$$\text{The differential normal force} = dN = pR_m b \, d\phi. \tag{10.32}$$

$$\text{The Axial component of } dN = dF = pR_m b \, d\phi \sin\alpha. \tag{10.33}$$

$$F = \int_0^{2\pi} pR_m b \, d\phi \sin\alpha = 2\pi p R_m b \sin\alpha = F_m \sin\alpha. \tag{10.34}$$

Thus torque T is given by

$$T = \frac{F_n \mu}{R_m b \sin\alpha}\left(\frac{R_o^3 - R_i^3}{3}\right) = F_n \mu\left[\frac{2}{3}\frac{(R_o^3 - R_i^3)}{(R_o^2 - R_i^2)}\right]. \tag{10.35}$$

$$\text{Thus } T = \frac{F\mu}{\sin\alpha}\left[\frac{2}{3}\frac{(R_o^3 - R_i^3)}{(R_o^2 - R_i^2)}\right]. \tag{10.36}$$

For uniform wear $p = c/r$ (see above)

$$T = \int_{R_i}^{R_o} p\mu 2\pi r\left(\frac{dr}{\sin\alpha}\right)r = 2\pi c\mu \frac{(R_o^2 - R_i^2)}{2\sin\alpha} \tag{10.37}$$

$$F_n = \int_{R_i}^{R_o} p 2\pi r\left(\frac{dr}{\sin\alpha}\right) = 2\pi c \frac{(R_o - R_i)}{\sin\alpha} \tag{10.38}$$

$$T = F_n \mu R_m. \tag{10.39}$$

The relationship of F_n to F can be obtained by first setting up the differential normal force on the differential area considered.

$$dF_n = p\left(\frac{dr}{\sin\alpha}\right) r\, d\phi \tag{10.40}$$

$$F = \int_{R_i}^{R_o}\int_{0}^{2\pi} p\left(\frac{dr}{\sin\alpha}\right) r\, d\phi \sin\alpha = 2\pi c(R_o - R_i) \tag{10.41}$$

$$\text{Thus torque } T = \frac{F\mu}{\sin\alpha}\frac{(R_o - R_i)}{2}. \tag{10.42}$$

Thus the only difference between a cone clutch and plate clutch is the effect of $\sin\alpha$ (as would be expected) and in designing a clutch we should consider the constant wear situation.

10.10 Brakes

A brake is a means of bringing a moving body to rest, or to a state of uniform motion, and holding it against the action of external forces or turning moments. There are three main types of brake:

- band brakes;
- drum brakes;
- disk brakes.

In the process of braking a body the energy removed is given up as heat. This generation of heat leads to a falling off of efficiency and thus heat must be dissipated by the most efficient means available to prevent *brake fade*. A brake has to develop the required torque for braking in a stable and controlled manner and must not reach temperatures high enough to impair its performance or damage its components. With continued use of brakes it is possible for the rate of heat generation to exceed the rate of heat dispersal, resulting in a rise in temperature of the friction material. In general a rise in temperature of a brake lining will result in a decreased coefficient of friction and a greater force must then be exerted on the brakes to maintain the same degree of braking. This should not be confused with the extra movement required in the case of drum brakes, when as a result of a rise in temperature the drum brake expands.

A measure of braking is known as the brake factor and may be defined:

$$\text{Brake Factor} = \frac{\text{Retarding Force}}{\text{Actuating Force}}. \qquad (10.42A)$$

10.10.1 The band brake

In a band brake (Figure 10.33), a flexible steel band lined with friction material is tightened against a rotating drum. Because of its self servo-action a band brake can be very powerful. Positive self-servo action occurs when the frictional force augments the actuating force so increasing the torque. The brake factor of a band brake increases rapidly with the coefficient of friction of the brake and the angle of wrap. Too much self-servo, makes the brake unstable and likely to grab and judder. If the drum rotates in the opposite sense such that the friction force opposes the actuating force (i.e. negative self-servo), the brake factor is very small.

10.10.2 The drum brake

Drum brakes (Figure 10.34) may be either external contracting or internal expanding or a combination of both. In the external form usually two shoes

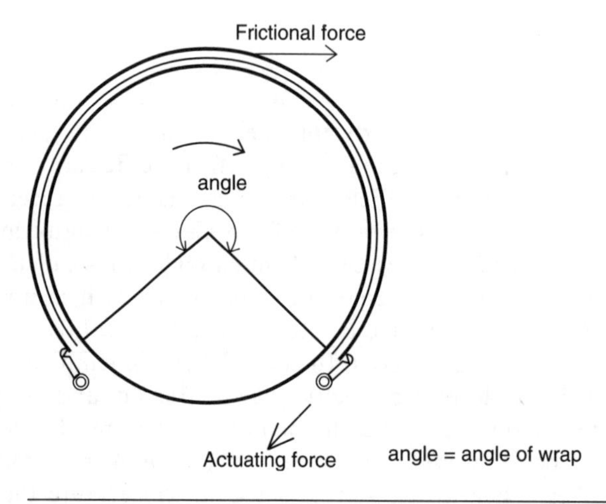

Figure 10.33 Band brake.

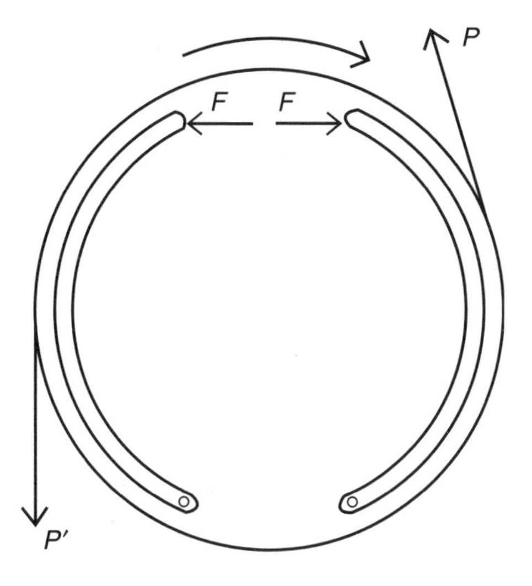

Figure 10.34 Drum brake.

are used pivoted such that the shoes contract when the brake is applied. In the case of the internal form again two shoes are common though more may be met. They are normally pivoted so as to expand when the brake is applied. When the direction of rotation of the drum is from the loaded end to the pivoted end the shoe is said to be a leading shoe, if opposite the shoe is a trailing shoe. With a leading shoe the frictional drag increases the effective applied force while with a trailing shoe the drag decreases. Shoes are lined with friction material usually over an arc of 900–1100 degrees.

10.10.3 The disk brake

In disk brakes, the drum of the drum brake is replaced by a disk and the shoes by a piece of friction material gripping the disk from opposite sides and held by a caliper (Figure 10.35). Multi-disk brakes are common in aircraft and large industrial plant.

Due to the *brake fade* problem the disk brake is rapidly replacing drum brakes in the automotive industry. Although disk brakes present their own problems they are capable of operating at a much greater temperature. Although the brake factor of this brake is low it is extremely stable and less affected by high temperature.

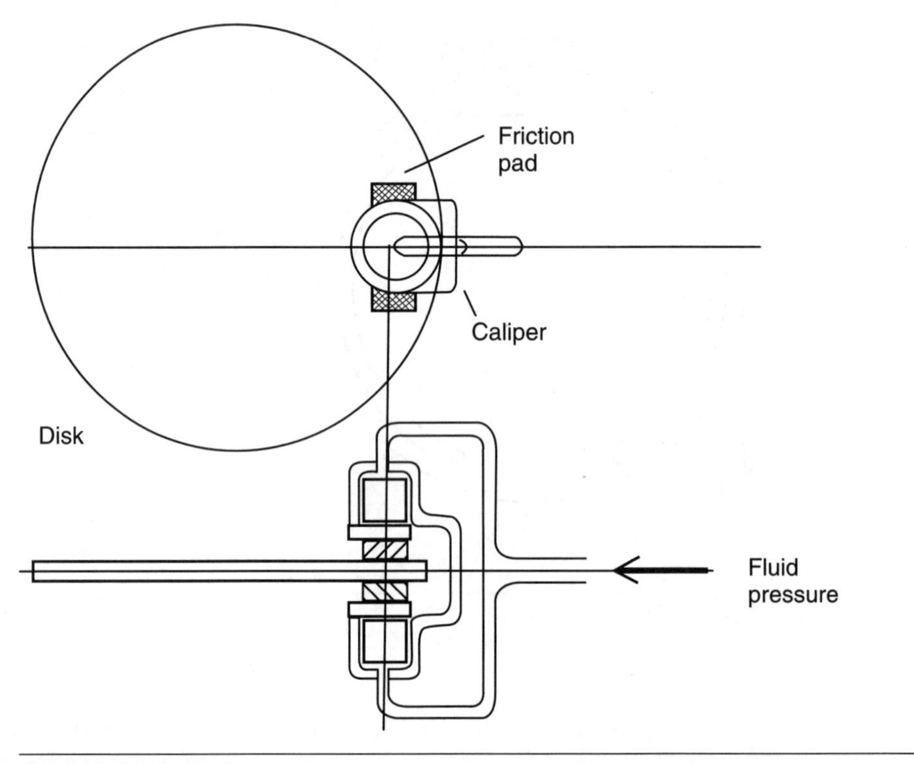

Figure 10.35 Disk brake.

10.10.4 Brake selection

The following lists the factors that need to be considered when choosing a brake type:

- torque capacity;
- cost;
- stability;
- temperature operating range;
- brake factor;
- humidity;
- dirt;
- vibration;
- water.

Problems

Q10.1 Explain the terms: (a) mechanism; (b) kinematic chain.

Q10.2 Explain what is meant by the four-bar chain.

Q10.3 For a cam with straight flanks and roller-end follower, derive expressions for: (a) displacement; (b) velocity; and (c) acceleration.

Q10.4 For a cam with curved flanks and flat-end follower, derive expressions for: (a) displacement; (b) velocity; and (c) acceleration.

Q10.5 For a circular cam and flat-end follower, derive expressions for: (a) displacement, (b) velocity; and (c) acceleration.

Q10.6 For a circular cam with straight flanks and roller-end follower, derive expressions for: (a) displacement, (b) velocity; and (c) acceleration.

Further reading

[1] Bolton, W. (1993) *Mechanical Science*, Blackwell Scientific Publications.
[2] Bolton, W. (1995) *Mechatronics: Electronic Control Systems in Mechanical Engineering*, Essex: Longman.
[3] Norton, R.L. (1992) *Design of Machinery,* McGraw-Hill.

CHAPTER 11

Interfacing microcontrollers with actuators

Chapter objectives

When you have finished this chapter you should be able to:

- understand the practical aspects of general purpose three-state transistors;
- understand how to interface microcontrollers with relays;
- understand how to interface microcontrollers with solenoids;
- understand how to interface microcontrollers with stepper motors;
- understand how to interface microcontrollers with permanent magnet motors;
- understand how to interface microcontrollers with sensors;
- be able to deal with power supplies requirements for a mechatronic system;
- understand how to interface microcontrollers with a DAC.

11.1 Introduction

This chapter presents the practical steps to be taken in interfacing microcontrollers with actuators, which are essential components of most mechatronic systems. Microcontrollers on their own are of no use in mechatronics unless they are interfaced with actuators that perform specific tasks in the real world. A proper knowledge of how to interface microcontrollers with actuators is essential. Therefore, this chapter is the bridge between theory and practice and should be utilized as the 'cook book' for putting most of the theories covered in the preceding chapters into practice. We first cover the basic gates such as AND, OR, NAND, NOT, latches, drivers, and decoders, which utilize general-purpose transistors. Then we discuss how to practically interface microcontrollers with relays, solenoids, stepper motors, permanent magnet motors, sensors, power supplies, and ADCs/DACs, and communication using the RS 232 and RS 485 protocols.

11.2 Interfacing with general-purpose three-state transistors

The two basic logic families of integrated circuits (ICs) are made from the transistor-transistor-logic (TTL), based on the bipolar technology, and the complementary metal-oxide semiconductor (CMOS), based on metal-oxide semiconductor field-effect transistor (MOSFET) technology. A family of logic gates such AND, OR, NAND, NOT, latches, drivers, and decoders, are produced from ICs having a number of transistors made with the same technology and having the same electrical characteristics, together with some resistors. It is important for those working in the area of mechatronics to be familiar with the differences in requirements between the TTL and CMOS families especially when there is the need to interconnect these transistors.

11.2.1 The 74LS373 octal latch

The 74LS373 octal D-type transparent latch shown in Figure 11.1 is an 8-bit latch, which feature tri-state outputs designed specifically for driving highly capacitive or relatively low impedance loads. They are particularly suitable for implementing buffer registers, I/O ports, bidirectional bus drivers and working registers.

Signal	Pin	Pin	Signal
$\overline{OC}$	1	20	Vcc
1Q	2	19	8Q
1D	3	18	8D
2D	4	17	7D
2Q	5	16	7Q
3Q	6	15	6Q
3D	7	14	6D
4D	8	13	5D
4Q	9	12	5Q
GND	10	11	C+

Figure 11.1 74LS373 pin connections.

The eight latches of the 74LS373 are transparent D-type latches. While the *enable* is taken as *low*, the Q outputs will be latched at the levels that were set up at the D inputs.

A buffered output control ($\overline{OC}$) can be used to place the eight outputs in either a normal logic state (*high* or logic *low* levels) or a high-impedance state. In the high impedance state an increased drive provides the capability to drive the bus lines in a bus organized system without the need for interface or pull up components.

The output control $\overline{OC}$ does not affect the internal options of the latches. Old data can be retained or new data can be entered while the outputs are off.

11.2.2 The 74LS244 octal buffer and line driver

The 74LS244 octal buffer and line driver shown in Figure 11.2 has tri-state outputs. The octal buffer and line driver is designed specifically to improve both the performance and density of tri-state memory address drivers, clock drivers and bus-oriented receivers and transmitters. The device provides the choice of selected combinations of inverting outputs, symmetrical G (active-low input control) inputs and complementary G and g inputs.

11.2.3 The 74ALS138 decoder

The 74ALS138 decoder (Figure 11.3) is designed to be used in high performance memory decoding or data routing applications requiring very short propagation delay times. In high performance memory systems, these decoders can be used to minimize the effects of system decoding. When employed with high speed memories utilizing a fast enable circuit, the delay times of these decoders and the enable time of the memory are usually less than the typical access time of the

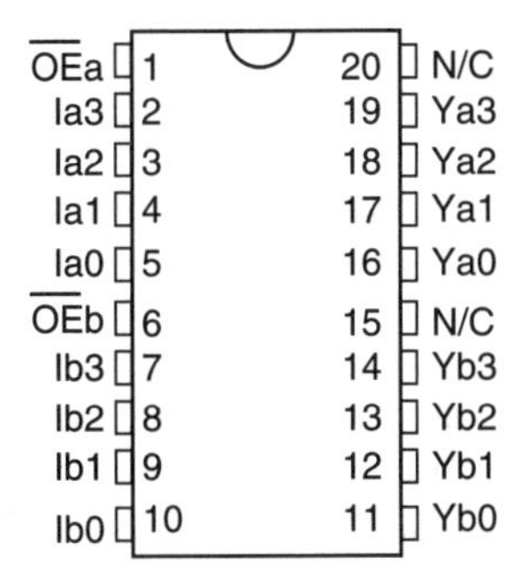

Figure 11.2 74LS244 pin connections.

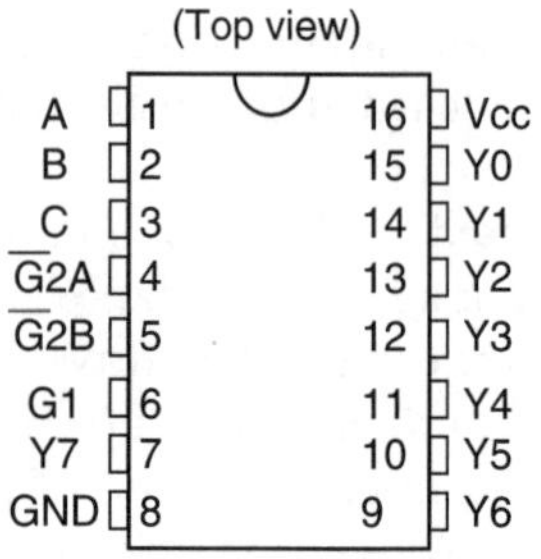

Figure 11.3 74ALS138 pin connections.

memory. The conditions at the binary select inputs and the three enable inputs select one of eight input lines. Two active *low* and one active *high* enable inputs reduce the need for external gates or inverters when expanding. Similarly a 24-line decoder can be implemented without external inverters and a 32-line decoder requires only one inverter. An *enable* input can be used as a data input for demultiplexing applications.

11.2.4 The 74LS00 quad NAND gate

The 74LS00 is a quadruple two-input positive NAND gate.

11.3 Interfacing relays

5 V relays are available, but with the limited current driving ability of the PIC, they will not trigger. Hence, the Darlington-pair driver IC (ULN2003A) is used to provide this current capability. This driver IC can be connected to a higher current source and it will switch the flow of high current to the motors as required. However, if very high current is used, it is always a good idea to use diodes to prevent the e.m.f. from the de-energizing of the relay harming the PIC. Figure 11.4 shows a block diagram of how to interface PIC microcontrollers with a relay.

In order to amplify the output of a PIC or the interfacing board of a PC to drive solenoids, a ULN2003A seven Darlington-pair array driver has to be used. The ULN2003A has seven Darlington-pairs per package and each has an output current and output voltage of up to 500 mA and 50 V respectively.

The relay is a 12 double-pole, double-throw (DPDT) type, which has a coil resistance of 200 Ω and a nominal power of 500 mW. It also has contact voltage and contact current ratings of 30 V d.c. and 10 A, respectively.

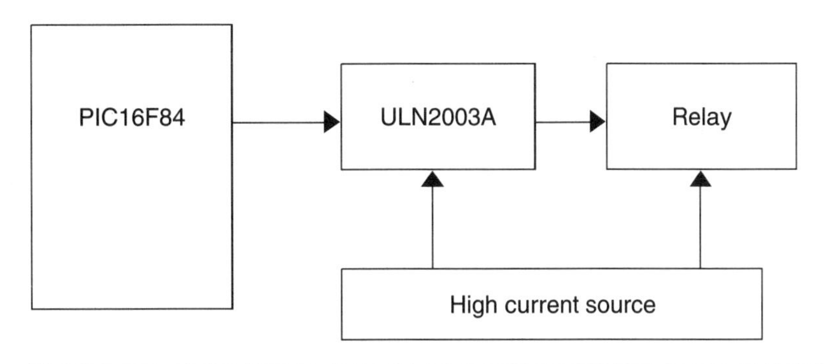

Figure 11.4 Block diagram for interfacing relays (high current peripherals).

Table 11.1 OMRON coil ratings

Rated voltage	*Coil resistance*	*Coil inductance armature OFF*	*Coil inductance armature ON*	*Must operate voltage ON*	*Must release voltage*	*Power consumption*	*Maximum voltage VDC*
12	275	1.15 H (ref. value)	2.29 H (ref. value)	8.4 V (70% of rated voltage)	1.8 V (70% of rated voltage)	Approx. 0.53 W	110% of rated voltage

Table 11.1 shows the datasheet containing information about the relay. The power consumption is approximately 0.53 W and the coil resistance is 275 Ω, hence we can deduce the current rating of the relay coil as:

$$I = \sqrt{\frac{P}{R}}$$
$$\therefore I = \sqrt{\frac{500\,\text{mW}}{200\,\Omega}} = 50\,\text{mA}. \quad (11.1)$$

11.4 Interfacing solenoids

Solenoids are useful devices, for example for clamping workpieces on a worktable during drilling or milling processes.

When a microcontroller or a PC interface is used to drive a solenoid, one of the outputs of the PIC microcontroller or the interfacing board of the PC has to be amplified in order to control the operation of the clamping solenoids. The mode of operation is a simple on/off control.

Let us consider a 10 W, 6 V d.c. 100% duty cycle solenoid used in a clamping device. Often suppliers give no further datasheet information, so in order to obtain the rated current requirement of a solenoid its coil resistance has to be first measured experimentally using a digital multimeter. The rated current is obtained as follows:

$$P = I^2 R$$
$$I = \sqrt{\frac{P}{R}}\,. \tag{11.2}$$

For example, if there are two solenoids, the total current consumption is $2I$ and often this current cannot be supplied by a PIC microcontroller or a PC interfacing board directly. In order to amplify the output of the interfacing board to drive the solenoids a ULN2003A seven Darlington-pair array driver and a 12 V d.c. single pole double-throw high power relay (already discussed) have to be used. Figure 11.5 shows a block diagram representation of how the voltages and currents vary from a PC-based interfacing board to the solenoids.

A circuit diagram of the amplifier circuit to drive the clamping solenoids is shown in Figure 11.6.

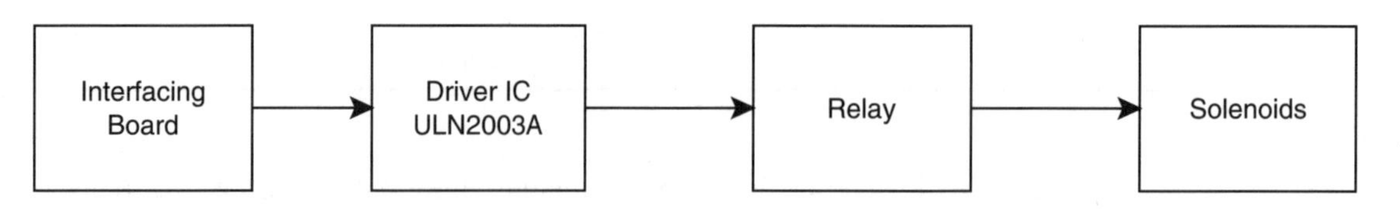

Figure 11.5 Block diagram for interfacing solenoids.

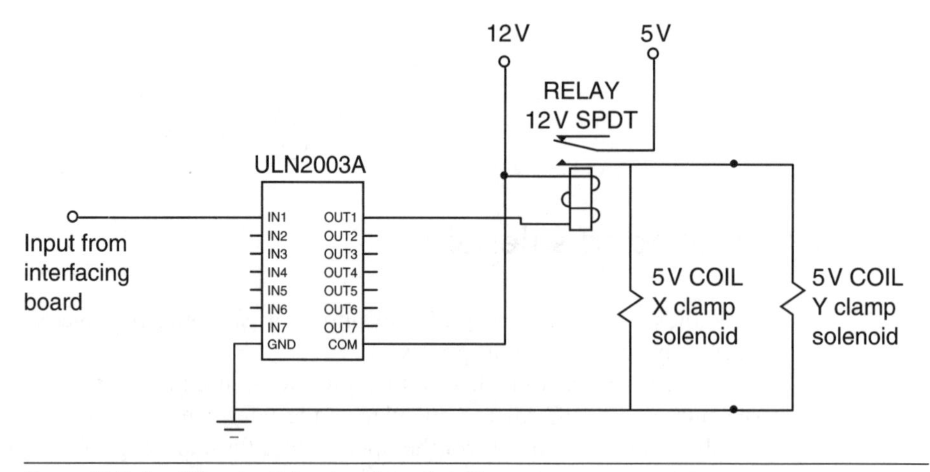

Figure 11.6 Amplifier circuit for clamping solenoids.

11.5 Interfacing stepper motors

Hybrid stepper motors (four-phase) are capable of delivering much higher working torque and stepping rates (1.8 degrees) than permanent magnet, servo and wiper motors. They also maintain a high detent torque even when not energized. This feature is good for positional integrity. Commonly used stepper motors are directly compatible with proprietary stepper motor drive boards.

11.5.1 Methodology for stepper motor control design

The successful operation of a stepper motor requires the following elements as shown in Figure 11.7.

- **Control unit:** A typical control unit is microprocessor based, a PC, or a microcontroller.
- **Power supply:** A 15 V/6 A power supply is suitable for the stepper motor drive card. For small-to-medium loads, motors and driver cards typically require a voltage supply of 15 V. 5 V and 12 V regulators are used to step down the voltages for the different sensors and for the operational amplifiers and bootloader board, respectively.
- **Drive card:** This converts the signals from the control unit into the required stepper motor sequence. Typical driver cards are unipolar 2 A and bipolar 3.5 A stepper motor drive boards. The unipolar driver card is used for motors with current ratings less than 2 A while the bipolar driver card is used for motors with current ratings less than 3.5 A.
- **Stepper motor:** Normally, each stepper motor requires a separate driver card.

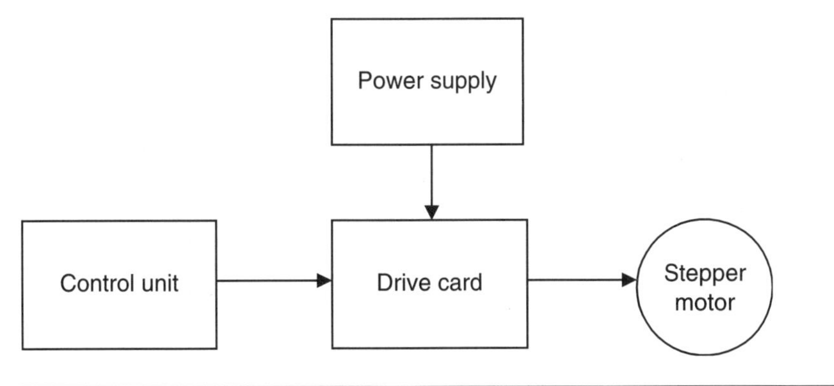

Figure 11.7 Stepper motor control system.

11.5.1.1 *Motor drive methods*

The motor drive methods used for the control of the hybrid stepper motors are:

- unipolar stepper motor drive;
- bipolar stepper motor drive.

The normal way of driving a four-phase hybrid stepper motor is shown in Figures 11.8 and 11.9.

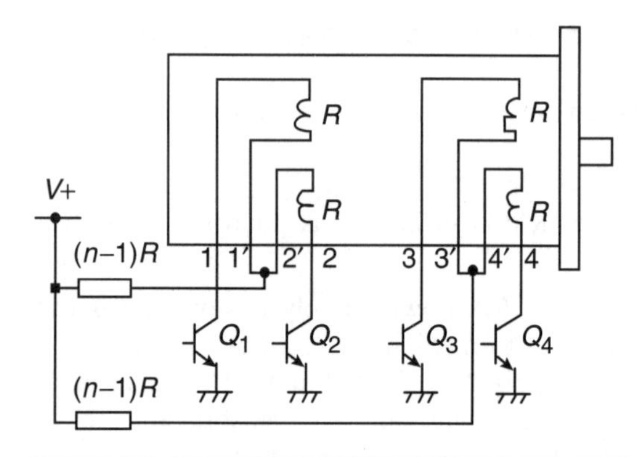

Figure 11.8 Unipolar drive circuit. (Courtesy the RS Catalog, 2001).

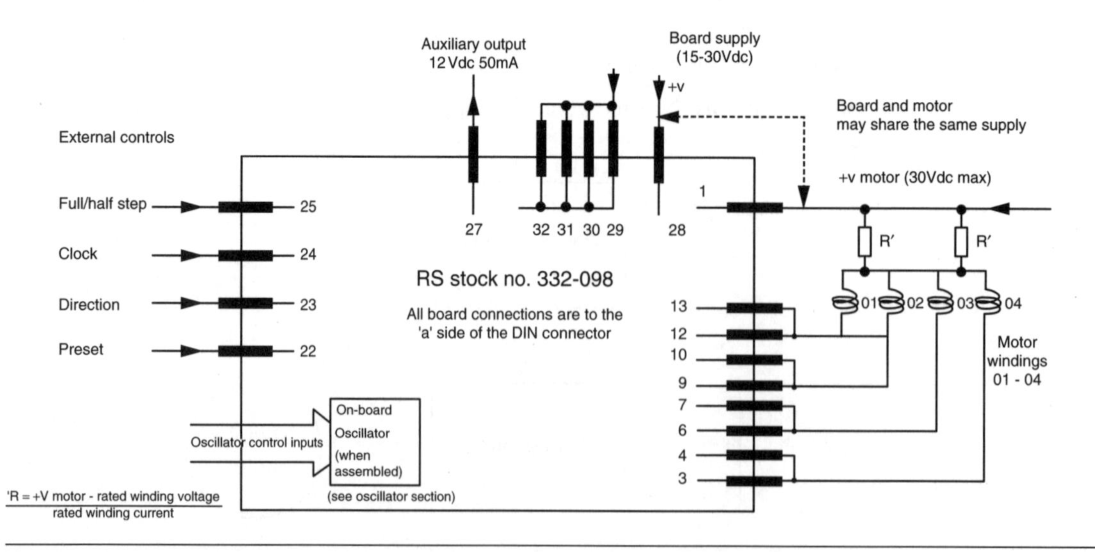

Figure 11.9 Unipolar stepper motor drive board connections. (Courtesy the RS Catalog, 2001).

11.6 Interfacing permanent magnet motors

The use of d.c. motors is common in mechatronics applications. In Chapter 6, we discussed that the limited source/drain current of a PIC can cause design difficulties. This problem is also encountered when interfacing a PC with motors or other actuators that draw heavy current. Motors in particular, require a lot of current. For example, all automotive windshield wiper motors operate at three distinct speeds in both directions. The usual rotation speed is 40 revolutions per minute. A maximum of 4 A d.c. is drawn during startup at 12 V. However, the normal working current consumption is as low as 1.32 A. The speed reduces to 30 revolutions per minute for rotation in the opposite direction. The PIC microcontroller is not able to interface with these motors without a proper interface. Two methods of interfacing a PC or microcontrollers with PM motors are:

- use of relay circuits;
- use of H-bridge circuits.

11.6.1 Relay circuits

One of the best proven methods of driving PM motors is using a Darlington-pair driver IC with a relay. As in the earlier examples, this driver IC can be connected to a higher current source and it will switch the flow of high current to the motors as required. However, if very high current is used, it is always recommended that diodes are used to prevent the e.m.f. from the de-energizing of the relay harming the PIC. Figure 11.10 shows a block diagram of how to interface a PIC microcontroller with a PM motor using a relay.

Figure 11.11 shows the pin connections for a typical 12 V d.c. DPDT relay. In this example, pin 8 is connected to the PIC microcontroller, thereby acting as a

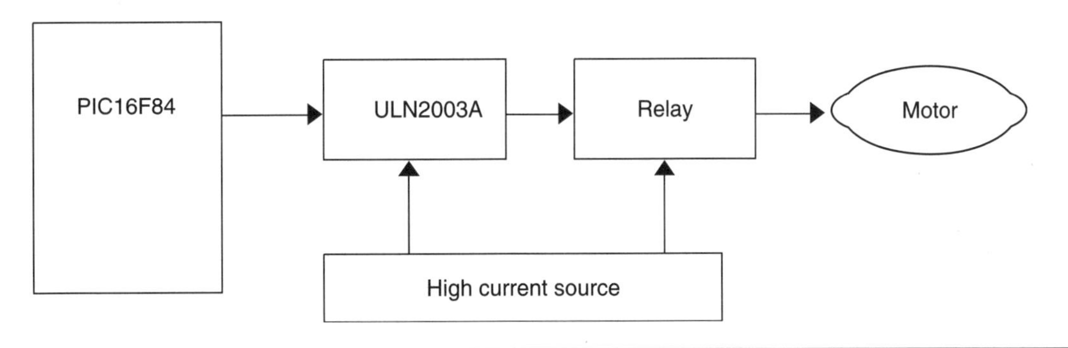

Figure 11.10 Block diagram for powering up high current peripherals.

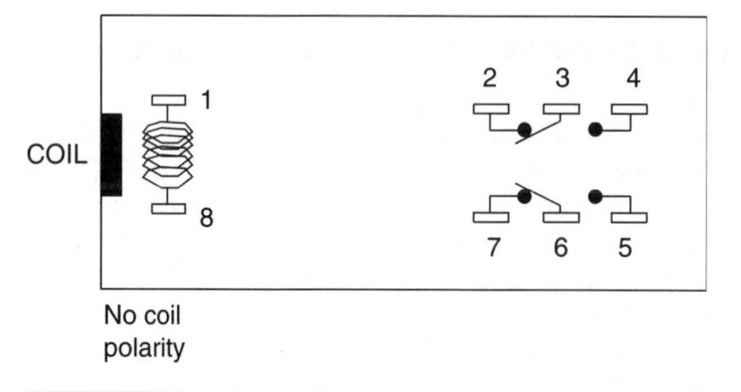

Figure 11.11 Pin connections for a typical 12 d.c. DPDT relay.

switch. Pin 1 is connected to ground. Pins 2 and 3; and 3 and 4 are respectively continuous. These are considered to be high switch because pin 3 is connected to 12 V and it toggles between pins 2 and 4. Moreover, pins 5 and 6; and 6 and 7 are respectively continuous. These are considered to be low switch because pin 6 is connected to the ground and it toggles between pins 5 and 7. This means that pin 4 is positive while pin 5 is negative; so also are pins 2 and 7, respectively. We can use pins 4 and 5 as outputs (or pins 2 and 7). Connecting pin 4 to the positive terminal of a d.c. motor and pin 5 to the negative terminal will make the motor rotate in a clockwise direction. Another pair of 12 V d.c. DPDT relays is used, this time we connect pin 4 to the negative of the d.c. motor and pin 5 to the positive terminal. This will make the motor rotate in a counter-clockwise direction. Care must be taken to ensure that the two pins 8 of the two relays do not become high at the same time as this will result in a shut and the motor will be damaged with the possibility of physical harm. Table 11.2 shows the truth table. It is observed that the last condition (both pins connected from PIC to pin 8 of each relay) is not allowed. One other point to note is that there is some delay in the relay switching from 3 to 2 and then to 4. A delay of 100 ms should be catered for in the program.

$$R = \frac{+V_{\text{motor}} - \text{Rated winding voltage}}{\text{Rated winding current}} \tag{11.3}$$

Table 11.2 Truth table for relay circuit.

Case	*Pin A0 (PIC)*	*Pin A1 (PIC)*	*Description*
1	0	0	0
2	1	0	CW
3	0	1	CCW
4	1	1	NA

$$R = \frac{+15\,\text{V} - 5\,\text{V}}{1\,\text{A}} = 10\,\Omega \tag{11.4}$$

$$P = I^2 R = (1)^2 \times 10 = 10\,\text{W} \tag{11.5}$$

$$\frac{1}{R} = \frac{1}{R_1} + \frac{1}{R_2} = \frac{1}{18} + \frac{1}{18} = \frac{1}{9}$$

$$\therefore R = 9\,\Omega; \quad P = 5\,\text{W} + 5\,\text{W} = 10\,\text{W} \tag{11.6}$$

Table 11.1 shows the datasheet containing information about the relay, and the current rating of the relay coil can be calculated:

$$I = \sqrt{\frac{P}{R}}$$

$$\therefore \; I = \sqrt{\frac{500\,\text{mW}}{200\,\Omega}} = 50\,\text{mA}. \tag{11.7}$$

Therefore, two relays must be used. From this calculation it can be deduced that the ULN2003A will be able to drive the relay coil of the 12 V d.c. DPDT relay which requires 50 mA compared to 500 mA which a UNL2003A can provide.

11.6.2 H-bridge circuits

In Chapter 9, we were interested on how to control speed, but in this chapter we are interested in the compatibility of interfacing a microcontroller and a motor which is an actuator.

Figure 9.23 shows a simple H-bridge circuit using pulsed-width modulated (PWM) controlled voltages consists of four switching transistors.

A PM motor speed control system that is more compatible with microcontrollers is shown in Figure 11.12. A PWM voltage is applied to a single input, and the motor direction is controlled with a HIGH/LOW voltage on a second input. Using an H-bridge circuit is a very efficient method because it requires only a single unipolar power supply.

11.7 Interfacing sensors

Usually, circuits employing sensors are easy to implement since most sensors operate on 5 V. In this section, we discuss how to interface some useful sensors such

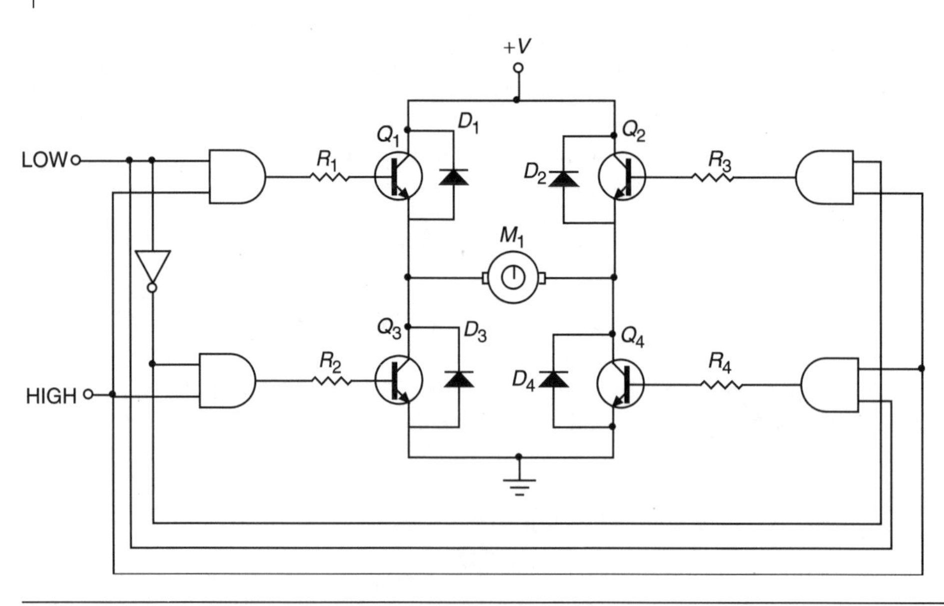

Figure 11.12 A microcontroller-compatible H-bridge circuit.

as the diode/phototransistor pair, the photoreflector sensor, the infrared sensor, and line tracers.

11.7.1 The diode/phototransistor pair

A diode/phototransistor pair (e.g. OP140/OP550) is useful in designing an encoder for PM motor speed control. In this case a rotating disk with slots is attached to the motor and located between the diode/phototransistor pair. The rotating disk will generate pulses which are then transmitted to a microcontroller. A diode/phototransistor pair biasing circuit is shown in Figure 11.13. The circuit provides a cost-effective means for easy interfacing with a microcontroller in a mechatronic system.

11.7.2 Photoreflector sensors

Two photoreflector sensors are used to locate an object. These sensors are useful when placed on the tips of a pick-and-place gripper, as shown in Figure 11.14, to give accurate positioning. While the pick-and-place robot is approaching the object, the sensor output is high indicating that the object is in front if it. The distance between the sides of the gripper is bigger than the length of the object to be picked. Therefore, only one sensor can go high, or both the photoreflective sensors

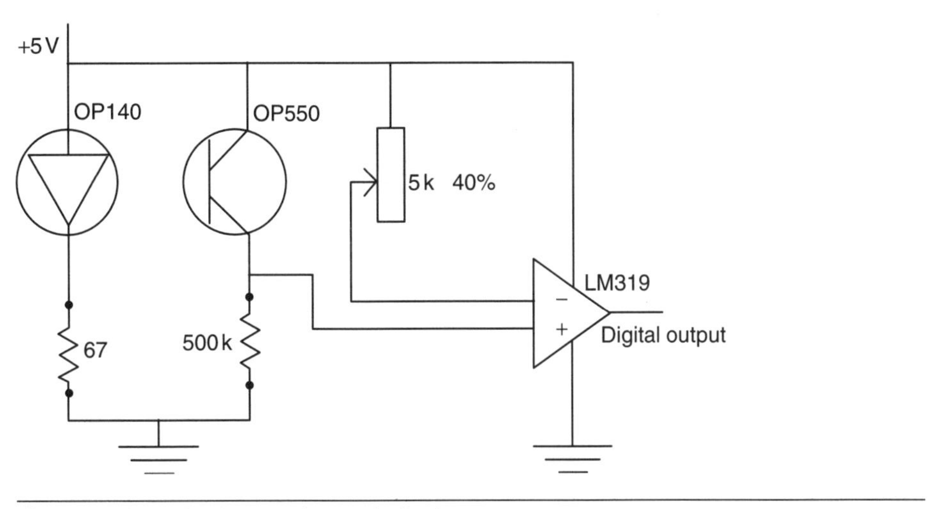

Figure 11.13 Diode/phototransistor pair circuit.

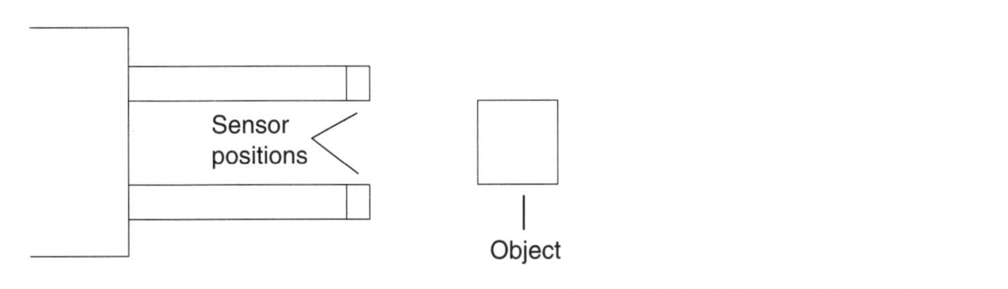

Figure 11.14 Photoreflector sensor positions for a pick-and-place gripper.

stay low. If the left sensor goes high, the robot will rotate left until the sensor goes low. If the right sensor goes high, the robot will rotate right until the sensor goes low. When both the sensors are low, it indicates that the object is in the center of the gripper in readiness for pick up.

11.7.3 Obstacle detection

Common sensors used in mobile robots for detecting obstacles are the digital Sharp GP2Y0D02YK infrared (IR) sensor shown in Figure 11.15 and the analog Sharp IR GP2Y0A02YK. The digital device operates at 5 V and gives out a logic signal if it detects any object within 0.8 m of its field of view (a thin streak extending between its emitter and detector). This sensor's behavior is less influenced by the color of the object since it uses an optical triangle method of measurement. The analog device outputs analog voltages proportional to an object's distance.

Figure 11.15 The Sharp digital IR sensor.

11.7.4 Line tracing

Purpose-built line-tracing sensors, operating at 5 V are available which can distinguish between white and black surfaces. Over a white surface, a logic high signal is produced. The distance between the sensor and the surface is critical and is typically between 8 and 12 mm, hence false detection is possible where the surface may not be flat. A useful approach would be to attach a mechanism to the sensor in order to maintain the distance of the line tracer within the specified region.

11.8 Interfacing with a DAC

The following illustrates an example relating to a PC-based (microcontroller) CNC drilling machine. The fundamental architecture involves digital signals fed from the interface card using the output pins that are required to be in analog form, within a range of 0–5 V. The analog signal is linked to a pulse width modulator.

The DAC selected is an AD557. It has a range of 0–2.56 V and is 8-bit. It is specifically designed for interfacing microprocessor-based applications. The drill motor speed has a maximum of 2000 rpm.

Since the DAC produces a maximum of only 2.56 V, an op amp been used to step up the signal to 5 V. The drill speed resolution is given by:

$$\text{Speed} = \frac{\text{Range}}{2^n - 1} = \frac{2000 - 0}{2^8 - 1} = 7.843\,\text{rpm}. \tag{11.8}$$

This means each bit corresponds to 7.843 revolutions per minute. This is a very reasonable resolution. The circuit of the DAC and the op amp is shown in Figure 11.16.

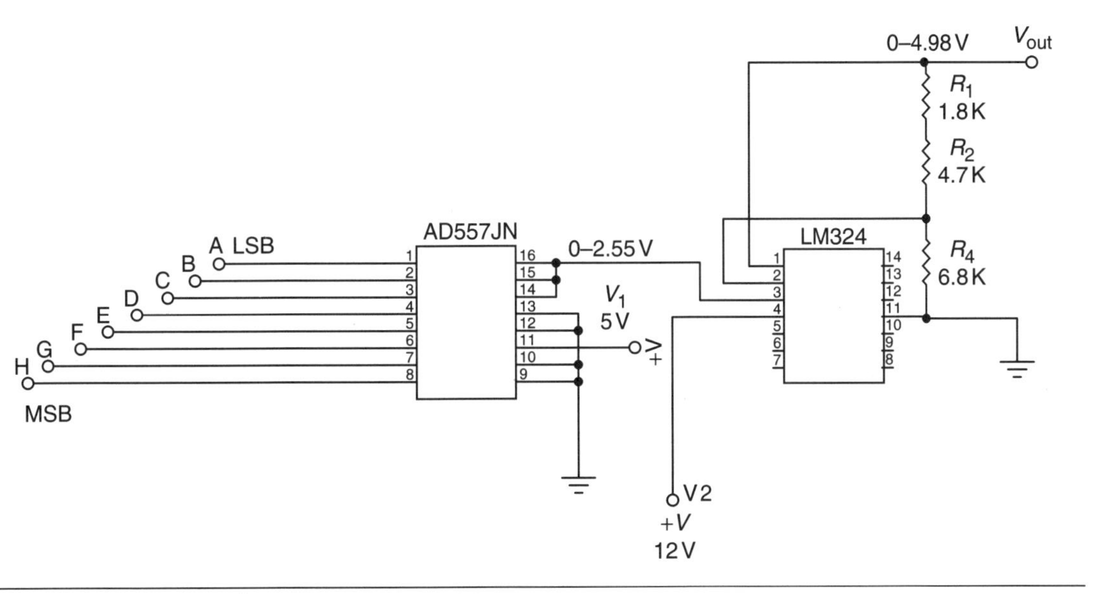

Figure 11.16 DAC and op amp circuit for a CNC drill.

11.9 Interfacing power supplies

This section presents the design aspects of a power supply unit and then discusses the purpose of each component. The section also discusses the ways in which power quality can be improved and even made flexible for a wide range of voltages. A power system is designed from the characteristics of the primary energy source. All over the world, electrical energy is available in the form of alternating current rated between 100–120 V or 240 V. The a.c. frequency is 50 Hz. Mechatronic applications deal with electronics, where smaller ratings of direct current voltages are required. TTL ICs operate on a 5 V d.c. supply, stepper cards operate on a 12 V d.c. supply, and stepper motors operate on a 15–24 V d.c. supply. This means that a power supply unit has to be used to convert the a.c. voltage to a d.c. voltage at the 100/120 V or 240 V level, before stepping down to the level compatible with the ICs.

Power supplies are discussed in Chapter 3.

A typical power supply circuit is shown in Figure 11.17. The line voltage is first stepped down with a transformer from the line input to a level close to the desired 5 V for the operation of TTL ICs. This means, for a 240 V a.c. input, the output of the transformer is precisely 24 V a.c. A transformer with a different turns ratio can be used for different requirements, but generally, a 10:1 transformer is suitable for most applications.

The stepped-down a.c. signal is then rectified. The rectified waveform is a series of positive half cycles, with very high ripples unsuitable for use in most

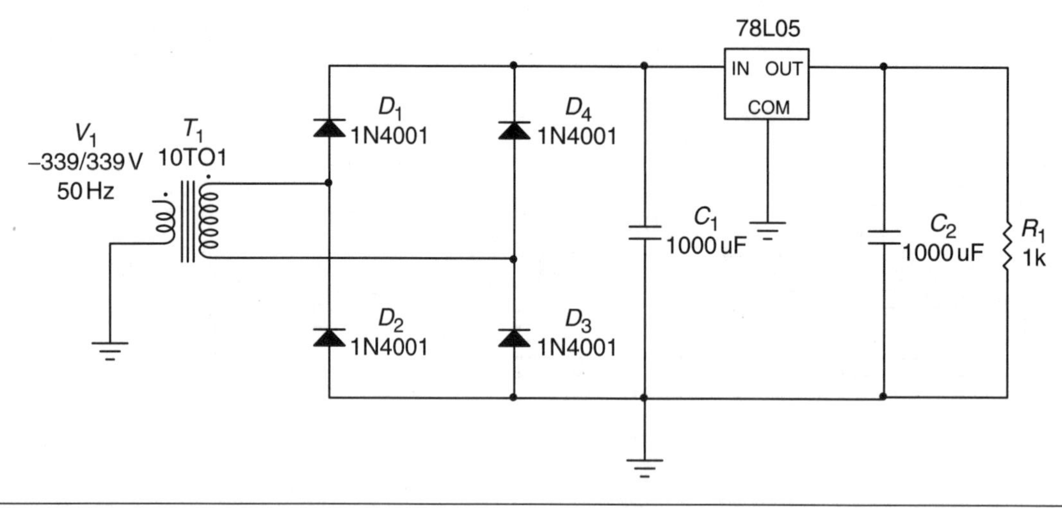

Figure 11.17 A regulated 5 V d.c. power supply.

electronics required in mechatronic applications. A large capacitor is then used to filter (smooth the ripple) the rectified waveform to make it more stable for use in electronic circuits.

If a 100 μF capacitor is used, there is a substantial deterioration in the quality of the signal current from the rectifier. If a 1000 μF capacitor filters the voltage, the filtering is negligible. If a 6000 μF capacitor is used, most of the noise is successfully filtered.

In practice, the supply voltage will change if the load resistance or a.c. line voltage varies, so most power supplies are regulated using a linear voltage regulator. The voltage regulator trims down the voltage to whatever level is required and maintains that voltage during use of the circuit.

Finally, an additional filter capacitor is placed in parallel to the load. The purpose of this filter is to trap voltage spikes during transition of the voltage from 0 V to the maximum voltage, V_{max}. Without this capacitor the load will feel a surge on switch on. The value of capacitor can be determined by the user on the delay that is required. A 100 μF capacitor has a response delay of 4.5 ms, whereas a 630 μF has a delay 20 ms. Most power supply units not using delicate electronics would not normally use the additional filter capacitor, but for applications such as microcontrollers or PC controller boards, this additional filter makes the power supply safer. The regulated voltage has a minimum of 0 V initially ($t = 0$ s). However, as t approaches 20 ms, the voltage levels off to 5 V after a small overshoot.

One disadvantage of using linear voltage regulators is their low current ratings (~100 mA). This restricts its use, for example they cannot be used to supply regulated voltage to electric motors.

11.10 Interfacing with RS 232 and RS 485

The communication standards generally used within mechatronic systems are the RS 232 and the RS 485 serial line standards. To communicate with a PC, standard transmitter and receiver chips are required.

11.11 Compatibility at an interface

In this section, we summarize the requirements for interfacing microcontrollers (in particular, the PIC) with some useful actuators commonly used in mechatronic systems. Generally, the PIC microcontroller cannot interface with an actuator that requires more than a 5 V input. Therefore, a major decision to be made in a mechatronic system is to consider the interface requirements between a PIC microcontroller and the actuator. Some of these requirements, which are useful as design guides are listed in Table 11.3.

Generally, PIC microcontrollers can easily interface with actuators that are designed primarily around TTL and CMOS electronic ICs. This accounts for why PIC microcontrollers do not need any special requirement when interfaced with general purpose tri-state transistors, MOSFETs, sensors, liner-tracers, and ADCs/DACs.

Problems

Q11.1 Why are PIC microcontrollers easily interfaced with general purpose three-state transistors?

Table 11.3 Requirements for connecting PIC microcontrollers with some actuators

Actuator	*Requirements for interfacing with PIC*	*Parameters*
3-state transistors	No special requirement is needed	5 V
Relays	ULN2003A is required	High current
Solenoids	ULN2003A is required	High current
Stepper motors	LM324 – Unipolar/bipolar driver	15–24 V
PM motors	ULN2003A – relay	12–24 V
MOSFETs	No special requirement is needed	5 V
Sensors, liner-tracers	No special requirement is needed	5 V
ADCs and DACs	No special requirement is needed	5 V
Power supply	MC78XX/LM78XX is required	120/240 V

Q11.2 Why are PIC microcontrollers easily interfaced with sensors such as IR sensors, line-tracers, etc?

Q11.3 Why are PIC microcontrollers easily interfaced with ADCs and DACs?

Q11.4 What special requirements are required when interfacing PIC microcontrollers with relays? Describe a design for this purpose.

Q11.5 What special requirements are required when interfacing PIC microcontrollers with solenoids? Describe a design for this purpose.

Q11.6 What special requirements are required when interfacing PIC microcontrollers with stepper motors? Describe a design for this purpose.

Q11.7 What special requirements are required when interfacing PIC microcontrollers with permanent magnet motors? Describe a design for this purpose.

Q11.8 What special requirements are required when designing a power supply for PIC microcontrollers? Describe a design for this purpose (from 240 V through 12 V to 5 V that the PIC requires).

Further reading

[1] Cathey, J.J. (2001) *Electric Machines: Analysis and Design Applying MATLAB*, McGraw-Hill.
[2] El-Sharkawi, M.A. (2000) *Fundamentals of Electric Drives*, Brooks/Cole Publishers.
[3] Fitzgerald, A.E., Kingsley Jr, C. and Umans, S.D. (2003) *Electric Machinery* (6th. ed.), McGraw-Hill.
[4] Stiffler, A.K. (1992) *Design with Microprocessors for Mechanical Engineers*, McGraw-Hill.

CHAPTER 12

Control theory: modeling

Chapter objectives

When you have finished this chapter you should know:

- the role of Laplace transformations in obtaining the transfer function of a system;
- how to describe qualitatively the transient response of first- and second-order systems;
- how to find transfer functions for an electrical network and mechanical systems;
- how to find a mathematical model, known as a state-space representation, for line, time-invariant systems;
- how to convert between transfer functions and state-space models;
- how to reduce a block diagram of multiple subsystems to a single block representing the transfer function from input to output.

12.1 Introduction

In the control system design process, the first step is to transform the requirements into a physical system, followed by translating a qualitative description of the system into a functional block diagram that describes the component parts of the system. Thereafter, the physical system is transformed into a schematic diagram. Once the schematic is drawn, the control system designer uses physical laws (Kirchoff's law for electrical systems and Newton's law for mechanical systems) together with simplifying assumptions, to model the system mathematically. The two methods for mathematical modeling that we discuss in this chapter

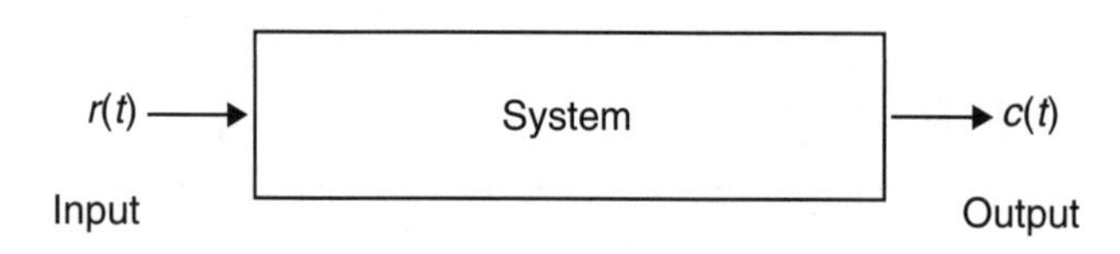

Figure 12.1 Block diagram of a control system.

are: (a) transfer function in the frequency domain; and (b) state-space representation in the time domain.

12.2 Modeling in the frequency domain

In control systems, we would prefer a mathematical representation, which has separate and distinct parts: namely input, output, and system, as shown in Figure 12.1. Since a system represented by differential equations is difficult to model as a block diagram, we use the Laplace transform, which is very useful for representing the input, output, and system, separately. The approach is based on converting a system's differential equation to a transfer function. Consequently, we obtain a mathematical model that algebraically relates a representation of the output to the representation of the input.

12.2.1 Laplace transforms

Laplace transforms simplify the representation of physical systems. The Laplace transform is defined as

$$L[f(t)] = F(s) = \int_{0-}^{\infty} f(t)e^{-st}\mathrm{d}t \tag{12.1}$$

where $s = \sigma + \mathrm{j}\omega$ is a complex number. Consequently, knowing $f(t)$ and the integral in Equation 12.1, we can find $F(s)$, which is the Laplace transform of $f(t)$.

Using Equation 12.1, it is possible to derive a Laplace transform table relating $F(s)$ to $f(t)$ as shown in Table 12.1.

The inverse Laplace transform, which allows us to determine $f(t)$ given $F(s)$ is defined as

$$L^{-1}[F(s)] = \frac{1}{2\pi\mathrm{j}}\int_{\sigma-\mathrm{j}\omega}^{\sigma+\mathrm{j}\infty} F(s)\mathrm{e}^{st}\mathrm{d}s = f(t)u(t) \tag{12.2}$$

where $u(t) = \begin{cases} 1 & \text{if} \quad t > 0 \\ 0 & \text{if} \quad t > 0 \end{cases}$ is the unit step function.

Table 12.1 Laplace transform table

Item no.	*f(t)*	*F(s)*
1	$\delta(t)$	1
2	$u(t)$	$\frac{1}{s}$
3	$tu(t)$	$\frac{1}{s^2}$
4	$t^n u(t)$	$\frac{n!}{s^{n+1}}$
5	$e^{-at}u(t)$	$\frac{1}{s+a}$
6	$\sin \omega t u(t)$	$\frac{\omega}{s^2+\omega^2}$
7	$\cos \omega t u(t)$	$\frac{s}{s^2+\omega^2}$

12.2.1.1 *Partial fraction expansion*

The inverse Laplace transform of a complicated function can be found if we are able to convert the complicated function to simpler functions for which we know the Laplace transform. If $F(s) = N(s)/D(s)$, where $N(s)$ is the numerator and $D(s)$ is the denominator, one possibility is that $N(s) \geq D(s)$ and the other is that $N(s) < D(s)$. For the first case we need to successively divide the numerator by the denominator until we have a function similar to the second case (we have a remainder whose numerator is of the order less than that of the denominator).

See website for downloadable MATLAB code to solve Cases 1,2 and 3

Case 1: Roots of the denominator of $F(s)$ are real and distinct

As an illustration, let us carry out the partial-fraction expansion of

$$F(s) = \frac{3}{(s+2)(s+3)}. \tag{12.3}$$

This is the case where $N(s) < D(s)$ and hence we can write the partial-fraction expansion as a sum of terms where each factor of the original denominator forms the denominator of each term, and constants, called *residues*, from the numerators as in the following form:

$$F(s) = \frac{3}{(s+2)(s+3)} = \frac{K_1}{(s+2)} + \frac{K_2}{(s+3)}. \tag{12.4}$$

To determine K_1, we multiply through by $(s+2)$, and to determine K_2, we multiply through by $(s+3)$. Let us now determines these parameters:

$$\frac{3(s+2)}{(s+2)(s+3)} = \frac{K_1(s+2)}{(s+2)} + \frac{K_2(s+2)}{(s+3)}, \tag{12.5}$$

$$\text{yielding } \frac{3}{(s+3)} = K_1 + \frac{K_2(s+2)}{(s+3)}. \tag{12.6}$$

$$\text{When } s = -2, \frac{3}{1} = K_1, \text{ leading to } K_1 = 3. \tag{12.6A}$$

Multiplying through by $(s+3)$:

$$\frac{3(s+3)}{(s+2)(s+3)} = \frac{K_1(s+3)}{(s+2)} + \frac{K_2(s+3)}{(s+3)}, \tag{12.7}$$

$$\text{yielding } \frac{3}{(s+2)} = \frac{K_1(s+3)}{(s+2)} + K_2. \tag{12.8}$$

$$\text{When } s = -3, \frac{3}{-1} = K_2, \text{ leading to } K_2 = -3. \tag{12.8A}$$

The partial fraction expression for the given function becomes

$$F(s) = \frac{3}{(s+2)} - \frac{3}{(s+3)}. \tag{12.9}$$

Hence, using $F(s)$ from Table 12.1 and the expression in Equation 12.9, we find the inverse Laplace transform $f(t)$ as

$$f(t) = 3(e^{-2t} - e^{-3t})u(t). \tag{12.10}$$

Case 2: Roots of the denominator of $F(s)$ are real and repetitive

In this case, there is a multiplying factor, n, leading to additional terms up to n. As an illustration, let us carry out the partial fraction expansion of

$$F(s) = \frac{3}{(s+2)(s+3)^2}. \tag{12.11}$$

$N(s) < D(s)$ is still the case except that the denominator root at -3 is raised to the power of 2, so that additional terms consisting of these denominator factors will be required:

$$F(s) = \frac{3}{(s+2)(s+3)} = \frac{K_1}{(s+2)} + \frac{K_2}{(s+3)^2} + \frac{K_3}{(s+3)}. \qquad (12.12)$$

To determine K_1, we multiply through by $(s+2)$, and to determine K_2, we multiply through by $(s+3)$. Let us now determine these parameters:

$$\frac{3(s+2)}{(s+2)(s+3)} = \frac{K_1(s+2)}{(s+2)} + \frac{K_2(s+2)}{(s+3)^2} + \frac{K_3(s+2)}{(s+3)} \qquad (12.13)$$

$$\text{yielding } \frac{3}{(s+3)} = K_1 + \frac{K_2(s+2)}{(s+3)^2} + \frac{K_3(s+2)}{(s+3)}. \qquad (12.14)$$

$$\text{When } s = -2, \frac{3}{1} = K_1, \text{ leading to } K_1 = 3. \qquad (12.14A)$$

Multiplying through by $(s+3)^2$:

$$\frac{3(s+3)^2}{(s+2)(s+3)^2} = \frac{K_1(s+3)^2}{(s+2)} + \frac{K_2(s+3)^2}{(s+3)^2} + \frac{K_3(s+3)^2}{(s+3)} \qquad (12.15)$$

$$\text{yielding } \frac{3}{(s+2)} = \frac{K_1(s+3)^2}{(s+2)} + K_2 + K_3(s+3). \qquad (12.16)$$

$$\text{When } s = -3, \frac{3}{-1} = K_2, \text{ leading to } K_2 = -3. \qquad (12.16A)$$

To determine K_3, we differentiate Equation 12.16 with respect to s:

$$\frac{-3}{(s+2)^2} = \frac{K_1(s+3)s}{(s+2)} + K_3. \qquad (12.17)$$

$$\text{When } s = -3, \frac{-3}{(-1)^2} = -3\ , \text{ leading to } K_2 = -3. \qquad (12.17A)$$

The partial fraction expression for the given function becomes

$$F(s) = \frac{3}{(s+2)} - \frac{3}{(s+3)^2} - \frac{3}{(s+3)}. \quad (12.18)$$

Hence, using the $F(s)$ from Table 12.1 and the expression in Equation 12.18, we find the inverse Laplace transform, $f(t)$ to be

$$f(t) = 3(e^{-2t} - te^{-3t} - e^{-3t})u(t). \quad (12.19)$$

Case 3: Roots of the denominator of $F(s)$ are complex or imaginary

As an illustration, let us carry out the partial fraction expansion of

$$F(s) = \frac{2}{s(s^2+3s+4)}. \quad (12.20)$$

$N(s) < D(s)$ is still true, but the roots of one of the denominator terms are complex. The term having the complex roots will take the following form:

$$\frac{2}{s(s^2+3s+4)} = \frac{K_1}{s} + \frac{K_2 s + K_3}{(s^2+3s+4)}. \quad (12.21)$$

To determine K_1, we multiply through by s, and to determine K_2 and K_3, we multiply through by $s(s^2+3s+4)$. Let us now determine these parameters:

$$\frac{2s}{s(s^2+3s+4)} = \frac{K_1 s}{s} + \frac{(K_2 s + K_3)s}{(s^2+3s+4)}, \quad (12.22)$$

$$\text{yielding } \frac{2}{(s^2+3s+4)} = K_1 + \frac{(K_2 s + K_3)s}{(s^2+3s+4)}. \quad (12.23)$$

$$\text{When } s = 0, \frac{2}{4} = K_1, \text{ leading to } K_1 = \frac{1}{2}. \quad (12.23A)$$

Multiplying right through with $s(s^2+3s+4)$:

$$\frac{2(s^2+3s+4)s}{s(s^2+3s+4)} = \frac{K_1(s^2+3s+4)s}{s} + \frac{K_2 s + K_3(s^2+3s+4)s}{(s^2+3s+4)}, \quad (12.24)$$

$$\text{yielding } 2 = K_1(s^2+3s+4) + (K_2 s + K_3)s. \quad (12.25)$$

Rearranging terms:

$$2 = s^2(K_1 + K_2) + s(3K_1 + K_3) + 4K_1$$
$$2 = s^2(\tfrac{1}{2} + K_2) + s(\tfrac{3}{2} + K_3) + 2. \qquad (12.25A)$$

Balancing terms:

$$(\tfrac{1}{2} + K_2) = 0$$
$$(\tfrac{3}{2} + K_3) = 0, \qquad (12.25B)$$

$$\text{leading to } K_2 = -\tfrac{1}{2};\ K_3 = -\tfrac{3}{2}$$

The partial fraction expression for the given function becomes

$$F(s) = \frac{\tfrac{1}{2}}{s} - \frac{1}{2}\frac{(s+3)}{(s^2 + 3s + 4)}. \qquad (12.26)$$

Completing the squares in the denominator and rearranging the numerator of the second term, we have

$$F(s) = \frac{\tfrac{1}{2}}{s} - \frac{1}{2}\frac{(s + \tfrac{3}{2}) + \sqrt{\tfrac{9}{7}}\left(\sqrt{\tfrac{7}{4}}\right)}{(s + \tfrac{3}{2})^2 + \left(\sqrt{\tfrac{7}{4}}\right)^2}. \qquad (12.27)$$

Noting that

$$L\left[Ae^{-at}\cos\omega t + Be^{-at}\sin\omega t\right] = \frac{A(s+a) + B\omega}{(s+a)^2 + \omega^2} \qquad (12.27A)$$

(combination of expressions from Table 12.1 and using the expression in Equation 12.27), we find the inverse Laplace transform, $f(t)$, as

$$f(t) = \tfrac{1}{2} - \tfrac{1}{2}e^{-\tfrac{3}{2}t}\left(\cos\sqrt{\tfrac{7}{4}}t + \sqrt{\tfrac{9}{7}}\sin\sqrt{\tfrac{7}{4}}t\right) \qquad (12.28)$$

12.2.2 The transfer function

The transfer function algebraically relates a system's output to its input; this function allows separation of the input, the system, and the output into three separate and distinct parts as shown in Figure 12.2. The ratio of the output transform, $C(s)$, divided by the input transform, $R(s)$, gives the gain, $G(s)$.

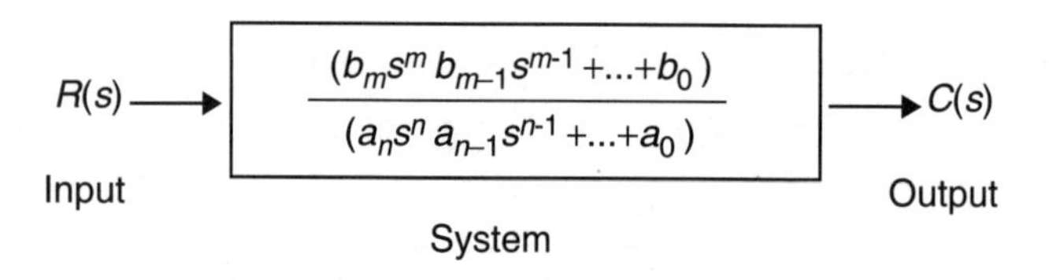

Figure 12.2 Block diagram of a transfer function.

EXAMPLE 12.1

Find the transfer function represented by

$$\frac{d^3c(t)}{dt} + 2\frac{d^2c(t)}{dt} + 5\frac{dc(t)}{dt} = 3r(t). \tag{12.28A}$$

Solution

Taking the Laplace transform of both sides, assuming zero initial conditions, leads to

$$s^3C(s) + 2s^2C(s) + 5sC(s) = 3R(s). \tag{12.28B}$$

The transfer function is

$$G(s) = \frac{C(s)}{R(s)} = \frac{3}{(s^3 + 2s^2 + 5s)} = \frac{3}{s(s^2 + 2s + 5)}. \tag{12.28C}$$

12.2.3 Electrical network transfer function

12.2.3.1 *Passive elements*

The three passive elements considered in basic electrical networks are resistors, inductors, and capacitors. Table 12.2 shows these passive elements and the relationships between voltage and current, voltage and charge, and current and voltage.

The steps involved in solving complex electrical networks (having multiple loops and nodes) using mesh analysis, are as follows:

1. Replace passive element values with their impedances.
2. Replace all sources and time variables with their Laplace transform.
3. Assume a transform current and a current direction in each mesh.
4. Write Kirchoff's voltage law around each mesh.

Table 12.2 Voltage–current, and current–voltage relations for passive elements

Elements	*Voltage–current*	*Voltage–charge*	*Current–voltage*	*Impedance*
Resistor	$v(t) = Ri(t)$	$v(t) = R\frac{dq(t)}{dt}$	$i(t) = \frac{1}{R}v(t)$	R
Inductor	$v(t) = L\frac{di(t)}{dt}$	$v(t) = L\frac{d^2q(t)}{dt}$	$i(t) = \frac{1}{L}\int_0^t v(\tau)d\tau$	Ls
Capacitor	$v(t) = \frac{1}{C}\int_0^t i(\tau)d\tau$	$v(t) = \frac{1}{C}q(t)$	$i(t) = C\frac{dv(t)}{dt}$	$\frac{1}{Cs}$

5. Solve the simultaneous equations for the output.
6. Form the transform function.

EXAMPLE 12.2

See website for downloadable MATLAB code to solve this problem

For the network shown in Figure 12.3, determine the transfer function, $I_2(s)/V(s)$.

Solution

The first step is to convert the network into Laplace transforms for impedances and current variables (assuming zero initial conditions) as shown in Figure 12.3(b).

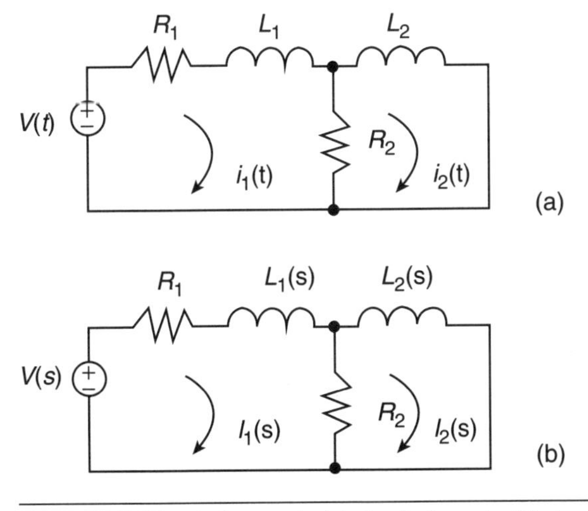

Figure 12.3 Electrical network: (a) circuit elements; (b) circuit impedances.

Mesh equations:

$$R_1I_1(s) + L_1sI_1(s) + R_2(I_1(s) - I_2(s)) = V(s) \tag{12.29}$$

$$L_2sI_2(s) + R_2(I_2(s) - I_1(s)) = 0 \tag{12.30}$$

leading to

$$(R_1 + L_1s + R_2)I_1(s) - R_2I_2(s) = V(s) \tag{12.31}$$

$$-R_2I_1(s) + (R_2 + L_2s)I_2(s) = 0 \tag{12.32}$$

Using Cramer's rule since this is a (2×2) matrix, we can solve for the current values and further analyze the circuit.

$$I_2(s) = \frac{\begin{vmatrix} (R_1 + L_1s + R_2) & V(s) \\ -R_2 & 0 \end{vmatrix}}{\Delta}, \tag{12.33}$$

$$\text{where } \Delta = \begin{vmatrix} (R_1 + L_1s + R_2) & -R_2 \\ -R_2 & (R_2 + L_2s) \end{vmatrix}. \tag{12.33A}$$

The solution to the problem of Example 12.2 is

$$G(s) = \frac{R_2}{L_1L_2s^2 + (R_2L_2 + R_2L_1 + R_1L_2)s + R_1R_2}. \tag{12.33B}$$

12.2.3.2 *Operational amplifiers*

The principles covered in Chapter 5 are directly applied here and there is no need for repetition. We merely apply impedances for the elements used.

Inverting op amps

As previously derived, for inverting op amps:

$$G(s) = \frac{V_o(s)}{V_i(s)} = \frac{-Z_2(s)}{Z_1(s)}. \tag{12.34}$$

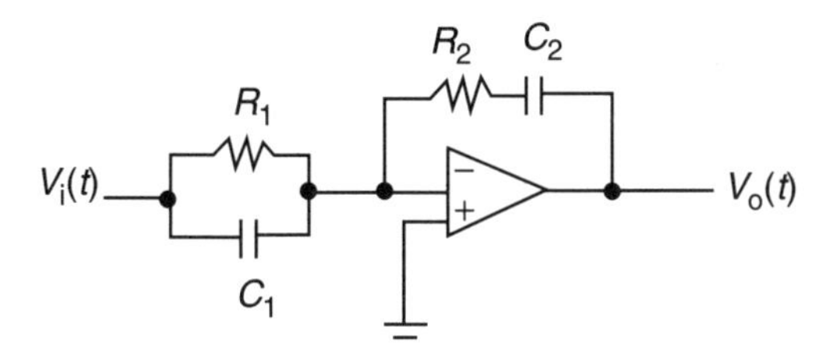

Figure 12.4 An op amp used as a PID controller in Example 12.3.

Non-inverting op amps

As previously derived, for non-inverting op amps:

$$G(s) = \frac{V_o(s)}{V_i(s)} = 1 + \frac{Z_2(s)}{Z_1(s)}. \tag{12.35}$$

EXAMPLE 12.3

See website for downloadable MATLAB code to solve this problem

In an op amp used as a PID controller, shown in Figure 12.4, the values of the electrical elements are $C_1 = 4\,\mu\text{F}$; $C_2 = 0.1\,\mu\text{F}$; $R_1 = 400\,\text{k}\Omega$; $R_2 = 300\,\text{k}\Omega$.

Determine the transfer function, $V_o(s)/V_i(s)$.

Solution

$$Z_1(s) = \frac{R_1 \times {}^1\!/_{C_1 s}}{R_1 + {}^1\!/_{C_1 s}} = \frac{1}{C_1 s + {}^1\!/_{R_1}} = \frac{1}{4 \times 10^{-6} s + {}^1\!/_{400\times10^3}} = \frac{400 \times 10^3}{1.6s + 1}$$

$$Z_2(s) = R_1 + {}^1\!/_{C_1 s} = 300 \times 10^3 + {}^1\!/_{0.1\times10^{-6} s} = 300 \times 10^3 + \frac{10^7}{s}$$

$$\frac{V_o(s)}{V_i(s)} = -\left(\frac{300 \times 10^3 + \frac{10^7}{s}}{400 \times 10^3/(1.6s+1)}\right) = \frac{(300 \times 10^3 + 10^7)(1.6s+1)}{400 \times 10^3 \times s} \tag{12.35A}$$

$$\frac{V_o(s)}{V_i(s)} = -\left(\frac{480\,000s^2 + 300\,000s + 16 \times 10^6 s + 10^7}{400 \times 10^3 s}\right)$$

$$= -1.2\frac{s^2 + 34.06s + 20.83}{s}$$

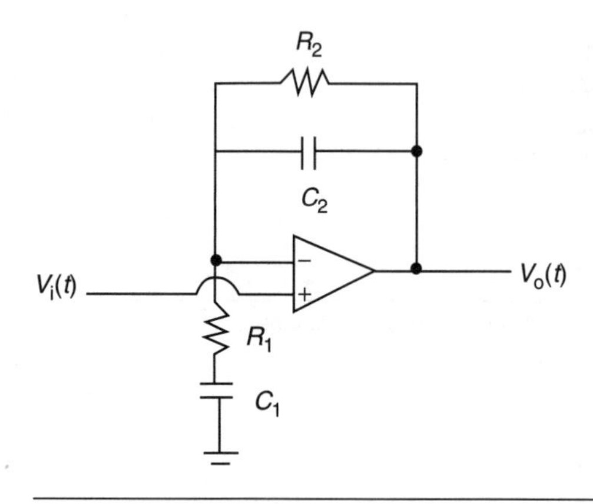

Figure 12.5 An op amp used as a PID controller in Example 12.4.

EXAMPLE 12.4

See website for downloadable MATLAB code to solve this problem

In an op amp used as a PID controller, shown in Figure 12.5, the values of the electrical elements are $C_1 = 4\,\mu F$; $C_2 = 0.1\,\mu F$; $R_1 = 400\,k\Omega$; $R_2 = 300\,k\Omega$.

Determine the transfer function, $V_o(s)/V_i(s)$.

Solution

Notice that in this case, the electrical elements in the input are in series while the output elements are in parallel (opposite to Example 12.3).

$$Z_1(s) = R_1 + {}^1\!/_{C_1 s} = 400 \times 10^3 + {}^1\!/_{4\times10^{-6}s} = 400 \times 10^3 + \frac{2.5 \times 10^7}{s}$$

$$Z_2(s) = \frac{R_2 \times {}^1\!/_{C_2 s}}{R_2 + {}^1\!/_{C_2 s}} = \frac{1}{C_2 s + {}^1\!/_{R_2 s}} = \frac{1}{0.1 \times 10^{-6} s + {}^1\!/_{300\times10^3}} = \frac{300 \times 10^3}{0.03s + 1} \tag{12.35B}$$

Tidying up leads to

$$G(s) = \frac{V_o(s)}{V_i(s)} = 1 + \frac{Z_2(s)}{Z_1(s)} = \frac{(s + 58.6)(s + 0.3555)}{(s + 33.33)(s + 0.625)}. \tag{12.35C}$$

12.2.4 Mechanical system transfer function

12.2.4.1 *Translational*

The three primary elements considered in basic mechanical systems are springs, dampers, and mass. Table 12.3 shows these primary elements and the relationships between force and velocity, and force and displacement.

Table 12.3 Force–velocity/displacement relations for mechanical elements

Elements	*Force–velocity*	*Force–displacement*	*Impedance*
Spring	$f(t) = K\int_0^t v(\tau)d\tau$	$f(t) = Kx(t)$	K
Viscous damping	$f(t) = f_v v(t)$	$f(t) = f_v \frac{dx(t)}{dt}$	$f_v s$
Inertia	$f(t) = M\frac{dv(t)}{dt}$	$f(t) = L\frac{d^2x(t)}{dt}$	Ms^2

Table 12.4 Torque–velocity/displacement relations for mechanical rotational elements

Elements	*Torgue–angular velocity*	*Torgue–angular displacement*	*Impedance*
Spring	$T(t) = K\int_0^t \omega(\tau)d\tau$	$T(t) = K\theta(t)$	K
Viscous damping	$T(t) = f_v D\omega(t)$	$T(t) = D\frac{d\theta(t)}{dt}$	Ds
Inertia	$T(t) = J\frac{d\omega(t)}{dt}$	$T(t) = J\frac{d^2\theta(t)}{dt}$	Js^2

12.2.4.2 *Rotational*

The three primary elements considered in basic mechanical rotational systems are again springs, dampers, and mass. Table 12.4 shows these primary elements and the relationships between torque and velocity, and torque and displacement.

12.2.4.3 *Gear system*

For a gear system the displacements for meshed gears are given as,

$$r_1\theta_1 = r_2\theta_2 \tag{12.36}$$

or

$$\frac{\theta_2}{\theta_1} = \frac{r_1}{r_2} = \frac{N_1}{N_2}. \tag{12.37}$$

The rotational energy is the torque times the angular displacement, given as

$$T_1\theta_1 = T_2\theta_2. \tag{12.38}$$

Solving this equation for the torque ratio,

$$\frac{T_2}{T_1} = \frac{\theta_1}{\theta_2} = \frac{N_2}{N_1}. \tag{12.39}$$

Reflected rotational mechanical impedance

The reflected rotational mechanical impedance through gear trains is given as

$$\text{mechanical impedance} = \left(\frac{N_d}{N_s}\right)^2, \tag{12.40}$$

where N_d is the number of teeth of gear on destination shaft and N_s is the number of teeth of gear on source shaft.

Hence, from Equation 12.39, the input torque T_1 can be reflected to the output shaft as

$$T_2 = T_1\left(\frac{N_2}{N_1}\right). \tag{12.41}$$

The equation of motion is therefore

$$(Js^2 + Ds + K)\theta_2(s) = T_1(s)\left(\frac{N_2}{N_1}\right). \tag{12.42}$$

Hence we have two cases: a rotational system driven by gears (Case 1 shown in Figure 12.6), and a rotational mechanical system driven by gears (Case 2 shown in Figure 12.7).

EXAMPLE 12.5

See website for downloadable MATLAB code to solve this problem

In the mechanical rotational system shown in Figure 12.8, the values of the mechanical elements properties are:

$J_1 = 2$ kg m^2; $J_2 = 3$ kg m^2;

$D_1 = 2$ N m s rad^{-1}; $D_2 = 4$ N m s rad^{-1};

$k_1 = 1$ N m rad^{-1}; $k_2 = 2$ N m rad^{-1};

$N_1 = 1000$ rpm; $N_2 = 2000$ rpm.

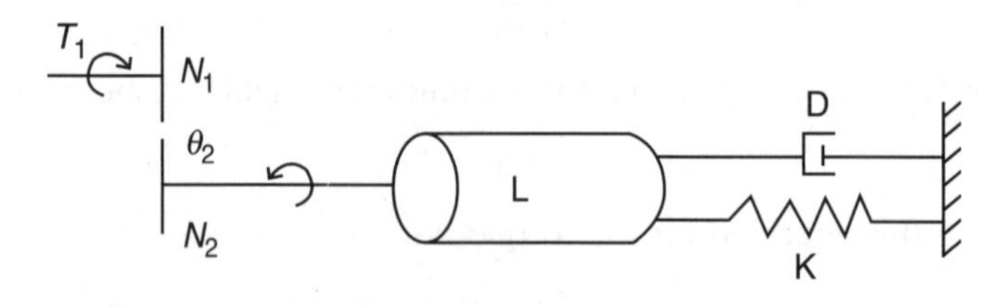

Figure 12.6 A rotational system driven by gears.

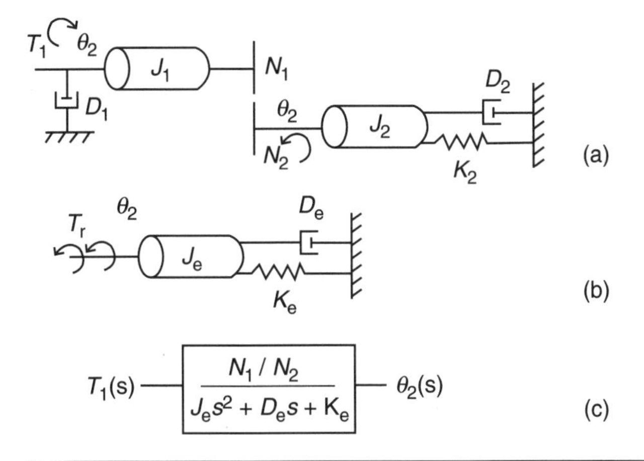

Figure 12.7 A rotational mechanical system driven by gears.

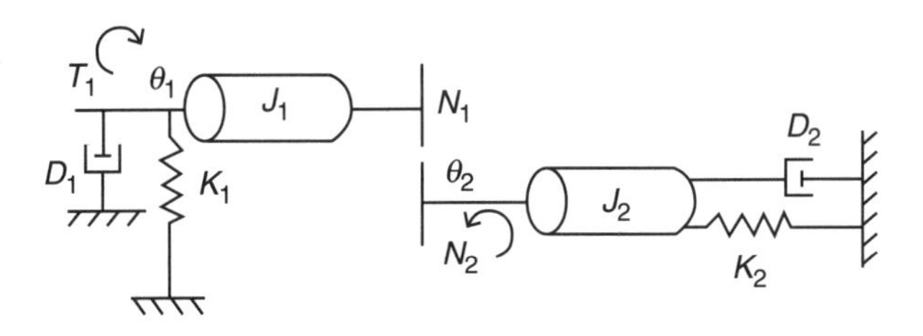

Figure 12.8 A mechanical rotational system.

Determine the transfer function, $G(s) = \theta_2(s)/T(s)$.

Solution

From Equation 12.42,

$$\left(Js^2 + Ds + K\right)\theta_2(s) = T_1(s)\left(\frac{N_2}{N_1}\right)$$

$$\therefore \quad G(s) = \frac{\theta_2(s)}{T_1(s)} = \frac{\left({}^{N_2}\!/\!_{N_1}\right)}{\left(Js^2 + Ds + K\right)}$$

$$J_e = J_1\left({}^{N_2}\!/\!_{N_1}\right)^2 + J_2 = 2 \times 4 + 3 = 11 \tag{12.42A}$$

$$D_e = D_1\left({}^{N_2}\!/\!_{N_1}\right)^2 + D_2 = 2 \times 4 + 4 = 12$$

$$K_e = K_1\left({}^{N_2}\!/\!_{N_1}\right)^2 + K_2 = 1 \times 4 + 2 = 6$$

$$\therefore \quad G(s) = \frac{\theta_2(s)}{T(s)} = \frac{\left({}^{N_2}\!/\!_{N_1}\right)}{\left(Js^2 + Ds + K\right)} = \frac{2}{11s^2 + 12s + 6}.$$

Figure 12.9 Legged robot. (Courtesy of the Lami Laboratoire du Microinformatique, Ecole Polytechnique Fédérale de Lausanne: PTI.)

2.3 Modeling in the time domain

Modeling systems which use linear, time-invariant differential equations and subsequent transfer functions are limited in applications that involve time-varying systems. The time-domain approach is therefore a unified method for modeling and analyzing a wide range of systems including non-linear systems that vary with time. Examples include missiles in operation, and legged robots that move on unfriendly terrain such as the one shown in Figure 12.9.

The steps involved in state space representation are:

1. Select a particular *subset* of all possible variables, referred to as *state variables*.
2. For an nth-order system, write n simultaneous, first-order differential equations in terms of the state variables, referred to as *state equations*.
3. Knowing the initial condition of all the *state variables*; at t_0 as well as the system input for $t \geq t_0$, solve the simultaneous differential equations for the state variables for $t \geq t_0$.
4. Algebraically combine the state variables with the system's input and find all of the other system variables for $t \geq t_0$. This algebraic equation is referred to as the *out equation*.
5. Consider the state equations and the output equations as a viable representation of the system. This representation is referred to as a *state-space representation*.

12.3.1 The general state-space representation

A system is represented in state space by the following set of equations:

$$\dot{x} = Ax + Bu \tag{12.43}$$

$$y = Cx + Du, \tag{12.44}$$

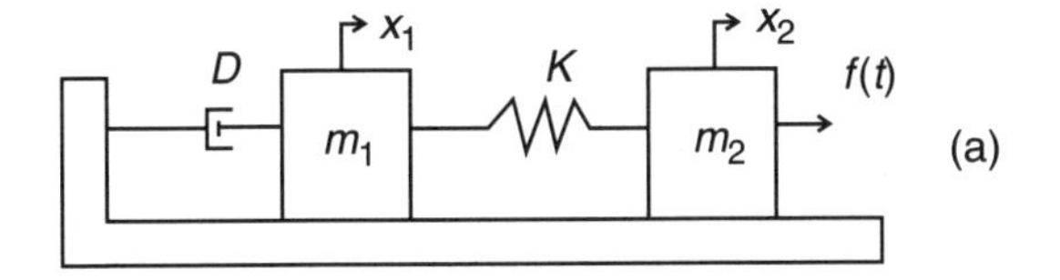

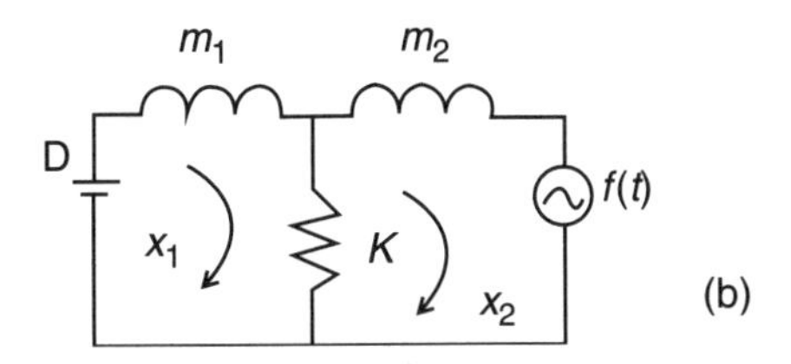

Figure 12.10 A mechanical system.

for $t \geq t_0$ and initial conditions, $x(t_0)$, where

$\boldsymbol{x}$ is the state vector; $\dot{x}$ is the derivative of the state vector with respect to time; $\boldsymbol{y}$ is the output vector; $\boldsymbol{u}$ is the input or control vector; A is the system matrix; B is the input matrix; C is the output matrix; and D is the feed-forward matrix.

Equation 12.43 is the *state equation*, and Equation 12.44 is the *output equation*.

As an example, Figure 12.10 shows a mechanical system.

The equations that describe the system dynamics are:

$$m_1\ddot{x}_1 + D\dot{x}_1 + K(x_1 - x_2) = 0 \tag{12.45}$$

$$m_2\ddot{x}_2 + K(x_2 - x_1) = f(t). \tag{12.46}$$

Here, we let $x_1, v_1, x_2, v_2, v_1 = \dot{x}_1;\ v_2 = \dot{x}_2;\ \dot{v}_1 = \ddot{x}_1;\ \dot{v}_2 = \ddot{x}_2$, be the state variables. (12.46A)

$$\begin{bmatrix} \dot{x}_1 \\ \dot{v}_1 \\ \dot{x}_2 \\ \dot{v}_2 \end{bmatrix} = \begin{bmatrix} 0 & 1 & 0 & 0 \\ -K/M_1 & -D/M_1 & K/M_1 & 0 \\ 0 & 0 & 0 & 1 \\ K/M_2 & 0 & -K/M_2 & 0 \end{bmatrix} \begin{bmatrix} x_1 \\ v_1 \\ x_2 \\ v_2 \end{bmatrix} + \begin{bmatrix} 0 \\ 0 \\ 0 \\ 1/M_2 \end{bmatrix} f(t). \tag{12.46B}$$

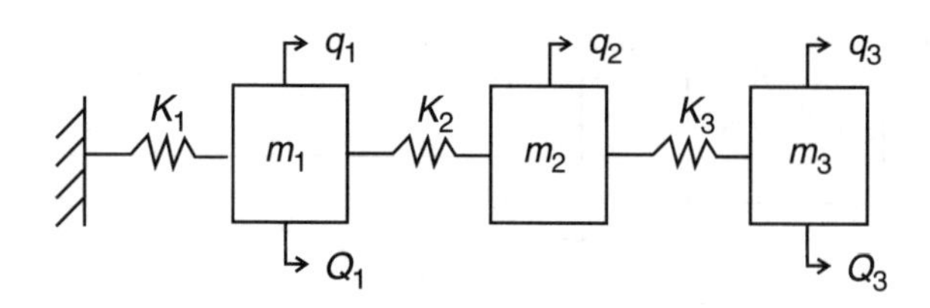

Figure 12.11 A translational mechanical system.

12.3.2 Application of state-space representation

EXAMPLE 12.6

See website for downloadable MATLAB code to solve this problem

Figure 12.11 shows a translational mechanical system, represented in state space, where $x_3(t)$ is the output. Represent it in state space.

$m_1 = 4$ kg; $m_2 = 2$ kg; $m_3 = 1$ kg;

$K_1 = 3$ N m rad^{-1}; $K_2 = 1$ N m rad^{-1}; $K_3 = 1$ N m rad^{-1}; $Q_1 = F \sin\Omega t$;

$Q_3 = 2F \sin \Omega t, (\Omega^2 = 2K_3/m_3)$.

Solution

The translational mechanical system of Figure 12.11 has the following equivalent electrical circuit shown in Figure 12.12:

The equations that describe the system dynamics are:

$$4\ddot{x}_1 + 3x_1 + 1(x_1 - x_2) = Q_1 \tag{12.47}$$

$$2\ddot{x}_1 + 1(x_2 - x_1) + 1(x_2 - x_3) = 0 \tag{12.48}$$

$$1\ddot{x}_1 + 1(x_3 - x_2) = Q_3 \tag{12.49}$$

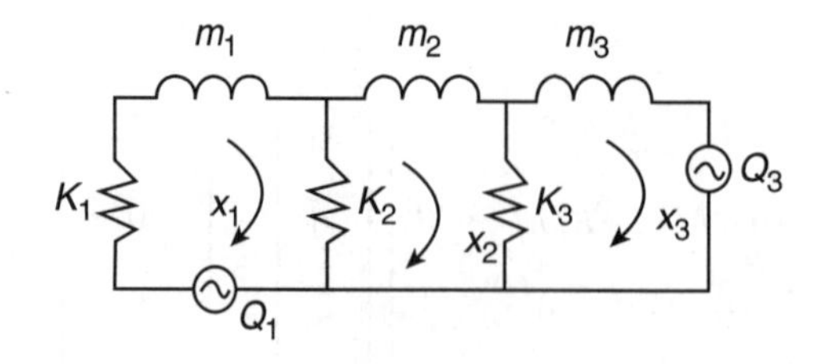

Figure 12.12 The equivalent electrical circuit of Figure 12.11.

Tidying up terms, we have

$$4\ddot{x}_1 + 4x_1 - x_2 = Q_1 \tag{12.50}$$

$$2\ddot{x}_1 - x_1 + 2x_2 - x_3 = 0 \tag{12.51}$$

$$1\ddot{x}_1 + 0x_1 - x_2 + x_3 = Q_3 \tag{12.52}$$

Here, we let x_1, v_1, x_2, v_2, x_3, v_3; $v_1 = \dot{x}_1$; $v_2 = \dot{x}_2$; $v_3 = \dot{x}_3$; $\dot{v}_1 = \ddot{x}_1$; $\dot{v}_2 = \ddot{x}_2$; $\dot{v}_3 = \ddot{x}_3$, be the state variables. (12.52A)

We can now set up six equations (two times the number of degrees for the problem) as follows:

$$\begin{aligned}
\dot{x}_1 &= v_1 \\
\ddot{x}_1 &= -x_1 + 0.25x_2 + Q_1 \\
\dot{x}_2 &= v_2 \\
\ddot{x}_2 &= 0.5x_1 - x_2 + 0.5x_3 \\
\dot{x}_3 &= v_3 \\
\ddot{x}_3 &= 0x_1 + x_2 - x_3 + Q_3
\end{aligned} \tag{12.52B}$$

The equations that describe the system dynamics are:

$$\begin{bmatrix} x_1 \\ v_1 \\ x_2 \\ v_2 \\ x_3 \\ v_3 \end{bmatrix} = \begin{bmatrix} 0 & 1 & 0 & 0 & 0 & 0 \\ -1 & 0 & 0.25 & 0 & 0 & 0 \\ 0 & 0 & 0 & 1 & 0 & 0 \\ 0.5 & 0 & -1 & 0 & 0.5 & 0 \\ 0 & 0 & 0 & 0 & 0 & 1 \\ 0 & 0 & 1 & 0 & -1 & 0 \end{bmatrix} \begin{bmatrix} x_1 \\ v_1 \\ x_2 \\ v_2 \\ x_3 \\ v_3 \end{bmatrix} + \begin{bmatrix} 0 \\ 1 \\ 0 \\ 0 \\ 0 \\ 2 \end{bmatrix} F \sin(\Omega t) \tag{12.52C}$$

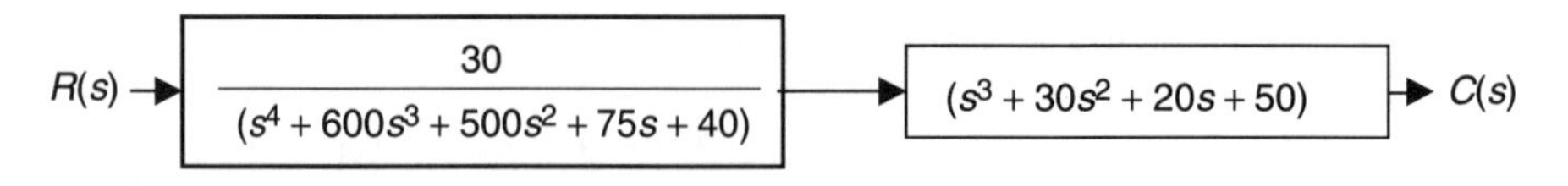

Figure 12.13 Splitting the function into two blocks.

12.4 Converting a transfer function to state space

An advantage of a state space is that it can be used to simulate a physical system on the digital computer. If we want to simulate a system represented as a transfer function, it is necessary to first convert it to a state space before simulation.

EXAMPLE 12.7

See website for downloadable MATLAB code to solve this problem

Find the state-space representation of the transfer function given as:

$$G(s) = \frac{30\left(s^3 + 30s^2 + 20s + 50\right)}{\left(s^4 + 600s^3 + 500s^2 + 75s + 40\right)}. \quad (12.52D)$$

Solution

We split the function into two blocks as shown in Figure 12.13.
From the background theory given, we have for the denominator:

$$\ddddot{c} + 600\,\dddot{c} + 500\ddot{c} + 75\dot{c} + 40c = 30r$$

$$\begin{aligned} x_1 &= c & \dot{x}_1 &= x_2 \\ x_2 &= \dot{c} & \dot{x}_2 &= x_3 \\ x_3 &= \ddot{c} & \dot{x}_3 &= x_4 \\ x_4 &= \dddot{c} \end{aligned} \quad (12.52E)$$

resulting in

$$\begin{bmatrix} \dot{x}_1 \\ \dot{v}_1 \\ \dot{x}_2 \\ \dot{v}_2 \end{bmatrix} = \begin{bmatrix} 0 & 1 & 0 & 0 \\ 0 & 0 & 1 & 0 \\ 0 & 0 & 0 & 1 \\ -40 & -75 & -500 & -600 \end{bmatrix} \begin{bmatrix} x_1 \\ v_1 \\ x_2 \\ v_2 \end{bmatrix} + \begin{bmatrix} 0 \\ 0 \\ 0 \\ 30 \end{bmatrix}. \quad (12.52F)$$

For the numerator,

$$C(s) = \left(b_3s^3 + b_2s^2 + b_1s + b_0\right)X_1(s) = \left(s^3 + 30s^2 + 20s + 50\right)X_1(s) \quad (12.52G)$$

Taking the Laplace transform (with zero initial conditions),

$$c = \dddot{x}_1 + 30\ddot{x}_1 + 20\dot{x}_1 + 50x_1$$

$$\begin{aligned} x_1 &= x_1 \\ \dot{x}_1 &= x_2 \\ \ddot{x}_1 &= x_3 \\ \dddot{x}_1 &= x_4 \end{aligned} \qquad (12.52H)$$

hence,

$$y = c(t) = x_4 + 30x_3 + 20x_2 + 50x_1$$

$$y = \begin{bmatrix} b_0 & b_1 & b_2 & b_3 \end{bmatrix} \begin{bmatrix} x_1 \\ x_2 \\ x_3 \\ x_4 \end{bmatrix} = \begin{bmatrix} 50 & 20 & 30 & 1 \end{bmatrix} \begin{bmatrix} x_1 \\ x_2 \\ x_3 \\ x_4 \end{bmatrix}. \qquad (12.52I)$$

The graphical representation of this problem is given in Figure 12.14.

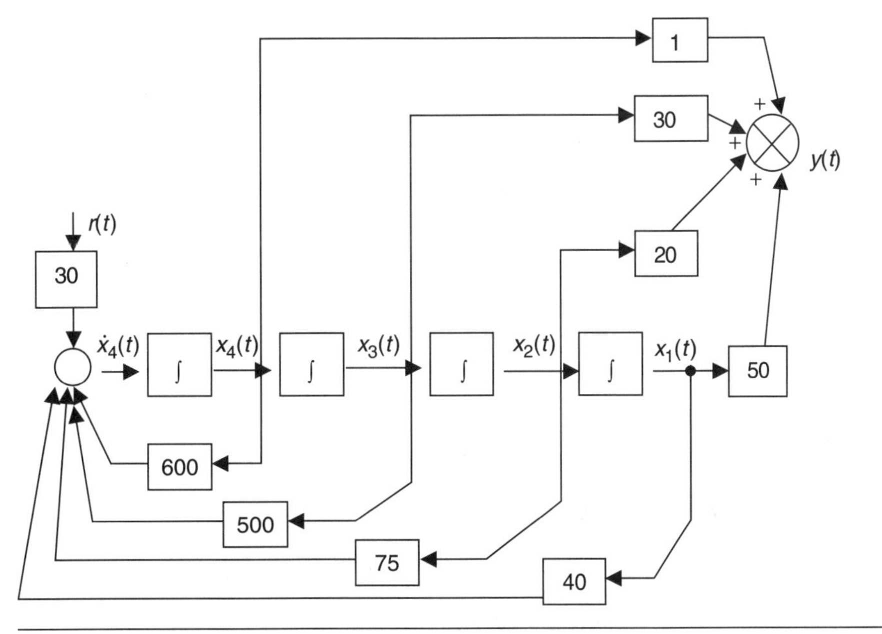

Figure 12.14 Diagrammatical state-space representation for Example 12.7.

12.5 Converting a state-space representation to a transfer function

Given the state and output equations (see Section 12.3.1)

$$\dot{x} = Ax + Bu \tag{12.53a}$$

$$y = Cx + Du, \tag{12.53b}$$

take the Laplace transform assuming zero initial conditions:

$$sX(s) = AX(s) + BU(s) \tag{12.54a}$$

$$Y(s) = CX(s) + DU(s) \tag{12.54b}$$

Solving for $X(s)$ in Equation 12.54a,

$$(sI - A)X(s) = BU(s) \tag{12.55}$$

$$X(s) = (sI - A)^{-1}BU(s), \tag{12.56}$$

where I is the identity matrix.

Substituting $X(s)$ in Equation 12.54b gives

$$Y(s) = C(sI - A)^{-1}BU(s) + DU(s) \tag{12.57a}$$

or

$$Y(s) = \left(C(sI - A)^{-1}B + D\right)U(s). \tag{12.57b}$$

The expression $C(sI - A)^{-1}B + D$ is referred to as the transfer matrix because it relates the output vector, $Y(s)$, to the input vector, $U(s)$. So we can rewrite Equation 12.57b in terms of the transfer matrix as

$$T(s) = \frac{Y(s)}{U(s)} = C(sI - A)^{-1}B + D. \tag{12.58}$$

12.6 Block diagrams

In general, many systems are composed of subsystems. These subsystems take any of the following forms: cascade, parallel, and closed-loop. We now present

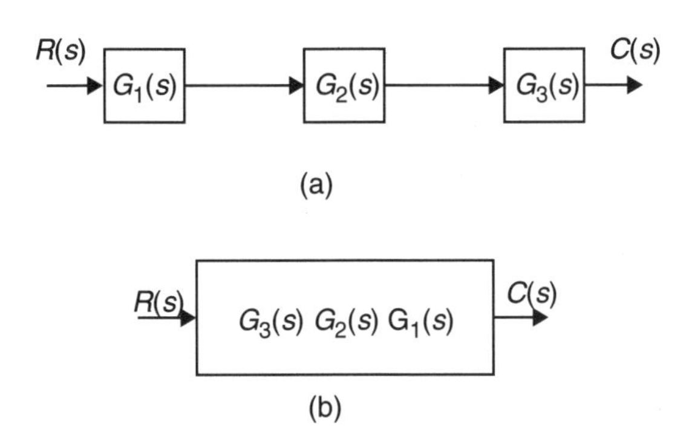

Figure 12.15 Cascade form block diagram.

these briefly and then move on to block algebra, which is necessary when interconnecting subsystems.

12.6.1 Cascade form

In a cascade form, each intermediate signal value is derived from the product of the input and the transfer function as shown in Figure 12.15.

12.6.2 Parallel form

In a parallel form, the parallel subsystems have a common input and an output formed by the algebraic sum of the outputs from all the subsystems as shown in Figure 12.16.

12.6.3 Feedback form

A general feedback control system would typically consist of an input, the plant and a controller, feedback, and an output, as shown in Figure 12.17. Due to the importance of feedback control systems, we will derive the transfer function that represents the system from its input to its output.

We can write the following for Figure 12.17:

$$E(s) = R(s) \pm C(s)H(s). \tag{12.59}$$

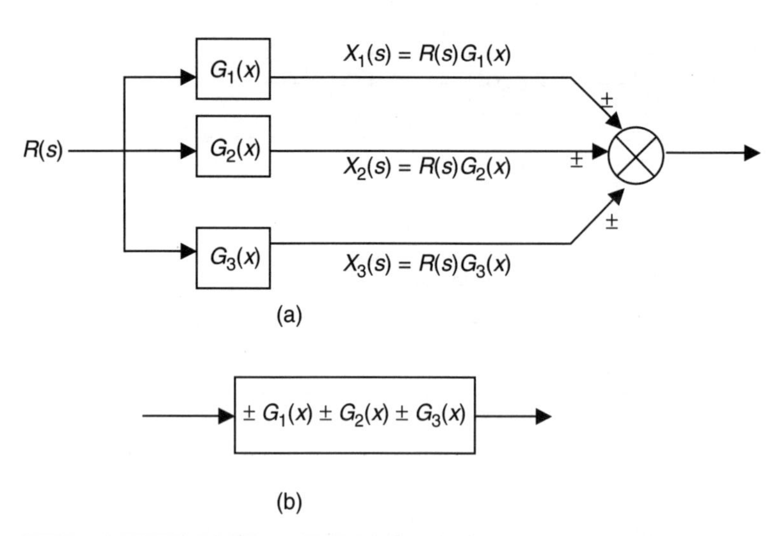

Figure 12.16 Parallel form block diagram.

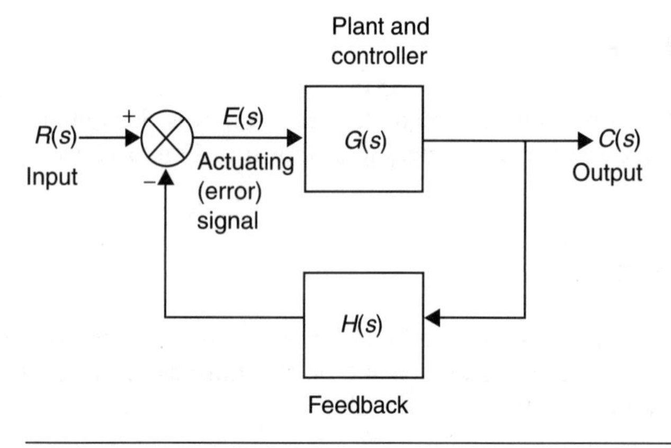

Figure 12.17 Simplified feedback control system.

But $C(s) = E(s)G(s)$

$$E(s) = \frac{C(s)}{G(s)}. \tag{12.60}$$

Putting Equation 12.60 into 12.59, we have

$$\frac{C(s)}{G(s)} = R(s) \pm C(s)H(s). \tag{12.61}$$

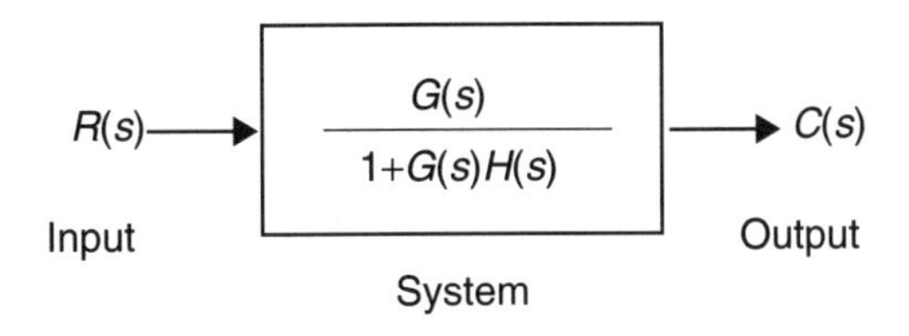

Figure 12.18 Equivalent transfer function of the closed-loop system.

Since $C(s)/R(s) = Ge(s)$, we then have

$$\frac{C(s)}{R(s)G(s)} = 1 \pm \frac{C(s)H(s)}{R(s)} \tag{12.62}$$

$$\frac{Ge(s)}{G(s)} = 1 \pm Ge(s)H(s) \tag{12.63}$$

Tidying up Equation 12.63 leads to

$$Ge(s) = G(s) \pm Ge(s)G(s)H(s) \tag{12.64}$$

$$Ge(s) \pm Ge(s)G(s)H(s) = G(s) \tag{12.65}$$

$$Ge(s)[1 \pm G(s)H(s)] = G(s) \tag{12.66}$$

$$\therefore\ Ge(s) = \frac{G(s)}{[1 \pm G(s)H(s)]}. \tag{12.67}$$

The equivalent transfer function is given in Figure 12.18.

12.6.4 Block diagram algebra

When multiple subsystems are interconnected, a few more schematic elements must be added to the block diagram. These elements include summing junctions and pickoff points. We will consider the cases of moving a block to the right and left of summing junctions.

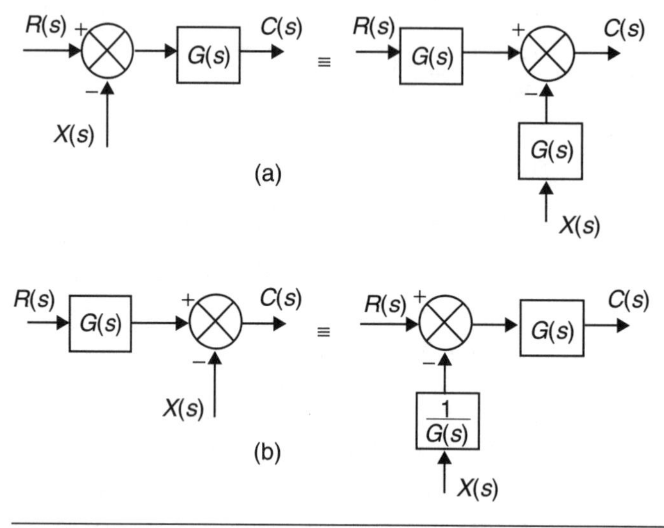

Figure 12.19 Moving a block to the (a) left of, and (b) right of a summing junction.

12.6.4.1 *Moving a block to the right and left of summing junctions*

Figure 12.19 shows the operations involved in moving a block to the right and left of summing junctions. When a block moves from the left to the right of a summing junction, the transfer function remains unchanged. However, when a block moves from the right to the left of a summing junction, the transfer function on the side of the feedback is in inverse form.

12.6.4.2 *Moving a block to the right and left of pickoff points*

Figure 12.20 shows the operations involved in moving a block to the right and the left of pickoff points. When a block moves from the left to the right of a pickoff point, the transfer function is inversed on all other links except the link from where movement is made. However, when a block moves from the right to the left of a pickoff point, the transfer function on each link to the right of the pickoff point remain unchanged.

EXAMPLE 12.8

See website for downloadable MATLAB code to solve this problem

Reduce the system shown in Figure 12.21 to a single transfer function.

Solution

Step 1: Add the two cascaded blocks and obtain the closed-loop transfer function as shown in Figure 12.22.

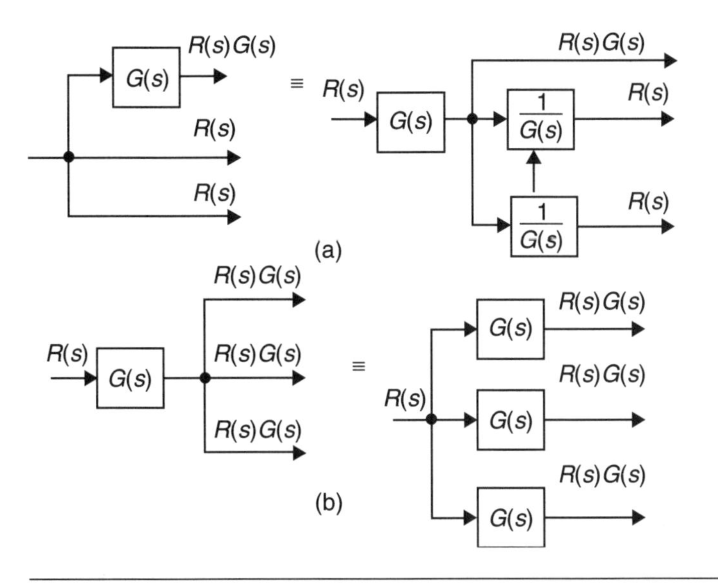

Figure 12.20 Moving a block to the (a) left of, and (b) right of a pickoff point.

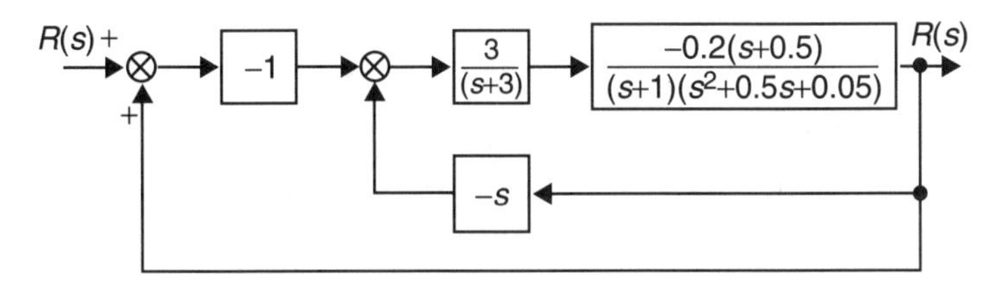

Figure 12.21 A feedback control system for Exercise 12.8.

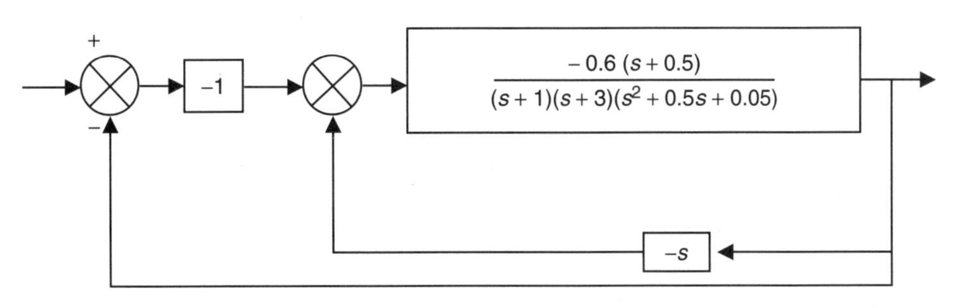

Figure 12.22 Adding the two cascaded blocks of Figure 12.21.

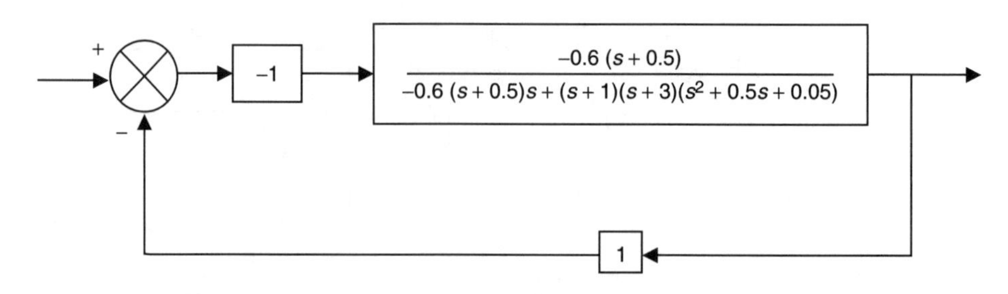

Figure 12.23 Equivalent block diagram of adding the two cascaded blocks of Figure 12.21.

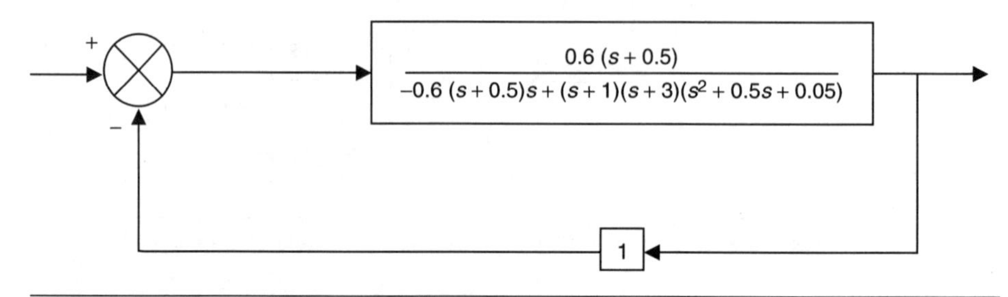

Figure 12.24 Dealing with the cascaded blocks of Figure 12.22.

Step 2: Deal with the cascades (see Figure 12.24).

Step 3: Deal with the closed-loop transfer function (see Figure 12.25).

To deal with unity feedback, with feed-forward being of the form a/b (i.e. $H(s) = 1$ and $G(s) = a/b$), we can show that

$$Ge(s) = \frac{a/b}{[1 + a/b \times 1]} = \frac{a}{a+b}. \tag{12.67A}$$

There are three approaches for reducing block diagrams:

- Solution via series, parallel, and feedback command.
- Solution via algebraic operations.
- Solution via append and connect command.

EXAMPLE 12.9

Reduce the system shown in Figure 12.26 to a single transfer function.

Solution

Step 1: Reduce the parallel pair consisting of $G_3(s)$ and unity, and push $G_1(s)$ to the right of the summing junction in order to have

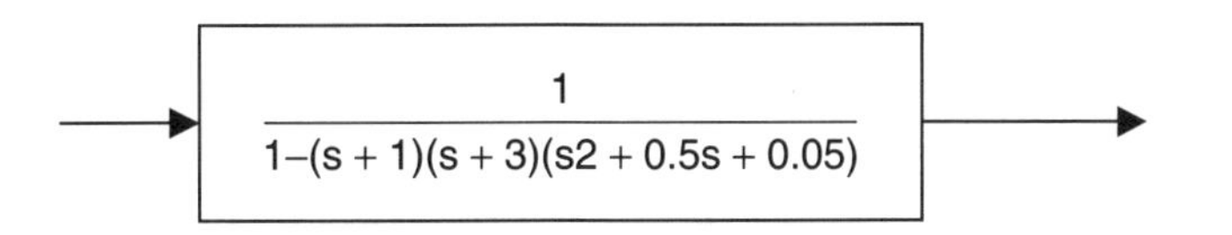

Figure 12.25 Closed-loop transfer function of Figure 12.22.

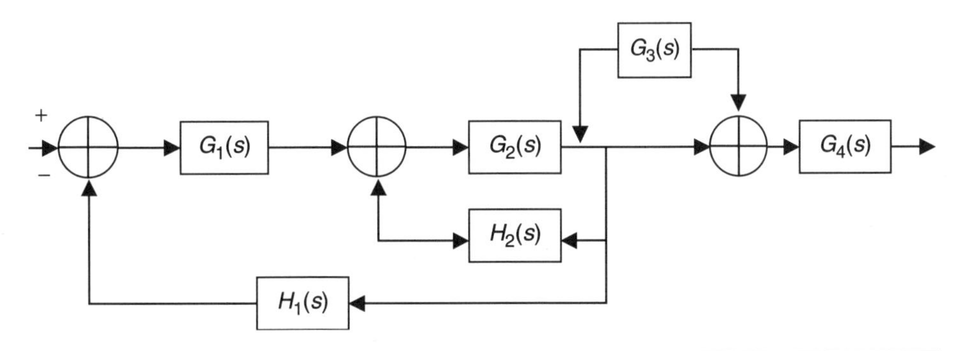

Figure 12.26 Control system for Example 12.9.

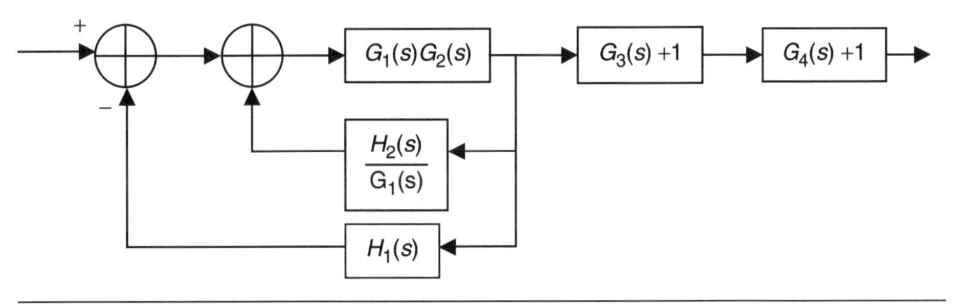

Figure 12.27 Intermediate system of Figure 12.26.

parallel subsystems in the feedback. Figure 12.27 shows the resulting intermediate system.

Step 2: Collapse the summing junctions into one, and add the two parallel feedback elements together, and add the two cascaded blocks before the output. Figure 12.28 shows the resulting intermediate system.

Step 3: Obtain the closed-loop transfer function as shown in Figure 12.29.

Step 4: Multiply the two cascaded blocks to obtain the final result shown in Figure 12.30.

+
−
G1(s)G2(s)
(G3(s) + 1)(G4(s))
H2(s)/H1(s) + H1(s)

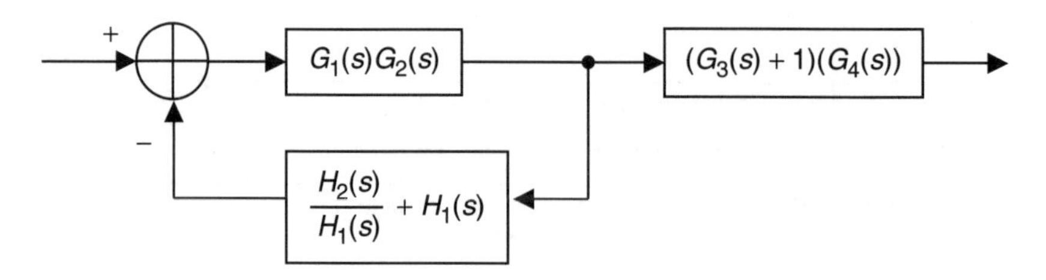

Figure 12.28 The resulting intermediate system of Figure 12.27.

R(s)
G1(s)G2(s) / 1 + G2(s)H2(s) + G2(s)G2(s)H2(s)
(G3(s) + 1)(G4(s))
C(s)

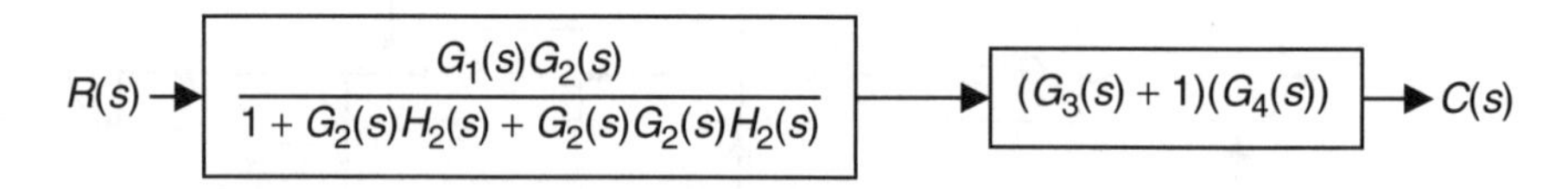

Figure 12.29 The further reduced intermediate system of Figure 12.28.

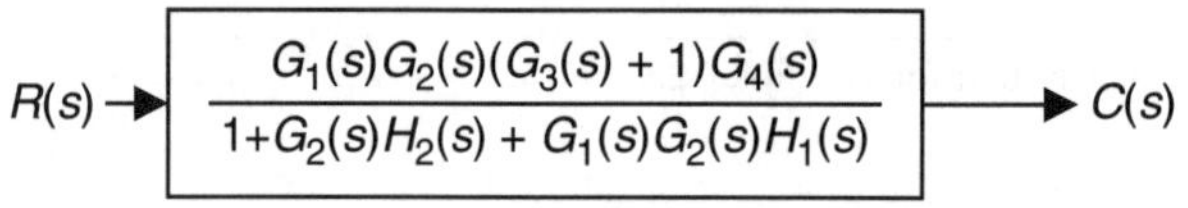

R(s)
G1(s)G2(s)(G3(s) + 1)G4(s) / 1+G2(s)H2(s) + G1(s)G2(s)H1(s)
C(s)

Figure 12.30 Closed-loop transfer function of Figure 12.29.

Problems

Q12.1 Carry out the partial fraction expansion, and hence find the inverse Laplace transform $f(t)$ of

$$F(s) = \frac{5}{(s+3)(s+4)}. \tag{12.67B}$$

Q12.2 Carry out the partial fraction expansion, and hence find the inverse Laplace transform $f(t)$ of

$$F(s) = \frac{5}{(s+3)(s+4)^2}. \tag{12.67C}$$

Q12.3 Find the transfer function represented by

$$\frac{d^4c(t)}{dt} + 2\frac{d^3c(t)}{dt} + 3\frac{d^2c(t)}{dt} + 6\frac{dc(t)}{dt} = 4r(t). \tag{12.67D}$$

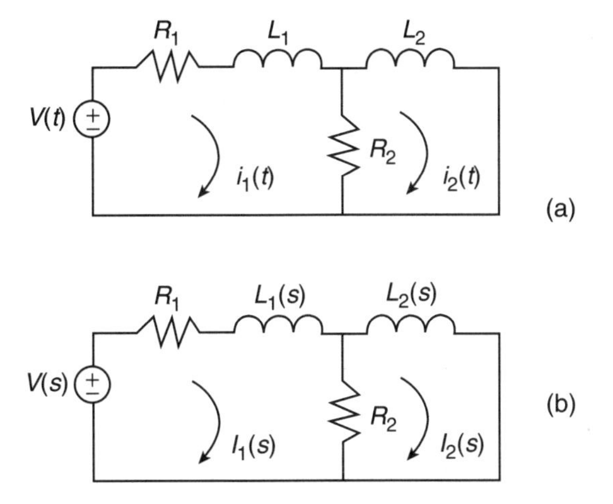

Figure 12.31 Electrical network: (a) circuit elements; (b) circuit impedances, for Q12.4.

Q12.4 For the network shown in Figure 12.31, determine the transfer function, $I_1(s)/V(s)$.

Q12.5 In an op amp used as a PID controller, shown in Figure 12.4, the values of the electrical elements are $C_1 = 8\,\mu F$; $C_2 = 0.2\,\mu F$; $R_1 = 800\,k\Omega$; $R_2 = 600\,k\Omega$.
Determine the transfer function, $V_o(s)/V_i(s)$.

Q12.6 In an op amp used as a PID controller, shown in Figure 12.5, the values of the electrical elements are $C_1 = 8\,\mu F$; $C_2 = 0.2\,\mu F$; $R_1 = 800\,k\Omega$; $R_2 = 600\,k\Omega$.
Determine the transfer function, $V_o(s)/V_i(s)$.

Q12.7 In the mechanical rotational system shown in Figure 12.8, the values of the mechanical elements properties are:

$J_1 = 4\,\text{kg m}^2$; $J_2 = 6\,\text{kg m}^2$;

$D_1 = 4\,\text{N m s rad}^{-1}$; $D_2 = 8\,\text{N m s rad}^{-1}$;

$k_1 = 2\,\text{N m rad}^{-1}$; $k_2 = 4\,\text{N m rad}^{-1}$;

$N_1 = 2000\,\text{rpm}$; $N_2 = 4000\,\text{rpm}$.

Determine the transfer function, $G(s) = \theta_2(s)/T(s)$.

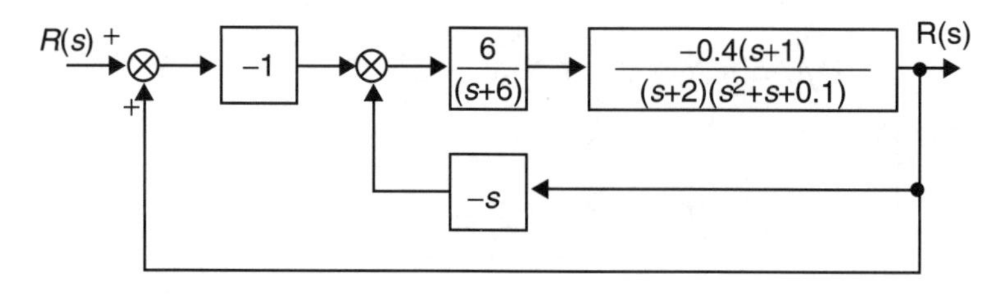

Figure 12.32 A feedback control system for Q12.10.

Q12.8 Figure 12.11 shows a translational mechanical system, represented in state space, where $x_3(t)$ is the output. Represent it in state space.

$$m_1 = 8\text{ kg};\quad m_2 = 4\text{ kg};\quad m_3 = 2\text{ kg};$$

$$K_1 = 6\text{ N m rad}^{-1};\quad K_2 = 2\text{ N m rad}^{-1};\quad K_3 = 2\text{ N m rad}^{-1};$$

$$Q_3 = 4F\sin\,\Omega t, (\Omega = 4K_2).$$

Q12.9 Find the state-space representation of the transfer function given as

$$G(s) = \frac{60\left(s^3 + 60s^2 + 40s + 100\right)}{(s^4 + 1200s^3 + 1000s^2 + 150s + 80)}. \tag{12.67E}$$

Q12.10 Reduce the system shown in Figure 12.32 to a single transfer function.

Q12.11 Repeat Questions 1–8 using MATLAB®.

Further reading

[1] Kuo, B.C. (1987) *Automatic Control Systems* (5th. ed.), Englewood Cliffs, NJ: Prentice-Hall.

[2] Nise, N.S. (2004) *Control Systems Engineering* (4th. ed.), New York: John Wiley & Sons.

[3] Ogata, K. (1990) *Modern Control Engineering* (2nd. ed.), Englewood Cliffs, NJ: Prentice-Hall.

[4] Philips, C.L. and Nagel, H.T. (1984) *Digital Control Systems Analysis and Design*, Englewood Cliffs, NJ: Prentice-Hall.

Internet resources

- LAMI robots (six feet): http://diwww.epfl.ch/lami/robots/robots.html

CHAPTER 13

Control theory: analysis

Chapter objectives

When you have finished this chapter you should be able to:

- understand poles and zeros as valuable tools for analysis and design of control systems;
- find the step response of a first-order control system;
- find the step response of a second-order control system;
- use the Routh-Hurwitz criterion to determine the stability of control systems;
- determine steady-state errors for a unity feedback system;
- determine steady-state errors for a non-unity feedback system.

13.1 Introduction

This chapter demonstrates the applications of the system representation by evaluating the transient response from the system model. The concept of poles and zeros, which is a valuable analysis and design tool is first discussed, then we show how to analyze our models in order to find the step response of first- and second-order control systems.

13.2 System response

Nowadays, it is possible to implement continuous control systems directly using computers. If the computer is fast enough (a close enough approximation

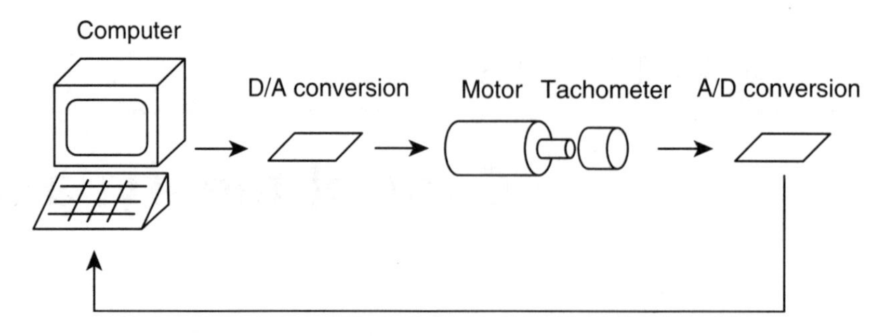

Figure 13.1 A computer-controlled motor speed control system.

to continuous), then the system generally works as predicted by continuous theory. Consider for example, a motor speed control system: normally, a d.c. motor's speed will not be constant if there are load variations, and feedback from a tachometer can be used to reduce speed fluctuations as shown in Figure 13.1.

However, as the sampling speed varies the system's performance changes. For example, at a high sampling rate, we might get a response such as the one shown in Figure 13.2(a). As the sampling rate is reduced, we might get the response shown in Figure 13.2(b). Finally, at a rate slow enough, the system might go unstable and the response would be depicted as shown in Figure 13.2(c).

13.2.1 Poles and zeros of a transfer function

The poles of a transform function are the values of the Laplace transform variable, s, that cause the transfer function to be infinite; they are the roots of the denominator of the function that are common to the roots of the numerator.

The zeros of a transform function are the values of the Laplace transform variable, s, that cause the transfer function to be zero; they are the roots of the numerator of the function that are common to the roots of the denominator.

The poles and zeros have certain characteristics:

- A pole of the input function generates the form of the *forced response.*
- A pole of the transfer function determines the form of the *natural response*.
- A pole on the real axis generates an exponential response of the form $e^{-\alpha t}$, where α is the pole location on the real axis.
- The poles and zeros generate the amplitudes for both the forced and natural responses.

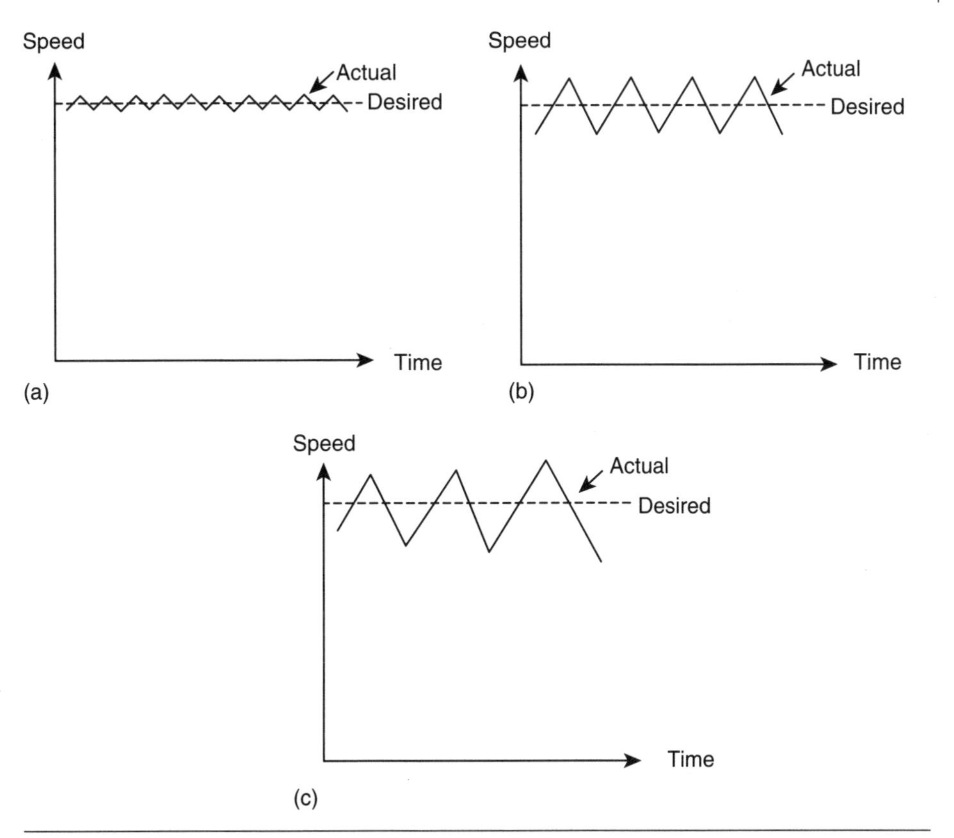

Figure 13.2 Variance between actual and desired response with different sampling rates: (a) high; (b) reduced; (c) 'extra-reduced' showing instability.

13.3 Dynamic characteristics of a control system

Most dynamic systems can be modeled as linear ordinary differential equations with constant coefficients. The general way of expressing the dynamics of linear systems is:

$$\sum_{n=0}^{N} A_n \frac{\mathrm{d}^n X_{\mathrm{out}}}{\mathrm{d}t^n} = \sum_{m=0}^{M} B_m \frac{\mathrm{d}^m X_{\mathrm{in}}}{\mathrm{d}t^m}, \tag{13.1}$$

where X_{out} is the output variable, X_{in} is the input variable, A_n and B_m are constant coefficients of the system's behavior; and N and M independently define the order of the system.

It should be mentioned that many mechatronic systems exhibit non-linear behavior and therefore, cannot be accurately modeled as linear systems.

Fortunately, a non-linear system may often exhibit linear behavior over a specified range of inputs and hence can be adequately approximated using a linear model over this range; a process known as linearization in dynamic system response. By varying the values of N and M, it is possible to examine in some detail different orders of dynamic systems.

13.4 Zero-order systems

A zero-order dynamic system is obtained by setting $N = M = 0$ in Equation 13.1. This leads to the following equation:

$$A_0 X_{out} = B_0 X_{in}, \tag{13.2}$$

leading to

$$X_{out} = \frac{B_0}{A_0} X_{in} = H X_{in}, \tag{13.3}$$

where as we have already discussed in amplifiers, H is the *gain* or *sensitivity* of the system.

A potentiometer used to measure displacement is a typical example of a zero-order dynamic system, where the output is related to the input as follows:

$$V_{out} = \frac{R_x}{R_p} V_s = \left(\frac{V_s}{L}\right) X_{in} = H X_{in}, \tag{13.4}$$

where R_p is the maximum resistance of the potentiometer, R_x is the current resistance of the wiper position, and L is the maximum amount of wiper travel.

13.5 First-order systems

A first-order dynamic system is obtained by setting $N = 1$ and $M = 0$ in Equation 13.1. This leads to the following equation:

$$A_1 \frac{dX_{out}}{dt} + A_0 X_{out} = B_0 X_{in}, \tag{13.5}$$

leading to

$$\frac{A_1}{A_0} \frac{dX_{out}}{dt} + X_{out} = \frac{B_0}{A_0} X_{in} = H X_{in}, \tag{13.6}$$

again, as we have already discussed in amplifiers, H is the *gain* or *sensitivity* of the system.

The ratio on the right-hand side of Equation 13.6 is known as the *time constant*:

$$\tau = \frac{A_1}{A_o}. \tag{13.7}$$

Consequently the first-order equation becomes

$$\tau \frac{dX_{out}}{dt} + X_{out} = HX_{in}. \tag{13.8}$$

There are several input models such as step, impulse, and sinusoidal functions. Let us apply the step function, although any of the other types could be considered. The mathematical expression for a step function is

$$X_{in} = \begin{cases} 0 & t < 0 \\ A_{in} & t \geq 0 \end{cases}, \tag{13.9}$$

with the initial condition being

$$X_{out}(0) = 0. \tag{13.10}$$

The standard approach to solving linear differential equations is to assume a solution of the form

$$X_{out} = Ce^{\lambda t}. \tag{13.11}$$

It is an easy matter to show that the characteristic equation (considering only the left-hand side of Equation 13.8) is given as

$$\tau\lambda + 1 = 0. \tag{13.12}$$

And the *homogenous* or *transient solution* is (since $\lambda = -1/\tau$):

$$X_{out_h} = Ce^{-t/\tau}. \tag{13.13}$$

We also now have the *particular* or *steady state solution* (considering only the left-hand side of Equation 13.8) to be:

$$X_{out_p} = HA_{in}. \tag{13.14}$$

The *general solution* (considering both Equation 13.13 and Equation 13.14) becomes:

$$X_{out}(t) = X_{out_h} + X_{out_p} = Ce^{-t/\tau} + HA_{in}, \tag{13.15}$$

which when we apply the initial condition of Equation 13.15 becomes

$$X_{out}(0) = 0 = C + HA_{in} \tag{13.16}$$

$$\therefore C - HA_{in}, \tag{13.17}$$

leading to

$$X_{out}(t) = HA_{in}(1 - e^{-t/\tau}). \tag{13.18}$$

A monostable vibrator (one-shot) containing a capacitor, used to generate a one-shot pulse is a typical example of a first-order dynamic system. Indeed, most *RC* circuits are first-order dynamic systems.

Let us approach the first-order systems from the point of view of transfer functions. For the first order system, $G(s) = a/(s + a)$, and considering the input of a unit step, where $R(s) = 1/a$, the output becomes

$$C(s) = R(s)G(s) = \frac{a}{s(s + a)}. \tag{13.19}$$

The inverse Laplace transform for the step response gives

$$c(t) = c_f(t) + c_n(t) = 1 - e^{-at} \tag{13.20}$$

Figure 13.3 shows the plot of Equation 13.20. Notice that the input pole at the origin generated the forced response, $c_f(t) = e^{-0t} = 1$, and the system pole at $-a$, the natural response $c_n(t) = -e^{-at}$.

When $t = 1/a$,

$$e^{-at}\Big|_{t=1/a} = e^{-1} = 0.37 \tag{13.21}$$

or

$$c(t)\Big|_{t=1/a} = 1 - e^{-at}\Big|_{t=1/a} = 1 - 0.37 = 0.63. \tag{13.22}$$

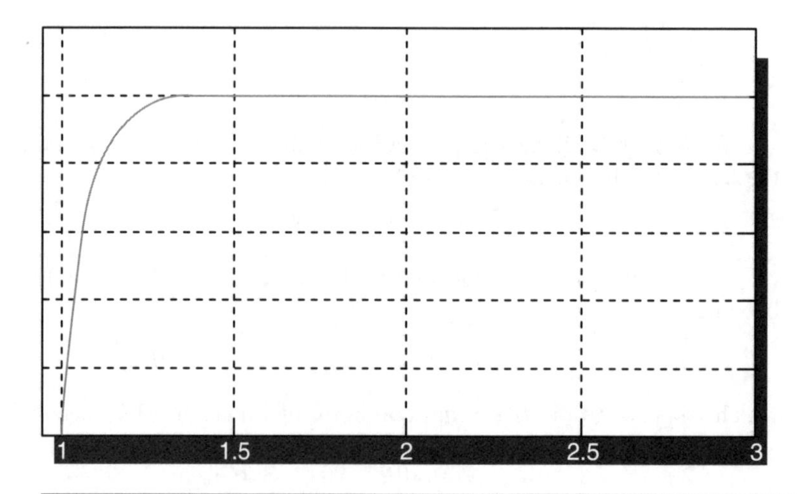

Figure 13.3 First-order system response to a unit step input.

Equations 13.20, 13.21, and 13.22 form the basis for the definition of the three transient response performance specifications: time constant, rise time, and settling time.

- The time constant, $1/a$, is the time for $e^{-\alpha t}$ to decay to 37% of its initial value or the time it takes the step response to rise to 63% of its final value.
- The rise time, T_r, is the time for the waveform to go from 0.1 to 0.9 of its final value. This is the same as letting $c(t)=0.9$ and 0.1, respectively, and solving for time, t, which gives

$$T_r = \frac{2.31}{a} - \frac{0.11}{a} = \frac{2.2}{a}. \quad (13.23)$$

- The settling time, T_s, is the time for the response to reach, and stay within, 2% of its final value. This is the same as letting $c(t)=0.98$ and solving for time, t, which gives

$$T_s = \frac{4}{a}. \quad (13.24)$$

13.6 Second-order systems

See website for additional material on using the Symbolic Math Toolbox

A second-order dynamic system is obtained by setting $N=2$ and $M=0$ in Equation 13.1. There are several responses that we need to consider: overdamped, underdamped, undamped, and critically-damped.

13.6.1 An overdamped response

Let us consider for this system that the input is a step function $R(s)=1/s$ and that,

$$G(s) = \frac{15}{(s^2 + 15s + 15)}. \quad (13.24A)$$

Concentrating on the denominator and finding the roots, we have

$$s = \left(-15 \pm \sqrt{15^2 - 4 \times 15}\right)/2 = \frac{(-15 \pm 12.85)}{2}$$
$$= -7.5 \pm 6.42 = -1.077 \quad \text{or} \quad -13.922 \quad (13.24B)$$

leading to

$$C(s) = \frac{15}{s(s^2 + 15s + 15)} = \frac{15}{s(s + 1.077)(s + 13.922)}. \tag{13.25}$$

The pole at the origin comes from the unit step input and two real poles come from the system. The input pole results in a constant forced response, while each of the two system poles on the real axis generate an exponential natural response, the frequency of which is equal to the pole location. Hence,

$$c(t) = K_1 + K_2 e^{-1.077t} + K_3 e^{-13.922t}. \tag{13.26}$$

13.6.2 An underdamped response

Let us consider for this system that the input is a step function $R(s) = 1/s$ and that,

$$G(s) = \frac{15}{(s^2 + 5s + 15)}. \tag{13.26A}$$

Concentrating on the denominator and finding the roots, we have

$$s = \left(-5 \pm \sqrt{5^2 - 4 \times 15}\right)/2 = \frac{(-5 \pm j5.916)}{2} = -2.5 \pm 2.968j \tag{13.26B}$$

leading to

$$C(s) = \frac{15}{s(s^2 + 5s + 15)} = \frac{15}{s(s + 2.5 - 2.968j)(s + 2.5 + 2.968j)} \tag{13.27}$$

$$c(t) = K_1 + e^{-2.5t}(K_2 \cos 2.958t + K_3 \sin 2.958t). \tag{13.28}$$

13.6.3 An undamped response

Let us consider for this system that the input is a step function $R(s) = 1/s$ and that,

$$G(s) = \frac{15}{(s + 15)}. \tag{13.28A}$$

Concentrating on the denominator and finding the roots, we have

$$s = \pm\sqrt{15}j \tag{13.28B}$$

leading to

$$C(s) = \frac{15}{s(s+15)} \tag{13.28C}$$

$$c(t) = 1 - \cos\sqrt{15}t. \tag{13.29}$$

13.6.4 A critically-damped response

Let us consider for this system that the input is a step function $R(s) = 1/s$ and that,

$$G(s) = \frac{15}{(s^2 + 10s + 25)}. \tag{13.29A}$$

Concentrating on the denominator and finding the roots, we have $s = -5$ (twice), leading to

$$C(s) = \frac{15}{s(s+5)^2} \tag{13.29B}$$

$$c(t) = 1 - 5e^{-5t} - e^{-5t}. \tag{13.30}$$

Figure 13.4 shows undamped, underdamped, and overdamped second-order systems.

13.7 General second-order transfer function

The general second order transfer function takes the form

$$G(s) = \frac{{\omega_n}^2}{s^2 + 2\zeta\omega_n s + {\omega_n}^2} \tag{13.31}$$

The roots of the denominator are

$$s_{1,2} = \left(-2\zeta\omega_n \pm \sqrt{4\zeta^2{\omega_n}^2 - 4{\omega_n}^2}\right)\Big/2 = -\zeta\omega_n \pm \omega_n\sqrt{\zeta^2 - 1}. \tag{13.32}$$

Several damping cases arise from Equation 13.32 and these are summarized in Table 13.1.

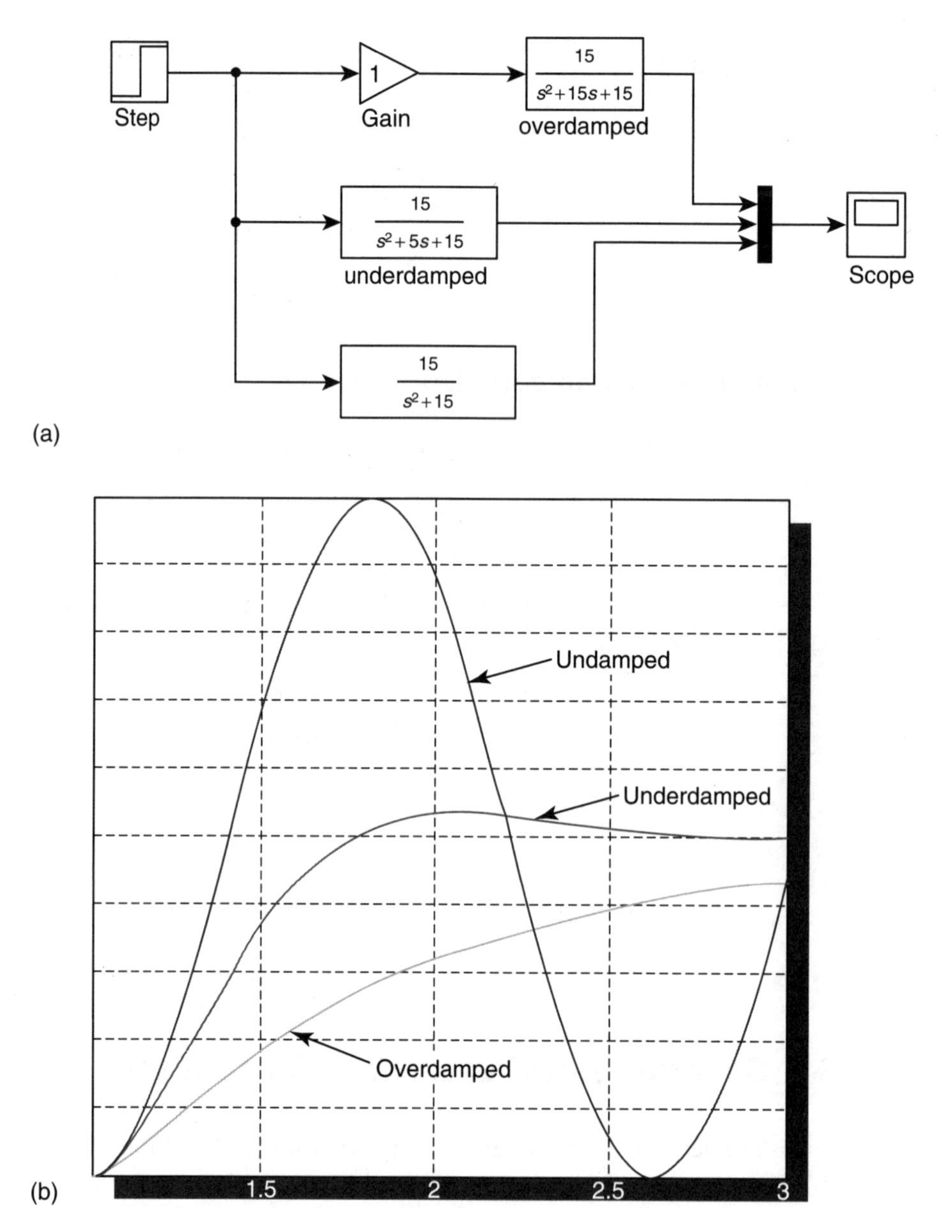

Figure 13.4 The undamped, underdamped, and overdamped responses for a step input.

Table 13.1 Damping cases for second-order system

Case	*Expression*	*Description*
$\zeta = 0$	$\pm\omega_n$	Undamped
$0 < \zeta < 1$	$\pm j\omega_n\sqrt{1-\zeta^2}$	Underdamped
$\zeta = 1$	$-\omega_n$	Critically-damped
$\zeta > 1$	$-\zeta\omega_n \pm \omega_n\sqrt{\zeta^2 - 1}$	Overdamped

EXAMPLE 13.1

For the transfer function given, determine the values of ζ and ω_n:

$$G(s) = \frac{64}{s^2 + 6s + 64}. \tag{13.32A}$$

Solution

Comparing the given transfer function and Equation 13.31, we find that

$$\omega_n{}^2 = 64$$
$$\therefore \omega_n = \sqrt{64} = 8. \tag{13.32B}$$

Also, $2\zeta\omega_n = 6$, so

$$\zeta = \frac{6}{2 \times 8} = 0.375$$

EXAMPLE 13.2

Categorize the following transfer functions as underdamped, critically-damped, or overdamped.

$$\text{(a) } G(s) = \frac{25}{s^2 + 10s + 25}; \quad \text{(b) } G(s) = \frac{36}{s^2 + 10s + 36}; \quad \text{(c) } G(s) = \frac{16}{s^2 + 10s + 16}. \tag{13.32C}$$

Solution

(a) Comparing the given transfer function and Equation 13.31, we find that: $\omega_n{}^2 = 25$, so $\omega_n = 5$. Also, $2\zeta\omega_n = 10$, so $\zeta = 1$, implying critical damping.

(b) Comparing the given transfer function and Equation 13.31, we find that ${\omega_n}^2 = 36$, so $\omega_n = 6$. Also, $2\zeta\omega_n = 10$, so $\zeta = 0.833$, implying underdamping.

(c) Comparing the given transfer function and Equation 13.31, we find that ${\omega_n}^2 = 16$, so $\omega_n = 4$. Also, $2\zeta\omega_n = 10$, so $\zeta = 1.25$, implying overdamping.

13.7.1 Underdamped second-order systems

Underdamping is often encountered in physical problems, so let us consider the step response, $R(s) = 1/s$, for the general second-order system.

$$C(s) = R(s)G(s) = \frac{{\omega_n}^2}{s(s^2 + 2\zeta\omega_n s + {\omega_n}^2)} \tag{13.33}$$

leading to

$$C(s) = \frac{{\omega_n}^2}{s(s^2 + 2\zeta\omega_n s + {\omega_n}^2)} = \frac{K_1}{s} + \frac{K_2 s + K_3}{(s^2 + 2\zeta\omega_n s + {\omega_n}^2)}\,. \tag{13.34}$$

For the underdamped case, where the poles are at

$$\pm j\omega_n\sqrt{1-\zeta^2}, \tag{13.35}$$

using partial fraction method, we have

$$C(s) = \frac{1}{s} + \frac{(s + \zeta\omega_n) + \frac{\zeta}{\sqrt{1-\zeta^2}}\omega_n\sqrt{1-\zeta^2}}{(s^2 + \zeta\omega_n)^2 + {\omega_n}^2(1-\zeta^2)} \tag{13.35A}$$

Taking the Laplace transform gives the following result,

$$\begin{aligned} c(t) &= 1 - e^{\zeta\omega_n t}\left(\cos\omega_n\sqrt{1-\zeta^2}t + \frac{\zeta}{\sqrt{1-\zeta^2}}\sin\omega_n\sqrt{1-\zeta^2}t\right) \\ &= 1 - \frac{\zeta}{\sqrt{1-\zeta^2}}e^{\zeta\omega_n t}\left(\cos\omega_n\sqrt{1-\zeta^2}t - \phi\right), \end{aligned} \tag{13.36}$$

where

$$\phi = \tan^{-1}\left(\frac{\zeta}{\sqrt{1-\zeta^2}}\right). \tag{13.36A}$$

We now define four other parameters associated with the underdamped response for the second-order system. These parameters are rise time, peak time, percentage overshoot, and settling time.

13.7.1.1 *Rise time*

The rise time, T_r, is the time for the waveform to go from 0.1 to 0.9 of its final value. This is the same as letting $c(t) = 0.9$ and 0.1, respectively, and solving for $\omega_n t$. Subtracting the two values of $\omega_n t$ gives us the normalized rise time, $\omega_n T_r$, for a given damping ratio, ζ. Solving Equation 13.36 for these conditions is difficult and hence the use of a computer is advisable. Table 13.2 shows the normalized rise time for values of damping ratios. We shall see later how to use these values to determine rise time.

13.7.1.2 *Peak time*

The peak time, T_p, is the time required to reach the first, or maximum, peak. It is obtained by differentiating Equation 13.36 with respect to time and equating the differential to zero, then solving for the time.

$$\dot{c}(t) = \frac{\omega_n}{\sqrt{1-\zeta^2}} e^{-\zeta\omega_n t} \sin \omega_n \sqrt{1-\zeta^2} t = 0 \tag{13.37}$$

Solving this equation for time gives

$$\omega_n \sqrt{1-\zeta^2} t = n\pi, \tag{13.38}$$

Table 13.2 Normalized rise time

ζ	T
0.1	1.104
0.2	1.203
0.3	1.321
0.4	1.463
0.5	1.638
0.6	1.854
0.7	2.126
0.8	2.467
0.9	2.883

yielding (for $n=1$)

$$T_p = \frac{\pi}{\omega_n\sqrt{1-\zeta^2}}. \tag{13.39}$$

13.7.1.3 ***Percentage overshoot***

The percentage overshoot, $\%OS$, is the amount that the waveform overshoots the steady-state, or final, value at the peak time, expressed as a percentage of the steady-state value.

$$\%OS = \frac{c_{max} - c_{final}}{c_{final}} \times 100. \tag{13.40}$$

The maximum value, c_{max}, (when $t = T_p$) is substituted in Equation 13.36. Since the maximum value of cosine is 1, then

$$c_{max} = c(T_p) = 1 + e^{-\left(\zeta\pi/\sqrt{1-\zeta^2}\right)}. \tag{13.41}$$

We take $c_{final} = 1$ for a unit step, and then obtain the expression for percentage overshoot as

$$\%OS = e^{-\left(\zeta\pi/\sqrt{1-\zeta^2}\right)} \times 100. \tag{13.42}$$

Since the percentage overshoot is a function of the damping factor, it is possible to express the damping factor in terms of the percentage overshoot from Equation 13.42. Rearranging, we get

$$e^{-\left(\zeta\pi/\sqrt{1-\zeta^2}\right)} = \frac{\%OS}{100} = z. \tag{13.43}$$

Taking the log of both sides of this equation yields

$$-\zeta\pi/\sqrt{1-\zeta^2} = \ln(z). \tag{13.44}$$

Squaring both sides of the equation gives

$$-\zeta^2\pi^2/1 - \zeta^2 = \ln^2(z). \tag{13.45}$$

Solving for the damping factor yields

$$\zeta = \frac{-\ln(z)}{\sqrt{\pi^2 + \ln^2(z)}}. \tag{13.46}$$

Notice that we have included a negative sign to cater for z, which is normally less than 1.

13.7.1.4 *Settling time*

The settling time, T_s, is the time for the response to reach, and stay within, 2% of its final value. This is the same as letting $c(t) = 0.98$ and solving for time, t. In other words,

$$e^{-\zeta\omega_n t}\frac{1}{\sqrt{1-\zeta^2}} = 0.02. \tag{13.47}$$

Solving for the settling time,

$$T_s = \frac{-\ln\left(0.02\sqrt{1-\zeta^2}\right)}{\zeta\omega_n}. \tag{13.48}$$

For the range of $0.1 \leq \zeta \leq 0.9$, we can approximate the numerator of Equation 13.48 as 4 since it varies between 3.92 and 4.74, hence

$$T_s = \frac{4}{\zeta\omega_n}. \tag{13.49}$$

EXAMPLE 13.3

For the transfer function

$$G(s) = \frac{150}{s^2 + 20s + 150}, \tag{13.49A}$$

determine: (a) the peak time; (b) the percentage overshoot; (c) the settling time; and (d) the rise time.

Solution

$$\omega_n = \sqrt{150} = 12.2474$$

$$\zeta = \frac{20}{2\omega_n} = \frac{20}{2 \times 12.2474} = 0.8165 \quad \text{(underdamped)}. \tag{13.49B}$$

(a) The peak time is

$$T_p = \frac{\pi}{\omega_n\sqrt{1-\zeta^2}} = \frac{\pi}{12.2474\sqrt{1-0.8165^2}} = 0.444s. \tag{13.49C}$$

(b) The percentage overshoot is

$$\%OS = e^{-\left(\zeta\pi/\sqrt{1-\zeta^2}\right)} \times 100 = \exp\left(\frac{-0.8165 \times \pi}{\sqrt{1-0.8165^2}}\right) \times 100 = 1.176. \tag{13.49D}$$

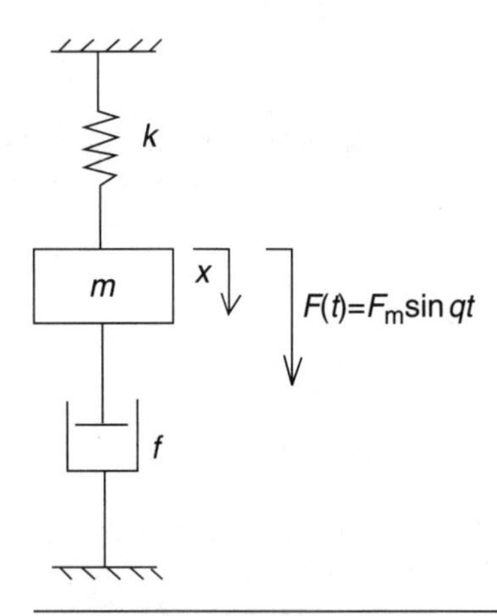

Figure 13.5 A second-order dynamic system.

(c) The settling time

$$T_s = \frac{4}{\zeta\omega_n} = \frac{4}{0.8165 \times 12.2474} = 0.4s\,. \tag{13.49E}$$

(d) The rise time is obtained using $\zeta = 0.8165$ in Table 13.2, for which the normalized rise time is $T_r = 2.5s$ (by interpolation) and the actual rise time

$$T_r = \frac{\hat{T}_r}{\omega_n} = \frac{2.5}{12.2474} = 0.204s. \tag{13.49F}$$

Consequently, for the underdamped second-order system, we can obtain peak time, percentage overshoot, settling time, and rise time without plots.

As an illustration of the second-order dynamic system, let us consider Figure 13.5, which is a mechanical mass-spring-damper system. This is a forced damped vibration system since the applied force is not natural but it has an amplitude and follows a sinusoidal waveform.

The governing differential equation is:

$$m\ddot{x} = -kx - f\dot{x} + F_m \sin qt. \tag{13.50}$$

We define

$$y_m = \frac{F_m}{k};\ 2\zeta\omega = \frac{f}{m} \text{ and } \omega^2 = \frac{k}{m}. \tag{13.51}$$

We now rewrite Equation 13.50 as

$$\ddot{x} + 2\zeta\omega\dot{x} + \omega^2 x = \frac{F_m}{k}\frac{k}{m}\sin qt \tag{13.52}$$

leading to

$$\ddot{x} + 2\zeta\omega\dot{x} + \omega^2 x = \omega^2 y_{\mathrm{m}} \sin qt. \tag{13.53}$$

Let us define

$$p = \omega\sqrt{1 - \zeta^2}. \tag{13.54}$$

Then for this class of problem, we choose a complementary function (transient solution) of the form

$$x_{\mathrm{cf}} = C \sin qt + E \cos qt. \tag{13.55}$$

Hence applying Equation 13.55 to Equation 13.53 we have

$$\begin{aligned} -q^2 C \sin qt - q^2 E \cos qt + 2\zeta\omega q\, C \cos qt \\ - 2\zeta\omega q E \sin qt + \omega^2 C \sin qt + \omega^2 E \cos qt = \omega^2 y_{\mathrm{m}} \sin qt \end{aligned} \tag{13.56}$$

We can now equate terms for sine and cosine accordingly:

$$\sin : -q^2 C - 2\zeta\omega q E + \omega^2 C = \omega^2 y_{\mathrm{m}} \tag{13.57}$$

$$\cos : -q^2 E + 2\zeta\omega q C + \omega^2 E = 0. \tag{13.58}$$

Hence,

$$(\omega^2 - q^2)C - 2\zeta\omega q E = \omega^2 y_{\mathrm{m}} \tag{13.59}$$

$$2\zeta\omega q C + (\omega^2 - q^2)E = 0. \tag{13.60}$$

From Equation 13.60, we have

$$E = \frac{-2\zeta\omega q C}{(\omega^2 - q^2)}. \tag{13.61}$$

Substituting in Equation 13.59, we now have

$$(\omega^2 - q^2)C + 2\zeta\omega q \frac{2\zeta\omega q C}{(\omega^2 - q^2)} = \omega^2 y_{\mathrm{m}}. \tag{13.62}$$

Consequently, from Equation 13.62 the expressions for C and E are obtained as

$$C = \frac{\omega^2(\omega^2 - q^2) y_{\mathrm{m}}}{\left[(\omega^2 - q^2)^2 + 4\zeta^2\omega^2 q^2\right]} \tag{13.63}$$

$$E = \frac{(-2\zeta\omega q)}{(\omega^2 - q^2)} \left\{ \frac{\omega^2(\omega^2 - q^2)y_m}{\left[(\omega^2 - q^2)^2 + 4\zeta^2\omega^2 q^2\right]} \right\} = \left\{ \frac{(-2\zeta\omega q)\omega^2 y_m}{\left[(\omega^2 - q^2)^2 + 4\zeta^2\omega^2 q^2\right]} \right\} \quad (13.64)$$

The *general solution* PI, obtained by substituting Equations 13.63 and 13.64 into Equation 13.55 becomes

$$\mathrm{PI} = \frac{\omega^2(\omega^2 - q^2)y_m \sin qt}{\left[(\omega^2 - q^2)^2 + 4\zeta^2\omega^2 q^2\right]} + \left\{ \frac{(-2\zeta\omega q)\omega^2 y_m \cos qt}{\left[(\omega^2 - q^2)^2 + 4\zeta^2\omega^2 q^2\right]} \right\} \quad (13.65)$$

$$= \left\{ \frac{\omega^2 y_m}{\left[(\omega^2 - q^2)^2 + 4\zeta^2\omega^2 q^2\right]} \right\} \left[(\omega^2 - q^2)\sin qt - 2\zeta\omega q \cos qt\right], \quad (13.66)$$

$$= \left\{ \frac{\omega^2 y_m}{\left[(\omega^2 - q^2)^2 + 4\zeta^2\omega^2 q^2\right]} \right\} \left\{ \sqrt{\left[(\omega^2 - q^2)^2 + 4\zeta^2\omega^2 q^2\right]} \times \sin(qt - \psi) \right\}. \quad (13.67)$$

Finally, we have

$$\mathrm{PI} = \frac{\omega^2 y_m \sin(qt - \psi)}{\sqrt{\left[(\omega^2 - q^2)^2 + 4\zeta^2\omega^2 q^2\right]}} = \frac{y_m \sin(qt - \psi)}{\sqrt{\left\{\left[1 - \left(\frac{q^2}{\omega^2}\right)\right]^2 + 4\zeta^2\left(\frac{q^2}{\omega^2}\right)\right\}}}. \quad (13.68)$$

We now define first, the amplitude ratio:

$$\mu = \frac{1}{\sqrt{\left\{\left[1 - \left(\frac{q^2}{\omega^2}\right)\right]^2 + 4\zeta^2\left(\frac{q^2}{\omega^2}\right)\right\}}} \quad (13.69)$$

and then the phasor:

$$\psi = \tan^{-1}\left[\frac{2\zeta q/\omega}{\left(1 - q^2/\omega^2\right)}\right] \quad (13.70)$$

We now plot the frequency response of the second-order system as a function of the damping ratio ζ as shown in Figure 13.6. When the input frequency, q, is equal to the natural frequency ω (i.e. $q/\omega = 1$) resonance occurs. We note from Equation 13.69 that when $\zeta = 0$ (no damping), the amplitude ratio is maximum.

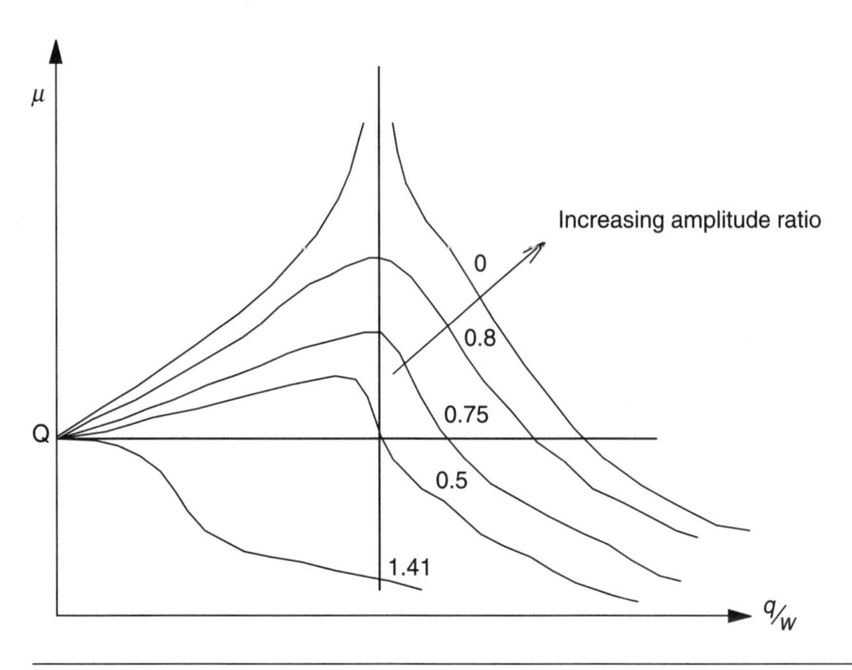

Figure 13.6 Second-order system amplitude response.

13.7.1.5 *The maximum amplitude ratio*

The maximum amplitude ratio is obtained by taking the squares of both sides of Equation 13.52, differentiating the denominator with respect to ω and equating to zero:

$$\frac{\mathrm{d}}{\mathrm{d}(q^2/\omega^2)}\left\{\left[1-\left(\frac{q^2}{\omega^2}\right)\right]^2+4\zeta^2\left(\frac{q^2}{\omega^2}\right)\right\}=-2\left(1-\left(\frac{q^2}{\omega^2}\right)\right)+4\zeta^2=0 \qquad (13.71)$$

leading to

$$q_r=\omega\sqrt{1-2\zeta^2} \quad \text{at resonance.} \qquad (13.72)$$

But

$$p=\omega\sqrt{1-\zeta^2}, \qquad \text{hence} \quad \omega>p>q_r. \qquad (13.72\text{A})$$

We can now obtain the maximum value of the amplitude ratio from Equations 13.72 and 13.69 as

$$\mu_{\max}=\frac{1}{\sqrt{\left\{[1-(1-2\zeta^2)]^2+4\zeta^2(1-2\zeta^2)\right\}}}=\frac{1}{2\zeta\sqrt{(1-\zeta^2)}}. \qquad (13.73)$$

This maximum value together with the frequency ratio, $q/\omega = 1$, give $1/2\zeta$ and help us to draw the graph of the amplitude ratio, μ, against the frequency ratio, q/ω.

Although we have successfully determined the frequency response, the method is unwieldy. Not only that, it is difficult to know what form of solution to assume. We now present the Operator *D*-Method, which is very user friendly.

13.7.2 Operator-*D* method

In this method, we use the following notations:

$$\ddot{x} = D^2 x; \quad \dot{x} = Dx; \quad \sin qt = I_{\mathrm{m}} e^{jqt} \tag{13.73A}$$

Let us restart from Equation 13.53, which is reproduced below:

$$\ddot{x} + 2\zeta\omega\dot{x} + \omega^2 x = \omega^2 y_{\mathrm{m}} \sin qt. \tag{13.73B}$$

Applying the Operator-*D* notations, we have

$$(D^2 + 2\zeta\omega D + \omega^2)x = \omega^2 y_{\mathrm{m}} I_{\mathrm{m}} e^{jqt}. \tag{13.74}$$

We can immediately write the *general solution* PI as:

$$\mathrm{PI} = x = \frac{\omega^2 y_{\mathrm{m}}}{(D^2 + 2\zeta\omega D + \omega^2)} I_{\mathrm{m}} e^{jqt} \tag{13.75}$$

$$x = I_{\mathrm{m}} \frac{\omega^2 y_{\mathrm{m}}}{(-q^2 + 2\zeta\omega q\mathrm{j} + \omega^2)} e^{jqt} \tag{13.76}$$

$$= \frac{I_{\mathrm{m}} \omega^2 y_{\mathrm{m}} \left[(\omega^2 - q^2) - 2\zeta\omega q\mathrm{j}\right] e^{jqt}}{\left[(\omega^2 - q^2)^2 + 4\zeta^2\omega^2 q^2\right]} \tag{13.77}$$

$$= \frac{I_{\mathrm{m}} \omega^2 y_{\mathrm{m}} \left[(\omega^2 - q^2) - 2\zeta\omega q\mathrm{j}\right](\cos qt + \mathrm{j} \sin qt)}{\left[(\omega^2 - q^2)^2 + 4\zeta^2\omega^2 q^2\right]} \tag{13.78}$$

$$= \frac{\omega^2 y_{\mathrm{m}} \left[(\omega^2 - q^2) \sin qt - 2\zeta\omega q \cos qt\right]}{\left[(\omega^2 - q^2)^2 + 4\zeta^2\omega^2 q^2\right]} \tag{13.79}$$

and

$$\mathrm{PI} = x = \frac{\omega^2 y_{\mathrm{m}} \sin(qt - \psi)}{\sqrt{[(\omega^2 - q^2)^2 + 4\zeta^2\omega^2 q^2]}} = \frac{y_{\mathrm{m}} \sin(qt - \psi)}{\sqrt{\left\{\left(1 - \left[\frac{q^2}{\omega^2}\right]\right)^2 + 4\zeta^2\left(\frac{q^2}{\omega^2}\right)\right\}}} \tag{13.80}$$

$$\psi = \tan^{-1}\left[\frac{2\zeta^{q}/_{\omega}}{\left(1 - {}^{q^2}/_{\omega 2}\right)}\right], \tag{13.81}$$

which is exactly the same solution as with the previous method. As demonstrated using this example, the Operator-D method is quite powerful and takes a much shorter time to analysis the response equation.

EXAMPLE 13.4

A mass of 25 kg is suspended from a spring with stiffness 500 000 N m^{-1}. The damping is negligible. The mass is initially resting in its equilibrium position when a fluctuating force of amplitude 10 000 N and frequency 100 rad s^{-1} is suddenly applied to it. Determine:

(a) the amplitude of the steady state vibration; and
(b) the arbitrary constants.

Sketch the displacement–time curve for the first few cycles of the oscillation.

Solution

The system is represented in Figure 13.7.
The governing differential equation is given as:

$$\begin{aligned} m\ddot{x} &= -kx + F_{\mathrm{m}} \sin qt \\ (\ddot{x} + \omega^2)x &= \omega^2 y_{\mathrm{m}} \sin qt. \end{aligned} \tag{13.81A}$$

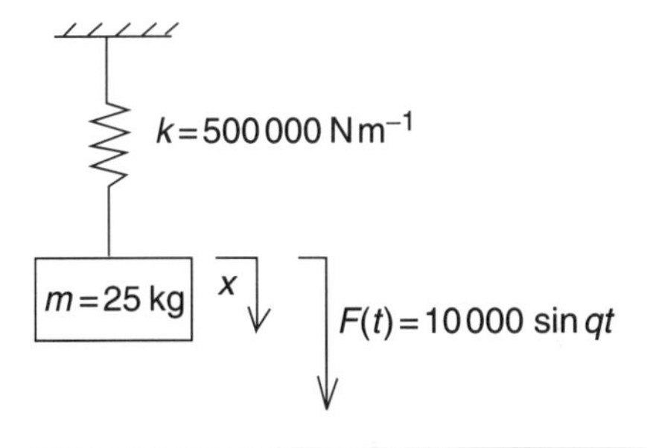

Figure 13.7 System for Example 13.4.

Applying the Operator-D notation, we have

$$(D^2 + \omega^2)x = \omega^2 y_m I_m e^{jqt} \tag{13.81B}$$

Equating the left-hand side to zero and solving for D, we obtain the complementary function:

$$(D^2 + \omega^2) = 0;$$
$$\therefore D = \pm j\omega. \tag{13.81C}$$
$$\text{Hence } x_{cf} = A \sin \omega t + B \cos \omega t.$$

We can immediately write the *general solution* PI as:

$$\text{PI} = x = \frac{\omega^2 y_m}{(D^2 + \omega^2)} I_m e^{jqt}$$

$$x = I_m \frac{\omega^2 y_m}{(-q^2 + \omega^2)} e^{jqt}$$

$$= \frac{I_m \omega^2 y_m [(\omega^2 - q^2)] e^{jqt}}{[(\omega^2 - q^2)^2]}$$

$$= \frac{I_m \omega^2 y_m [(\omega^2 - q^2)](\cos qt + j \sin qt)}{[(\omega^2 - q^2)^2]}$$

$$= \frac{\omega^2 y_m [(\omega^2 - q^2) \sin qt]}{[(\omega^2 - q^2)^2]} = \frac{\omega^2 y_m \sin qt}{(\omega^2 - q^2)}. \tag{13.81D}$$

$$\text{Hence, } x = A \sin \omega t + B \cos \omega t + \frac{y_m \sin qt}{(1 - q^2/\omega^2)}$$

$$\mu = \frac{1}{\left|1 - \left(\frac{q^2}{\omega^2}\right)\right|}.$$

$$\text{Now, } \omega = \sqrt{\frac{k}{m}} = \sqrt{\frac{500\,000}{25}} = 141.2 \text{ rad s}^{-1}.$$

$$\text{And } \mu = \frac{1}{\left|1 - \left(\frac{100^2}{141.2^2}\right)\right|} = 2.$$

$$\text{Now, } y_m = \frac{F_m}{k} = \frac{10\,000}{500\,000} = 0.02\, m.$$

This value is at steady-state condition.

Using the equation for the final equation for displacement, at time $t = 0$, $x_o = 0$, hence $B = 0$.

$$\dot{x} = \omega A \cos \omega t - \omega B \sin \omega t + \frac{q y_m \cos qt}{(1 - q^2/\omega^2)}.$$

$\dot{x} = 0$ at $t = 0$ since the system is resting at equilibrium position initially.

$$\therefore 0 = \omega A + \frac{q y_m}{(1 - q^2/\omega^2)}$$

$$\therefore A = \frac{-q/\omega}{(1 - q^2/\omega^2)} y_m = \frac{-\sqrt{2}/2}{1 - 2/4} y_m = -\sqrt{2} y_m. \tag{13.81E}$$

Resulting in

$$x = -\sqrt{2} \sin \omega t + \frac{y_m \sin qt}{(1 - 2/4)} = y_m \left(2 \sin q100t - \sqrt{2} \sin 141.2t\right) \tag{13.81F}$$

where $y_m = 0.02$ (as already calculated). This equation describes the response of the dynamic system being considered in this example; the equation is used to draw the graph of x as a function of $\sin 100t$ and $\sin 141.2t$ with the appropriate amplitudes.

13.8 Systems modeling and interdisciplinary analogies

An ordinary linear differential equation is known to model any linear system, relating the output response of the system to the input, whether electrical, mechanical, hydraulic, or thermal. The analogies between these several engineering disciplines have been developed over time. The mechanical analogy used is the mobility analogy in which the physical analogies are sacrificed in favor of creating equivalent mathematical relationships, which hold in network analysis. Table 13.3 summarizes the interdisciplinary analogies that exist between different systems. Figure 13.8 shows a simple analog mechanical system and a simple electrical system. Their respective governing differential equations are given in Equations 13.82 and 13.83. The analogy between these systems is of practical importance because each of them can be independently analyzed and substituted for each system, for example, for a mechanical system, it is now possible to use the mesh circuit theory to write the governing differential equations.

Table 13.3 Interdisciplinary analogies

General	*Electrical*	*Mechanical Translational/Rotational*		*Hydraulic (Acoustic)*	*Thermal*
Flow variable (through variable)	Current, $I = dq/dt$ [A]	Velocity, v [m s^{-1}]	Angular velocity, [rad s^{-1}]	Volume (fluid) flow, Q [m^3 s^{-1}]	Heat flow, q [J s^{-1}]
Displacement (q)	Charge (q)	Displacement (x)	Angular displacement	Volume, V	
Potential variable (across variable)	Voltage, V [V]	Force, F [N]	Torque, T [N m]	Pressure drop, p [P]	Temperature difference, T [°C]
Integrating element (delay component)	Inductance, L [H] Faraday's law: $I = \int V dt/L$	Mass (i.e., inertia), m [kg] Newton's second law of motion $F = m\,dv/dt$	Polar moment of inertia, $T = Jd/dt$	Inertance, M $G = \int p\,dt/M$ e.g., for pipe: $M = \rho L/A$	Not Applicable
Proportional element (dissipative component)	Resistance, $R = \rho L/A$ [Ω] Ohm's law: $I = V/R$	Viscous friction (e.g., dashpot or damper) $F = BvB =$ damping constant	Viscous friction $T = DwD =$ damping factor	Fluid resistance, $RG = p/R$	Heat transfer resistance, $Rq = T/RR_{convect} = 1/(hA)$

Differentiating element (accumulative component)	Capacitance, $C = \varepsilon A/d$ [F] $I = C\,dV/dt$	Elasticity Hooke's law: $F = k \int v\,dt k =$ spring constant (stiffness)	Elasticity (e.g., torsion bar or coil spring) $T = k \int dt k =$ torsional spring constant	Fluid capacitance, $CG = C\,dp/dt$	Thermal heat capacity, $mc_p q = mc_p\,dT/dt$
Other variables	Charge, $q = \int I\,dt$ [C]	Displacement, $x = \int v\,dt$	Angle, $\beta = \int dt$	Flow velocity, $u = G/A$ [ms^{-1}] Volume, $V = \int G\,dt$ [m^2]	Heat, $Q = \int q\,dt$ [J]
Junction/node law $\sum$(flow) = 0	Kirchhoff's current law $\sum I = 0$	d'Alembert's principle $\sum F = 0$	Second law of rotational mechanical systems $\sum T = 0$	$\sum G = 0$	$\sum q = 0$
Closed loop law $\sum$(potential) = 0	Kirchhoff's voltage law $\sum V = 0$	Continuity of space law $\sum v = 0$	$\sum = 0$	$\sum p = 0$	$\sum T = 0$
Power = (potential)(flow)	$P = IV$ [W]	$P = Fv$	$P = T$	$P = Gp$	$P = Tq$ [W°C]
Kinetic energy	$E_k = \frac{1}{2} L I^2$ [J]	$E_k = \frac{1}{2} m v^2$	$E_k = \frac{1}{2} J$	$E_k = \frac{1}{2} M G^2$	Not applicable
Potential energy	$E_p = \frac{1}{2} q^2/C$ [J]	$E_p = \frac{1}{2} k x^2$	$E_p = \frac{1}{2} k \beta^2$	$E_p = \frac{1}{2} \left[\int G\,dt\right]^2/C$	Not applicable

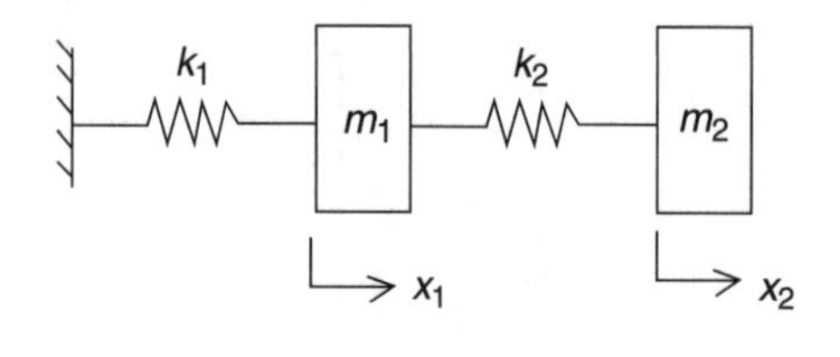

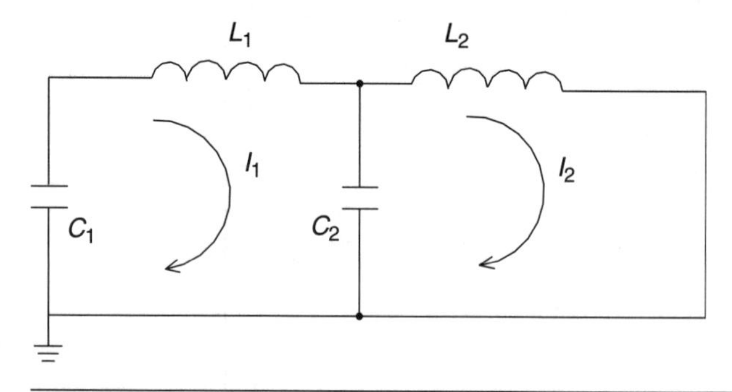

Figure 13.8 Analog mechanical and electrical systems.

$$\begin{aligned}(m_1\ddot{x}_1 + k_1x_1 + k_2x_1) - k_2x_2 = 0 \\ (m_2\ddot{x}_2 + k_2x_2) - k_2x_1 = 0\end{aligned} \tag{13.82}$$

$$\begin{aligned}\left(L_1\dot{i}_1 + 1/C_1q_1 + C_2q_1\right) - 1/C_2q_2 = 0 \\ \left(L_2\dot{i}_2 + 1/C_2q_2\right) - 1/C_2q_1 = 0.\end{aligned} \tag{13.83}$$

13.9 Stability

Stability is the most important system specification. An unstable system cannot be designed for a specific transient response or steady-state error requirement. There are many definitions for stability, depending upon the type of system or the point of view. In this section we limit ourselves to linear, time-invariant systems.

We have already discussed that we can control the output of a system if the steady-state response consists of only the forced response. But the total response of a system is the sum of the forced and natural responses, or

$$c(t) = c_{\text{forced}}(t) + c_{\text{natural}}(t). \tag{13.84}$$

We now present the following definitions of *stability*, *instability*, and *marginal stability*.

13.9.1 Stable systems

Let us focus on the natural response definitions of stability. Recall from our study of system poles that poles in the left half-plane (lhp) yield either pure exponential decay or damped sinusoidal natural responses. These natural responses decay to zero as time approaches infinity. Thus, if the closed-loop system poles are in the left half of the s-plane and hence have a negative real part, the system is stable. That is, stable systems have closed-loop transfer functions with poles only in the left half-plane. Here are some definitions of stable systems:

- A linear, time-invariant system is stable if the natural response approaches zero as time approaches infinity.
- A system is stable if every bounded input yields a bounded output.

13.9.2 Unstable systems

Poles in the right half-plane (rhp) yield either pure exponentially increasing or exponentially increasing sinusoidal natural responses. These natural responses approach infinity as time approaches infinity. Thus, if the closed-loop system poles are in the right half of the s-plane and hence have a positive real part, the system is unstable. Also, poles of multiplicity greater than one on the imaginary axis lead to the sum of responses of the form $At^n \cos(\omega t + \varphi)$, where $n = 1, 2, \ldots$, which also approaches infinity as time approaches infinity. Thus, unstable systems have closed-loop transfer functions with at least one pole in the right half-plane and/or poles of multiplicity greater than one on the imaginary axis. Here are some definitions of unstable systems:

- A linear, time-invariant system is unstable if the natural response grows without bound as time approaches infinity.
- A system is unstable if any bounded input yields an unbounded output.

13.9.3 Marginally stable systems

A system that has imaginary axis poles of multiplicity 1 yields pure sinusoidal oscillations as a natural response. These responses neither increase nor decrease in amplitude. Thus, marginally stable systems have closed-loop transfer functions with only imaginary axis poles of multiplicity 1 and poles in the left half-plane. Here is a definition of a marginally stable system:

- A linear, time-invariant system is marginally stable if the natural response neither decays nor grows but remains constant or oscillates as time approaches infinity.

13.10 The Routh-Hurwitz stability criterion

The Routh-Hurwitz stability criterion for stability is a method that yields stability information without the need to solve for the closed-loop system poles. Using this method, we can determine how many closed-loop system poles are in the left half-plane, in the right half-plane and on the $j\omega$-axis. An important observation is that we say *how many*, not *where*. We can determine the number of poles in each section of the s-plane, but we cannot find their exact coordinates.

The method requires two steps:

1. Generate a data table called a *Routh-table*.
2. Interpret the Routh table to tell how many closed-loop system poles are in the left half-plane, the right half-plane, and on the $j\omega$-axis.

13.10.1 Generating a Routh table

Referring to the equivalent closed-loop transfer function shown in Figure 13.9, we focus our attention on the denominator since we are interested in the system poles. First create the Routh table shown in Table 13.4. We commence by labeling the rows with powers of s from the highest power of the denominator of the closed-loop transfer function to s^0. Next start with the coefficient of the highest power of s in the denominator and list, horizontally in the first row, every other coefficient. In the second row, list horizontally, starting the next highest power of s, every coefficient that was skipped in the first row. The remaining entries are filled in as follows. Each entry is a negative determinant of entries in the previous

$$R(s) \rightarrow \boxed{\dfrac{N(s)}{a_5s^5 + a_4s^4 + a_3s^3 + a_2s^2 + a_1s + a_0}} \rightarrow C(s)$$

Figure 13.9 A closed-loop transfer function.

Table 13.4 Initial template for Routh table

s^5	a_5	a_3	a_1
s^4	a_3	a_2	a_0
s^3			
s^2			
s^1			
s^0			

Table 13.5 Completed Routh table

s^5	a_5	a_3	a_1
s^4	a_4	a_2	a_0
s^3	$-\dfrac{\begin{vmatrix} a_5 & a_3 \\ a_4 & a_2 \end{vmatrix}}{a_4} = b_1$	$-\dfrac{\begin{vmatrix} a_5 & a_1 \\ a_4 & a_0 \end{vmatrix}}{a_4} = b_2$	$-\dfrac{\begin{vmatrix} a_5 & 0 \\ a_4 & a_2 \end{vmatrix}0}{a_4} = b_3 = 0$
s^2	$-\dfrac{\begin{vmatrix} a_4 & a_2 \\ b_1 & b_2 \end{vmatrix}}{b_1} = c_1$	$-\dfrac{\begin{vmatrix} a_4 & a_0 \\ b_1 & 0 \end{vmatrix}}{b_1} = c_2$	$-\dfrac{\begin{vmatrix} a_4 & 0 \\ b_1 & 0 \end{vmatrix}}{b_1} = c_3 = 0$
s^1	$-\dfrac{\begin{vmatrix} b_1 & b_2 \\ c_1 & c_2 \end{vmatrix}}{c_1} = d_1$	$-\dfrac{\begin{vmatrix} b_1 & b_3 \\ c_1 & c_3 \end{vmatrix}}{c_1} = d_2$	$-\dfrac{\begin{vmatrix} b_1 & 0 \\ c_1 & 0 \end{vmatrix}}{c_1} = d_3 = 0$
s^0	$-\dfrac{\begin{vmatrix} c_1 & c_2 \\ d_1 & d_2 \end{vmatrix}}{d_1} = e_1$	$-\dfrac{\begin{vmatrix} c_1 & c_3 \\ d_1 & d_3 \end{vmatrix}}{d_1} = e_2$	$-\dfrac{\begin{vmatrix} c_1 & 0 \\ d_1 & 0 \end{vmatrix}}{d_1} = e_3 = 0$

two rows divided by the entry in the first column directly above the calculated row. The left-hand column of the determinant is always the first column of the previous *two rows,* and the right-hand column is the elements of the column above and to the right. The table is complete when all of the rows are completed down to s^0. Table 13.5 is the completed Routh table for Figure 13.9.

EXAMPLE 13.5

Generate the Routh table for the system shown in Figure 13.10.

Solution

The first step is to find the equivalent closed-loop system because we want to test the denominator of this function, not the given forward transfer function, for pole location. Using the feedback formula, we obtain the equivalent system as Figure 13.10(b). We will apply the Routh-Hurwitz criterion to the denominator, $(s^3 + 9s^2 + 26s + 81)$. First label the rows with powers of s from s^3 down to s^0 in

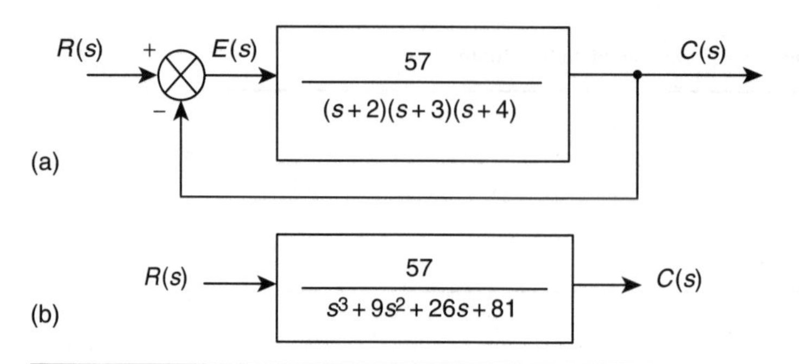

Figure 13.10 Closed-loop transfer function for Example 13.5.

Table 13.6 Completed Routh table for Example 13.5

s^3	1	26	0
s^2	(9) 1	(81) 9	0
s^1	$-\dfrac{\begin{vmatrix}1 & 26\\1 & 9\end{vmatrix}}{1}=17$	$-\dfrac{\begin{vmatrix}1 & 0\\1 & 0\end{vmatrix}}{1}=0$	
s^0	$-\dfrac{\begin{vmatrix}1 & 9\\17 & 0\end{vmatrix}}{17}=9$	$-\dfrac{\begin{vmatrix}1 & 0\\17 & 0\end{vmatrix}}{17}=0$	

a vertical column, as shown in Table 13.6. Next form the first row of the table, using the coefficients of the denominator of the closed-loop transfer function.

We commence with the coefficient of the highest power and skip every other power of s. We now form the second row with the coefficients of the denominator skipped in the previous step. Subsequent rows are formed with determinants as shown.

For convenience any row of the Routh table can be multiplied by a positive constant without changing the values of the rows below. This can be proved by examining the expressions for the entries and verifying that any multiplicative constant from a previous row cancels out. In the second row of Table 13.6, for example, the row was multiplied by 1/9. We see later that care must be taken not to multiply the row by a negative constant.

13.10.2 Interpreting a Routh table

The basic Routh table applies to systems with poles in the left and right half-planes. Systems with imaginary poles (and the kind of Routh table that results) will be discussed in the next section. Simply stated, the Routh-Hurwitz criterion declares that the number of roots of the polynomial that are in the right half-plane is equal to the number of sign changes in the first column.

If the closed-loop transfer function has all poles in the left half of the s-plane, the system is stable. Thus, a system is stable if there are no sign changes in the first column of the Routh table. For example, Table 13.6 has two sign changes in the first column. Thus, the system of Figure 13.10 is stable since no poles exist in the right half-plane.

Now that we have described how to generate and interpret a Routh table, let us look at two special cases that can arise.

13.10.2.1 *Zero only in the first column*

If the first element of a row is zero, division by zero would be required to form the next row. Two methods are normally used: (a) the epsilon method; and (b) the reciprocal-roots method.

In the first method, to avoid this zero-row phenomenon, an epsilon, ε, is assigned to replace the zero in the first column. The value of ε is then allowed to approach zero from either the positive or the negative side, after which the signs of the entries in the first column can be determined.

In the second method, we show that the polynomial we are looking for, the one with the reciprocal roots, as simply the original polynomial with its coefficients written in reverse order. For example,

$$s^n + a_{n-1}s^{n-1} + \ldots + a_1 s + a_0 = 0 \tag{13.85}$$

If s is replaced by $1/d$, then d will have roots which are the reciprocal of s. Making this substitution gives,

$$\left(\frac{1}{d}\right)^n + a_{n-1}\left(\frac{1}{d}\right)^{n-1} + \ldots + a_1\left(\frac{1}{d}\right) + a_0 = 0. \tag{13.86}$$

Factoring out $\left(\frac{1}{d}\right)^n$,

$$\left(\frac{1}{d}\right)^n\left[1 + a_{n-1}\left(\frac{1}{d}\right)^{-1} + \ldots + a_1\left(\frac{1}{d}\right)^{1-n} + a_0\left(\frac{1}{d}\right)^{-n}\right] = 0 \tag{13.86A}$$

$$\left(\frac{1}{d}\right)^n\left[1 + a_{n-1}d + \ldots + a_1 d^{n-1} + a_0 d^n\right] = 0. \tag{13.87}$$

Table 13.7 Partial Routh table for Example 13.6

s^5	1	5	6
s^4	(2) 1	(10) 5	(4) 2

Thus, the polynomial with reciprocal roots is a polynomial with the coefficient written in reverse order.

EXAMPLE 13.6

See website for downloadable MATLAB code to solve this problem

Determine the stability of the closed-loop transfer function

$$T(s) = \frac{5}{s^5 + 2s^4 + 5s^3 + 10s^2 + 6s + 4} \tag{13.88}$$

Solution

Filling the first two rows of the Routh table gives Table 13.7. Here the pivot for the third row contains a zero. So we form a polynomial with its coefficients written in the reverse order.

First write a polynomial that is the reciprocal roots of the denominator of Equation (13.88). This polynomial is formed by writing the denominator in reverse order. Hence,

$$D(s) = \frac{5}{4s^5 + 6s^4 + 10s^3 + 5s^2 + 2s + 1}. \tag{13.89}$$

We form the Routh table as shown in Table 13.8 using Equation 13.89. Since there are two sign changes, the system is unstable and has two right-half plane poles. Notice that Table 13.8 does not have a zero in the first column.

13.10.2.2 *Entire row is zero*

We now look at the second special case. Sometimes while making a Routh table, we find that an entire row consists of zeros because there is an even polynomial that is a factor of the original polynomial. This case must be handled differently from the case of a zero in only the first column of a row. Let us look at an example that demonstrates how to construct and interpret the Routh table when an entire row of zeros is present.

EXAMPLE 13.7

Determine the number of right-half plane poles in the closed-loop transfer function

$$T(s) = \frac{30}{s^5 + 5s^4 + 5s^3 + 25s^2 + 6s + 30}. \tag{13.90}$$

Table 13.8 Completed Routh table for Example 13.6

s^5	(4) 2	(10) 5	(2) 1
s^4	6	5	1
s^3	$-\frac{\begin{vmatrix} 2 & 5 \\ 6 & 5 \end{vmatrix}}{6} = 3.33$	$-\frac{\begin{vmatrix} 2 & 1 \\ 6 & 1 \end{vmatrix}}{6} = 0.66$	$-\frac{\begin{vmatrix} 2 & 0 \\ 6 & 0 \end{vmatrix}}{6} = 0$
s^2	$-\frac{\begin{vmatrix} 6 & 5 \\ 3.33 & 0.66 \end{vmatrix}}{3.33} = 3.81$	$-\frac{\begin{vmatrix} 6 & 1 \\ 3.33 & 0 \end{vmatrix}}{3.33} = 1$	
s^1	$-\frac{\begin{vmatrix} 3.33 & 0.66 \\ 3.81 & 1 \end{vmatrix}}{3.81} = 0.214$	$-\frac{\begin{vmatrix} 3.33 & 0 \\ 3.81 & 0 \end{vmatrix}}{3.81} = 0$	
s^0	$-\frac{\begin{vmatrix} 3.81 & 1 \\ 0.214 & 0 \end{vmatrix}}{0.214} = 1$		

Table 13.9 Partial Routh table for Example 13.7

s^5	1	5	6
s^4	(5) 1	(25) 5	(30) 6
s^3	$-\frac{\begin{vmatrix} 1 & 5 \\ 1 & 5 \end{vmatrix}}{1} = 0$	$-\frac{\begin{vmatrix} 1 & 6 \\ 1 & 6 \end{vmatrix}}{1} = 0$	0

Solution

Start by forming the Routh table for the denominator of Equation 13.90 (see Table 13.9). In the second-row we multiply through by 1/5 for convenience. We stop at the third row, since the entire row consists of zeros, and use the following procedure.

First we return to the row immediately above the row of zeros and form an auxiliary polynomial, using the entries in that row as coefficients. The polynomial will start with the power of s in the label column and continue by skipping every other power of s. Thus, the polynomial formed for this example is

$$P(s) = s^4 + 5s^2 + 6. \tag{13.91}$$

Table 13.10 Completed Routh table for Example 13.7

s^5	1	5	6
s^4	(5) 1	(25) 5	(30) 6
s^3	4	10	0
s^2	$-\dfrac{\begin{vmatrix}1 & 5\\ 4 & 10\end{vmatrix}}{4} = 2.5$	$-\dfrac{\begin{vmatrix}1 & 6\\ 4 & 0\end{vmatrix}}{4} = 6$	
s^1	$-\dfrac{\begin{vmatrix}4 & 10\\ 2.5 & 6\end{vmatrix}}{2.5} = 0.4$	$-\dfrac{\begin{vmatrix}4 & 0\\ 2.5 & 0\end{vmatrix}}{2.5} = 0$	
s^0	$-\dfrac{\begin{vmatrix}2.5 & 6\\ 0.4 & 0\end{vmatrix}}{0.4} = 6$		

Next we differentiate the polynomial with respect to s and obtain

$$P(s) = 4s^3 + 10s + 0. \tag{13.92}$$

Finally, we use the coefficients of Equation 13.92 to replace the row of zeros. Again, for convenience the third row is multiplied by 1/4 after replacing the zeros.

The remainder of the table is formed in a straightforward manner by following the standard form shown in Table 13.10 which shows that all entries in the first column are positive. Hence, there are no right-half plane poles.

13.10.3 Stability design using the Routh-Hurwitz criterion

EXAMPLE 13.8

See website for downloadable MATLAB code to solve this problem

Find the gain range, K (assume $K > 0$) for the system shown in Figure 13.11 to cause the system to be: (a) stable; (b) unstable; and (c) marginally stable.

Solution

The first step is to find the equivalent closed-loop system because we want to apply the Routh-Hurwitz criterion to the denominator $(s^3 + 15s^2 + 50s + K)$ of this function, not the given forward transfer function, for pole location. First label the rows with powers of s from s^3 down to s^0 in a vertical column, as

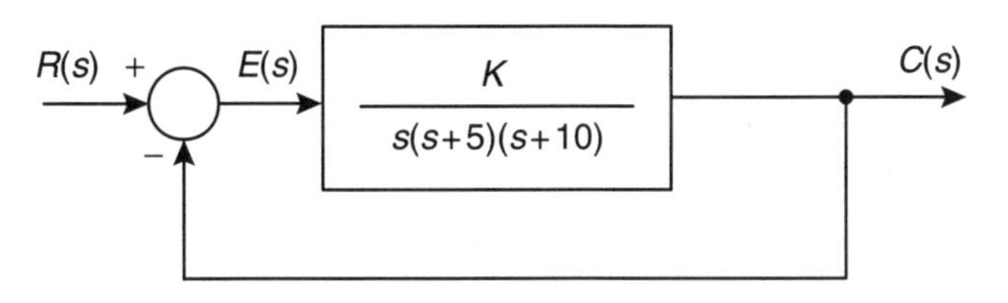

Figure 13.11 A system to be examined for stability in Example 13.8.

Table 13.11 Completed Routh table for example 13.8

s^3	1	50
s^2	15	K
s^1	$-\dfrac{\begin{vmatrix} 1 & 50 \\ 15 & K \end{vmatrix}}{15} = \dfrac{750 - K}{15}$	0
s^0	K	0

shown in Table 13.11. Next form the first row of the table, using the coefficients of the denominator of the closed-loop transfer function.

$$T(s) = \frac{K}{s^3 + 15s^2 + 50s + K}. \tag{13.93}$$

We consider the three possible cases.

- **$K > 750$:** All elements in first column are positive except the row s^1. There are two sign changes showing that the system is unstable.
- **$K < 0$:** All elements in first column are positive and there is no sign change showing that the system is stable, with three poles in the left-halfplane.
- **$K < 750$:** We have an entire row of zeros, which signify $j\omega$ poles.

Let us explore the last case further. First we return to the row immediately above the row of zeros and form an auxiliary polynomial, using the entries in that row as coefficients. The polynomial will start with the power of s in the label column and continue by skipping every other power of s. Thus, the polynomial formed for this example is

$$P(s) = 15s^2 + 750. \tag{13.94}$$

Next we differentiate the polynomial with respect to s and obtain

$$P(s) = 30s + 0. \tag{13.95}$$

Table 13.12 Completed Routh table for Example 13.8 with $K = 750$

s^3	1	50
s^2	(15) 1	(750) 50
s^1	30	0
s^0	$-\dfrac{\begin{vmatrix} 1 & 50 \\ 30 & 0 \end{vmatrix}}{30} = 50$	0

Finally, we use the coefficients of Equation 13.95 to replace the row of zeros. For convenience we multiply the second row entries by 1/15.

The remainder of the table (Table 13.12) is formed in a straightforward manner by following the standard form. Since there is no sign change from the even polynomial, s^2, down to the bottom of the table, the even polynomial has its two roots on the jω-axis of unit multiplicity. The system is therefore marginally stable.

13.11 Steady-state errors

The steady-state error is the difference between the input and the output for a prescribed test input as $t \to \infty$. Since we are concerned with steady-state error after the steady state has been reached our discussions are limited to stable systems, where the natural response approaches zero as $t \to \infty$. The control engineer must first check that the system is stable while performing steady-state error analysis and design.

13.11.1 Steady-state error for unity feedback systems

The steady-state error can be calculated from either a system's close-loop transfer function, $T(s)$, or the open-loop transfer function, $G(s)$, for a unity feedback system. Let us deal with these two cases.

13.11.1.1 *Steady-state error in terms of T(s)*

Consider the feedback control system shown in Figure 13.12(a)

$$E(s) = R(s) - C(s). \tag{13.96}$$

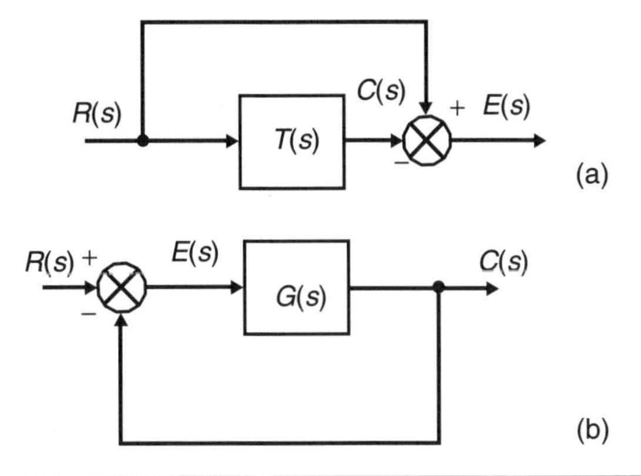

Figure 13.12 Steady-state error: (a) in terms of $T(s)$; (b) in terms of $G(s)$.

But

$$C(s) = R(s)T(s), \tag{13.97}$$

hence

$$E(s) = R(s) - R(s)T(s) = R(s)[1 - T(s)]. \tag{13.98}$$

From the final value theorem,

$$e(\infty) = \lim_{t\to\infty} e(t) = \lim_{s\to 0} sE(t). \tag{13.98A}$$

EXAMPLE 13.9

For a unit step input, and closed-loop transfer function,

$$T(s) = \frac{4}{s^2 + 6s + 8}, \tag{13.98B}$$

determine the steady-state error of the system.

Solution

$$R(s) = \tfrac{1}{s}; \quad 1 - T(s) = 1 - \frac{4}{s^2 + 6s + 8} = \frac{s^2 + 6s + 4}{s^2 + 6s + 8}$$

$$E(s) = \frac{1}{s} \times \frac{s^2 + 6s + 4}{s^2 + 6s + 8} \tag{13.98C}$$

$$e(\infty) = \lim_{s\to 0} s\,E(s) = \lim_{s\to 0} s\frac{s}{s} \times \frac{s^2 + 6s + 4}{s^2 + 6s + 8} = {}^4\!/_8 = {}^1\!/_2.$$

13.11.1.2 *Steady-state error in terms of G(s)*

Consider the feedback control system shown in Figure 13.12(b)

$$E(s) = R(s) - C(s). \tag{13.99}$$

But

$$C(s) = E(s)G(s), \tag{13.100}$$

hence

$$E(s) = R(s) - E(s)G(s). \tag{13.101}$$

Rearranging

$$E(s)[1 + G(s)] = R(s) \tag{13.102}$$

$$E(s) = \frac{R(s)}{1 + G(s)}. \tag{13.103}$$

From the final value theorem,

$$e(\infty) = \lim_{t\to\infty} e(t) = \lim_{s\to 0} sE(t). \tag{13.103A}$$

We now apply this theorem,

$$e(\infty) = \lim_{s\to 0} sE(s) = \lim_{s\to 0} \frac{sR(s)}{1 + G(s)}. \tag{13.104}$$

Let us consider three signals: unit input, ramp input, and parabolic input.

Step input

In this case $R(s) = 1/s$, hence

$$e(\infty) = e_{\text{step}}(\infty) = \lim_{s\to 0} \frac{s(1/s)}{1 + G(s)} = \frac{1}{1 + \lim\limits_{s\to 0} G(s)}. \tag{13.105}$$

For zero steady-state error, $\lim_{s\to 0} G(s) = \infty$. If there is one integrator in the forward path, then $\lim_{s\to 0} G(s) = \infty$ and the steady-state error is zero.

Ramp input

In this case $R(s) = 1/s^2$, hence

$$e(\infty) = e_{\text{step}}(\infty) = \lim_{s\to 0} \frac{s(1/s^2)}{1 + G(s)} = \lim_{s\to 0} \frac{1}{s + sG(s)} = \frac{1}{\lim\limits_{s\to 0} sG(s)}. \tag{13.106}$$

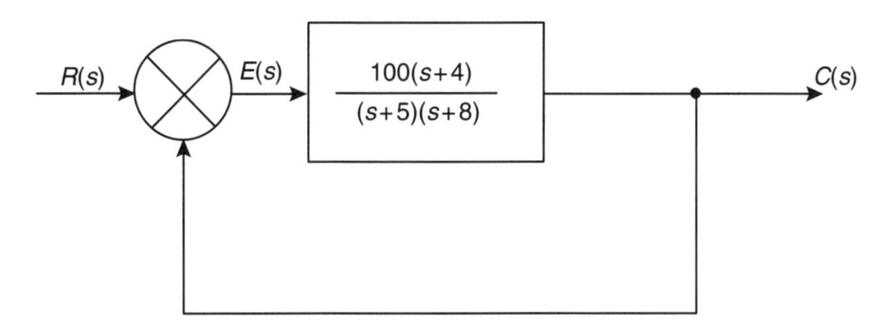

Figure 13.13 A control system for Example 13.10.

For zero steady-state error, $\lim_{s\to 0} sG(s) = \infty$. If there is one integrator in the forward path, then $\lim_{s\to 0} sG(s) = \infty$ and the steady-state error is zero.

Parabolic input

In this case, $R(s) = 1/s^3$, hence

$$e(\infty) = e_{\text{step}}(\infty) = \lim_{s\to 0} \frac{s(1/s^3)}{1 + G(s)} = \lim_{s\to 0} \frac{1}{s^2 + s^2 G(s)} = \frac{1}{\lim\limits_{s\to 0} s^2 G(s)}. \tag{13.107}$$

For zero steady-state error, $\lim_{s\to 0} s^2 G(s) = \infty$. If there is one integrator in the forward path, then $\lim_{s\to 0} s^2 G(s) = \infty$ and the steady-state error is zero.

Let us consider the effect of the absence or presence of an integrator in the feed-forward path using some examples.

EXAMPLE 13.10

Figure 13.13 shows a control system. The function $u(t)$ is the unit step. Determine the steady-state error for inputs of: (a) $10u(t)$; (b) $10tu(t)$; (c) $10t^2u(t)$.

Solution

(a) The input $R(s) = L[10u(t)] = 10/s$, hence for step input,

$$e(\infty) = e_{\text{step}}(\infty) = \frac{1}{1 + \lim\limits_{s\to 0} G(s)}$$

$$\lim_{s\to 0} G(s) = \lim_{s\to 0} \frac{100(s+4)}{(s+5)(s+8)} = \frac{100 \times 4}{5 \times 8} = 10 \tag{13.107A}$$

$$e(\infty) = \frac{10}{1+10} = \frac{10}{11}.$$

(b) The input $R(s) = L[10tu(t)] = 10/s^2$, hence for ramp input,

$$e(\infty) = e_{\text{step}}(\infty) = \frac{1}{\lim\limits_{s\to 0} sG(s)}$$

$$\lim_{s\to 0} G(s) = \lim_{s\to 0} \frac{s100(s+4)}{(s+5)(s+8)} = \frac{s \times 100 \times 4}{5 \times 8} = 0 \tag{13.107B}$$

$$e(\infty) = \frac{10}{0} = \infty.$$

(c) The input $R(s) = L\left[10t^2u(t)\right] = (10 \times 2!)/s^3$.

For parabolic input,

$$e(\infty) = e_{\text{step}}(\infty) = \frac{1}{\lim\limits_{s\to 0} s^2G(s)}$$

$$\lim_{s\to 0} G(s) = \lim_{s\to 0} \frac{s^2100(s+4)}{(s+5)(s+8)} = \frac{s^2 \times 100 \times 4}{5 \times 8} = 0 \tag{13.107C}$$

$$e(\infty) = \frac{20}{0} = \infty.$$

EXAMPLE 13.11

Figure 13.14 shows a control system. The function $u(t)$ is the unit step. Determine the steady-state error for inputs of: (a) $10u(t)$; (b) $10tu(t)$; (c) $10t^2u(t)$.

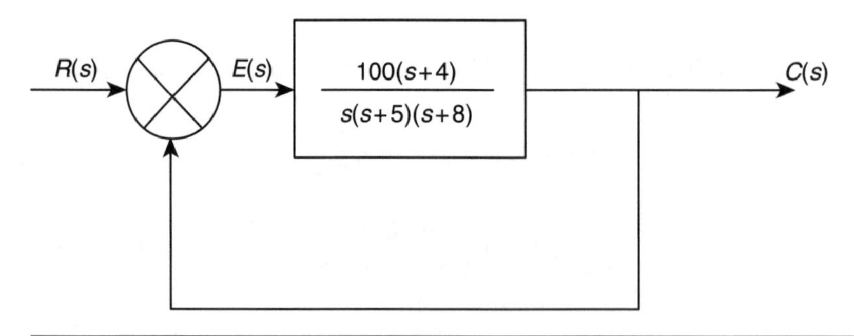

Figure 13.14 A control system for Example 13.11.

Solution

(a) The input $R(s) = L[10u(t)] = 10/s$, hence for step input,

$$e(\infty) = e_{\text{step}}(\infty) = \frac{1}{1 + \lim_{s\to 0} G(s)}$$

$$\lim_{s\to 0} G(s) = \lim_{s\to 0} \frac{100(s+4)}{s(s+5)(s+8)} = \frac{100 \times 4}{0 \times 5 \times 8} = \infty \tag{13.107D}$$

$$e(\infty) = \frac{10}{\infty} = 0.$$

(b) The input $R(s) = L[10tu(t)] = 10/s^2$, hence for ramp input,

$$e(\infty) = e_{\text{step}}(\infty) = \frac{1}{\lim_{s\to 0} sG(s)}$$

$$\lim_{s\to 0} G(s) = \lim_{s\to 0} \frac{s100(s+4)}{s(s+5)(s+8)} = \frac{100 \times 4}{5 \times 8} = 10 \tag{13.107E}$$

$$e(\infty) = \frac{10}{0+10} = 1.$$

(c) The input $R(s) = L\left[10t^2u(t)\right] = (10 \times 2!)/s^3$.

For parabolic input,

$$e(\infty) = e_{\text{step}}(\infty) = \frac{1}{\lim_{s\to 0} s^2G(s)}$$

$$\lim_{s\to 0} G(s) = \lim_{s\to 0} \frac{s^2100(s+4)}{s(s+5)(s+8)} = \frac{s^2 \times 100 \times 4}{s \times 5 \times 8} = 0 \tag{13.107F}$$

$$e(\infty) = \frac{20}{0} = \infty.$$

13.11.2 Static error constants and system type

Static error constants can be used to specify the steady-state error characteristics of control systems. (Steady-state error is an important design consideration for a DVD camcorder.)

13.11.2.1 *Static error constants*

We have derived the following steady-state error relationships:

For a step input, $u(t)$,

$$e(\infty) = e_{\text{step}}(\infty) = \frac{1}{1 + \lim_{s \to 0} G(s)}. \tag{13.107G}$$

For a ramp input, $tu(t)$,

$$e(\infty) = e_{\text{step}}(\infty) = \frac{1}{\lim_{s \to 0} sG(s)}. \tag{13.107H}$$

For a parabolic input, $\frac{1}{2}t^2u(t)$,

$$e(\infty) = e_{\text{step}}(\infty) = \lim_{s \to 0} \frac{s(1/s^3)}{1 + G(s)} = \lim_{s \to 0} \frac{1}{s^2 + s^2G(s)} = \frac{1}{\lim_{s \to 0} s^2G(s)} \tag{13.107I}$$

The steady-state error is determined by the terms in the denominator. These are referred to as the static error constants (Table 13.13).

13.11.3 Steady-state error through static error constants

We can determine the steady-state error through the static error constants. Let us consider some examples.

EXAMPLE 13.12 For the control system shown in Figure 13.15, determine:

(a) the static error constants;

(b) the expected error for the standard step, ramp, and parabolic input.

Table 13.13 Static error constants

Error constants	*Definitions*
Position constant, K_p	$K_p = \lim_{s \to 0} G(s)$
Velocity constant, K_v	$K_v = \lim_{s \to 0} sG(s)$
Acceleration constant, K_{pa}	$K_a = \lim_{s \to 0} s^2G(s)$

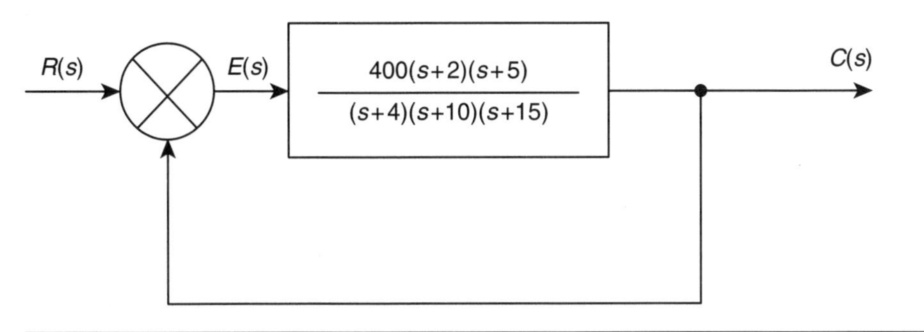

Figure 13.15 A control system for Example 13.12.

Solution

(a)

$$K_p = \lim_{s\to 0} G(s) = \frac{400 \times 2 \times 5}{4 \times 10 \times 15} = 6.67$$

$$K_v = \lim_{s\to 0} sG(s) = 0 \qquad (13.107J)$$

$$K_a = \lim_{s\to 0} s^2 G(s) = 0.$$

(b) For a step input, $R(s) = 1/s$

$$e(\infty) = \frac{1}{1+K_p} = \frac{1}{1+1.67} = 0.15. \qquad (13.107K)$$

For a ramp input, $R(s) = 1/s^2$

$$e(\infty) = \frac{1}{K_v} = \infty. \qquad (13.107L)$$

For a parabolic input, $R(s) = 1/s^3$

$$e(\infty) = \frac{1}{K_a} = \infty. \qquad (13.107M)$$

EXAMPLE 13.13

For the control system shown in Figure 13.16, determine:

(a) the static error constants;

(b) the expected error for the standard step, ramp, and parabolic inputs.

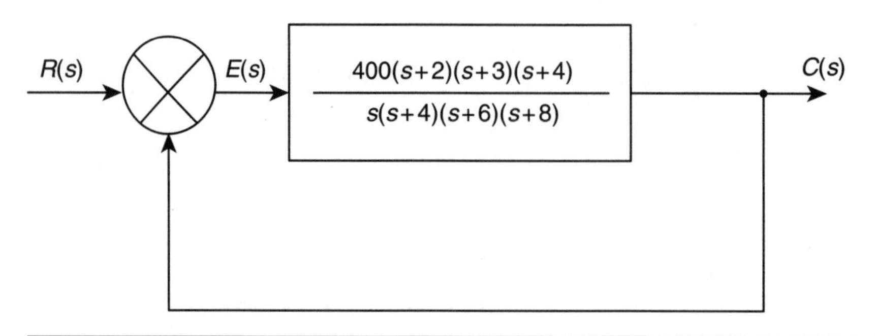

Figure 13.16 A control system for Example 13.13.

Solution

(a)

$$K_p = \lim_{s \to 0} G(s) = \frac{400 \times 2 \times 3 \times 4}{0 \times 4 \times 6 \times 8} = \infty.$$

$$K_v = \lim_{s \to 0} sG(s) = \frac{400 \times 2 \times 3 \times 4}{4 \times 6 \times 8} = 50 \qquad (13.107N)$$

$$K_a = \lim_{s \to 0} s^2 G(s) = 0.$$

(b) For a step input, $R(s) = 1/s$

$$e(\infty) = \frac{1}{1 + K_p} = \frac{1}{\infty} = 0. \qquad (13.107P)$$

For a ramp input, $R(s) = 1/s^2$

$$e(\infty) = \frac{1}{K_v} = \frac{1}{50} = 0.02. \qquad (13.107Q)$$

For a parabolic input, $R(s) = 1/s^3$

$$e(\infty) = \frac{1}{K_a} = \frac{1}{0} = \infty. \qquad (13.107R)$$

EXAMPLE 13.14 For the control system shown in Figure 13.17, determine:

(a) the static error constants;

(b) the expected error for the standard step, ramp, and parabolic inputs.

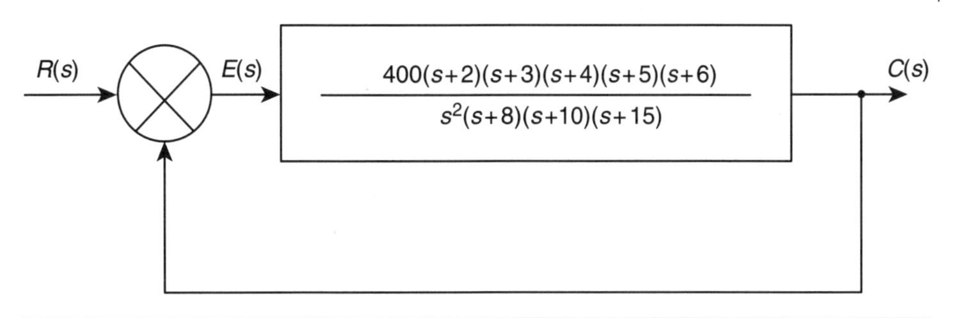

Figure 13.17 A control system for Example 13.14.

Solution

(a)

$$K_p = \lim_{s \to 0} G(s) = \infty$$

$$K_v = \lim_{s \to 0} sG(s) = \infty \tag{13.107S}$$

$$K_a = \lim_{s \to 0} s^2 G(s) = \frac{400 \times 2 \times 3 \times 4 \times 5 \times 6}{8 \times 10 \times 15} = 240.$$

(b) For a step input, $R(s) = 1/s$

$$e(\infty) = \frac{1}{1 + K_p} = \frac{1}{\infty} = 0. \tag{13.107T}$$

For a ramp input, $R(s) = 1/s^2$

$$e(\infty) = \frac{1}{K_v} = \frac{1}{\infty} = 0. \tag{13.107U}$$

For a parabolic input, $R(s) = 1/s^3$

$$e(\infty) = \frac{1}{K_a} = \frac{1}{240} = 4.16 \times 10^{-3}. \tag{13.107V}$$

13.11.3.1 *System type*

From our discussions so far, we have identified three types of system that are related to the error constants; Table 13.14 summarizes these.

Table 13.14 Steady-state errors (SSEs) and types

			Type 0		*Type 1*		*Type 2*	
Input	*Name*	*SSE formula*	*SSE constant*	*Error*	*SSE constant*	*Error*	*SSE constant*	*Error*
$u(t)$	step	$\frac{1}{1+K_p}$	$K_p = C$	$\frac{1}{1+K_p}$	$K_p = \infty$	0	$K_p = \infty$	0
$tu(t)$	ramp	$\frac{1}{K_v}$	$K_v = 0$	∞	$K_v = C$	$\frac{1}{K_v}$	$K_v = \infty$	0
$\frac{1}{2}t^2 u(t)$	parabola	$\frac{1}{K_a}$	$K_a = 0$	∞	$K_a = 0$	∞	$K_a = C$	$\frac{1}{K_a}$

13.11.4 Steady-state error specifications

From the above, we now know that the damping ratio, ζ, settling time, T_s, peak time, T_p, and overshoot, %OS are used as parameters for finding the transient response of a control system. For steady-state errors, the position constant, K_p, the velocity constant, K_v, and the acceleration constant K_a, are used. Let us consider some examples.

EXAMPLE 13.15

A control system has $K_p = 500$. Determine:

(a) whether the system is stable;

(b) the system type;

(c) the input test signal; and

(d) the error that can be expected for the input.

Solution

(a) The system is stable.

(b) The system is of Type 0 since only Type 0 system has a finite value of K_p.

(c) The input test signal is a step.

(d) The steady-state error is

$$e(\infty) = \frac{1}{1 + K_p} = \frac{1}{1 + 500} = 1.996 \times 10^{-3}. \qquad (13.107W)$$

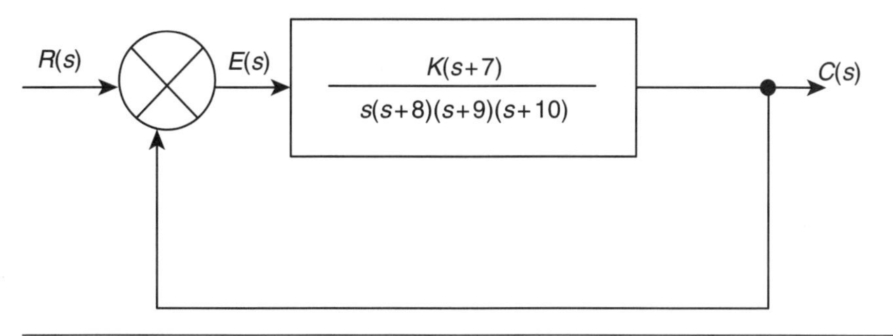

Figure 13.18 A control system for Example 13.16.

EXAMPLE 13.16

The control system shown in Figure 13.18 has a 15% error in the steady state. Determine the value of K.

Solution

The system is of Type 1, and hence the error state in the problem must apply to a ramp. Only a ramp yields a finite error in a Type 1 system. Hence,

$$e(\infty) = {}^{1}\!/_{K_p} = 0.15 \quad (15\%)$$

$$K_v = \frac{1}{0.15} = 6.67 = \lim_{s\to 0} sG(s) = \lim_{s\to 0} \frac{s \times K \times 6}{s \times 8 \times 9 \times 10} \qquad (13.107\text{X})$$

$$\therefore K = \frac{6.672 \times 8 \times 9 \times 10}{6} = 800.$$

EXAMPLE 13.17

A unity feedback system has the following forward transfer function:

$$G(s) = \frac{K(s+15)}{s^2(s+10)(s+12)}. \qquad (13.107\text{Y})$$

Determine the value of K to yield a 10% error in the steady state.

Solution

The system is of Type 3, and hence the error state in the problem must apply to a parabola. Only a parabolic form yields a finite error in a Type 3 system.

Hence,

$$e(\infty) = {}^{1}\!/\!_{K_a} = 0.1 \quad (10\%)$$

$$K_v = \frac{1}{0.1} = 10 = \lim_{s \to 0} sG(s) = \lim_{s \to 0} \frac{s^2 \times K \times 15}{s^2 \times 10 \times 12} \tag{13.107Z}$$

$$\therefore K = \frac{10 \times 10 \times 12}{15} = 80.$$

13.11.5 Steady-state error for non-unity feedback system

In practice, non-unity feedback systems are often encountered. A general feedback system shown in Figure 13.19(a) consists of the transducer, $G_1(s)$, controller and plant, $G_2(s)$, and feedback, $H_1(s)$. Pushing the input transducer past the summing junction results in the general non-unity feedback system shown in Figure 13.19(b), where

$$G(s) = G_1(s)G_2(s) \tag{13.108}$$

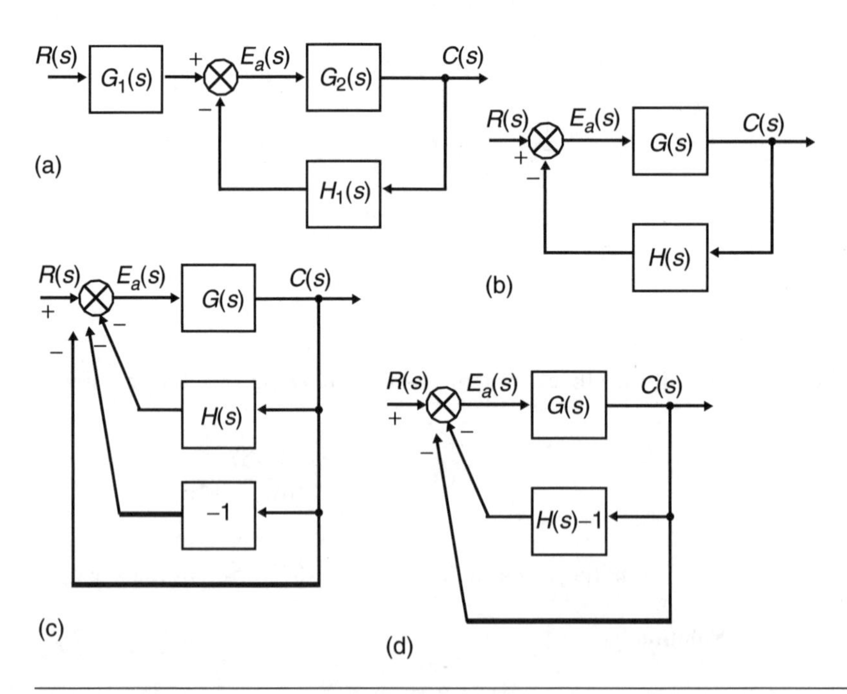

Figure 13.19 Transforming a general non-unity feedback system into an equivalent unity feedback system.

and

$$H(s) = \frac{H_1(s)}{G_1(s)}. \tag{13.109}$$

In this case we have the actuating signal,

$$E_a(s) = R(s) - H(s). \tag{13.110}$$

We add and subtract unity feedback as shown in Figure 13.19(c). Then we combine $H(s)$ with the negative feedback to obtain $\overline{H}(s) = H(s) - 1$ as shown in Figure 13.19(d). Finally, we combine the feedback system consisting of $G(s)$ and $\overline{H}(s) = H(s) - 1$, being an equivalent forward path and a unity feedback. The equivalent transfer function becomes

$$G_e(s) = \frac{G(s)}{1 + G(s)[H(s) - 1]} = \frac{G(s)}{1 + G(s)H(s) - G(s)}. \tag{13.111}$$

EXAMPLE
13.18

See website for downloadable MATLAB code to solve this problem

Figure 13.20 shows a feedback system. Assume that the input and output units are the same. For a unit step input, determine:

(a) the system type;

(b) the error constant associated with the system type; and

(c) the steady-state error.

Solution

$$G_e(s) = \frac{G(s)}{1 + G(s)[H(s) - 1]} = \frac{G(s)}{1 + G(s)H(s) - G(s)}, \tag{13.111A}$$

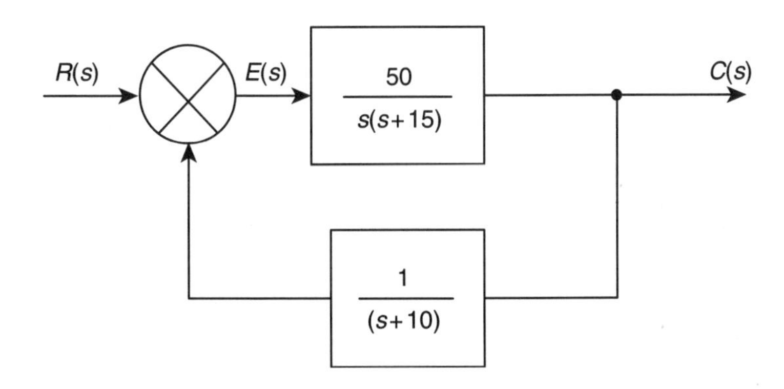

Figure 13.20 A control system for Example 13.18.

where

(a)

$$G(s) = \frac{50}{s(s+15)}; \quad H(s) = \frac{1}{(s+10)}$$

$$H(s) - 1 = \frac{1}{(s+10)} - 1 = \frac{1-s-10}{(s+10)} = \frac{-s-9}{(s+10)}$$

$$G(s)[H(s)-1] = \frac{50}{s(s+15)} \frac{(-s-9)}{(s+10)} = \frac{50(-s-9)}{s(s+10)(s+15)}$$

$$1 + G(s)[H(s)-1] = 1 + \frac{50(-s-9)}{s(s+10)(s+15)} = \frac{s(s+10)(s+15)+50(-s-9)}{s(s+10)(s+15)}$$

$$= \frac{s^3+25s^2+100s-450}{s(s+10)(s+15)}.$$

$$\therefore G_e(s) = \frac{50}{s(s+15)} \frac{s(s+10)(s+15)}{s^3+25s^2+100s-450} = \frac{50(s+10)}{s^3+25s^2+100s-450}. \tag{13.111B}$$

Since there is no pure integrator the system is Type 0. Hence,

(b)

$$K_p = \lim_{s\to 0} G_e(s) = \lim_{s\to 0} \frac{50(s+10)}{s^3+25s^2+100s-450} = \frac{-50\times 10}{-450} = -1.11. \tag{13.111C}$$

(c)

$$e(\infty) = \frac{1}{1+K_p} = \frac{10}{1-1.11} = -9.1. \tag{13.111D}$$

13.11.6 Sensitivity

Sensitivity is the degree to which changes in a control system parameters affect system transfer functions, and consequently performance. Sensitivity is inversely proportional to the system performance.

Sensitivity, S, is the ratio of the fractional change in the function to the fractional change in the parameter as the fraction change of the parameter tends to zero.

$$S = \lim_{\Delta P\to 0} \frac{\Delta F/F}{\Delta P/P} = \lim_{\Delta P\to 0} \frac{P\,\Delta F}{F\,\Delta P} \tag{13.112}$$

or

$$S = \frac{P\,\delta F}{F\,\delta P}. \tag{13.113}$$

Problems

First-order systems

Q13.1 A system has a transfer function, $G(s) = 20/(s+20)$. Determine: (a) the time constant, T_c; (b) the rise time, T_r; and (c) the settling time, T_s.

Second-order systems

Q13.2 For the transfer function

$$G(s) = \frac{100}{s^2 + 5s + 100}. \tag{13.113A}$$

determine the values of ζ and ω_n.

Q13.3 Categorize the following transfer functions as underdamped, critically-damped, or overdamped.

(a)
$$G(s) = \frac{64}{s^2 + 16s + 64}; \tag{13.113B}$$

(b)
$$G(s) = \frac{49}{s^2 + 10s + 49}; \tag{13.113C}$$

(c)
$$G(s) = \frac{9}{s^2 + 10s + 9}. \tag{13.113D}$$

Q13.4 For the following transfer functions, determine:

(i) the peak time;

(ii) the percentage overshoot;

(iii) the settling time; and

(iv) the rise time.

(a)
$$G(s) = \frac{225}{s^2 + 25s + 225}; \tag{13.113E}$$

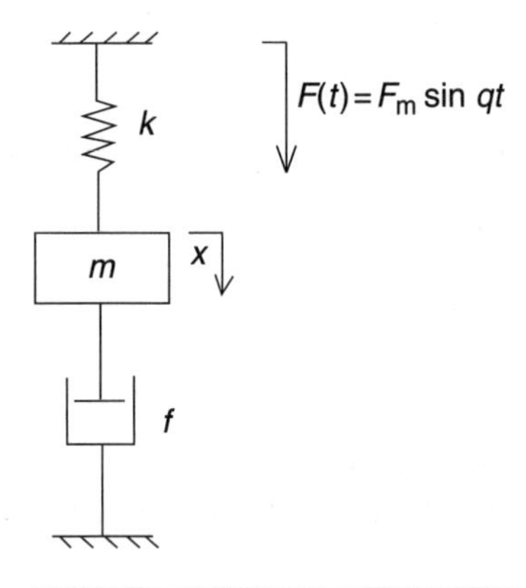

Figure 13.21 System for Q13.5(b).

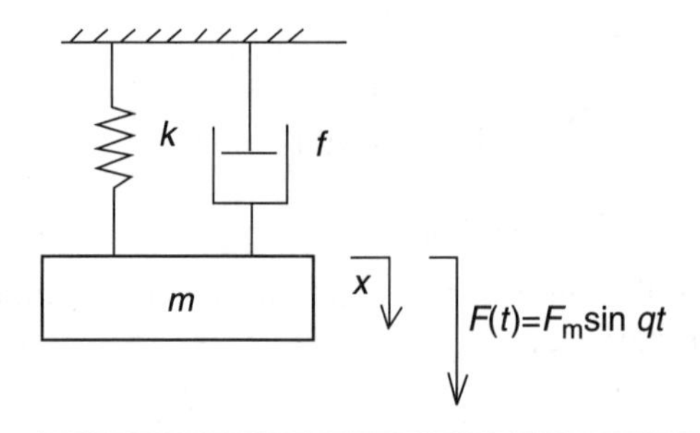

Figure 13.22 System for Q13.5(c).

(b)
$$G(s) = \frac{169}{s^2 + 16s + 169}; \tag{13.113F}$$

(c)
$$G(s) = \frac{120}{s^2 + 12s + 120}; \tag{13.113G}$$

(d)
$$G(s) = \frac{950}{s^2 + 50s + 950}. \tag{13.113H}$$

Q13.5 For the mass-spring-damper systems shown in (a) Figure 13.5, (b) Figure 13.21, (c) Figure 13.22, (d) Figure 13.23 and (e) Figure 13.24, for which the mass (m) is 2500 kg, the elastic constant (k) is 250 N m, the damping constant (ζ) is 40 000 Ns m^{-1}, the amplitude of a sinusoidally fluctuating force is 10 000 N, and the circular frequency (q) is 10 rad s^{-1}. For part (e) the radius of swing (a) is 50 mm.

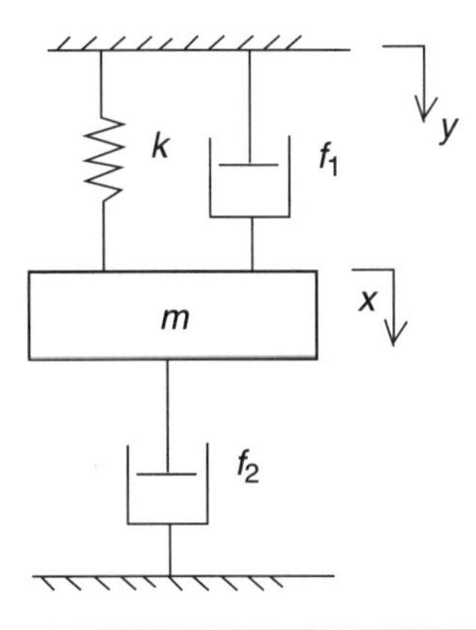

Figure 13.23 System for Q13.5(d).

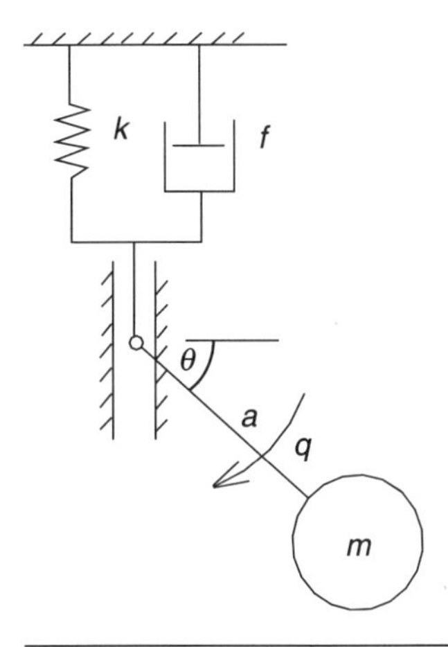

Figure 13.24 System for Q13.5(e).

(i) What is the equation for the steady-state response of $x(t)$?

(ii) What is the amplitude ratio μ?

(iii) Sketch the second-order system amplitude response.

Q13.6 Draw analogous electrical systems for Q13.5(a–e) with all elements and flows labeled.

Q13.7 How would simulation software help in solving problems Q13.5(a–e)? Students can use MATLAB® to solve problems 13.1–13.4.

Further reading

[1] Beckwith, T.G., Buck, N.L. and Marangoni, R.D. (1982) *Mechanical Measurements* (3rd. ed.), Reading, MA: Addison-Wesley.
[2] Doeblin, E. (1990) *Measurement Systems Applications and Design* (4th. ed.), New York: McGraw-Hill.
[3] Dorf, R.C. and Bishop, R.H. (2001) *Modern Control Systems* (9th. ed.), Prentice Hall.
[4] Figliola, R. and Beasley, D. (1995) *Theory and Design of Mechanical Measurements* (2nd. ed.), New York: John Wiley.
[5] Nise, N. (2004) *Control Systems Engineering* (4th. ed.), New York: John Wiley & Sons.

Internet resources

- http://www.engin.umich.edu/group/ctm/freq/nyq.html
- http://www.engin.umich.edu/group/ctm/examples/motor2/freq2.html
- http://www-me.mit.edu/Sections/RLocus/2.11-rationale.html

CHAPTER 14

Control theory: graphical techniques

Chapter objectives

When you have finished this chapter you should be able to:

- create a root locus;
- use the locus to understand the closed-loop system behavior given an open-loop system and a feedback controller;
- calculate the root locus gain at any point on the locus;
- plot frequency response;
- use frequency response to analyze stability.

14.1 Introduction

The root locus, Bode plots, and Nyquist plots techniques are graphical methods for understanding the performance of closed-loop control systems. In the root locus technique the input is of the form of step, impulse and ramp. However, for frequency response (Bode and Nyquist plots) sinusoidal inputs are considered.

14.2 Root locus

Root locus is a graphical technique used to describe qualitatively the performance of a closed-loop system as various parameters are changed. It is a plot of the closed-loop poles of a transfer function as one gain in the transfer function is varied. This gain is normally the control gain although it could be a parameter variation in the plant. The root locus provides stability, accuracy, sensitivity, and transient information and is useful for system analysis and design. Before going into describing the fundamentals of root locus, sketching rules, and interpretation, we first present complex numbers and their representation as vectors.

14.2.1 Vector representation of complex numbers

The *s-plane* or *complex plane* is a two-dimensional space defined by two orthogonal axes: the real number axis and the imaginary number axis. A point in the s-plane represents a complex number. Each complex number, s, has both a real component, typically represented by sigma, and an imaginary component, typically represented by omega.

$$s = \sigma + j\omega. \tag{14.1}$$

Any point in the complex plane has an angle (or phase) and magnitude defined, respectively, as

$$\angle s = \tan^{-1}\left(\frac{\omega}{\sigma}\right) \qquad |s| = \sqrt{\sigma^2 + \omega^2}. \tag{14.2}$$

Graphically, each complex number, s, is plotted in the s-plane.

14.2.2 Properties of root locus

In mathematical terms, given a forward-loop transfer function,

$$KG(s), \tag{14.3}$$

where K is the root locus gain, and the corresponding closed-loop transfer function is

$$\frac{KG(s)}{1 + KG(s)}. \tag{14.3A}$$

The root locus is the set of paths traced by the roots of

$$1 + KG(s) = 0 \tag{14.4}$$

as K varies from zero to infinity. As K changes, the solution to this equation also changes. This equation is called the *characteristic equation*. The roots to the equation are the poles of the forward-loop transfer function. The equation defines where the poles will be located for any value of the *root locus gain*, K. In other words, it defines the characteristics of the system behavior for various values of controller gain.

The root locus is a graphical procedure for determining the poles of a closed-loop system given the poles and zeros of a forward-loop system. Graphically, the locus is the set of paths in the complex plane traced by the closed-loop poles as the root locus gain is varied from zero to infinity.

14.2.2.1 ***Angle criterion***

The angle criterion is used to determine the departure angles for the parts of the root locus near the open-loop poles and the arrival angles for the parts of the root locus near the open-loop zeros. When used with the magnitude criterion, the angle criterion can also be used to determine whether or not a point in the s-plane is on the root locus.

The angle criterion on the root locus is defined as

$$\angle KG(s) = -180^\circ \tag{14.5}$$

Note that $+180^\circ$ could be used rather than -180°. The use of -180° is just a convention. Since $+180^\circ$ and -180° are the same angle, they both produce the same result.

The angle criterion is a direct result of the definition of the root locus; it is another way to express the locus requirements. The root locus is defined as the set of roots that satisfy the characteristic equation

$$1 + KG(s) = 0 \tag{14.5A}$$

or, equivalently,

$$KG(s) = -1. \tag{14.6}$$

Taking the phase of each side of the equation yields the angle criterion.

14.2.2.2 ***Angle of departure***

The *angle of departure* is the angle at which the locus leaves a pole in the s-plane. The *angle of arrival* is the angle at which the locus arrives at a zero in the s-plane.

By convention, both types of angles are measured relative to a ray starting at the origin and extending to the right along the real axis in the *s*-plane. Both arrival and departure angles are found using the angle criterion.

14.2.2.3 *Break points*

Break points occur on the locus where two or more loci converge or diverge. Break points often occur on the real axis, but they may appear anywhere in the *s*-plane. The loci that approach/diverge from a break point do so at angles spaced equally about the break point. The angles at which they arrive/leave are a function of the number of loci that approach/diverge from the break point.

14.2.2.4 *Characteristic equation*

The *characteristic equation* of a system is based upon the transfer function that models the system. It contains information needed to determine the response of a dynamic system. There is only one characteristic equation for a given system.

14.2.3 Root locus plots

This section outlines the steps to create a root locus and illustrates the important properties of each step in the process. This chapter treats how to sketch a root locus given the forward-loop poles and zeros of a system. Using these steps, the locus is detailed enough to evaluate the stability and robustness properties of the closed-loop controller. Let us now consider the steps involved in root locus plots.

Step 1: Open-loop roots

We start with the forward-loop poles and zeros. Since the locus represents the path of the roots (specifically, paths of the closed-loop poles) as the root locus gain is varied, we start with the forward-loop configuration, that is the location of the roots when the gain of the closed-loop system is zero. Each locus starts at a forward-loop pole and ends at a forward-loop zero. If the system has more poles than zeros, then some of the loci end at zeros located infinitely far from the poles.

Step 2: Real axis crossings

Many root loci have paths on the real axis. The real axis portion of the locus is determined by applying the following rule:

On the real axis, for $K > 0$, the root locus exists to the left of an odd number of finite open-loop poles and/or finite open-loop zeros.

If an odd number of forward-loop poles and forward-loop zeros lie to the right of a point on the real axis that point belongs to the root locus. We note that the real axis section of the root locus is determined entirely by the number of forward-loop poles and zeros and their relative locations. Since the final root locus is always symmetric about the real axis, the real axis part is somewhat easy.

Step 3: Asymptotes

The asymptotes indicate where the poles will go as the gain approaches infinity. For systems with more poles than zeros, the number of asymptotes is equal to the number of poles minus the number of zeros, $N_a = (\text{number of poles}) - (\text{number of zeros})$. In some systems, there are no asymptotes; when the number of poles is equal to the number of zeros, then each locus is terminated at a zero rather than asymptotically to infinity.

The asymptotes are symmetric about the real axis, and they stem from a point defined by the relative magnitudes of the open-loop roots. This point is called the *centroid*. Note that it is possible to draw a root locus for systems with more zeros than poles, but such systems do not represent physical systems. In these cases, one can think of some of the poles being located at infinity.

The root loci approach straight lines as asymptotes as the locus approaches infinity. Moreover, the equation of the asymptotes is given by the real-axis intercept, σ_a, and angle, θ_a, as follows:

$$\sigma_a = \frac{\sum \text{finite poles} - \sum \text{finite zeros}}{\text{number of finite poles} - \text{number of finite zeros}} \tag{14.7}$$

$$\theta_a = \frac{(2k+1)\pi}{\text{number of finite poles} - \text{number of finite zeros}}, \tag{14.8}$$

where $k = 0, \pm 1, \pm 2, \pm 3$, and the angle is given in radians with respect to the positive extension of the real axis.

Step 4: Break points

Break points occur where two or more loci join then diverge. Although they are most commonly encountered on the real axis, they may also occur elsewhere in the complex plane. Each break point is a point where a double (or higher order) root exists for some value of K. Three methods for determining break points (differentiation, transition, and max–min gain) are discussed.

Method 1: Differentiation method
Mathematically, given the root locus equation

$$1 + KG(s)H(s) = 0, \tag{14.9}$$

since $KG(s)H(s) = -1$, we solve for K and differentiate it with respect to s and find its optimal value by equating the differential to zero. This means that $K = -1/G(s)H(s)$. We note that the transfer function $G(s)$ consists of a numerator, $A(s)$, and denominator, $B(s)$, (i.e. $G(s) = A(s)/B(s)$), then the break points can be determined from the roots of

$$\frac{dK}{ds}\frac{B(s)\overline{A}(s) - \overline{B}(s)A(s)}{B^2(s)} = 0. \tag{14.10}$$

If K is real and positive at a value s that satisfies this equation, then the point is a break point. There will always be an even number of loci around any break point; for each locus that enters the locus, there must be one that leaves.

Method 2: Transition method

This second method is a variation on the differential calculus method, in which *break-away* and *break-in points* satisfy the relationship

$$\sum_{i=1}^{m} \frac{1}{\sigma + z_i} = \sum_{j=1}^{n} \frac{1}{\sigma + p_j}, \tag{14.11}$$

where z_i and p_j are the negatives of the zero and pole values, respectively of $G(s)H(s)$.

Method 3: Maximum–minimum gain method

In this third method, we find the maximum gain between poles and minimum gain between zeros. The real-axis value that gives the *maximum gain* between poles corresponds to the *break-away* point, while the real-axis value that gives the *minimum gain* between zeros corresponds to the *break-in* point.

Step 5: Angles of departure/arrival

The angle criterion determines which direction the roots move as the gain moves from zero (angles of departure, at the forward-loop poles) to infinity (angles of arrival, at the forward-loop zeros). An angle of departure/arrival is calculated at each of the complex forward-loop poles and zeros.

Step 6: Axis crossings

The points where the root locus intersects the imaginary axis indicate the values of K at which the closed-loop system is marginally stable. The closed-loop system will be unstable for any gain for which the locus is in the right-half plane of the complex plane.

If the root locus crosses the imaginary axis from left to right at a point where $K = K_0$ and then stays completely in the right-half plane, then the closed-loop system is unstable for all $K > K_0$. Therefore, knowing the value of K_0 is very useful. Some systems are particularly awkward when their locus dips back and forth across the imaginary axis. In these systems, increasing the root locus gain will cause the system to go unstable initially and then become stable again.

Step 7: Sketch the locus

The complete root locus can be drawn by starting from the forward-loop poles, connecting the real axis section, break points, and axis crossings, then ending at either the forward-loop zeros or along the asymptotes to infinity. If the hand-drawn locus is not detailed enough to determine the behavior of your system, then one may want to use MATLAB® or some other computer tool to calculate the locus exactly.

EXAMPLE 14.1

See website for downloadable MATLAB code to solve this problem

Sketch the root locus for the system shown in Figure 14.1.

Solution

First calculate the asymptotes, given by the real-axis intercept, σ_a, as

$$\sigma_a = \frac{(0 - 1 - 2 - 6) - (-4)}{4 - 1} = \frac{-5}{3} = -1.667. \tag{14.11A}$$

Then, find the angles, θ_a, that intercept at the asymptotes, as follows:

$$\theta_a = \frac{(2k+1)\pi}{4-1} = \frac{\pi}{3},\ \pi,\ \frac{5\pi}{3}, \tag{14.11B}$$

where $k = 0, \pm 1, \pm 2, \pm 3$, and the angle is given in radians with respect to the positive extension of the real axis.

Figure 14.2 shows the complete root locus and the asymptotes that have just been calculated.

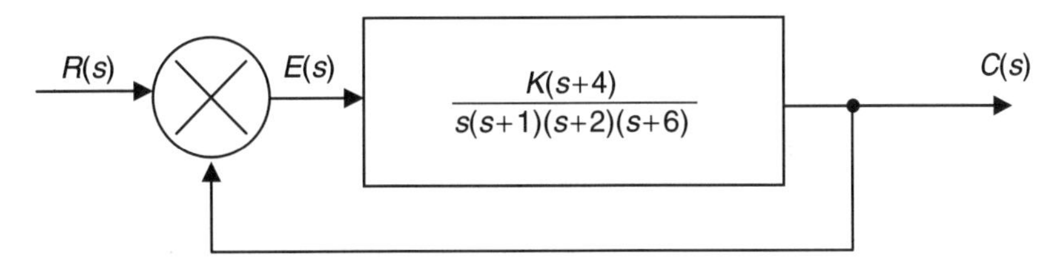

Figure 14.1 A control system for Example 14.1.

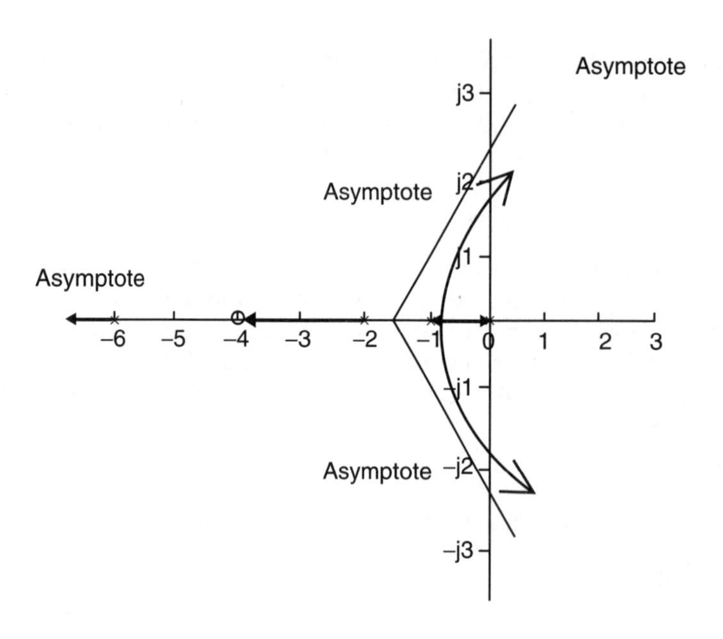

Figure 14.2 Root locus and asymptotes of Figure 14.1.

EXAMPLE 14.2

Sketch the root locus for the system shown in Figure 14.3.

Solution

First calculate the asymptotes, given by the real-axis intercept, σ_a, as

$$\sigma_a = \frac{(-2-4-5)-(0)}{3-0} = \frac{-11}{3} = -3.667. \tag{14.11C}$$

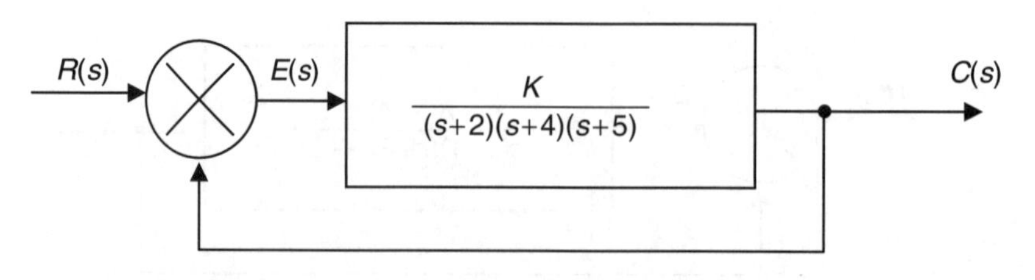

Figure 14.3 A control system for Example 14.2.

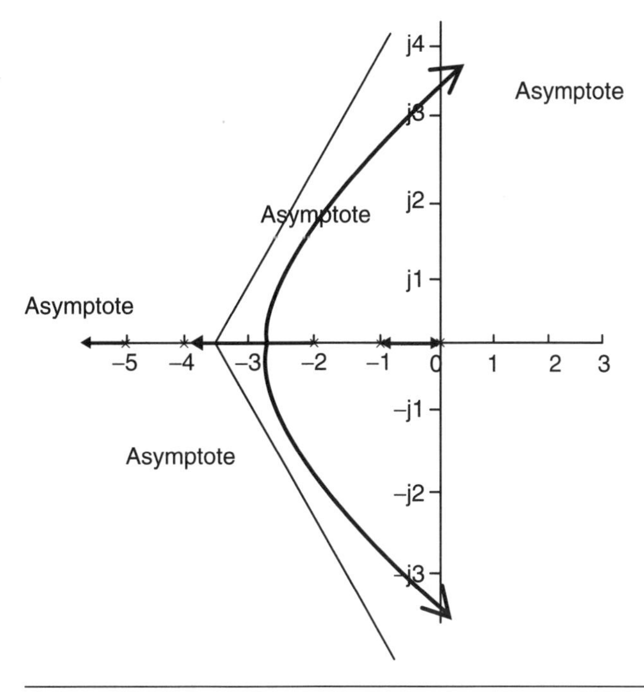

Figure 14.4 Root locus and asymptotes for Figure 14.3.

Then, find the angles θ_a, that intercept at the asymptotes, as follows:

$$\theta_a = \frac{(2k+1)\pi}{3-0} = \frac{\pi}{3},\ \pi,\ \frac{5\pi}{3}, \tag{14.11D}$$

where $k=0, \pm 1, \pm 2, \pm 3$, and the angle is given in radians with respect to the positive extension of the real axis.

Figure 14.4 shows the complete root locus and the asymptotes that have just been calculated.

EXAMPLE 14.3

Sketch the root locus for the system shown in Figure 14.5.

Solution

First calculate the asymptotes, given by the real-axis intercept, σ_a, as

$$\sigma_a = \frac{(-1-3)-(-4-6)}{2-2} = \frac{6}{0} = \infty. \tag{14.11E}$$

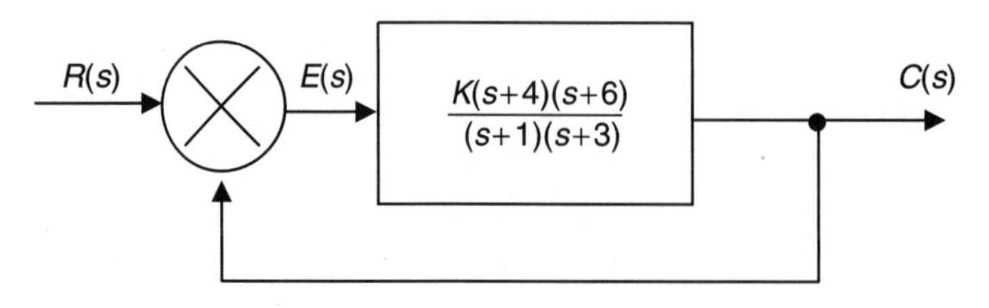

Figure 14.5 A control system for Example 14.3.

Then, find the angles θ_a, that intercept at the asymptotes, as follows:

$$\theta_a = \frac{(2k+1)\pi}{2-2} = \infty, \tag{14.11F}$$

where $k = 0, \pm 1, \pm 2, \pm 3$, and the angle is given in radians with respect to the positive extension of the real axis.

Figure 14.6 shows the complete root locus and the asymptotes that have just been calculated.

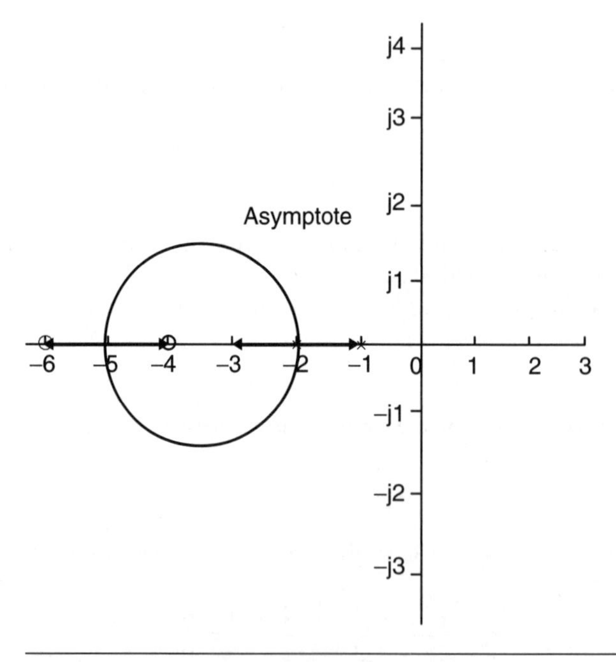

Figure 14.6 Root locus and asymptotes for Figure 14.5.

14.3 Frequency response techniques

The frequency response is a representation of the system's response to sinusoidal inputs at varying frequencies. The output of a linear system to a sinusoidal input is a sinusoid of the same frequency but with a different magnitude and phase. The *frequency response* is defined as the magnitude and phase differences between the input and output sinusoids. In this chapter, we will see how we can use the open-loop frequency response of a system to predict its behavior in closed-loop.

The frequency response method may be less intuitive than the root locus method. However, it has certain advantages, especially in real-life situations such as modeling transfer functions from physical data. The frequency response of a system can be viewed two different ways: via (1) the Bode plot or via (2) the Nyquist diagram. Both methods display the same information; the difference lies in the way the information is presented. We will study both methods in this chapter.

To plot the frequency response, we create a vector of frequencies (varying between zero and infinity) and compute the value of the plant transfer function at those frequencies. If $G(s)$ is the open-loop transfer function of a system and ω is the frequency vector, we then plot $G(\mathrm{j}\omega)$ against the frequency ω. Since $G(\mathrm{j}\omega)$ is a complex number, we can plot both its magnitude and phase (the Bode plot) or its position in the complex plane (the Nyquist plot).

Consider a mechanical system whose input force is sinusoidal and its steady-state output response is also sinusoidal and at the same frequency as the input. Then we can represent the input and output as phasors, $M_i(\omega)\angle\varphi_i(\omega)$ and $M_o(\omega)\angle\varphi_o(\omega)$, respectively; where M_i and M_o are the amplitudes of the sinusoids, and φ_i and φ_o are the phase angles of the sinusoids as shown in Figure 14.7.

We can then write the output steady-state sinusoid as

$$M_o(\omega)\angle\phi_o(\omega) = M_i(\omega)M(\omega)\angle[\phi_i(\omega) + \phi(\omega)]. \tag{14.12}$$

From this, the system's function is found to be

$$M(\omega) = \frac{M_o(\omega)}{M_i(\omega)} \tag{14.13}$$

and

$$\phi(\omega) = \phi_o(\omega) - \phi_i(\omega). \tag{14.14}$$

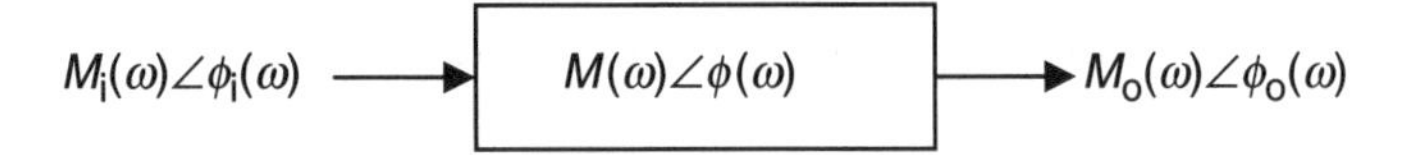

Figure 14.7 Transfer function in phasor form.

We note that

$$M_G = |G(j\omega)| \tag{14.15}$$

and

$$\phi(\omega) = \angle G(j\omega). \tag{14.16}$$

In other words, the frequency response is

$$G(j\omega) = G(s)|_{s\to j\omega}. \tag{14.17}$$

Let us now discuss these two frequency response methods.

14.3.1 Bode plots

A Bode plot is the representation of the magnitude and phase of the open-loop transfer function of a system $G(j\omega)$, where the frequency vector ω contains only positive frequencies. In other words, it is the log-magnitude and phase frequency response curve as functions of $\log(\omega)$. They can be approximated as straight lines, simplifying the method.

Consider for example the following general transfer function:

$$G(s) = \frac{K(s+z_1)(s+z_2)(s+z_3)\ \dots\ (s+z_k)}{s^m(s+p_1)(s+p_2)(s+p_3)\ \dots\ (s+p_n)}. \tag{14.18}$$

The overall magnitude frequency response is the product of the magnitude frequency response of each term, given as

$$|G(j\omega)| = \frac{|K||(s+z_1)||(s+z_2)||(s+z_3)|\ \dots\ |(s+z_k)|}{|s^m||(s+p_1)||(s+p_2)||(s+p_3)|\ \dots\ |(s+p_n)|}\Bigg|_{s\to j\omega}. \tag{14.19}$$

When we convert the magnitude frequency response into decibels (dB), noting that a decibel is defined as $20\log_{10}|G(j\omega)|$, we now have

$$\begin{aligned} 20\log|G(j\omega)| = {} & 20\log|K| + 20\log|(s+z_1)| + \dots + |(s+z_k)| + \\ & - 20\log|s^m| - 20\log|(s+p_1)| - \dots - (s+p_n) - \dots \Big|_{s\to j\omega} \end{aligned} \tag{14.20}$$

Consequently, we can build the Bode plot of a system by adding together the Bode plots of the magnitudes of the constituent elements. Similarly, the phase plot of a system is obtained by adding together the phase plots of the magnitudes of the constituent elements.

By using a number of basic elements, the Bode plot for a wide range of systems can be easily obtained. Let us present the Bode plots for some basic elements.

14.3.1.1 *Bode plot for transfer function of a constant gain, G(s) = K*

The frequency response function is given as

$$G(s) = K. \tag{14.21}$$

The magnitude plot

The frequency response, obtained by replacing s by jω is given as

$$|G(\mathrm{j}\omega)| = |K| = K. \tag{14.22}$$

The magnitude frequency response in decibels, is given as

$$20\log|G(\mathrm{j}\omega)| = 20\log(K). \tag{14.23}$$

The magnitude plot is therefore a straight line of constant magnitude depending on the value of K.

Phase plot

The phase is zero.

Figure 14.8 shows the Bode plot.

14.3.1.2 *Bode plot for transfer function of a step function, G(s) = 1/s*

The frequency response function is given as

$$G(s) = \frac{1}{S}. \tag{14.24}$$

The magnitude plot

The frequency response, obtained by replacing s by jω is given as

$$G(\mathrm{j}\omega) = 1/\mathrm{j}\omega = -\mathrm{j}/\omega. \tag{14.25}$$

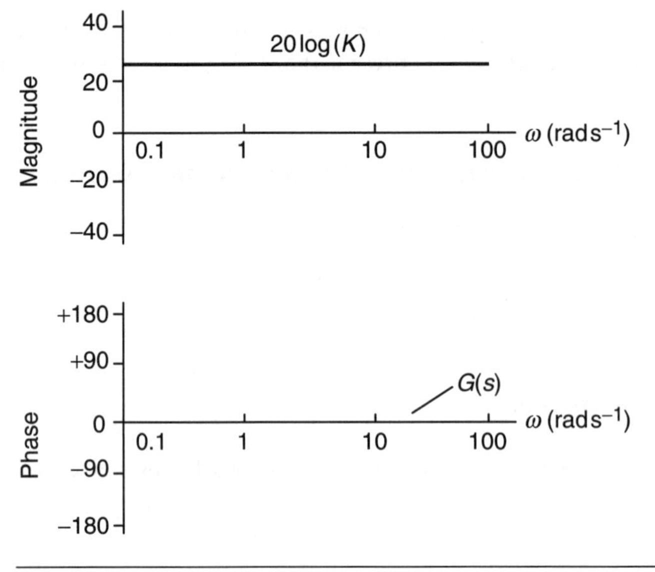

Figure 14.8 Bode plot for $G(s) = K$.

The magnitude frequency response in decibels, is given as

$$20\log\left|G(-\mathrm{j}/\omega)\right| = 20\log(1/\omega) = -20\log(\omega). \tag{14.26}$$

Examining the following cases:

$$\omega \to 0.1 : M_{\mathrm{G}} = 20\,\mathrm{dB},$$

$$\omega \to 1 : M_{\mathrm{G}} = 0\,\mathrm{dB}, \text{and}$$

$$\omega \to 10 : M_{\mathrm{G}} = -20\,\mathrm{dB},$$

leading to a straight line passing through 20 dB, 0 dB, and −20 dB.

Phase plot

The phase angle is given as

$$\tan\phi = \frac{-1/\omega}{0} \to \infty. \tag{14.26A}$$

Hence, for all situations, $\theta_{\mathrm{G}} = -90°$.
Figure 14.9 shows the Bode plot.

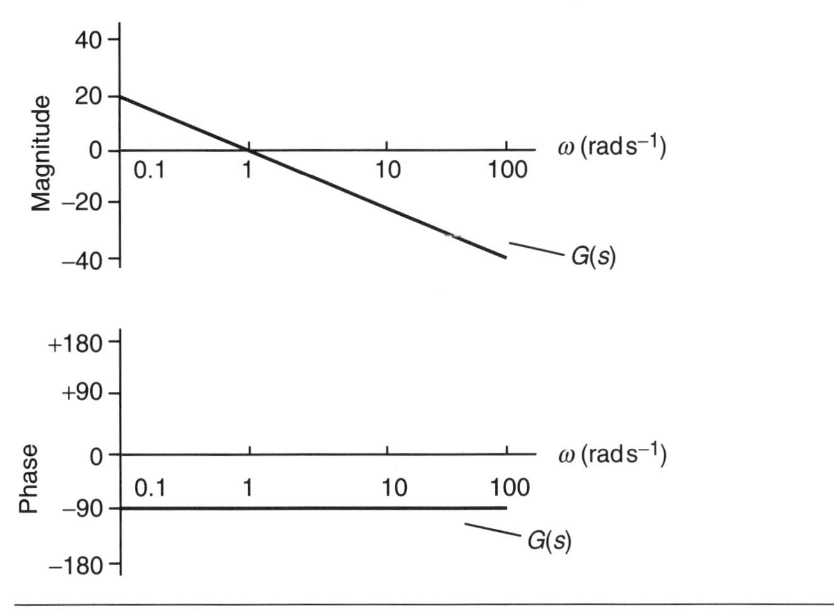

Figure 14.9 Bode plot for $G(s) = 1/s$.

14.3.1.3 *Bode plot for transfer function of a first-order system*

The frequency response function is given as

$$G(s) = \frac{1}{(\tau s + 1)}. \tag{14.27}$$

The magnitude plot

The frequency response, obtained by replacing s by $j\omega$ is given as

$$|G(j\omega)| = \left|\frac{1}{(j\tau\omega + 1)}\right| = \frac{1}{\sqrt{\tau^2\omega^2 + 1}}. \tag{14.28}$$

The magnitude frequency response in decibels, is given as

$$20\log|G(j\omega)| = 20\log\left(\frac{1}{\sqrt{\tau^2\omega^2 + 1}}\right) = -20\log\sqrt{\tau^2\omega^2 + 1}. \tag{14.29}$$

Examining the following cases:

$$\omega \to 0 : 20\log|G(j\omega)| = 20\log|1| = 0,$$
$$\omega \to 1/\tau : 20\log|G(j\omega)| = -20\log\sqrt{2} = -3,$$
$$\omega \to 10/\tau : 20\log|G(j\omega)| = -20\,\text{dB (this occurs when } \omega \to 10\omega_n = 10 \times 6 = 60.$$

Phase plot

The phase angle is given as

$$\tan\phi = -\omega\tau. \tag{14.29A}$$

Examining the following cases:

$$\omega \to 0 : \theta_G = 0°,$$
$$\omega \to 1/\tau : \theta_G = -45°,$$
$$\omega \to 10/\tau : \theta_G = -90°.$$

Figure 14.10 shows the Bode plot.

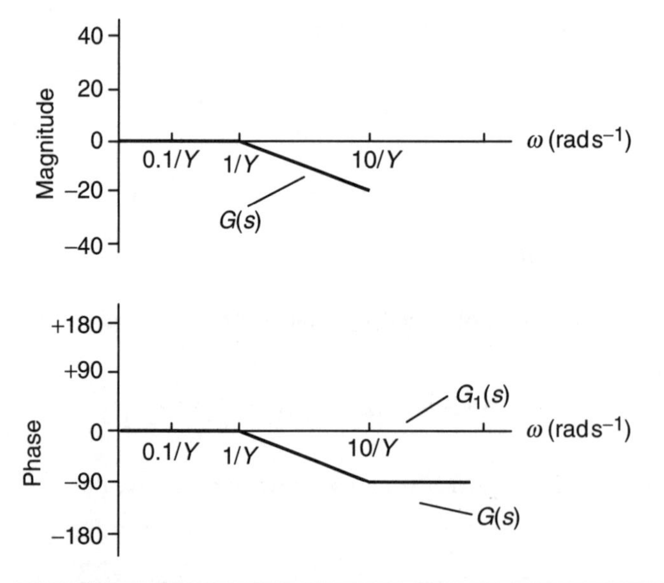

Figure 14.10 Bode plot for first-order system.

14.3.1.4 ***Bode plot for transfer function of a second-order system***

The frequency response function is given as

$$G(s) = \frac{\omega_n^2}{s^2 + 2\xi\omega_n s + \omega_n^2}. \tag{14.30}$$

The magnitude plot

The frequency response, obtained by replacing s by $j\omega$ is given as

$$G(j\omega) = \frac{\omega_n^2}{-\omega^2 + 2j\xi\omega_n\omega + \omega_n^2} = \frac{1}{1 - \left(\frac{\omega}{\omega_n}\right)^2 + 2j\xi\left(\frac{\omega}{\omega_n}\right)}. \tag{14.31}$$

The magnitude frequency response in decibels, is given as

$$20\log\frac{1}{\sqrt{\left(1-\left(\frac{\omega}{\omega_n}\right)^2\right)^2 + \left(2\xi\left(\frac{\omega}{\omega_n}\right)\right)^2}} = -20\log\sqrt{\left(1-\left(\frac{\omega}{\omega_n}\right)^2\right)^2 + \left(2\xi\left(\frac{\omega}{\omega_n}\right)\right)^2}. \tag{14.32}$$

The magnitude plot is therefore a straight line of constant magnitude depending on the value of K.

Phase plot

$$\tan\phi = \frac{2\xi\left(\frac{\omega}{\omega_n}\right)}{1 - \left(\frac{\omega}{\omega_n}\right)^2} \tag{14.32A}$$

Examining the following cases:

$\omega/\omega_n \to 0 : |G(j\omega)| = 1; 20\log|G(j\omega)| = 0.$

$\omega/\omega_n \to 1 : 20\log|G(j\omega)| = -20\log(2\zeta)$; this is a function of the damping factor.

$\omega/\omega_n \to 10 : 20\log|G(j\omega)| = -20\log(\omega/\omega_n)^2 = -40\,\text{dB}.$

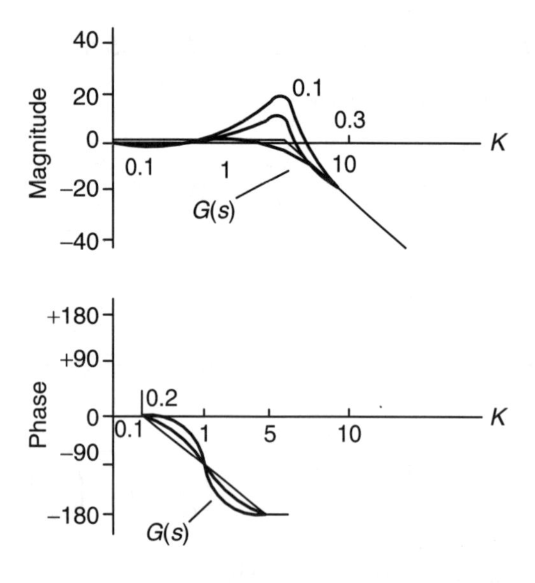

Figure 14.11 Bode plot for second-order system: $K = \omega/\omega_n$.

Figure 14.11 shows the Bode plot.

Table 14.1 summarizes Bode plots.

EXAMPLE 14.4

Sketch the Bode plot of the system shown in Figure 14.12, where

$$G(s) = \frac{3}{s(s^2 + 4s + 25)}. \tag{14.32B}$$

Table 14.1 Guidelines for Bode plots

	M_G	θ_G
$G(s) = K$	$2\log_{10}(K)$	0
$G(s) = 1/s$	$\omega \to 0.1 : M_G = 20\,\text{dB}$ $\omega \to 1 : M_G = 0$ $\omega \to 10 : M_G = -20\,\text{dB}$	$\omega \to 0 : \theta_G = -90°$ $\omega \to 1 : \theta_G = -90°$ $\omega \to 10 : \theta_G = -90°$
$G(s) = 1/(\tau s + 1)$	$\omega \to 0 : M_G = 0$ $\omega \to 1/\tau : M_G = -20\log(\omega\tau)$ $\omega \to 10/\tau : M_G = -20\,\text{dB}$	$\omega \to 0 : \theta_G = 0°$ $\omega \to 1/\tau : \theta_G = -45°$ $\omega \to 10/\tau : \theta_G = -90°$
$G(s) = \omega_n^2/\left(s^2 + 2\xi\omega_n s + \omega_n^2\right)$	$\omega/\omega_n \to 0 : M_G = 0$ $\omega/\omega_n \to 1 : M_G = -20\log(2\xi)$ $\omega/\omega_n \to 10 : M_G = -40\,\text{dB}$	$\omega/\omega_n \to 0.2 : \theta_G = 0°$ $\omega/\omega_n \to 1 : \theta_G = -90°$ $\omega/\omega_n \to 5 : \theta_G = -180°$

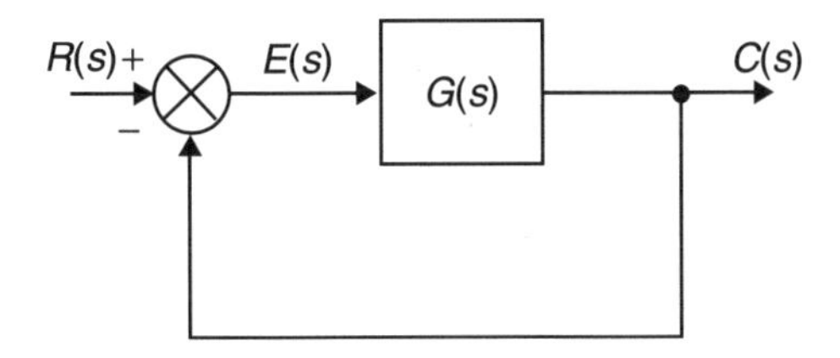

Figure 14.12 Closed-loop unity feedback system for Example 14.4.

Solution

There are three identifiable components:

- a constant gain, $G_1(s) = K = 3/25 = 0.12$;
- a step function, $G_2(s) = 1/s$;
- a second-degree function, $G_3(s) = 25/(s^2 + 4s + 25)$.

Magnitude

For $G_1(s), 20\log|G(j\omega)| = 20\log(0.12) = -18.42\,\text{dB}$.

For $G_2(s)$, using Table 14.1 (second case):

$$\omega \to 0.1 : M_G = 20\,\text{dB},$$
$$\omega \to 1 : M_G = 0\,\text{dB, and}$$
$$\omega \to 10 : M_G = -20\,\text{dB},$$

leading to a straight line passing through 20 dB, 0 dB, and −20 dB.

For $G_3(s)$, $\omega_n = \sqrt{25} = 5$ (break point) and $2\zeta\omega_n - 4$, giving $\zeta = 0.4$.

Let us now consider the following conditions using Table 14.1:

$\omega/\omega_n \to 0 : M_G = 0\,\text{dB}$,

$\omega/\omega_n \to 1 : M_G = -20\log(2\zeta) = -20\log(2 \times 0.4) = 1.94$; this occurs when $\omega \to \omega_n = 5$,

$\omega/\omega_n \to 10 : M_G = -40\,\text{dB}$, this occurs when $\omega \to 10\omega_n = 10 \times 5 = 50$.

Phase

For $G_1(s) = 0.12, \omega \to 0 : \theta_G = 0°$ for all situations.

For $G_2(s)$, $= \omega \to 0 : \theta_G = -90°$ for all situations.

For $G_3(s)$ (third case from Table 14.1),

$$\omega/\omega_n \rightarrow 0.2 : \theta_G = 0°;\ \text{this occurs when } \omega \rightarrow 0.2\omega_n = 0.2 \times 5 = 1$$
$$\omega/\omega_n \rightarrow 1 : \theta_G = 90°;\ \text{this occurs when } \omega \rightarrow \omega_n = 5$$
$$\omega/\omega_n \rightarrow 5 : \theta_G = 180°;\ \text{this occurs when } \omega \rightarrow 5\omega_n = 5 \times 5 = 25.$$

Figure 14.13 shows the Bode plot.

EXAMPLE 14.5

See website for downloadable MATLAB code to solve this problem

Draw the Bode plot for the transfer function

$$G(s) = \frac{4}{s(2s+1)(s^2+5s+36)} \tag{14.32C}$$

Solution

There are four identifiable components:

- a constant gain $G_1(s) = K = 4/36 = 0.111$;
- a step function, $G_2(s) = 1/s$;
- a first-degree function $G_3(s) = 1/(2s+1)$;
- a second-degree function, $G_4(s) = 36/(s^2+5s+6)$.

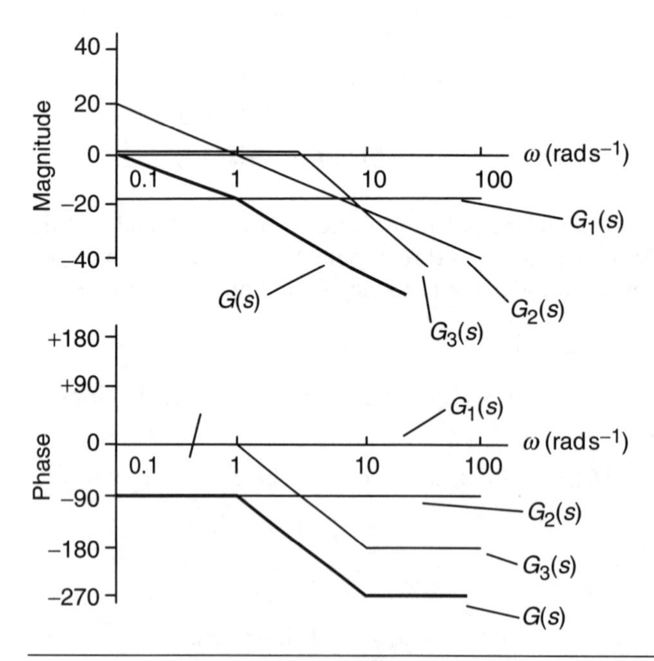

Figure 14.13 Bode plot for Example 14.4.

Magnitude

For $G_1(s), 20\log|G(j\omega)| = 20\log(0.111) = -19\,\text{dB}$.

For $G_2(s)$, using Table 14.1 (second case):

$$\omega \to 0.1 : M_G = 20\,\text{dB},$$
$$\omega \to 1 : M_G = 0\,\text{dB, and}$$
$$\omega \to 10 : M_G = -20\,\text{dB},$$

leading to a straight line passing through 20 dB, 0 dB, and −20 dB.

For $G_3(s)$, using Table 14.1 (third case):

$$\omega \to 0 : M_G = 0\,\text{dB},$$
$$\omega \to 1/\tau : M_G = -20\log(\omega\tau), \text{ when } \omega = 50 : M_G = -20\log(50/2) = -28,$$
$$\omega \to 10/\tau : M_G = -20\text{ dB, this occurs when } \omega \to 10\omega_n = 10 \times 6 = 60,$$

For $G_4(s)$, $\omega_n = \sqrt{36} = 6$ (break point) and $2\zeta\omega_n = 5$, giving $\zeta = 0.4167$.

Let us now consider the following conditions using Table 14.1:

$$\omega/\omega_n \to 0 : M_G = 0\,\text{dB},$$
$$\omega/\omega_n \to 1 : M_G = -20\log(2\zeta) = -20\log(2 \times 0.4167)$$
$$= 1.583; \text{ this occurs when } \omega \to \omega_n = 6,$$
$$\omega/\omega_n \to 10 : M_G = -40\,\text{dB, this occurs when } \omega \to 10\omega_n = 10 \times 6 = 60.$$

Phase

For $G_1(s) = 0.12, \omega \to 0 : \theta_G = 0°$ for all situations.

For $G_2(s), \omega \to 0 : \theta_G = 90°$ for all situations.

For $G_3(s)$ (third case from Table 14.1),

$$\omega \to 0 : \theta_G = 0°; \text{ this occurs when } \omega \to 0.1/2 = 0.05$$
$$\omega \to 1/\tau : \theta_G = -45°; \text{ this occurs when } \omega \to 1/2 = 0.5$$
$$\omega \to 10/\tau : \theta_G = -90°; \text{ this occurs when } \omega \to 10/2 = 5$$

For $G_4(s)$ (fourth case from Table 14.1),

$$\omega/\omega_n \to 0.2 : \theta_G = 0°; \text{ this occurs when } \omega \to 0.2\omega_n = 0.2 \times 6 = 1.2$$
$$\omega/\omega_n \to 1 : \theta_G = -90°; \text{ this occurs when } \omega \to \omega_n = 6$$
$$\omega/\omega_n \to 5 : \theta_G = 180°; \text{ this occurs when } \omega \to 5\omega_n = 5 \times 6 = 30$$

Figure 14.14 shows the Bode plot.

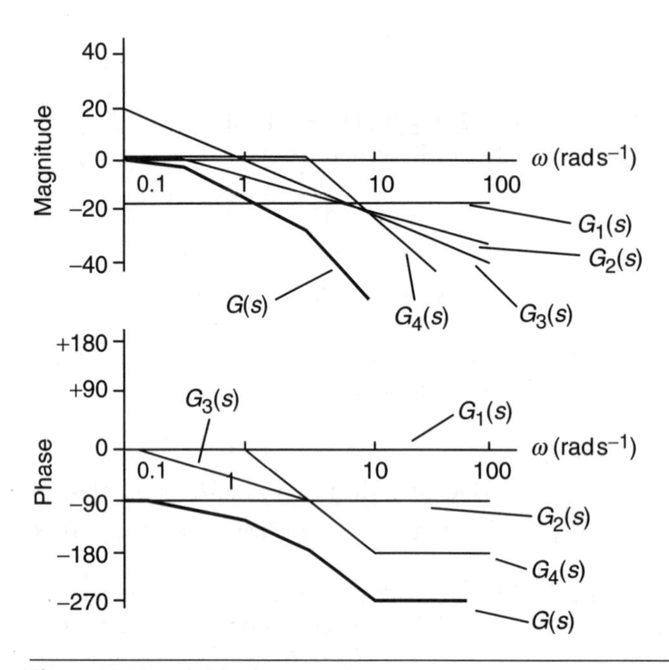

Figure 14.14 Bode plot for Example 14.5.

EXAMPLE 14.6

Determine the value of K for a system with the following open-loop transfer function which will give a gain margin of 4 dB.

$$G(s) = \frac{K}{s(2s+1)(4s+1)}. \tag{14.32D}$$

Solution

We substitute $s = \mathrm{j}\omega$ in the open-loop transfer function:

$$G(\mathrm{j}\omega) = \frac{K}{\mathrm{j}\omega(2\mathrm{j}\omega+1)(4\mathrm{j}\omega+1)}. \tag{14.32E}$$

Clearing terms in the denominator,

$$\mathrm{j}\omega(2\mathrm{j}\omega+1)(4\mathrm{j}\omega+1) = \left(-2\omega^2+\mathrm{j}\omega\right)(4\mathrm{j}\omega+1) = -6\omega^2+\mathrm{j}\omega\left(1-8\omega^2\right)$$

$$\therefore\ G(\mathrm{j}\omega) = \frac{K}{-6\omega^2+\mathrm{j}\omega(1-8\omega^2)} = \frac{-K\left[6\omega^2+\mathrm{j}\omega\left(1-8\omega^2\right)\right]}{36\omega^2+\omega^2(1-8\omega^2)^2}$$

$$\therefore\ \left|G(\mathrm{j}\omega)\right| = \frac{K}{-6\omega^2+\mathrm{j}\omega(1-8\omega^2)} = \frac{K}{\sqrt{36\omega^4+\omega^2(1-8\omega^2)^2}}. \tag{14.32F}$$

The phase angle is given as

$$\tan\phi = \frac{(1 - 8\omega^2)}{6\omega} \tag{14.32G}$$

For $\phi = 180°$, $\tan\phi = 0$; hence, $1 - 8\omega^2 = 0$ or $\omega = \frac{1}{\sqrt{8}} = 0.3536$.
The gain margin $= 20\log|G(j\omega)|$, hence

$$\begin{aligned} 4 &= -20\log\left(\frac{K}{\sqrt{36\omega^4 + \omega^2(1 - 8\omega^2)^2}}\right) \\ &= -20\log\left(\frac{K}{\sqrt{36(0.3536)^4 + 0}}\right) = -20\log(1.33K). \end{aligned} \tag{14.32H}$$

Tidying up leads to

$$\begin{aligned} -4/20 = -0.2 &= \log(1.33K) \\ 1.33K &= 10^{-0.2} = 0.631 \\ K &= \frac{0.631}{1.33} = 0.474. \end{aligned} \tag{14.32I}$$

14.3.2 Nyquist plots

The Nyquist criterion relates the stability of a closed-loop system to the open-loop frequency response and open-loop pole location. Consequently, knowledge of the open-loop system's frequency response gives us information regarding the stability of the closed-loop system. This concept is similar to the root locus, which starts with information about the open-loop system, its poles and zeros, and then finishes up with the transient and stability information for the closed-loop system. Although the Nyquist criterion will yield stability information, we will extend the concept to transient response and steady-state errors. Consequently, frequency response techniques are an alternative approach to the root locus.

14.3.2.1 *Derivation of the Nyquist criterion*

The Nyquist criterion can tell us how many closed-loop poles are in the right half-plane (rhp). We will examine:

- the relationship between the poles of $1 + G(s)H(s)$ and the poles of $G(s)H(s)$;

- the relationship between the zeros of $1 + G(s)H(s)$ and the poles of the closed-loop transfer function, $T(s)$;
- the concept of *mapping* points; and
- the concept of mapping *contours*.

Let us consider the system of Figure 14.15.
We let

$$G(s) = \frac{N_G}{D_G} \tag{14.33}$$

$$H(s) = \frac{N_H}{D_H} \tag{14.34}$$

$$G(s)H(s) = \frac{N_G N_H}{D_G D_H} \tag{14.35}$$

$$1 + G(s)H(s) = 1 + \frac{N_G N_H}{D_G D_H} = \frac{D_G D_H + N_G N_H}{D_G D_H} \tag{14.36}$$

$$T(s) = \frac{G(s)}{1 + G(s)H(s)} = \frac{N_G}{D_G} \frac{D_G D_H}{D_G D_H + N_G N_H} = \frac{N_G D_H}{D_G D_H + N_G N_H}. \tag{14.37}$$

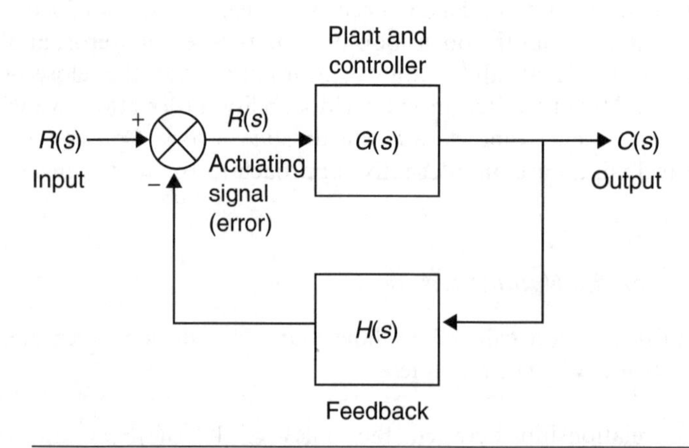

Figure 14.15 A closed-loop control system.

From Equations 14.36 and 14.37, we conclude that:

- the poles of $1+G(s)H(s)$ are the same G the poles of $G(s)H(s)$, the open-loop system; and
- the zeros of $1+G(s)H(s)$ are the same as the poles of $T(s)$, the closed-loop system.

EXAMPLE 14.7

See website for downloadable MATLAB code to solve this problem

Figure 14.16 shows the control system of a mobile robot. Sketch the Nyquist diagrams for the system.

Solution

$$G(s) = \frac{750}{(s+2)(s+4)(s+8)}\Big|_{s\to j\omega}. \tag{14.37A}$$

Now let us deal with the denominator of the transfer function.

$$\begin{aligned}
(j\omega+2)(j\omega+4)(j\omega+8) &= \left(-\omega^2+6j\omega+8\right)(j\omega+8) \\
&= \left(-j\omega^3-6\omega^2+8j\omega-8\omega^2+48j\omega+64\right) \\
&= \left(-14\omega^2+64\right)+j\left(-\omega^3+56\omega\right) \\
\therefore\; G(j\omega) &= \frac{750}{(-14\omega^2+64)+j(-\omega^3+56\omega)} \\
&= 750\frac{\left(-14\omega^2+64\right)-j\left(-\omega^3+56\omega\right)}{(-14\omega^2+64)^2+(-\omega^3+56\omega)^2}.
\end{aligned} \tag{14.37B}$$

For $\omega \to 0, G(j\omega) = 750/64$;
For the imaginary part $=0, 56\omega-\omega^3=0 \Rightarrow \omega = 7.48$.

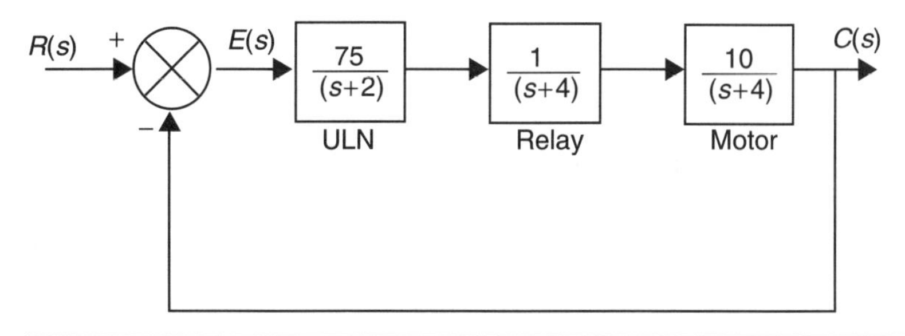

Figure 14.16 A control system for Example 14.7.

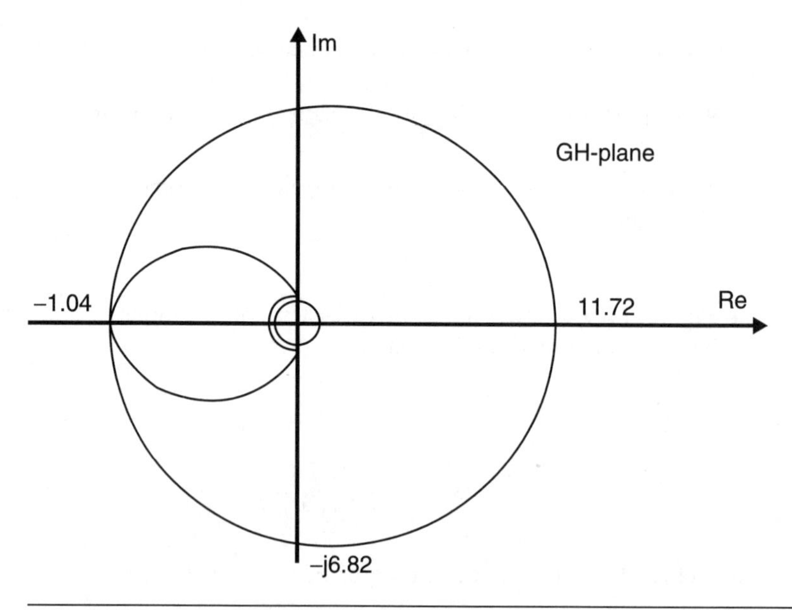

Figure 14.17 Nyquist plot for Example 14.7.

Substituting this value into the transfer function expression gives

$$G(j\omega) = 750\frac{(-14(7.48)^2+64)}{(-14(7.48)^2+64)^2} = 750\frac{1}{(-14(7.48)^2+64)} = -1.04. \qquad (14.37C)$$

Hence, the real value at the axis crossing is −1.04.
For the real part = 0, $-14\omega^2 + 64 = 0 \Rightarrow \omega = 2.138$.
Substituting this value into the transfer function expression gives

$$G(j\omega) = 750\frac{(-j)}{(56\omega - \omega^3)} = 750\frac{1}{(56(2.138)^2-(2.138)^3)} = -6.82. \qquad (14.37D)$$

Hence, the imaginary value at the axis crossing is −j6.82. The Nyquist plot is shown in Figure 14.17.

Further reading

[1] Beckwith, T.G., Buck, N.L. and Marangoni, R.D. (1982) *Mechanical Measurements* (3rd. ed.), Reading, MA: Addison-Wesley.
[2] Doeblin, E. (1990) *Measurement Systems Applications and Design* (4th. ed.), New York: McGraw-Hill.

[3] Dorf , R.C., and Bishop, R.H. (2001) *Modern Control Systems* (9th. ed.), Prentice Hall.
[4] Figliola, R. and Beasley, D. (1995) *Theory and Design of Mechanical Measurements*, (2nd. ed.), New York: John Wiley.
[5] Nise, N. (2000) *Control Systems Engineering* (3rd. ed.), New York: John Wiley.

Internet resources

- http://www.engin.umich.edu/group/ctm/freq/nyq.html
- http://www.engin.umich.edu/group/ctm/examples/motor2/freq2.html
- http://www-me.mit.edu/Sections/RLocus/2.11-rationale.html

CHAPTER 15

Robotic systems

Chapter objectives

When you have finished this chapter you should be able to:

- understand what robots are, and how they may be used;
- differentiate between mobile and stationary robots;
- understand basic definitions, configurations and components of robotic arms;
- understand forward as well as inverse (backward) transformations for robotic arms;
- understand the importance of resolutions, repeatability, and accuracy in the functioning of robotic arms;
- appreciate important aspects of robotic arm path planning.

Using MATLAB to analyze robotic manipulator path plannng

See website for a series of downloadable MATLAB codes which demonstrate aspects covered in this chapter, including forward and backward transformation for 2, 3 and 4 joint robot manipulators.

15.1 Types of robot

Robots are machines, which perform tasks similar to the human form. There are basically two types of robot: mobile and stationary. Mobile robots are free to move within a workspace. Stationary robots are fixed in one place (such as a robotic arm).

Typical applications of mobile robots are, nuclear accident cleanup, planetary exploration, automated guided vehicles in manufacturing factories, and mail delivery. Mobile robots are either guided or free roving.

Automated guided vehicles (AGVs) are typically wheeled robots that carry payloads within a factory. They are material handling systems for moving raw materials or partly finished goods from one workstation to another within a manufacturing system facility. AGVs navigate using one of the several methods: wires embedded in floors, light sources or reflectors, or colored tapes on the floor. The free-roving mobile robot does not necessarily follow any predefined path and, as such, needs to be able to move on more hostile surfaces than the factory floor. They typically have robust obstacle avoidance systems which include vision-type, camera-mounted systems for online capture and analysis of the environment. For many tasks, even within a friendly and mapped environment, the free-roving robot needs to possess some degree of autonomy enabling it to 'know' its current position, to plan its route to a destination, and to navigate to a destination. These 'skills' require robust route generation and path following algorithms. Hence, these robots are expensive due to the complex computational aspect of analysis. However, simpler systems include the use of basic sensors and algorithms for clearing obstacles. Research in the area of autonomous robots continues to be challenging and it is still in its infancy.

15.2 Robotic arm terminology

Here is some key terminology for robotic arms

- **Link:** The solid structural member of the arm
- **Joint:** The moving couplings between links. Joints can be either rotary (often driven by electric motors and chain/belt/gear transmissions, or by hydraulic cylinders and levers), or prismatic (slider joints in which a link is supported on a liner slider bearing, and linearly actuated by ball screws and motors or cylinders).
- **Degrees of freedom:** Each joint on the robot introduces a degree of freedom. Each degree of freedom can be a slider, rotary, or other type of actuator. Robots typically have five or six degrees of freedom. Six degrees of freedom are enough to allow the robot to reach all positions and orientations in three-dimensional space.
- **Orientation axes:** Roll, pitch and yaw are the common orientation axes used (Figure 15.1).
- **Position axes:** The tool, regardless of orientation, can be moved to a number of positions in space.
- **Tool center point (TCP):** The tool center point is located either on the robot, or on the tool as shown in Figure 15.2.
- **Work envelope/workspace:** The robot tends to have a fixed and limited geometry. The work envelope is the boundary of positions in space that the robot can reach. Figure 15.3 shows a typical work envelope.
- **Speed:** Speed refers either to the maximum velocity that is achievable by the tool center point, or by individual joints.

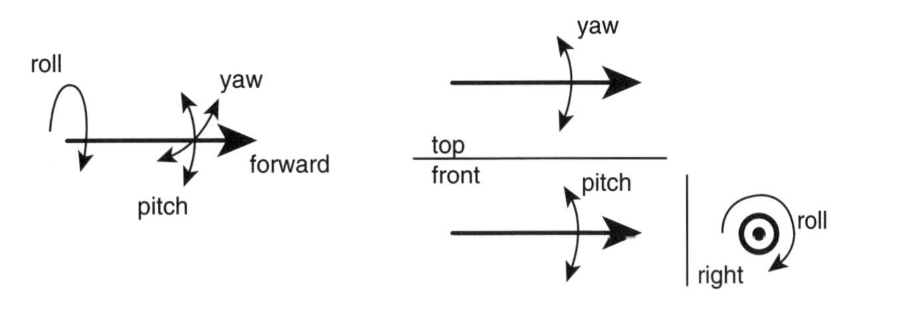

Figure 15.1 Movement in three dimensions.

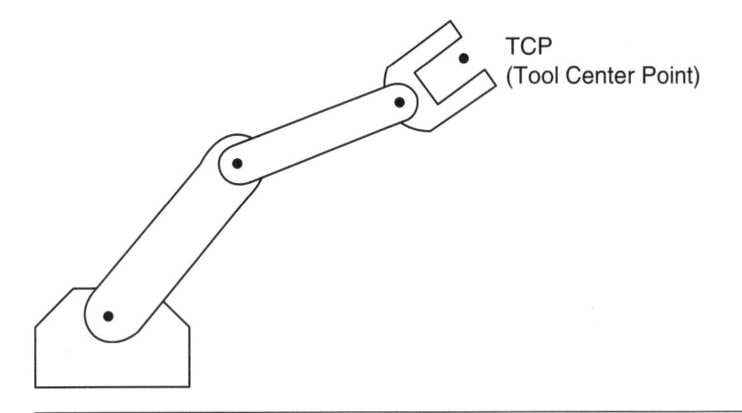

Figure 15.2 Tool center point.

- **Payload:** The payload is maximum mass the robot can lift before the robot fails, or there is a dramatic loss of accuracy.
- **Settling time:** A robot moves fast during a movement, but as it approaches the final working position, it slows down. The settling time is the time required for the robot to be within a given distance from the final position.
- **Coordinates:** The robot can move, therefore it is necessary to define positions.

15.3 Robotic arm configuration

Basic configurations include:

- the Cartesian/rectilinear/gantry (Figure 15.4);

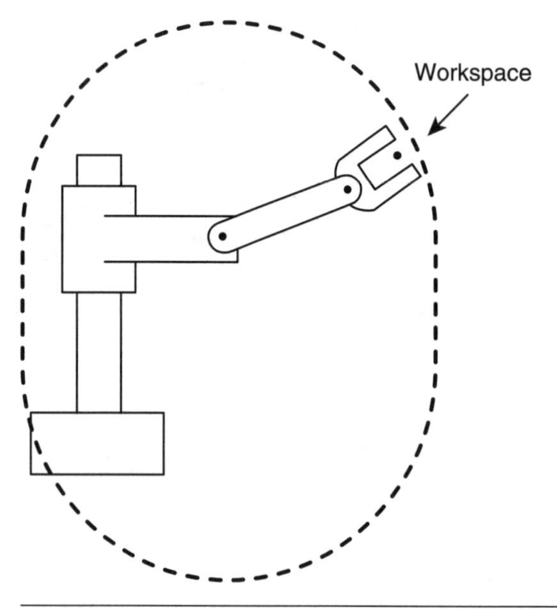

Figure 15.3 The work envelope.

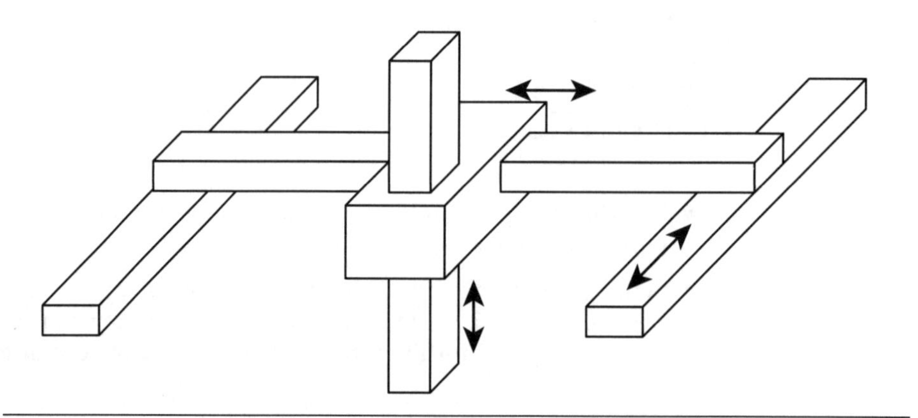

Figure 15.4 The Cartesian/rectilinear/gantry robotic arm.

- the cylindrical (Figure 15.5), having a revolute motion about a base, a prismatic joint for height, and a prismatic joint for radius;
- the spherical (Figure 15.6), having two revolute joints and one prismatic joint allow the robot to point in many directions, and then reach out some radial distance;

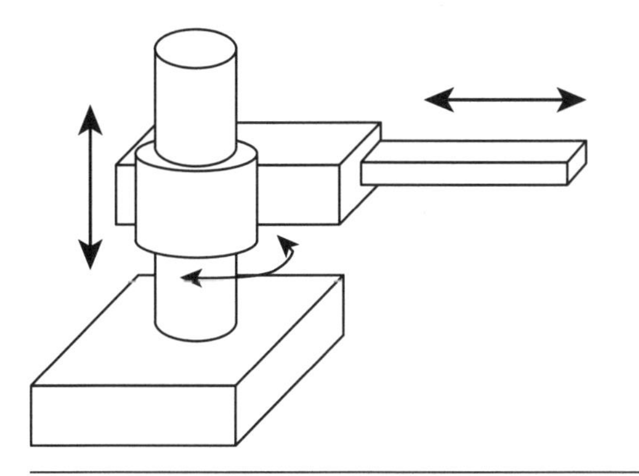

Figure 15.5 The cylindrical robotic arm.

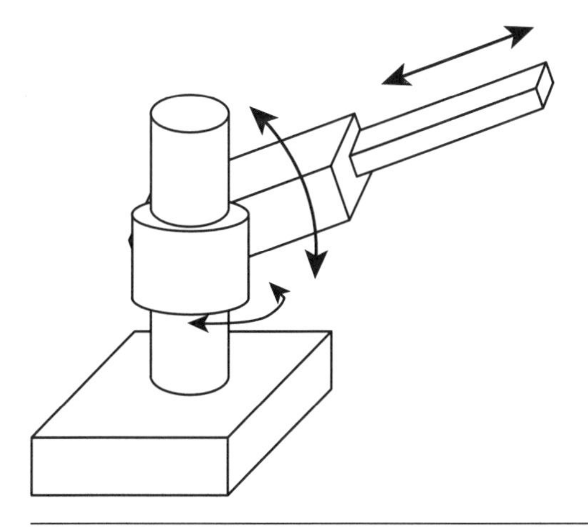

Figure 15.6 The spherical robotic arm.

- the articulated/jointed spherical/revolute (3R-type) (Figure 15.7), having three revolute joints to position the robot. This robot most resembles the human arm, with a waist, shoulder, elbow, wrist;
- the selective compliance arm for robotic assembly (SCARA) (Figure 15.8), conforming to cylindrical coordinates, but the radius and rotation is obtained by a two planar links with revolute joints.

Interested readers are referred to Craig (1989) for more details.

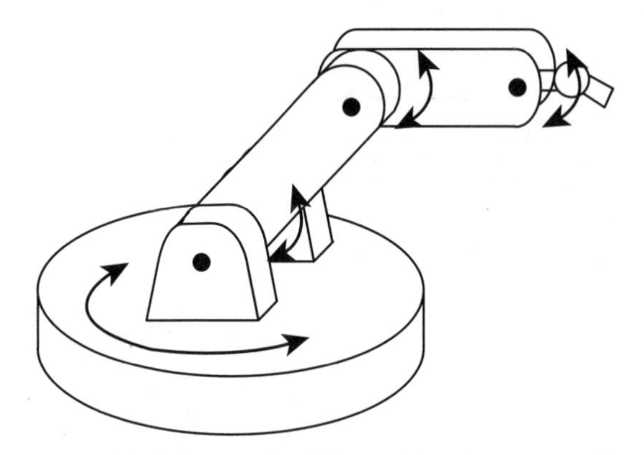

Figure 15.7 The articulated/jointed spherical/revolute robotic arm.

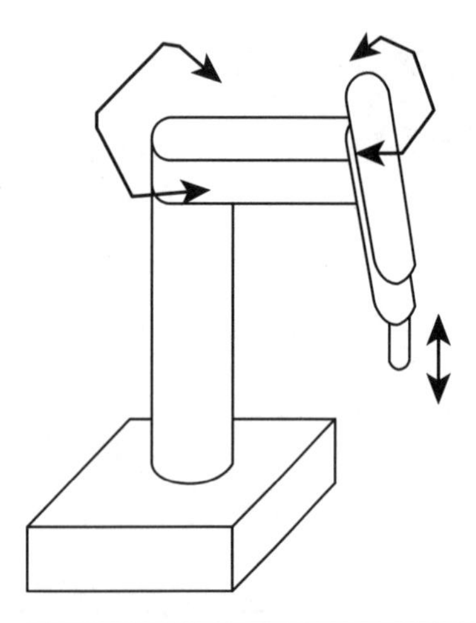

Figure 15.8 The SCARA robotic arm.

15.4 Robot applications

Some tasks which robotic arms can perform are:

- point-to-point
- manipulation
- path tracking
- operating
- telerobotics

- services; and
- biomedical

15.5 Basic robotic systems

The basic components of a robotic arm are shown in Figure 15.9.

- **Structure:** the mechanical structure (links, base, etc.) requires a great deal of mass to provide enough structural rigidity to ensure high accuracy for different payloads;
- **Actuators:** the stepper motors that drive the robot joints. This also includes mechanisms for transmission, locking, etc.;
- **Control computer:** to interface with the user, and control the robot's joints;
- **Sensors:** the sensors are of two types (proximity sensors to sense other objects, and force sensors to enable the robot to apply the exact amount of force when gripping an object;
- **End of arm tooling (EOAT):** provided by the user, and designed for specific tasks.

15.5.1 Robotic mechanical arm

The design of the robotic mechanical arm requires consideration of the linkages, as well as the static and dynamic considerations of the payload.

- **Design of linkages:** Linkage mechanisms connected at the joints are made to rotate using actuators (motors, chains/belts and gears). A knowledge of the

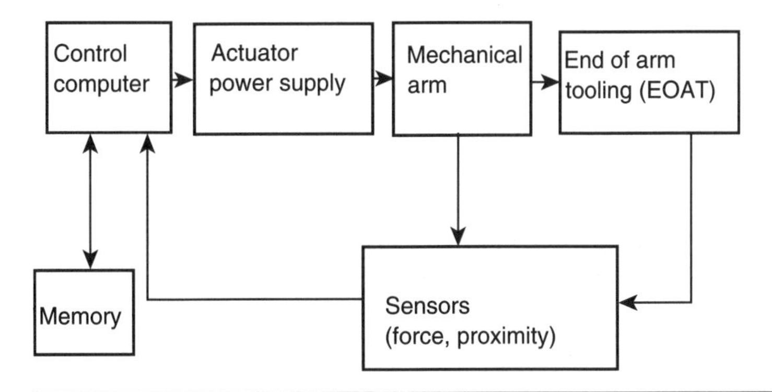

Figure 15.9 Basic components of a computer-controlled robotic arm.

linkage kinematics is required in the design of a robotic mechanical arm. Decisions need to be made regarding size (length, cross-sectional dimensions) of the arm, as well as its range (the maximum angle which the arm has to rotate).

- **Payload considerations:** Static and dynamic considerations cause positioning errors. *Static considerations* are gravity effects, drive gears and belts, joint play, and thermal effects. The most important *dynamic considerations* to be considered are the effects of acceleration.

15.5.1.2 *Static considerations*

Gravity effects cause downward deflection of the arm and support systems. The gravity effects consist of payload P, and robot link mass, w. Hence the total load on the robotic arm is given as

$$W_{\text{total}} = P + w. \tag{15.1}$$

Gravity effect due to payload

The deflection of the beam tip caused by the point load, P, is given as

$$\delta_{\text{payload}} = \frac{PL^3}{3EI}, \tag{15.2}$$

where E is Young's modulus, and the moment of inertia, $I = BH^3/12$.

Gravity effect due to robot link mass

The deflection of the beam tip caused by the uniformly distributed load, w, is given as

$$\delta_{\text{link mass}} = \frac{wL^3}{8EI}, \tag{15.3}$$

where E is Young's modulus, and the moment of inertia, $I = BH^3/12$.

The robotic arm deflection due to the payload and the robotic link mass is given as

$$\delta_{\text{total}} = \frac{PL^3}{3EI} + \frac{wL^3}{8EI}. \tag{15.4}$$

EXAMPLE 15.1

A robotic arm made from steel has a length of 1.524 m, a breadth of 0.102 m, and a height of 0.1524 m. The payload is 444.82 kg. The density of steel, ρ, is $7.87\,\mathrm{kg\,m^{-3}}$ and its Young's modulus, E, is 206.85 GPa. Determine the deflection of the robotic arm due to the payload and the robotic link mass.

Solution

The moment of inertia of the robotic arm is

$$I = \frac{BH^3}{12} = \frac{0.102(0.1524)^3}{12} = 3 \times 10^{-5}\,\mathrm{m}^4. \tag{15.5}$$

The robotic link mass is

$$w = \frac{\text{weight}}{\text{length}} = \frac{(\rho A l)g}{l} = \left(7.87 \times 10^3 \times 0.102 \times 0.1524\right)9.81 = 1200\,\mathrm{kg\,m^{-1}}. \tag{15.6}$$

Hence, for the parameters given, the total deflection is

$$\delta_{\text{total}} = \frac{PL^3}{3EI} + \frac{wL^3}{8EI} = \frac{(1.524)^3}{(206.85 \times 10^9 \times 3 \times 10^{-5})}\left[\frac{444.82}{3} + \frac{1200}{8}\right] = 170\,\mu\mathrm{m}. \tag{15.7}$$

Drive gears and belts drive play

Drive gears and belts often have noticeable amounts of slack (backlash) that cause positioning errors.

Joint play

Joint play (windup) occurs due to joint flexibility when long rotary members are used in a drive system and twist under load. When modeling the joint play (flexibility) of the robotic arm, we consider the angular twist of the joints, rotary drives, and shafts, under the load. The torsional stiffness, k, is defined as torque per radian twist, given as

$$k = \frac{T}{\theta} = \frac{G\Im}{l}, \tag{15.8}$$

where $\Im$ is the polar moment of inertia, $\Im = \pi D^3/32$.

Consequently, the twist, θ, is obtained as

$$\theta = \frac{TL}{G\Im} = \frac{32TL}{\pi D^4 G}. \tag{15.9}$$

Thermal effects

Temperature change leads to dimensional changes in the manipulator.

15.5.1.3 *Dynamic considerations*

The major considerations here are the effects of acceleration. Inertial forces can lead to deflection in structural members. These are normally only problems when a robot is moving very fast, or when a continuous path following is essential. However, during the design of a robot these factors must be carefully examined.

15.5.2 End-of-arm tooling (EOAT)

The best industrial robot is only as good as its end-of-arm tooling (EOAT). End-of-arm tooling is typically purchased separately, or custom built, and is very expensive.

15.5.2.1 *Classification of end-of-arm-tooling*

EOAT can be classified into grippers and tools. Grippers are either multiple/single or internal/external. Tools are subdivided into compliant, contact or non-contact.

There are at least seven methods used for gripping an object:

- grasp it
- hook it
- scoop it
- inflate around it
- attract it magnetically
- attract it by a vacuum
- stick to it

15.5.2.2 *Calculating gripper payload and gripping force*

Manufacturers usually identify the maximum payload that a manipulator can handle. If the manipulator can handle 28 kg (including a 5 kg wrist and a 3 kg gripper), then the gripper can handle only a 20 kg object. Other factors involved in payload calculations include:

- **Torque:** Torque exists when a part is picked up at a place other than its center of gravity.
- **Center of gravity:** Center of gravity is the point where its mass seems to be concentrated or the point where the part is balanced.
- **Coefficient of friction:** Coefficient of friction measures how efficiently the gripper holds the part.
- **Acceleration or deceleration:** Acceleration or deceleration is the rate of change of velocity of the part.
- **A safety factor:** The safety factor is a design factor to counteract unaccountable error or unforeseen factors. A typical safety factor is 2.

15.5.2.3 *EOAT design*

Typical design factors to be considered are shown in Table 15.1, while the design criteria are shown in Table 15.2.

15.5.2.4 *Gripper mechanisms*

Fingers are designed to:

- physically mate with the part for a good grip;
- apply enough force to the part to prevent slipping.

Movements of the fingers could be:

- pivoting (often uses pivotal linkages); or
- linear or translational movement (often uses linear bearings and actuators).

There are various types of gripper mechanisms available; these include: linkage actuation, gear and rack, cam, screw, rope and pulley, bladder, diaphragm, etc.

The following describes some grippers actuated using pneumatic/hydraulic cylinders; these can be replaced with other mechanical arrangements other than cylinders.

Table 15.1 End of tool design factors

Design factors	*Options to be considered*
Workpiece to be handled	■ part dimensions ■ mass ■ pre- and post-processing geometry ■ geometrical tolerances ■ potential for part damage
Actuators	■ mechanical ■ vacuum ■ magnet
Power source	■ electrical ■ pneumatic ■ hydraulic ■ mechanical
Range of gripping force	■ object mass ■ friction or nested grip ■ coefficient of friction between gripper and part ■ maximum accelerations during motion
Positioning	■ gripper length ■ robot accuracy and repeatability ■ part tolerances
Maintenance	■ number of cycles required ■ use of separate wear components ■ design for maintainability
Environment	■ temperature ■ humidity ■ dirt, corrosives, etc.
Temperature protection	■ heat shields ■ longer fingers ■ separate cooling system ■ heat resistant materials
Materials	■ strong, rigid, durable ■ fatigue strength ■ cost and ease of fabrication ■ coefficient of friction ■ suitable for environment
Other points	■ interchangeable fingers ■ design standards ■ use of mounting plate on robot ■ gripper flexible enough to accommodate product design change

Two finger gripper

A two-finger gripper is shown in Figure 15.10(a). As the pneumatic cylinder is actuated, the fingers move together and apart. A parallel finger actuator two finger gripper is shown in Figure 15.10(b). Figure 15.10(c) shows a mechanism that increases the holding force.

Table 15.2 End of tool design criteria

Design criteria	*Description*
Weight	low weight to allow larger payload, increase accelerations, decrease cycle time
Dimensions	minimum dimensions set by size of workpiece, and work area clearances
Range of parts	widest range of parts accommodated using inserts, and adjustable motions
Rigidity	rigidity to maintain robot accuracy and reduce vibrations
Force level	maximum force applied for safety, and to prevent damage to the work
Power source	power source should be readily available from the robot, or nearby
Maintenance	maintenance should be easy and fast
Safety	safety dictates that the work shouldn't drop when the power fails
Gripper centroid	ensure that part centroid is centered close to the robot to reduce inertial effects. For the worst case, ensure that it is between the points of contact
Holding pressure	holding pressures/forces/etc. are hard to control, try to hold parts with features or shapes
Compliance	compliance can help guide work into out-of-alignment conditions.
Sensors	sensors in the EOAT can check for parts not in the gripper, etc.
Work position	the gripper should tolerate variance in work position with part alignment features
Gripper changer	gripper changers can be used to make a robot multifunctional
Flexibility	multiple EOAT heads allow one robot to perform many different tasks without an EOAT change
Originality	Don't try to mimic human behavior
Speed of removal	design for quick removal or interchange of tooling by requiring a small number of tools (wrenches, screwdrivers, etc.)
Alignment	provide dowels, slots, and other features to lead to fast alignment when changing grippers
Fastener usage	use the same fasteners when possible
Shapes to avoid	eliminate sharp corners/edges to reduce wear on hoses, wires, etc.
Cable slackness	allow enough slack and flexibility in cables for full range of motion
Weight of material	use lightweight materials, and drill out frames when possible
Type of coatings	use hard coatings, or hardened inserts to protect soft gripper materials
Alternative designs	examine alternatives when designing EOAT
Design attention	the EOAT should be recognized as a potential bottleneck, and given extra design effort
Fixtures	use shear pins, and other devices to protect the more expensive components
Cleanliness	consider dirt, and use sealed bearings where possible
Location of weights	move as much weight away from the tip of the gripper towards the robot

In the two-fingered pneumatic actuated gripper shown in Figure 15.11, the fingers move outward for internal gripping.

Magnetic grippers

Obviously, magnetic grippers can be used only with ferrous materials. Electromagnets and permanent magnets are used. Electromagnets require a power supply and a controller. Polarity can be reversed on the magnet when it is

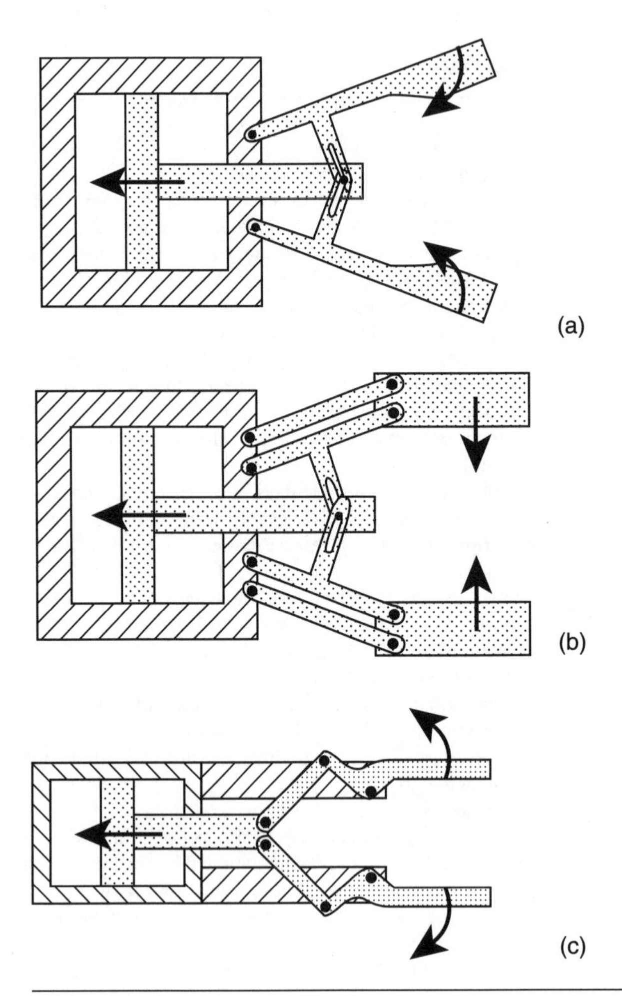

Figure 15.10 Two finger gripper: (a) non-parallel; (b) parallel; (c) parallel with increased grip.

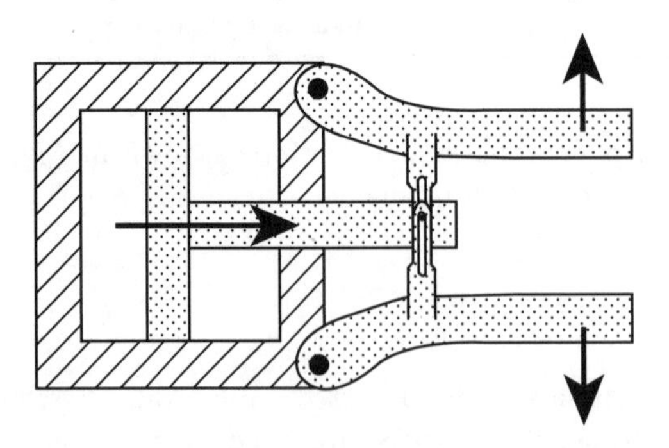

Figure 15.11 Internal two finger gripper.

put down to reverse residual magnetism. A mechanism is required to separate parts from a permanent magnet. They are good for environments that are sensitive to sparks.

Some of the advantages of magnetic grippers are:

- variation in part size can be tolerated;
- ability to handle metal parts with holes;
- pick up times are fast;
- requires only one surface for gripping;
- can pick up the top sheet from a stack.

Some of the disadvantages of magnetic grippers are:

- residual magnetism remains in the workpiece;
- possible side slippage.

Expanding grippers

Some parts have hollow cavities that can be used to advantage when grasping. Expanding grippers can also be used when gripping externally.

Other types of gripper

Most grippers for manipulation are sold with mounts so that fingers may be removed, and replaced. Gripper fingers can be designed to reduce problems when grasping.

15.6 Robotic manipulator kinematics

In robotic arm kinematics, we are interested in the forward transformation as well as inverse (backward) transformation. We will concentrate on the 3R articulated robot, although the basics apply to other robot classifications.

15.6.1 Forward transformation for three-axis planar 3R articulated robot

When the motors controlling the robotic shoulder, elbow, and wrist rotate, the gripper moves to a point in the world coordinate system. An important task is to

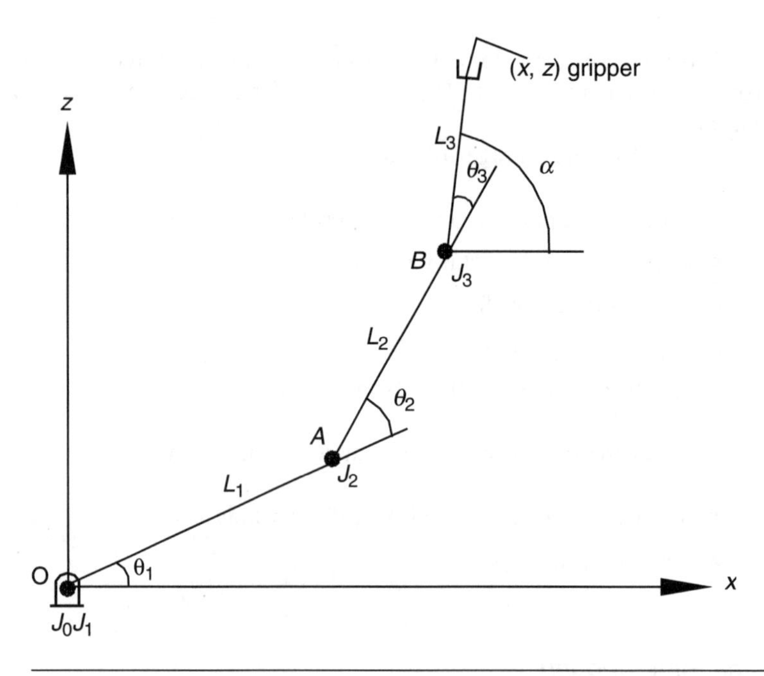

Figure 15.12 Robotic manipulator (arm) for backward analysis.

determine the position of the gripper when this happens. This problem is classified as a forward transformation problem.

Figure 15.12 shows the robotic arm, from which we compute the gripper position as:

$$x = L_1 \cos\theta_1 + L_2 \cos(\theta_1 + \theta_2) + L_3 \cos(\theta_1 + \theta_2 + \theta_3)$$

$$z = L_1 \sin\theta_1 + L_2 \sin(\theta_1 + \theta_2) + L_3 \sin(\theta_1 + \theta_2 + \theta_3)$$

$$\alpha = (\theta_1 + \theta_2 + \theta_3). \qquad (15.10)$$

15.6.2 Inverse transformation for three-axis planar 3R articulated robot

Here, we are interested in knowing what angles the joints will rotate for a given position of the gripper. The problem of determining the angles through which the robotic shoulder, elbow, and wrist rotate is known as inverse transformation. Mathematically this calculation is difficult, and there are often multiple solutions.

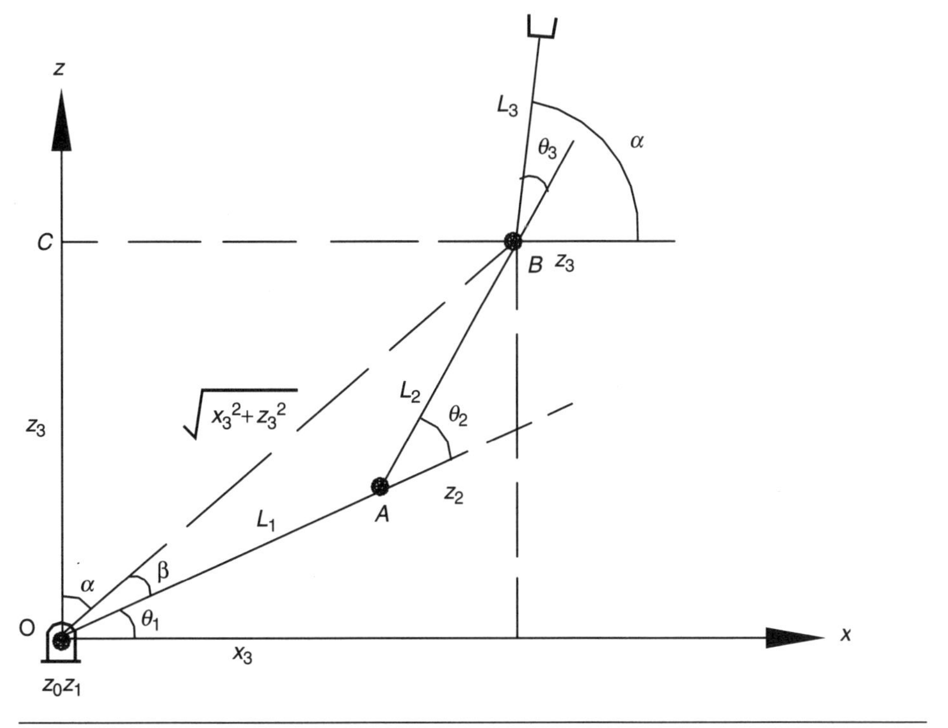

Figure 15.13 Robotic arm for backward analysis.

Figure 15.13 shows the robotic arm.
The following definitions are used for inverse transformation:

$$CB = x_3$$
$$OC = z_3$$
$$OB = \sqrt{x_3^2 + z_3^2}$$
$$x_3 = x - L_3 \cos\alpha$$
$$z_3 = z - L_3 \sin\alpha$$

From $\Delta OAB, OB^2 = AB^2 + OA^2 - 2AB \times OA \cos(180^\circ - \theta_2)$ leading to

$$x_3^2 + z_3^2 = L_1^2 + L_2^2 + 2L_1L_2 \cos\theta_2.$$

$$\cos\theta_2 = \frac{x_3^2 + z_3^2 - (L_1^2 + L_2^2)}{2L_1L_2}$$

$$\sin\theta_2 = \pm(1 - \cos\theta_2).$$

From ΔOCB, $\tan\alpha = \frac{CB}{OC} = \frac{x_3}{z_3}$

leading to

$$\sin\alpha = \frac{x_3}{\sqrt{x_3{}^2 + z_3{}^2}}.$$

From ΔOAB:

$$\frac{L_2}{\sin\beta} = \frac{\sqrt{x_3{}^2 + z_3{}^2}}{\sin(180 - \theta_2)} = \frac{\sqrt{x_3{}^2 + z_3{}^2}}{\sin\theta_2}.$$

From which,

$$\sin\beta = \frac{L_2 \sin\theta_2}{\sqrt{x_3{}^2 + z_3{}^2}}$$

$$\theta_1 = 90 - \alpha - \beta. \tag{15.11}$$

EXAMPLE 15.2

The links of a 3R robotic arm are $L_1 = 350$ mm, $L_2 = 250$ mm and $L_3 = 50$ mm. The gripper is at world coordinates given as $x = 300$ mm, $z = 400$ mm and $\alpha = 30°$. Determine the angles θ_1, θ_2 and θ_3, which the motor controlling the shoulder, elbow, and wrist have to be rotated.

Solution

$$x_3 = x - L_3 \cos\alpha = 300 - 50 \cos(30°) = 256.7$$

$$z_3 = z - L_3 \sin\alpha = 400 - 50 \sin(30°) = 375$$

$$\cos\theta_2 = \frac{x_3{}^2 + z_3{}^2 - (L_1{}^2 + L_2{}^2)}{2L_1L_2} = \frac{256.7^2 + 375^2 - (350^2 + 250^2)}{2 \times 350 \times 250} = 0.1229$$

$$\therefore \theta_2 = 82.9^{o}$$

$$\tan\alpha = \frac{x_3}{z_3} = \frac{256.7}{375} = 0.685$$

$$\therefore \alpha = 34.39^{o}$$

$$\sin\beta = \frac{L_2 \sin\theta_2}{\sqrt{x_3{}^2 + z_3{}^2}} = \frac{250 \times \sin 82.9^{o}}{\sqrt{256.7^2 + 375^2}} = 0.546$$

$$\beta = 33.09^\circ$$

$$\theta_1 = 90 - \alpha - \beta = 90 - 34.39 - 33.09 = 22.5^\circ$$

$$\theta_3 = \alpha - \theta_1 - \theta_2 = 30 - 82.9 - 22.5 = -75.4^\circ$$

So $\theta_1 = 22.5^\circ$, $\theta_2 = 82.9^\circ$ and $\theta_3 = -75.4^\circ$.

15.7 Robotic arm positioning concepts

15.7.1 Resolution

Resolution is based on a limited number of points that the robot can be commanded to reach out for.

15.7.2 Spatial resolution

Spatial resolution is the smallest increment of movement into which the robot can divide its work volume. Spatial resolution depends on two factors: the systems control resolution and the robots mechanical inaccuracies.

15.7.3 Electromechanical control resolution

This is the mechanical limit on the capacity to divide the range of each joint-link system into addressable points. It is designated CR_1.

15.7.4 Control resolution

This is determined by the robot's position control system and its feedback measurement system. The control resolution (CR_2) of a robot is given as $CR_2 = R/(2^n - 1)$, where R is the range of the joint and n is the number of bits in the bit storage register devoted to that particular joint.

The resolution of each joint-link is defined as the maximum between the mechanical and control resolutions, given as $CR = \max(CR_1, CR_2)$.

15.7.5 Repeatability

The robot mechanism will have some natural variance during a repetitive task. Repeatability is a measure of the error or variability when repeatedly reaching for a single position, or, put another way, it defines how close a robot will be to the same position as the same move it made before. It is the result of random errors only.

Repeatability is defined as $\pm 3\sigma$, where σ is the standard deviation.

15.7.6 Accuracy

Accuracy is determined by the resolution of the workspace, defining how close a robot gets to a desired position. It is a measure of the distance between a specified position and the actual position reached. Accuracy is defined as $(\mathrm{CR}/2) + 3\sigma$.

Accuracy is more important when performing off-line programming, because absolute coordinates are used.

15.7.7 Sources of errors

There are several possible sources of errors:

- Kinematic and calibration errors basically shift the points in the workspace resulting in an error *e*. Vendor specifications typically assume that calibration and modeling errors are zero.
- Random errors will prevent the robot from returning to the exact same location each time, and this can be shown with a probability distribution about each point.

The accuracy and repeatability are functions of:

- **Resolution:** the use of digital systems and other factors mean that only a limited number of positions are available. Thus user input coordinates are often adjusted to the nearest discrete position.
- **Kinematic modeling error:** the kinematic model of the robot does not exactly match the robot. As a result the calculations of required joint angles contain a small error.
- **Calibration errors:** the position determined during calibration may be slightly off, resulting in an error in calculated position.
- **Random errors:** problems arise as the robot operates. For example, friction, structural bending, thermal expansion, backlash/slip in transmissions, etc. can cause variations in position.

EXAMPLE 15.3

The range of one of the arms of a 3R articulated industrial robot is 30°. The controller used for this joint has an 8-bit storage capacity. The mean of the distribution of the mechanical errors is zero and the standard deviation is 0.5°. Determine: (a) the control resolution (CR_2); (b) the accuracy; and (c) the repeatability of the robot.

Solution

(a) The control resolution is:

$$\mathrm{CR} = \frac{R}{2^n - 1} = \frac{30}{2^8 - 1} = \frac{30}{256 - 1} = 0.118^{\circ}. \quad (15.12)$$

(b) The accuracy is:

$$\frac{\mathrm{CR}}{2} + 3\sigma = \frac{0.118}{2} + 3(0.5) = 1.559^{\circ}. \quad (15.13)$$

(c) The repeatability is

$$\pm 3\sigma = \pm 3(0.5) = \pm 1.5^{\circ}. \quad (15.14)$$

15.8 Robotic arm path planning

There are significant differences between the methods used to move a robot arm from point A to point B, or along a continuous path, and the routes to follow are infinite.

- **Slew motion:** This is the simplest form of motion. As the robot moves from A to point B, each axis of the manipulator travels as quickly as possible from its initial position to its final position. All axes begin moving at the same time, but each axis ends it motion in a length of time that is proportional to the product of its distance moved and its top speed (allowing for acceleration and deceleration). Slew motion usually results in unnecessary wear on the joints and often leads to unanticipated results in the path taken by the manipulator.
- **Joint interpolated motion:** This is similar to slew motion, except all joints start, and stop at the same time. This method demands only the speeds needed to accomplish any movement in the least time.
- **Straight-line motion:** In this method the tool of the robot travels in a straight line between the start and stop points. This can be difficult, and lead to rather erratic motions when the boundaries of the workspace are

approached. We note that straight-line paths are the only paths that will try to move the tool straight through space; all others will move the tool in a curved path.

Basic requirements are to develop a set of points from the start and stop points that minimize acceleration, and perform inverse kinematics to find the joint angles of the robot at the specified points.

Straight-line motion is not a very satisfactory means of achieving the desired movement.

15.8.1 Two methods for joint-space trajectory generation

Joint velocity, acceleration, and jerk are important in analysis, but we are fundamentally concerned with controlling joint angle.

15.8.1.1 *Third-order polynomial*

A third-order polynomial will give 'smooth' motion since there are continuous position and velocity. But *jerk* has infinite spikes at the start and end. The equation for a third-order polynomial that is useful is: $\theta(t) = at^3 + bt^2 + c$.

15.8.1.2 *Fifth-order polynomial*

To avoid jerk, we introduce a higher order polynomial having the same four 'smooth' motion joint constraints, plus two more constraints to avoid the infinite *jerk* spikes. A fifth-order polynomial fit gives a 'smooth' joint space trajectory generation, plus finite *jerk*, at joint i only. The equation for a fifth-order polynomial that is useful for path planning is $\theta(t) = at^5 + bt^4 + ct^3 + d$.

Not all terms are included. The end conditions are normally given so that the constants can be determined. MATLAB® can be used to solve high order polynomials.

EXAMPLE 15.4

The path planning of the gripper of a robotic manipulator is defined using the equation for a fifth-order polynomial given as $\theta(t) = at^5 + bt^4 + ct^3 + d$. The start and finish angles are 30° and 120°, respectively. Determine the gripper's: (a) angle; (b) speed; (c) acceleration; and (d) jerk. Sketch the graphs.

Solution

Differentiating the equation for angle results in the following:

$$\begin{aligned}
\theta(0) &= d = 30 \\
\theta(3) &= 243a + 81b + 27c + 30 = 120 \\
\dot{\theta}(3) &= 405a + 108b + 27c = 0 \\
\ddot{\theta}(3) &= 540a + 108b + 18c = 0 \\
\ddot{\theta}(1.5) &= 67.5a + 27b + 9c = 0
\end{aligned} \tag{15.15}$$

From the first three equations, we have the following matrix formulation, which can be easily solved using MATLAB®:

$$\begin{bmatrix} 234 & 81 & 27 \\ 405 & 108 & 27 \\ 540 & 108 & 18 \end{bmatrix} \begin{bmatrix} a \\ b \\ c \end{bmatrix} = \begin{bmatrix} 90 \\ 0 \\ 0 \end{bmatrix} \tag{15.16}$$

The solution of this problem is:

$$\begin{aligned}
\theta(t) &= -2.22t^5 - 16.67t^4 + 33.33t^3 + 30 \\
\dot{\theta}(t) &= 11.11t^4 - 66.67t^3 + 100t^2 \\
\ddot{\theta}(t) &= 44.44t^3 - 200t^2 + 200t \\
\dddot{\theta}(t) &= 133.33t^2 - 400t + 200
\end{aligned} \tag{15.17}$$

The last equation is jerk, which is important to control. These equations can now be graphed.

15.8.2 Computer control of robot paths (incremental interpolation)

Path planning is a simple process when the path planning methods already described are used before the movement begins. A simple real-time lookup table can be used. The path planner puts all of the values in a trajectory table. The online path controller will look up values from the trajectory table at predetermined time, and use these as set-points for the controller. The effect of the two tier structure is that the robot is always shooting for the next closest *knot-point* along the path. The scheme just described leads to errors between the planned and actual path, and lurches occur when the new set-points are updated for each servo motor. The quantization of the desired position requires a decision of what value to use, and

Table 15.3 Typical robot actuators

Type	*Advantages*
Pneumatics	■ simple, low maintenance ■ light, least expensive ■ low payload ■ easy to find fault ■ hard to do continuous control
Hydraulic	■ large payload ■ high power/weight ratio ■ leakage ■ noisy
Electric motors	■ feedback compatible ■ computer compatible ■ EOAT compatible ■ quiet, clean ■ low power/weight ratio

this value is fixed for a finite time. The result is that the path will tend to look somewhat bumpy.

15.9 Actuators

Robots are normally controlled using microcomputers or microcontrollers. The output from a robot needs to be transformed into usable forms using actuators. There are a large number of power sources that may be used for robots.

Table 15.3 lists some advantages of typical actuators.

Problems

Forward transformation

Q15.1 The joints and links of a 3R manipulator have lengths $L_1 = 525$ mm, $L_2 = 425$ mm, and $L_3 = 50$ mm. The joint angles for the three links are $\theta_1 = 15°$, $\theta_2 = 30°$ and $\theta_3 = 45°$, respectively. Determine the values of x and z in the world space coordinates.

Q15.2 The joints and links of a 3R manipulator have lengths $L_1 = 425$ mm, $L_2 = 325$ mm, and $L_3 = 50$ mm. The joint angles for the three links are $\theta_1 = 15°$, $\theta_2 = 30°$ and $\theta_3 = 45°$, respectively. Determine the values of x and z in the world space coordinates.

Backward transformation

Q15.3 The world space coordinates for a 3R manipulator are $x = 400$ mm and $z = 450$ mm. The links have lengths $L_1 = 525$ mm, $L_2 = 425$ mm, and $L_3 = 50$ mm. $\alpha = 25°$. Determine the joint angles θ_1, θ_2, and θ_3.

Q15.4 The world space coordinates for a 3R manipulator are $x = 350$ mm and $z = 400$ mm. The links have lengths $L_1 = 500$ mm, $L_2 = 425$ mm, and $L_3 = 50$ mm. $\alpha = 25°$. Determine the joint angles θ_1, θ_2, and θ_3.

Control resolution, accuracy, and repeatability

Q15.5 The range of one of the arms of a 3R articulated industrial robot is 60°. The controller used for this joint has a 12-bit storage capacity. The mean of the distribution of the mechanical errors is zero and the standard deviation is 0.05°. Determine: (a) the control resolution (CR_2); (b) the accuracy; and (c) the repeatability of the robot.

Further reading

[1] Craig, J.J. (1989) *Introduction to Robotics: Mechanics and Control* (2nd. ed.), Addison Wesley.

[2] Groover, M.P. (2001) *Automation, Production Systems, and Computer-integrated Manufacturing*, Prentice Hall.

[3] Kumar, S. (2003) *Development of a Mobile Robot with Obstacle Avoidance System*, MSc Thesis, University of the South Pacific.

[4] Onwubolu, G.C., Narayan, S. and Sharan, R.V. (2004) Development of a microcontroller-based pick and place robot for FMS application (awaiting publication).

[5] Reddy, H. *et al.* (2003) Development of obstacle avoidance mobile robot platform using a low-end budget microcontroller PIC, *Proceedings of the 10th. Electronics New Zealand Conference*, University of Waikato, Hamilton, NZ, pp. 59–64.

[6] Sikking, L. and Carnegie, D. (2003) The development of an indoor navigation algorithm for an autonomous mobile robot, *Proceedings of the 10th. Electronics New Zealand Conference*, University of Waikato, Hamilton, NZ, pp. 83–87.

[7] Suzuki, S. *et al.* (1991) How to describe the mobile robot's sensor-based behavior, *Robotics and Autonomous Systems*, **7**, 227–237.

[8] Tomizawa, T., Ohya, A. and Yuta, S. (2002) Book browsing system using an autonomous robot teleoperated via the Internet, *Proceedings of IROS'02*.

CHAPTER 16

Integrated circuit and printed circuit board manufacture

Chapter objectives

When you have finished this chapter you should be able to:

- understand the production of electronic grade silicon;
- understand the single crystal growing method using the Czochralski process;
- understand the shaping of silicon into wafers;
- understand the film deposition process;
- understand lithography, photo-masking, and the etching process;
- understand ion implantation in the silicon gate manufacturing process;
- appreciate IC packaging;
- have knowledge of the PCB manufacturing processes.

16.1 Integrated circuit fabrication

An integrated circuit (IC) is a collection of electronic devices such as transistors, diodes, and resistors that have been fabricated and electronically interconnected on a small flat chip of semiconductor material. The basic properties of semiconductors and different semiconductor devices have been described in previous chapters. Many good books (such as Groover, 2002) are available to describe the production of electronic grade silicon (EGS) which is used in the fabrication process of ICs. An outline of the process is shown in Figure 16.1.

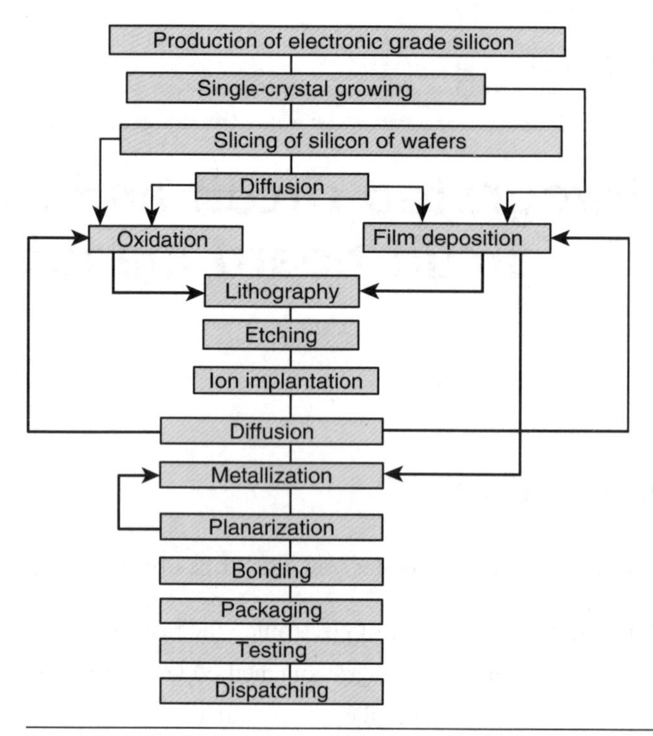

Figure 16.1 Fabrication process of integrated circuits.

16.1.1 Production of electronic grade silicon

Most ICs produced today are made from pure silicon. Silicon occurs naturally in the form of silicon dioxide, SiO_2, (in the form of quartzite), and this is the base material used in the production of pure silicon needed for a semiconductor.

Three steps are taken to produce electronic grade silicon (EGS):

1. The base material and a charge (consisting of coal, coke and wood chips) are heated in a submerged-electrode arc furnace to produce metallurgical grade silicon (MGS). MGS contains about 2% impurities. The MGS is then ground.
2. Hydrogen chloride is added to the MSG in a fluidized bed at temperatures around 300°C to form trichlorsilane ($SiHCl_3$) which is separated from the MSG by fractal distillation.
3. The trichlorsilane is then reduced to electronic grade silicon (EGS) using hydrogen.

16.1.2 **Growing a single crystal**

Semiconductors are fabricated from a wafer of silicon shaped as a disk. Wafers (varying from 75 mm to 150 mm in diameter, and less than 1 mm thick) are sliced from an ingot of single crystal silicon. The most widely used single-crystal growing method is the Czochralski process (Figure 16.2). Controlled amounts of impurities are added to molten polycrystalline silicon to achieve the required doping. Other contamination is avoided as this will adversely affect the silicon's electrical properties. A single silicon crystal ingot, 1 m long and 50–150 mm in diameter, can be fabricated using the Czochralski process.

The ingot is then shaped into a cylinder and sliced into wafers which are polished.

16.1.3 **Film deposition and oxidation**

Insulating and conducting films are used extensively in the fabrication of ICs. They are used for masking, for diffusion or implants, and for protection of the semiconductor surface. Film materials include poly-silicon, silicon nitride, silicon

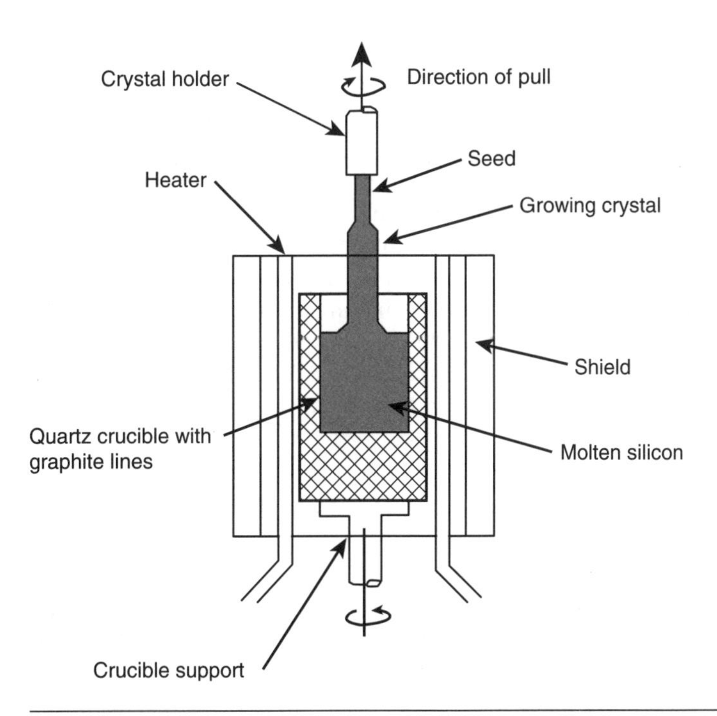

Figure 16.2 The Czochralski process. (Courtesy of Weste, N. and Eshraghian, K. (1988)).

dioxide, tungsten, titanium, and aluminum. A number of techniques are used to deposit the film, including evaporation, spattering, and chemical vapor deposition.

Silicon dioxide (SiO_2) is an insulator and since silicon surfaces have an extremely high affinity with oxygen, it can be easily produced on the surface of a silicon wafer.

16.1.4 Lithography

Lithography is the process by which the geometric patterns that define devices are transferred from a reticle to the silicon wafer surface (the substrate). Several lithographic techniques are used in semiconductor fabrication: photolithography, electron lithography, x-ray lithography, and ion lithography.

The reticle is the masking unit, and computer-aided design (CAD) techniques play a major role in its design and generation. The reticle is design at several times its final size and then reduced to the size required for the IC.

Lithography is used to achieve selective diffusion. To create different areas of silicon, containing different proportions of donors and acceptor impurities, further processing is needed. Lithography ensures that these areas are precisely placed and sized. The SiO_2 acts as a barrier (or mask) to doping impurities. Dopant atoms are able to pass into the wafer where there is an absence of SiO_2. Lithography is the process used for selectively removing the SiO_2. The SiO_2 is covered with an acid resistant coating except where diffusion windows are required. The SiO_2 is removed using an etching technique. The acid resistant coating is normally a photosensitive organic material known as photo-resist which can be polymerized by ultraviolet (UV) light. If the UV light is passed through a mask containing the desired pattern, the coating is polymerized where the pattern is to appear. The unpolymerized areas may then be removed with an organic solvent. Electron beam lithography (EBL) is used to avoid diffraction and achieve very high accuracy line widths.

Silicon can also be formed in an amorphous form commonly known as polycrystalline silicon or poly-silicon. This is used as an interconnection in ICs and as the gate electrode on MOS transistors. It has the ability to be used as a further mask to allow the precise definition of the source and drain electrodes.

16.1.5 Manufacturing issues

- **Wafer testing:** Several hundred ICs are produced on a wafer, and each is tested before the wafer is cut into separate devices. Computer-controlled equipment is used for these tests. The wafer is positioned on an X–Y moveable table beneath needle probes.
- **Chip separation:** A diamond-impregnated high precision saw is used to cut the wafer into individual devices.

- **Chip bonding:** This is the process of attaching an individual chip to its packaging to ensure reliability. Two commonly used methods are epoxy die bonding and eutectic die bonding.
- **Wire bonding:** Electrical connections are made between the contact pads on the chip surface and the package leads. Generally the connections are made using small diameter aluminum or gold wires. Gold has a better electrical conductivity than aluminum but is more expensive. Bonding techniques are ultrasonic bonding (with aluminum) and thermo-compression bonding (with gold).

16.1.6 IC packaging

After all the processing steps for the wafer have been completed, the wafer has to be transformed into individual chips or components. These steps are referred to as IC packaging, which includes the mechanical structure that holds and protects the circuitry.

IC packaging largely determines the overall cost of each completed IC, since the circuits are mass produced on a (EGS) wafer but then packaged individually. The chip size, number of external leads, operating conductors, heat dissipation, and power requirements all have to be taken into account when determining the package. There are two aspects normally considered in IC packaging: IC package design, and IC packaging processes.

Ceramics and plastics are used to encapsulate the IC chip. Plastic materials of two types: pre-molded packages and post-molded packages. In pre-molded packaging, the chip and lead-frame are connected to an enclosure base, which is molded prior to encapsulation. In post-molded packaging, an epoxy thermosetting plastic is transfer molded around the assembled chip and lead frame. The commonly used ceramic packaging material is alumina (Al_2O_3).

The advantages of using ceramics include:

- hermetic sealing;
- highly complex packages can be produced.

The disadvantages of using ceramics include:

- the high dielectric constant of alumina;
- poor dimensional control resulting from shrinking during heating.

The commonly used plastic packaging materials are epoxies, polyamides, and silicones.

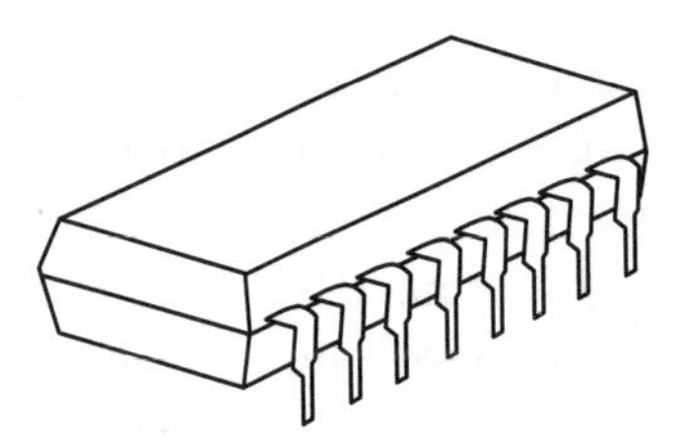

Figure 16.3 The IC dual-in-line package.

An advantage of using plastic is low cost. Disadvantages of using plastic include:

- no hermetic sealing;
- low reliability.

In nearly all practical applications, the IC is an important component in a larger electronic system. ICs must be attached to a PCB in one of two ways: pin-in-hole (PIH) technology in which the components have leads inserted through holes on the board and soldered on the underside, or surface-mount technology (SMT) in which the components are attached to the surface of the board. The dual in-line pack (DIP) (see Figure 16.3) is one of the most common packs for both PIH and SMT.

16.2 Printed circuit boards

A printed circuit board (PCB) is a laminated flat panel of insulator material (usually polymer composites reinforced with glass fabrics or paper) designed to provide the electrical interconnections between electronic components attached to it. The conducting paths are made of copper and are known as *tracks*. Other copper areas for attaching and electrically connecting components are also available and are called *lands*. Thin conducting paths on the surface of the board or sandwiched between layers of insulating material are used for interconnecting electronic components.

There are basically three principal types of PCB. They are

- the single-sided board, in which copper foil is only on one side of the insulating substrate;
- the double-sided board, in which copper foil is on both sides of the insulating substrate;
- the multilayer board, in which layers of copper foil and insulating substrates alternate.

The processes involved in PCB manufacture are

- production of starting boards;
- board preparation;
- circuit pattern imaging and etching;
- hole drilling;
- plating;
- testing;
- finishing.

16.2.1 Starting boards

Starting boards are the sheets of glass fiber and copper that are used to produce the final printed circuit. Copper foil is placed on one side (for single-sided PCBs), both sides (for double-sided PCBs), or in sandwiched layers within (for multi-layer PCBs) the glass fiber.

16.2.2 Board preparation

Starting boards are cut to fit the manufacturing equipment, and tooling holes are made for handling during the board fabrication. Alignment markers are particularly important in ensuring that holes later drilled through the board are coincident with the copper lands on both sides of and within the board's surface. Often the boards are bar-coded for identification. The board surface is thoroughly cleaned.

16.2.3 Circuit pattern imaging and etching

CAD packages are often used to produce the track layout of the PCB, although more basic techniques are possible for very simple board layouts. Photolithography is used to transfer the track pattern to the starting board. In this process the copper surfaces are covered with a light-sensitive resist. A photographic image of the track layout is placed and accurately aligned on the copper surface. The board and track image are then subjected to light exposing the photoresist. Dependent on the nature of the resist (positive or negative), the photographic image can be either a positive or negative image.

When the exposed board is immersed in a solution of etchant, the copper that is unprotected by the unexposed resist is removed leaving the track layout of circuit interconnections. The board is cleaned to remove the remaining photoresist.

16.2.4 Hole drilling

Once the tracks have been etched (and, in the case of multi-layer boards, the layers have been assembled) the boards need to be drilled for:

- component mounting holes (for non-surface mount device leads);
- plated-through holes (for when a conducting path is required to pass from one board layer to another);
- via holes (for when a wire connection is required to join a track on one board layer to a track on another layer);
- fastening holes (e.g. screw holes for physically mounting certain devices).

Computer numerically controlled (CNC) drills are often used to achieve the high accuracy required for these holes.

16.2.5 Plating

Holes that are to provide a conducting path through a board have to be plated.

16.2.6 Testing

Before a PCB is dispatched from the production line for use in electronic assembly, it must be tested to ensure that it meets the design specifications. Visual testing and continuity testing are two commonly used techniques.

16.2.7 Finishing

After testing, two finishing operations are necessary: the application of a thin solder layer on the track and land surfaces, and the application of a coating of solder resist to all areas of the board except the lands. The solder layer prevents oxidation of the copper. The solder resist ensures that the conducting solder only adheres to prescribed areas and so minimizes the risk of solder 'whiskers' causing erroneous connections.

16.2.8 Assembly

The assembly process is the act of positioning and fixing of electronic devices and other elements on the PCB. The steps are:

- **Placement:** In modern assembly plants most components are positioned and fixed automatically. If the component is not a surface mount device, the leads need to be preformed before they can be inserted into the board. After insertion, the leads are 'clinched' (to ensure the device is secure) and cropped before soldering.
- **Soldering:** Either hand or wave soldering can be used. Hand soldering has the advantages that it is localized, cheap, and has low energy consumption. Wave soldering, whereby the board is exposed to a wave of molten solder, is less labor intensive, and is favored for production lines. Re-flow soldering is used for surface mount devices.
- **Cleaning:** Contaminants that may cause degradation of the PCB assembly are removed.
- **Testing:** Further visual inspections are carried out at all assembly stages. After final assembly, electrical tests are performed to test specifications and test procedures to ensure that the electronic assembly functions correctly. These test may include 'burn-in' tests at elevated temperatures.
- **Reworking:** If a faulty board is found during testing, it may be possible to repair it (for instance, a broken track can be easily bridged).

16.2.9 Surface mounted devices

Surface mounted devices, unlike wired or 'legged' components, can be mounted on both sides of a PCB, and with a much higher packing density. However, the components are more expensive than conventional devices, are less easy to handle (due to their small size), and not all devices are available in the surface mount configuration.

Further reading

[1] Edwards, P.R. (1999) *Manufacturing Technology in the Electronics Industry*, London: Chapman & Hall.
[2] Groover, M.P. (2002) *Fundamentals of Modern Manufacturing: Materials, Processes, and Systems*, New York: John Wiley.
[3] Van Zant, P. (2000) *Microchip Fabrication*, New York: McGraw-Hill.
[4] Weste, N. and Eshraghian, K. (1988) *Principles of CMOS VLSI Design: A Systems Perspective*, Addison Wesley.

CHAPTER 17

Reliability

Chapter objectives

When you have finished this chapter you should be able to:

- understand the principles of reliability;
- understand how to deal with the reliability of series systems;
- understand how to deal with the reliability of parallel systems;
- understand how to deal with the reliability of generic series-parallel systems;
- understand how to deal with the reliability of major parallel systems;
- understand how to deal with the reliability of standby systems;
- appreciate common modes of failure;
- understand availability of systems with repair;
- understand the factors influencing failure rate;
- understand the practical applications of response surface methodology.

17.1 The meaning of reliability

The main difference between the quality of a device and the reliability of a device is that reliability involves a time factor. In a quality problem, the question may be asked: What is the probability of one defective device or one failure in a sample of ten parts? The parts are either good or defective at the time that they are examined. In a reliability problem, the question may be: What is the probability that the device will work for 100 hours without a failure?

Reliability is defined as the probability that a device continues to perform its intended function under given operating conditions and environments for a specified length of time. If however, as time goes on the device is not able to perform its intended function then it is no longer reliable; it has failed. The unreliability, F, of a device is defined as the probability that the device fails to perform its intended function under given operating conditions and environments for a specified length of time. Both reliability and unreliability vary with time. Reliability, $R(t)$, decreases with time, while unreliability, $F(t)$, increases with time and they are complementary:

$$R(t) + F(t) = 1. \tag{17.1}$$

The most basic method of achieving product reliability is through mature design. On new products, failure rates are determined under accelerated conditions and used to make reliability predictions. In complex assemblies, there may be hundreds of individual components that affect the reliability of the final product. Ideally, 100 percent reliability is desirable but that is not always possible to achieve. In products that affect human life, a high degree of reliability is absolutely necessary. These products have high quality components and are tested under extreme conditions. The reliability of a product, whether it is an airplane or a computer, is dependent on the quality of its components.

17.2 The life curve

Over many years, and across a wide variety of mechanical and electronic components and systems, people have calculated empirical population failure rates as units deteriorate over time. They repeatedly obtain a graph known as the *bath tub* curve (Figure 17.1). As can be seen, there are three phases in the life of a product.

The initial region that begins at time zero when a customer first begins to use the product is characterized by a high but rapidly decreasing failure rate. This region is known as the *early failure* or *burn in* period (phase 1). This decreasing failure rate typically lasts several weeks to a few months. During the early life failure or burn-in stage of a device, failures occur more frequently than during the operating or useful life phase which follows.

Next, the failure rate levels off and remains roughly constant for (hopefully) the majority of the useful life of the product. This long period of a level failure rate is known as the *useful life* or *stable failure* period. The constant failure rate level is called the *useful life rate* (phase 2). Most systems spend most of their life operating in this flat portion of the curve.

During the last part of the life of a device, the wear out phase (phase 3), the frequency of failure is again high and rises rapidly.

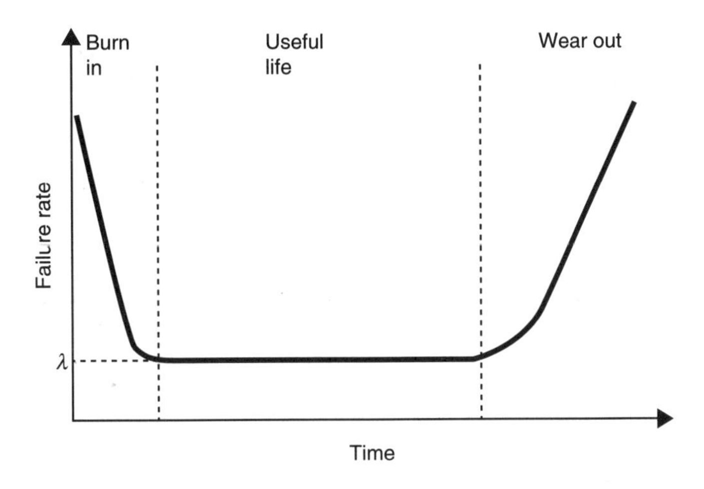

Figure 17.1 The reliability 'bath tub' curve.

Reliability calculations for a product or device can only be made in phase 2 as in phases 1 and 2 there is too much variation in the failure rate to make reliability predictions. A product usually only reaches a customer or end user after the initial problems (early failures) have occurred.

Failure rates and the subsequent reliability of devices are usually determined by a procedure called life testing. Life testing is the process of placing a device or unit of product under a specified set of test conditions and measuring the time it takes until failure. Life-test sampling plans are used to specify the number of units that are to be tested and for determining acceptability. The procedures for developing and using a life-test sampling plan are almost the same as those used for acceptance sampling. The producer and consumer's risks are specified, and an operating characteristic (OC) curve may be developed. The exponential distribution is used to find the probability of acceptance.

The following sections will define some of the concepts, terms, and models we need to describe, estimate and predict reliability.

17.3 Repairable and non-repairable systems

17.3.1 A non-repairable system

A non-repairable population is one for which individual items that fail are removed permanently from the population. While the system may be repaired by replacing failed units from either a similar or a different population, the members of the original population dwindle over time until all have eventually failed.

Defining T_1 as the survival time (or up time) for the ith failure, N as the number of items that fail during the period T, then the mean time to failure (MTTF) can be expressed as:

$$\text{Mean time to fail (MTTF)} = \frac{\text{total up time}}{\text{number of failures}} = \frac{1}{N}\sum_{i=1}^{N} T_i = \int_0^{\infty} R(t)\,\mathrm{d}t. \quad (17.2)$$

The mean time between failures (MTBF) is the average length of life of the devices being tested. It is the reciprocal of the failure rate.

$$\text{Mean failure rate (MFR)} = \frac{\text{number of failures}}{\text{total up time}} = \bar{\lambda} = \frac{N}{\sum_{i=1}^{N} T_i}. \quad (17.3)$$

If 1000 parts are placed on test and 20 failures are recorded between the sixth and seventh hour, then the failure rate, λ, is 20/1000 or 0.02 failures per hour. The mean time between failure for this example is $1/\lambda$, which is 1/0.02 or 50 hours.

17.3.2 A repairable system

A repairable system is one, which can be restored to satisfactory operation by any action, including part replacement or changes to adjustable settings. When discussing the rate at which failures occur during system operation time (and are then repaired) we will define a *rate of occurrence of failure* (ROCF) or *repair rate.* It is incorrect to talk about failure rates or hazard rates for repairable systems, as these terms apply only to the first failure times for a population of non-repairable components.

If there are N items or repairable products observed over a test interval, T, during which the down time, $T_{\mathrm{D}j}$, is associated with the jth failure, and there are N_{f} failures, then we have:

$$\text{Mean down time (MDT)} = \frac{\text{total down time}}{\text{number of failures}} = \frac{1}{N_{\mathrm{f}}}\sum_{i=1}^{N_{\mathrm{f}}} T_{\mathrm{Di}} \quad (17.4)$$

$$\text{Total up time (TUT)} = NT - \sum_{i=1}^{N_{\mathrm{f}}} T_{\mathrm{Di}} = NT - N_{\mathrm{f}}\text{MDT}. \quad (17.5)$$

The mean time between failures (MTBF) is defined as

$$\text{MTBF} = \frac{\text{total up time}}{\text{number of failures}} = \frac{NT - N_\text{f}\text{MDT}}{N_\text{f}}. \qquad (17.6)$$

The mean failure rate (MFR) is defined as

$$\text{MFR} = \frac{\text{number of failures}}{\text{total up time}} = \bar{\lambda} = \frac{N_\text{f}}{NT - N_\text{f}\text{MDT}}. \qquad (17.7)$$

17.3.3 Availability and unavailability

The availability of a device, A, is the fraction of the total test interval that it is performing its intended function under given operating conditions and environments. It is defined as:

$$\text{Availability} = \frac{\text{total up time}}{\text{test interval}} = A = \frac{\text{MTBF}}{\text{MTBF} + \text{MDT}} \qquad (17.8)$$

The unavailability of a device, U, is defined as:

$$\text{Unavailability} = \frac{\text{total down time}}{\text{test interval}} = U = \frac{\text{MDT}}{\text{MTBF} + \text{MDT}} \qquad (17.9)$$

Consequently,

$$A + U = 1. \qquad (17.10)$$

17.4 Failure or hazard rate models

A failure or hazard rate model can be any probability density function (PDF), $f(t)$, defined over the range of time from $t=0$ to $t=\infty$. While the bath tub curve represents the most general form of the time variation in instantaneous failure rate, simpler forms of $\lambda(t)$ satisfactorily represent the failure of many products and components. The most commonly used are the constant failure rate and the Weibull failure rate models.

17.4.1 The constant failure rate model

A study of the failure rate of a large range of aerospace industry products and components showed that 68 percent of all the products could be represented by $\lambda(t)$ characterized by a short and early failure region, and extended constant failure region and no wear-out region on the bath tub curve. Many electronic devices fall into this category.

The use of early failure region can either be reduced or eliminated using effective quality control techniques such as *burn in*. The high failure rate during the burn-in period accounts for parts with slight manufacturing defects not found during the manufacturer's testing. The constant failure rate during the useful life (phase 2) of a device is represented by λ. The failure rate is defined as the number of failures per unit time or the proportion of the sampled units that fail before some specified time.

$$\text{Failure rate} = \lambda = \frac{f}{n}, \tag{17.11}$$

where f is the total failures during a given time interval and n is the number of units or items placed on test. This means a constant failure rate model will be adequate for a large range of products and components:

$$\lambda(t) = \lambda(\xi) = \lambda = \text{constant} \tag{17.12}$$

$$R(t) = \exp\left(-\int_0^t \lambda(\xi)\,\mathrm{d}\xi\right) = \exp\left(-\lambda\int_0^t \mathrm{d}\xi\right) = \exp(-\lambda t) \tag{17.13}$$

and

$$F(t) = 1 - \exp(-\lambda t) \tag{17.14}$$

where $F(t)$ is the failure rate.

Alternatively, the exponential distribution formula is used to compute the reliability of a device or a system of devices in the useful life phase. The exponential formula has its roots in the Poisson formula. Instead of np, the product λt is used. The exponential is the Poisson formula with $x=0$. Reliability being the probability of zero failures in the specified time interval.

$$P(t) = \frac{\mathrm{e}^{-np}(np)^x}{x!} = \frac{\mathrm{e}^{-\lambda t}(\lambda t)^x}{x!}. \tag{17.15}$$

For $x=0$,

$$\text{Reliability } R(t) = P(0) = \mathrm{e}^{-\lambda t}. \tag{17.16}$$

Hence the mean time to failure is defined as

$$\text{MTTF} = \int_0^t R(t)\,\mathrm{d}t = \int_0^t \exp(-\lambda t)\,\mathrm{d}t = \left[\frac{-1}{\lambda}\exp(-\lambda t)\right]_0^\infty = \frac{-1}{\lambda}[0-1] = \frac{1}{\lambda}. \tag{17.17}$$

This means that MTTF is the reciprocal of failure rate in the constant failure rate case.

17.4.2 The Weibull failure rate model

The Weibull hazard rate function is given as

$$\lambda(t) = \frac{\beta}{\eta}\left(\frac{t - t_0}{\eta}\right)^{\beta-1} \quad \beta > 0, \tag{17.18}$$

where t_0 is the position of the origin, η is the scale parameter and β is the shape parameter.

The reliability time variation or distribution is given as

$$\begin{aligned} R(t) &= \exp\left\{-\int_{t_0}^{t} \lambda(\xi)\mathrm{d}\xi\right\} = \exp\left\{-\frac{\beta}{\eta^\beta}\right\}\int_{t_0}^{t}(\xi - t_0)^{\beta-1}\mathrm{d}\xi \\ &= \exp\left\{-\frac{\beta}{\eta^\beta}\left[\frac{1}{\beta}(\xi - t_0)^\beta\right]_{t_0}^{t}\right\} = \exp\left\{-\frac{1}{\eta^\beta}\left[(t - t_0)^\beta - 0\right]\right\}, \end{aligned} \tag{17.18A}$$

leading to

$$R(t) = \exp\left\{-\left(\frac{\xi - t_0}{\eta}\right)^\beta\right\}. \tag{17.19}$$

EXAMPLE 17.1

A sample of 450 devices was tested for 30 hours and five failures were recorded. The device is designed to operate for 1000 hours without failure. What is the reliability of the tested device?

Solution

$$\text{Failure rate} = \lambda = 5/(450)(30) = 5/13{,}500 = 0.0003704$$

$$\text{Reliability} = \mathrm{e}^{-\lambda t} = \mathrm{e}^{-(0.0003704)(1000)} = \mathrm{e}^{-0.3704} = 0.6905$$

The probability of a device operating for 1000 hours without a failure is 69.05%.

17.5 Reliability systems

Reliability theory developed apart from the mainstream of probability and statistics, and was used primarily as a tool to help nineteenth-century maritime and life insurance companies compute profitable rates to charge their customers. Even today, the terms *failure rate* and *hazard rate* are often used interchangeably. Once research began, scientists rapidly developed models in an attempt to explain their observations on reliability. Mathematically, these models can be broken down into two classes: series reliability and parallel reliability. More complex models can be built by combining the two basic elements of a reliability model. In a simple, parallel configuration, the system will work if at least one device works. The reliability calculations for these systems are an extension of basic probability concepts. There are other configurations in addition to the two basic systems, such as standby systems, switched systems, and combinations of each.

17.5.1 A series system

Many systems perform a task by having a single component perform a small part of it, and pass its result to another component in a serial fashion as shown in Figure 17.2. The new component then performs a small piece of the task, and continues passing it along, until the task is completed. This is how people typically write programs, and design hardware devices. It is exceedingly difficult and expensive to build a reliable series system.

For example, if one were to build a serial system with 100 components, each of which had a reliability of 0.999, the overall system reliability would be $0.999^{100} = 0.905$.

The reliability of series systems is modeled using the following equation:

$$R(t) = \prod_{i=1}^{N} R_i(t). \qquad (17.20)$$

Hence for a series system,

$$R_i = e^{-\lambda_i t} \qquad (17.21)$$

and

$$R_{\text{SYSTEM}} = e^{-\lambda_1 t} \times e^{-\lambda_2 t} \times e^{-\lambda_3 t} \ \ldots\ e^{-\lambda_I t} \ldots e^{-\lambda_M t}, \qquad (17.22)$$

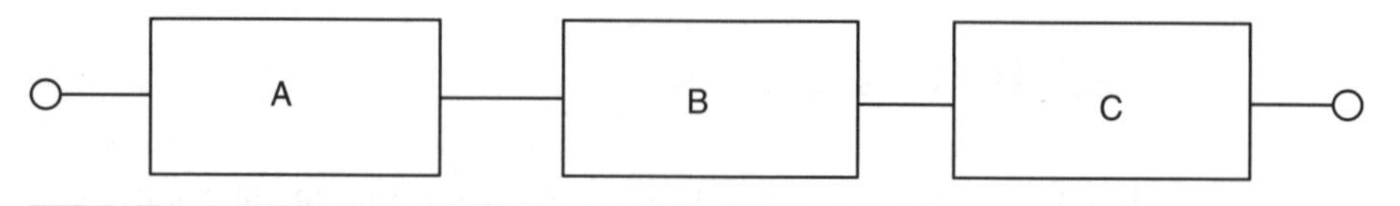

Figure 17.2 A series reliability system.

so that if $\lambda_{\mathrm{SYSTEM}}$ is the overall system failure rate,

$$R_{\mathrm{SYSTEM}} = \mathrm{e}^{-\lambda_{\mathrm{SYSTEM}1}t} = \mathrm{e}^{-(\lambda_1+\lambda_2+\lambda_3+\cdots+\lambda_I+\cdots+\lambda_M)t}$$
$$\lambda_{\mathrm{SYSTEM}} = \lambda_1 + \lambda_2 + \lambda_3 + \cdots + \lambda_i + \cdots + \lambda_m. \tag{17.23}$$

17.5.2 A parallel system

Figure 17.3 shows a system consisting of n (=2 for this example) individual elements in parallel.

The concept of redundancy can boost reliability. In the simplest case only one of the redundant components must be working to maintain the system's level of service. This is characterized by the following equation:

$$R(t) = 1 - \prod_{i=1}^{N}[1 - R_i(t)]. \tag{17.24}$$

All of the elements or systems are expected to be functioning properly. However, only one element/system is necessary to meet these requirements, the remainder merely increasing the reliability of the overall system (active redundancy). The important point to note here is that the overall system will only fail if every element/system fails. If one element/system survives the overall system survives.

Assuming that the reliability, f, of each element/system is independent of the reliability of the other elements, then the probability that the overall system fails is the probability that element/system 1 fails *and* element/system 2 fails

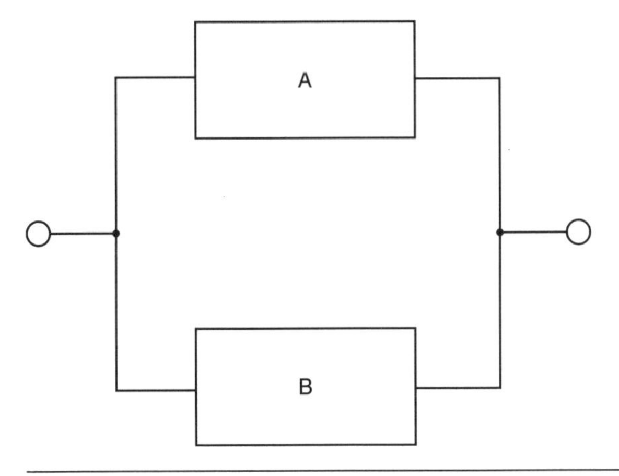

Figure 17.3 A parallel reliability system.

and element/system 2 fails, etc. The overall system unreliability is the product of the individual element/system unreliability:

$$F_{\text{SYSTEM}} = F_1 F_2 F_3 \ldots F_i \ldots F_m. \tag{17.25}$$

This means that the overall failure rate for a system is the product of the individual element or component failure rate.

Consider a system built with four identical modules. The system will operate correctly provided at least one module is operational. If the reliability of each module is 0.95, then the overall system reliability is: $1-(1-0.95)^4=$ 0.99999375.

In this way, a reliable system can be built despite the unreliability of its component parts, though the cost of such parallelism can be high.

17.5.3 A combination system

Figure 17.4 shows a combination system of components. This system consists of n (=2 for this example) identical subsystems in parallel and each subsystem consists of m (=2 for this example) elements in series. Models of more complex systems may be built by combining the simple serial and parallel reliability models.

In this case, R_{ji} is the reliability of the i element in the j subsystem, and

$$R_j = \prod_{i=1}^{M} R_{ji}. \tag{17.26}$$

The unavailability of the jth subsystem is:

$$F_j = 1 - \prod_{i=1}^{M} R_{ji}. \tag{17.27}$$

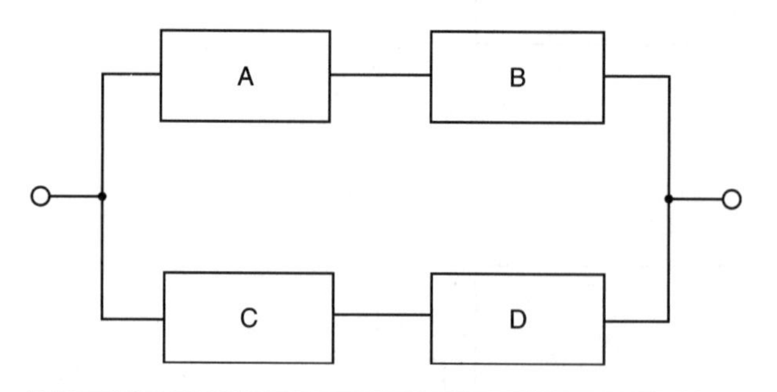

Figure 17.4 A series-parallel combination reliability system.

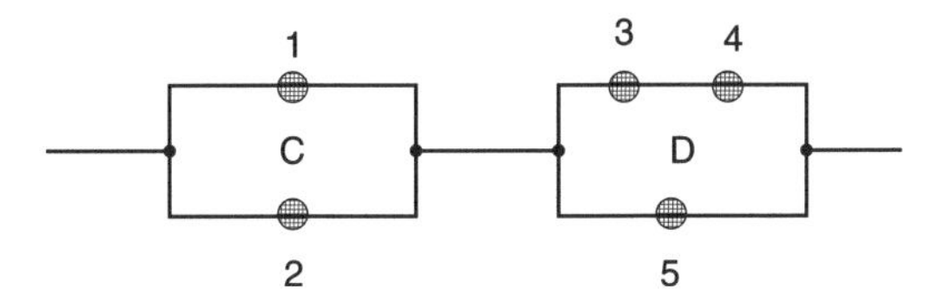

Figure 17.5 Another series-parallel reliability system.

The overall system unavailability is

$$F_{\text{OVERALL}} = \prod_{j=1}^{N} \left[1 - \prod_{i=1}^{M} R_{ji} \right]. \tag{17.28}$$

We first introduce a new term, which is useful when considering such combinational systems: the minimal path set.

A minimal path set is the smallest set of components whose functioning ensures the functioning of the system (see Ross, 1997). In Figure 17.5, the minimal path set is {1, 3, 4} {2, 3, 4} {1, 5} {2, 5}.

The total reliability of the system can be abstracted as the parallel reliability of the first half in series with the parallel reliability of the second half. For example, given that $R1 = 0.9$, $R2 = 0.9$, $R3 = 0.99$, $R4 = 0.99$, and $R5 = 0.87$, then $R(t) = [1 - (1 - 0.9)(1 - 0.9)][1 - (1 - 0.87)(1 - (0.99 \times 0.99))] = 0.987$.

Such a system might be built if system component 5 was extremely fast, but not very reliable. Components 3 and 4 are reliable but slow. So the system can race along using 5 until it fails. System service degrades until 5 can be reset and reintegrated into the system.

17.5.4 *N*-version modular redundancy

One of the classic, but expensive, methods for improving reliability, *n*-version modular redundancy, can be very effective when implemented correctly. Typically only the most mission critical systems, such as are found in the aerospace industry, will employ *n*-version modular redundancy.

The fundamental idea behind *n*-version modular redundancy is that of parallel reliability. These systems can also compensate for correctness issues stemming from faults injected during the design and specification phases of a project. The independent modules all perform the same task in parallel, and then use some voting scheme to determine what the correct answer is. This voting overhead means that *n*-modular redundant systems can only approach the theoretical limit of reliability for a fully parallel reliable system.

The reliability of an n-modular redundant system can be mathematically described as follows:

$$R_{M\ of\ N}(t) = \sum_{i=0}^{N-M} {}^{n}C_i R_m{}^{N-i}(t)[1 - R_m(t)]^i, \tag{17.29}$$

where

$${}^{n}C_i = \frac{N!}{(N-i)!i!}, \tag{17.29A}$$

N is the number of redundant modules, and M is the minimum number of modules required to be functioning correctly, disregarding voting arrangements.

EXAMPLE 17.2

Consider a five-module system requiring three correct modules, each with a reliability of 0.95. What is the reliability of the system?

Solution

$$R_{3\ of\ 5}(t) = {}^{5}C_0 R_m{}^5(t) + {}^{5}C_1 R_m{}^4(t)[1 - R_m(t)] + {}^{5}C_2 R_m{}^3(t)[1 - R_m(t)]^2. \tag{17.29B}$$

Expanding, we have

$$\begin{aligned} R_{3\ of\ 5}(t) &= R_m{}^5(t) + 5R_m{}^4(t)[1 - R_m(t)] + 10R_m{}^3(t)[1 - R_m(t)]^2 \\ &= R_m{}^5(t) + 5R_m{}^4(t) - 5R_m{}^5(t) + 10R_m{}^3(t)\left[1 - 2R_m(t) + R_m{}^2(t)\right] \\ &= R_m{}^5(t) + 5R_m{}^4(t) - 5R_m{}^5(t) + 10R_m{}^3(t) - 20R_m{}^4(t) + 10R_m{}^5(t) \\ &= 10R_m{}^3(t) - 15R_m{}^4(t) + 6R_m{}^5(t) \\ &= 10(0.95)^3 - 15(0.95)^4 + 6(0.95)^5 \\ &= 0.9988 \end{aligned} \tag{17.29C}$$

EXAMPLE 17.3

What is the reliability of a two element series system (Figure 17.6) in which $\lambda_A = 0.001$, $\lambda_B = 0.002$ and the mission time, $t = 50$ hours?

Solution

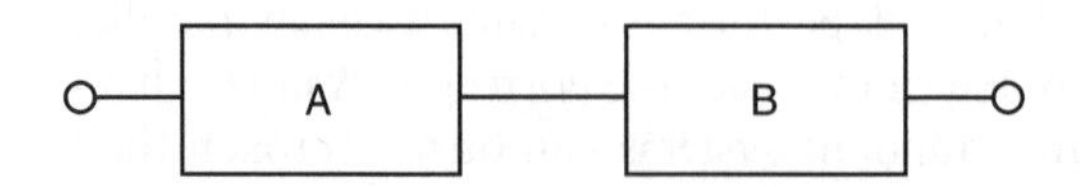

Figure 17.6 A series reliability system for Example 17.3.

The system elements are in series, so for the system to work, both devices must work. If one device fails, the system fails.

Let R_A be the reliability of device A, i.e. the probability that device A will work for at least 50 hours.

$$R_A = e^{-\lambda_A t} = e^{-(.001)(50)} = 0.9512. \tag{17.29D}$$

Let R_B be the reliability of device B, i.e. the probability that device B will work for at least 50 hours.

$$R_B = e^{-\lambda_B t} = e^{-(.002)(50)} = 0.9048. \tag{17.29E}$$

The reliability of the series system, R_S (the probability that the system will work for at least 50 hours)$= R_A \times R_B = 0.9512 \times 0.9048 = 0.860$.

EXAMPLE 17.4

What would be the reliability if the system elements of Example 17.3 were connected in parallel and the mission time remained 50 hours?

Solution

For the system to work, one or both devices must work. The system will fail when both devices fail.

From example 17.3, $R_A = 0.9512$ and $R_B = 0.9048$.

Let the probability that device A fails be $P_A = 1 - R_A$ and the probability that device B fails, $P_B = 1 - R_B$.

Then, the parallel system reliability,

$$R_P = R_A P_B + P_A R_B + R_A R_B.$$

$$R_P = (0.9512)(0.0952) + (0.0489)(0.9048) + (0.9512)(0.9048) = 0.9954.$$

An alternative solution is found using $R_P = 1 - P_A P_B$.

17.6 Response surface modeling

Response surface modeling (RSM) is a technique to determine and represent the cause and effect relationship between true mean responses and input control variables influencing the responses as a two- or three-dimensional hyper surface. In other words, RSM is the procedure for determining the relationship between various parameters with various operating criteria and exploring the effect of these process parameters on the coupled responses. Workers in mechatronics need to understand the concept and application of RSM in order to deliver products that will perform as specified during operation.

The steps involved in the RSM technique are:

1. Design a set of experiments for adequate and reliable measurement of the true mean response of interest.
2. Determine the mathematical model with best fits.
3. Find the optimum set of experimental factors that produces a maximum or minimum value of response
4. Represent the direct and interactive effects of process variables on the best parameters through two-dimensional and three-dimensional graphs. The accuracy and effectiveness of an experimental program depends on careful planning and execution of the experimental procedure.

Given the equation

$$y_u = \phi(x_{1u}, x_{2u}, \ldots, x_{ku}) + \varepsilon_u, \tag{17.30}$$

where $u = 1, 2, 3, \ldots, n$ and represents the number of N observations in the factorial experiment, and x_{iu} represents the level of the ith factor in the uth experiment. The function φ is called the response surface. The residual ε_u measures the experimental error in the uth observation.

Polynomial curve fitting equations normally exist in *first* degree and *second* degree. They are also referred to as first-order or second-order polynomials. The first-order polynomials have the form

$$y_u = \beta_0 + \beta_1 x_{1u} + \beta_2 x_{2u} + \cdots + \beta_k x_{ku} + \varepsilon_u. \tag{17.31}$$

EXAMPLE 17.5

The overall objective is to investigate the axial force and torque, the material removal rate, and power consumption during operation of a CNC drilling machine.

The aluminum workpiece is 2.5 mm thick. The drill speeds are 270 rpm, 1135 rpm, and 2000 rpm, with feeds rates of 1.3 mm s^{-1}, 1.9 mm s^{-1}, and 3.8 mm s^{-1}. Drill bit sizes are 2.5 mm, 3.5 mm and 5.0 mm.

Trial runs were carried out by varying one of the process parameters whilst keeping the rest of them at constant values. The upper limit of a factor was coded as +1 and the lower limit as −1. The selected process parameters with their limits units and notations are given in Table 17.1. The response behavior of the CNC drilling machine is to be technically discussed.

Table 17.1 Process control parameters and their limits

Parameters	*Units*	*Notation*	*Limits*		
			−1	*0*	*+1*
Spindle speed	rev min^{-1}	N	270	1135	2000
Feed rate	mm s^{-1}	f	1.3	1.9	3.8
Drill bit diameter	mm	d	2.5	3.5	5

Solution

Investigation plan

1. Identifying the important process control variable.
2. Finding the upper and lower limits of the control variables, (namely drill speed (N), drill feed rate (f), and drill bit diameter (d)).
3. Develop the design matrix.
4. Conduct the experiments as per the design matrix.
5. Record the responses, (namely drilling force (P), and drilling torque (D)).
6. Develop mathematical models.
7. Calculate the coefficients of the polynomials.
8. Check the adequacy of the models developed.
9. Test the significance of the regression coefficients, recalculating the value of the significant coefficients and arriving at the final mathematical models.
10. Present the main effects and the significant interaction effects of the process parameters on the responses in two- and three-dimensional (contour) graphical form.
11. Analyze the results.

The CNC drilling machine is based on a second-order design. This is more specifically known as the quadratic response surface and its basic form for 2-x variables is given as

$$y_u = \beta_0 + \beta_1 x_{1u} + \beta_2 x_{2u} + \beta_{11} x_{1u}^2 + \beta_{22} x_{2u}^2 + \beta_{12} x_{1u} x_{2u} + \varepsilon_u. \tag{17.32}$$

The surface contains linear terms, squared terms and cross product terms. In order to make designs easier to deal with the criterion of rotatability is used.

Table 17.2 Components for rotatable designs

No. of x-variables k	Total N	Value of α
2	13	1.414
3	20	1.682
4	31	2
5	32	2
6	53	2.378

Let point (0, 0, ..., 0) represent the center of the region in which the relation between y and x is under investigation.

$$\hat{y}_u = b_0 + \sum_{i=1}^{k} b_i x_{iu} + \sum_{i=1}^{k} b_{ii} x_{iu} + \sum_{i<j}^{k} b_{ij} x_{iu} x_{ju}. \quad (17.33)$$

where $\hat{y}$ is the estimated response at a point fitted on a second-order surface. Rotatable designs for any number of k variables can be built up from components shown in Table 17.2.

There are three x variables, x_1, x_2, and x_3. Hence there are 20 experimental observations and the α value is 1.682.

The following independently controllable process parameters are identified to carry out the experiments: drill speed (N), drill feed rate (f), and drill bit diameter (d).

The mathematical relationship for correlating P and the considered process variables is

$$\begin{aligned} Y_{U(P)} = {} & 51.6045 - 6.7395N + 9.0524f + 11.9810d \\ & + 2.4695N^2 + 1.5858f^2 - 0.1815d^2 \\ & - 1.25Nf + 3.75Nd - 1.2fd. \end{aligned} \quad (17.34)$$

The mathematical relationship for correlating T and the considered process variables is

$$\begin{aligned} Y_{u(T)} = {} & 1.2258 + 0.01246N - 0.03647f - 0.04788d \\ & - 0.1785N^2 - 0.1785f^2 - 0.1919d^2 \\ & - 0.0908Nf + 0.0317Nd + 0.1542fd. \end{aligned} \quad (17.35)$$

Equations 17.34 and 17.35, therefore, represent the relation correlating P and T with the considered process variables. The observed process-parameters values are shown in Table 17.3.

Table 17.3 Design matrix and observed values of process parameters

S. no	*Design matrix*										*P* (N)	*T* (Nm)
1	1	−1	−1	−1	1	1	1	1	1	1	35	0.98
2	1	1	−1	−1	1	1	1	−1	−1	1	20	0.98
3	1	−1	1	−1	1	1	1	−1	1	−1	65	0.98
4	1	1	1	−1	1	1	1	1	−1	−1	40	0.49
5	1	−1	−1	1	1	1	1	1	−1	−1	65	0.49
6	1	1	−1	1	1	1	1	−1	1	−1	60	0.49
7	1	−1	1	1	1	1	1	−1	−1	1	85	0.98
8	1	1	1	1	1	1	1	1	1	1	80	0.78
9	1	−1.682	0	0	2.829	0	0	0	0	0	60	0.49
10	1	1.682	0	0	2.829	0	0	0	0	0	45	0.98
11	1	0	−1.682	0	0	2.829	0	0	0	0	45	0.98
12	1	0	1.682	0	0	2.829	0	0	0	0	65	0.49
13	1	0	0	−1.682	0	0	2.829	0	0	0	40	0.98
14	1	0	0	1.682	0	0	2.829	0	0	0	60	0.98
15	1	0	0	0	0	0	0	0	0	0	40	1.47
16	1	0	0	0	0	0	0	0	0	0	45	0.49
17	1	0	0	0	0	0	0	0	0	0	60	1.96
18	1	0	0	0	0	0	0	0	0	0	40	0.98
19	1	0	0	0	0	0	0	0	0	0	60	0.49
20	1	0	0	0	0	0	0	0	0	0	65	1.96

A spreadsheet package can be used to calculate the values of these coefficients for different responses (Table 17.4), The spreadsheet solutions are shown in Tables 17.5 and 17.6.

Table 17.3 shows the data for four x variables and the corresponding y values for axial load (y_P) and torque (y_T). For more information on, and application of, RSM to practical problems, readers are referred to Chapter 20.

Table 17.4 Estimated values of the model coefficients

Specimen number	*Coefficient*	*Axial Load (N)*	*Load torque (N-m)*
1	b_0	51.60452	1.225812
2	b_1	−6.73952	0.012464
3	b_2	9.052406	−0.03647
4	b_3	11.98104	−0.04788
5	b_{11}	2.469493	−0.17853
6	b_{22}	1.585827	−0.17853
7	b_{33}	−0.1815	−0.09193
8	b_{12}	−1.25	−0.09076
9	b_{13}	3.75	0.031743
10	b_{23}	−1.25	0.154243

Table 17.5 Spreadsheet of axial load solutions

	Coefficients	*Standard Error*	*t Stat*	*P-value*	*Lower 95%*	*Upper 95%*	*Lower 95.0%*	*Upper 95.0%*
Intercept	51.60452	4.292283	12.02263	2.87E−07	42.04072	61.16833	42.04072	61.16833
−1	−6.73952	2.847664	−2.36668	0.0395	−13.0845	−0.39453	−13.0845	−0.39453
−1	9.052406	2.847664	3.178889	0.009838	2.707415	17.3974	2.707415	17.3974
−1	11.98104	2.847664	4.207322	0.001808	5.636048	18.32603	5.636048	18.32603
1	2.469493	2.771755	0.890949	0.393888	−3.70636	8.645349	−3.70636	8.645349
1	1.585827	2.771755	0.572138	0.579862	−4.59003	7.761683	−4.59003	7.761683
1	−0.1815	2.771755	−0.06548	0.94908	−6.35736	5.994351	−6.35736	5.994351
1	−1.25	3.720842	−0.33595	0.743854	−9.54055	7.040553	−9.54055	7.040553
1	3.75	3.720842	1.007836	0.337297	−4.54055	12.04055	−4.54055	12.04055
1	−1.25	3.720842	−0.33595	0.743854	−9.54055	7.040553	−9.54055	7.040553

Table 17.6 Spreadsheet of torque solutions

	Coefficients	*Standard Error*	*t Stat*	*P-value*	*Lower 95%*	*Upper 95%*	*Lower 95.0%*	*Upper 95.0%*
Intercept	1.225812	0.219456	5.585694	0.00034	0.729369	1.722256	0.729369	1.722256
−1	0.012464	0.160903	0.077461	0.939952	−0.35152	0.376451	−0.35152	0.376451
−1	−0.03647	0.160903	−0.22666	0.825749	−0.40046	0.327516	−0.40046	0.327516
−1	−0.04788	0.160903	−0.29757	0.772788	−0.41187	0.316108	−0.41187	0.316108
1	−0.17853	0.143827	−1.24126	0.245887	−0.50389	0.146833	−0.50389	0.146833
1	−0.17853	0.143827	−1.24126	0.245887	−0.50389	0.146833	−0.50389	0.146833
1	−0.09193	0.143827	−0.63915	0.538649	−0.41729	0.233433	−0.41729	0.233433
1	−0.09076	0.223324	−0.40639	0.69395	−0.59595	0.414438	−0.59595	0.414438
1	0.031743	0.223324	0.142141	0.8901	−0.47345	0.536938	−0.47345	0.536938
1	0.154243	0.223324	0.690671	0.507192	−0.35095	0.659438	−0.35095	0.659438

Problems

Q17.1 A reliability test was carried out on 25 non-repairable light emitter diodes (LEDs). The times at which failures occurred (in units of 10^3 hours) were as follows:

0.5, 1.0, 1.5, 1.3, 2.0,
2.5, 3.0, 3.5, 3.5, 4.0,
4.5, 5.0, 5.5, 5.5, 6.0,
6.5, 7.0, 7.0, 7.5, 8.0,
8.1, 8.6, 9.0, 9.5, 10.0.

(a) Use the data to determine: (i) the mean time to failure; and (ii) the mean failure rate.

(b) Use the equation $R(t) = 1 - 10^{-4t}$, which models the reliability of the LEDs, to estimate: (i) the mean time to failure; and (ii) the mean failure rate. Hint: use $t = 10^4$ hours as upper limit for integration.

Q17.2 A batch of 100 repairable computers was tested over a 12-month period. Records show that there were 20 failures within the period and the corresponding down times in hours were as follows:

5, 5, 7, 8, 5, 7, 8, 10, 5, 5,
8, 5, 5, 5, 6, 5, 4, 8, 8, 5.

Determine: (a) the mean down time; (b) the mean time between failures; (c) the mean failure rate; and (d) the availability.

Q17.3 An electric motor has a constant failure rate of 0.1 per annum. Determine: (a) the reliability; and (b) the unreliability after 1, 2, 3, 4, and 5 years. What deduction can be made on how reliability relates to time?

Q17.4 A rural water supply system consists of a solar panel ($\lambda = 0.6$), a power unit ($\lambda = 0.8$), a pump ($\lambda = 0.08$), and tank ($\lambda = 0.08$) connected in series. Determine the probability of losing the water supply unit after 6 months of operation for the following:

(a) a single water supply system;

(b) three identical water supply systems in parallel;

(c) a system with two solar panels, two power units, and a single switch ($\lambda = 0.08$). The switch selects the transmitter output signal, which is neither highest nor lowest. The selected signal is then passed to the pump and tank.

(d) What are your observations with regard to these calculations?

Q17.5 Repeat Q17.4 for a solar panel ($\lambda = 0.8$), a power unit ($\lambda = 1.0$), a pump ($\lambda = 0.15$), a tank ($\lambda = 0.15$) and a display unit ($\lambda = 0.15$) connected in series. Determine the probability of losing the water supply unit after 6 months of operation for the following:

(a) a single water supply system;

(b) three identical water supply systems in parallel;

(c) a system with three solar panels, three power units, and one switch ($\lambda = 0.15$). The switch selects the transmitter output signal, which is neither highest nor lowest. The selected signal is then passed to the pump, tank, and display unit.

(d) What are your observations with regard to these calculations?

Q17.6 A rural temperature measurement system consists of a thermocouple ($\lambda = 0.9$), a transmitter ($\lambda = 0.1$), and a recorder ($\lambda = 0.1$) connected in series. Determine the probability of losing the temperature measurement system after 6 months of operation for the following:

(a) a single temperature measurement system;

(b) three identical temperature measurement systems in parallel;

(c) a system with three thermocouples and a single selector relay ($\lambda = 0.1$). The selector relay selects the thermocouple output signal, which is neither highest nor lowest. The selected signal is then passed to the recorder.

(d) What are your observations with regard to these calculations?

Q17.7 A decision is to be made whether to buy two, three, or four stepper motors for a mechatronic system. Each stepper motor will normally be working and can supply up to 50 percent of the total power required. The reliability of each stepper motor can be specified by a constant failure rate of 0.25 per year. The motors are to be simultaneously tested and proved at 6-month intervals. The required system reliability must be at least 0.99.

(a) Use the binomial expansion $(R + F)^n$ to decide how many stepper motors must be bought.

(b) Determine the MTTF for the system.

Q17.8 Repeat Q17.7 with a failure rate of 0.2 per year.

Further reading

[1] Bentley, J. (1999) *Introduction to Reliability and Quality Engineering* (2nd. ed.), Addison-Wesley.

[2] Bhattacharya, B. and Sorkhel, S.K. (1999) Investigation for controlled electro-chemical machining through a response surface methodology based approach. *Journal of Materials Processing Technology*, 200–7.

[3] Cocharan, W.G. and Fox, G.M. (1977) *Experimental Designs* (2nd. ed.), New Delhi: Asia Publishing House.
[4] DeVale, J. (1998) Traditional reliability. 18-896, *Dependable Embedded Systems*, Spring 1999, Carnegie Mellon University, pp. 1–15.
[5] Ross, S.M. (1997) *Introduction to Probability Models* (6th. ed.), Academic Press.
[6] Siewiorek, D.P. and Swarz, R.S. (1992) *Reliable Computer Systems: Design and Evaluation* (2nd. ed.), Digital Press.
[7] Storey, N. (1996) *Safety-critical Computer Systems*, Addison-Wesley Longman.
[8] Villemeur, A. (1992) *Reliability, Availability, Maintainability and Safety Assessment: Volume 1 – Methods and Techniques*, Wiley and Sons.

Internet resources

- http://www.ece.cmu.edu/~koopman/des_s99/mechanical/index.html
- http://www.ece.cmu.edu/~koopman/des_s99/electronic_electrical/index.html
- http://www.ece.cmu.edu/~koopman/des_s99/traditional_reliability/
- http://www.cqeweb.com/Chapters-HTML/Chap8_html/reliability.htm

CHAPTER 18

Case studies

Chapter objectives

When you have finished this chapter you should be able to:

- identify and recognize many engineering and consumer products as mechatronic systems;
- understand the integration of control in mechatronic systems through computers and digital electronics;
- appreciate how mechatronics pulls together knowledge from different areas in order to realize engineering and consumer products that are useful in everyday life.

18.1 Introduction

This chapter presents detailed case studies of complete mechatronic systems: a computer numerically controlled (CNC) drilling machine, mobile robots, and a robotic arm. These projects have been used as mechatronic projects in the engineering department at The University of the South Pacific.

18.2 Case study 1: A PC-based computer numerically controlled (CNC) drilling machine

The biggest change in the numerical control (NC) industry has been the introduction of the microprocessor and microcomputer. The use of microcomputers in NC systems reflects the current trend of using software to minimize the use of hardware or to replace it. NC has become increasingly attractive for manufacturing systems due to the reduced cost and increasing control flexibility achieved by replacing the hardware with software.

18.2.1 Design and manufacture of the CNC drilling machine

The PC is used as a separate, front-end interface for a custom-built machine control unit (MCU), which controls a drilling machine using actuators. The PC-based CNC drilling machine comprises several integrated technologies, ranging from a neural network optimizer for finding the best sequence of points to be drilled, a custom-built MCU, and an enhanced parallel port for communication to the drilling machine.

The framework for the design is shown in Figure 18.1. Neural network design has not been discussed in this book and so will not be elaborated on here. The interested reader will find more detail in Kohonen (1984) and Onwubolu (2002). The neural network optimizes the sequencing of the points to be drilled so that production costs can be minimized. Once the best sequence has been found, the PC software communicates it to the controller, which is connected to the printer port of the PC. The controller coordinates all the system actions. Its output is connected to an interface card that sends signals to the stepper motors. The stepper motors carry the worktable on which the workpiece is supported. Two stepper motors are required to provide movement, one in the x-direction and one in the y-direction. The stepper motors are synchronous 1.8° motors giving a half step angle of 0.9° per revolution. The linear displacement of the load for the gear wheel (with a radius of 5 mm), using gear ratio of 5:1 is 0.0157 mm. This means that one pulse of the stepper motor is equivalent to a 0.0157 mm movement of the load being carried. Consequently, the control resolution, which is the distance between two adjacent addressable points in the axis movement, is 0.0157 mm, and the accuracy is 0.05 mm. The drilling machine navigates in the z-direction to drill the points located directly below it. Load analysis shows that stepper motor x carries the weights of the workpiece and the belt x. Stepper motor y carries the weights of the x-components (workpiece and belt x), stepper motor x, base plate x, pulley x, shaft x, rail, and belt y. Table 18.1 shows the part specifications.

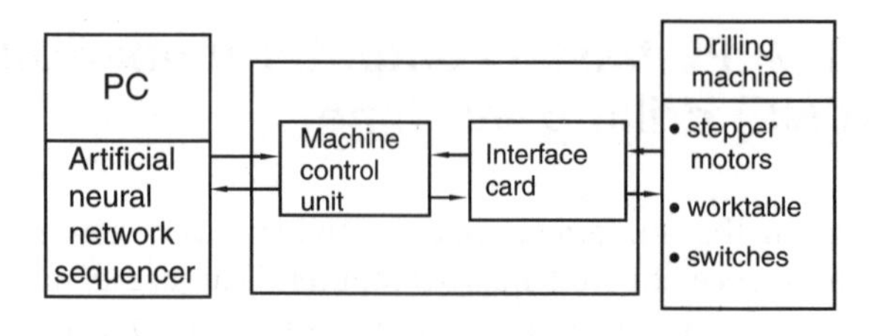

Figure 18.1 Framework for a PC-based CNC drilling machine.

Table 18.1 Drilling machine parts specification

Component	*Specification*
Shaft x	10 mm diameter × 350 mm
Shaft y	10 mm diameter × 350 mm
Load on stepper motor x	26.45 N
Load on stepper motor y	38.41 N
Load on motor z	12.75 N
Belt x	5 mm × 2 mm section × 700 mm long
Belt y	5 mm × 2 mm section × 700 mm long
Base plate	350 mm × 350 mm × 5 mm thickness
Worktable	320 mm × 300 mm × 2 mm thickness
Pulley	25 mm diameter

18.2.1.1 *The customized machine control unit (MCU)*

A conventional MCU is first discussed and then the custom-built MCU used in this example is described to show how advances in PC technology have revolutionized CNC.

The conventional MCU, shown in Figure 18.2(a), consists of:

- a central processing unit (CPU);
- memory;
- an input/output (I/O) interface;
- control for machine tool axes and spindle speed;
- sequence controls for other machine tool functions.

The components are interconnected with a system bus.

The control signals generated by the MCU must be converted to a form and power level suited to the particular position and velocity (feed rate) control systems used to drive the machine tool axes. These signals are responsible for the worktable position, the feed rate, and the spindle speed. In addition, the MCU performs auxiliary functions such as the control of a fixture clamping device, a tool changer and a tool storage unit, coolant use, and an emergency stop function.

The model of the custom-built MCU is shown in Figure 18.2(b). Since the CPU, memory, and the I/O interface are standard components in a PC, the MCU used in this case study comprises solely the machine tool and sequence control functions. The PC is used as a separate, front-end interface for the MCU. In this model the custom MCU is the hardware for interfacing the PC to the drilling machine through the PC's enhanced parallel port (EPP).

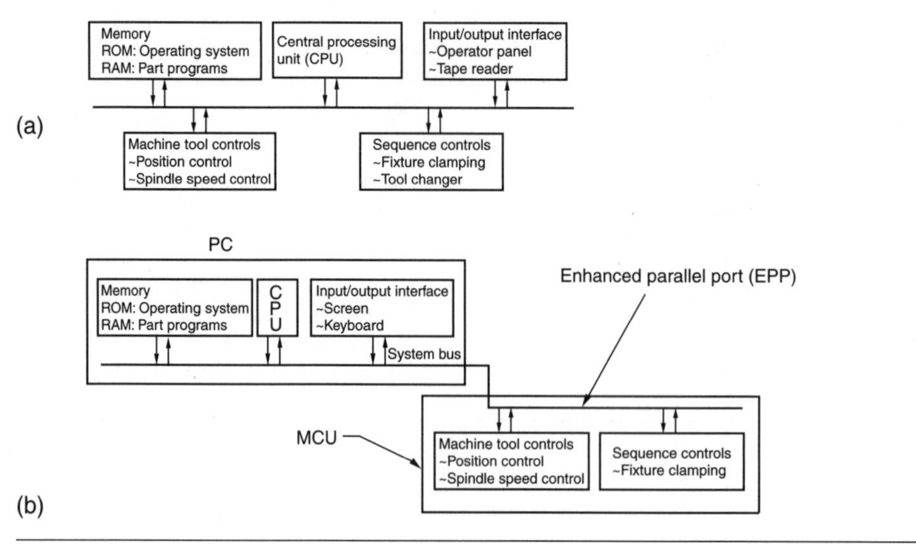

Figure 18.2 Features of (a) a conventional MCU; and (b) the MCU used in case study 1.

The advantages of the PC-based system are:

- simplification of the MCU;
- flexibility to execute a variety of software (shop-floor control, solid modeling, cutting tool management, and other CAM software), in addition to concurrently controlling the machine tool operation;
- ease of use compared with conventional CNC;
- no need for operating system software and machine interface software;
- reduced costs;
- simplified application software (only drilling coordinates are entered);
- updating machine interface software is easier.

The PC I/O hardware, the enhanced parallel port and its corresponding software, and the custom MCU, provide the data transfer interface between the PC and the drilling machine's actuators (stepper motors).

The CNC drill is shown in Figure 18.3.

18.2.1.2 *Graphical user interface*

A graphical user interface (GUI) is the communication tool between the operator and the computer controlling the system. It is displayed on the computer screen

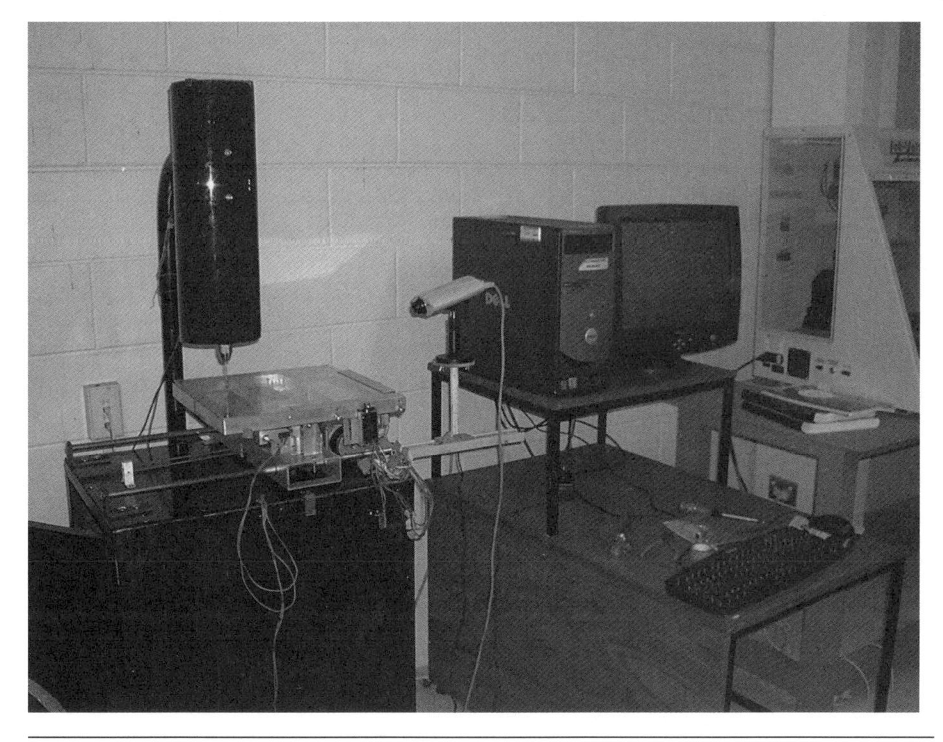

Figure 18.3 The CNC drill of case study 1.

and allows control parameters to be entered and measured system parameters to be displayed. The GUI for this case study was written in C++. The features of the GUI include graphic buttons for entering coordinates for drilling, file saving, running the neural network to optimize the sequence of points to be drilled, retrieving neural network output, selecting the workpiece thickness, and starting drilling operations.

18.2.2 Prediction and reduction of process times

Time is an important factor to consider in any manufacturing process and affects the cost of the product. Thus, an approximate method of predicting process times for the PC-based CNC drilling machine was developed. Factors influencing the processing time include the time to move the workpiece times and the actual drilling time. The process times for drilling different parts were predicted and compared with the actual times taken. The results obtained showed that the drilling prediction times were very close to the actual times taken.

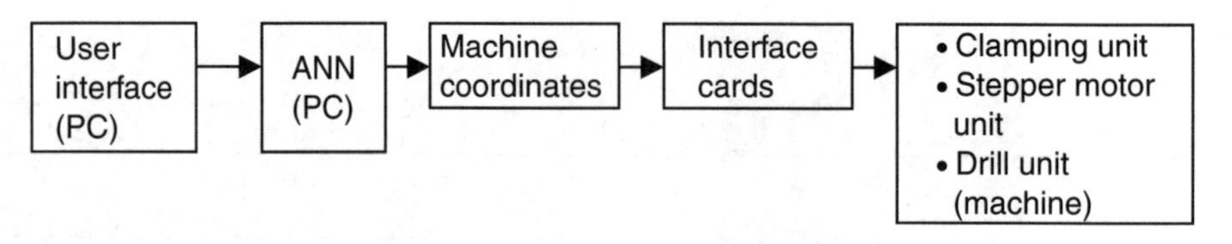

Figure 18.4 PC-base drilling machine architecture.

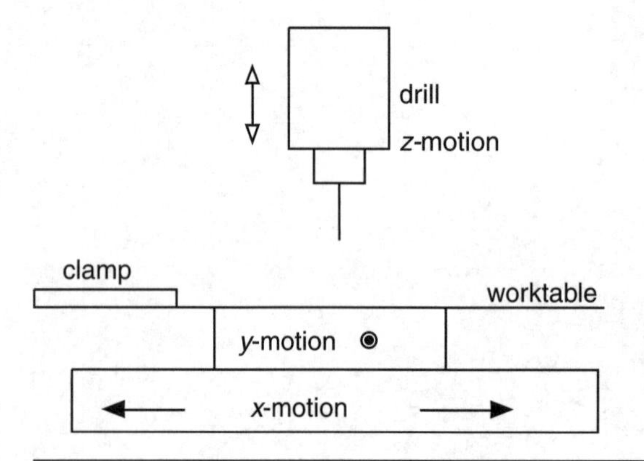

Figure 18.5 PC-based drilling machine layout.

A neuro-optimizer was included in the design to sequence the drilling order. This prediction model was useful to operation managers and engineers as it would also give more insight into the process time.

The system described here uses the PC's parallel port to communicate with the drilling machine. Figure 18.4 shows the system architecture.

The artificial neural network (ANN) software asks the user for the coordinates of the holes to be drilled in the workpiece. The ANN sorts these into an optimal order, which is then transformed into machine movements and communicated to the drilling machine. The drilling machine uses stepper motors to move the workpiece in the x- and y-directions simultaneously to the given coordinates and in the optimal sequence (Figure 18.5).

18.3 Case study 2: A robotic arm

This section describes the design and implementation of a five-axis robotic arm used for 'pick-and-place' in a flexible manufacturing system. The arm consisted of a base, three links and a gripper. The robot was equipped with sensors for gathering data, and hence information, about its environment. The joints were actuated using four-phase hybrid stepper motors. The arm was controlled by a C program uploaded to a PIC microcontroller which read the sensors and stepped the motors to the required positions. The use of a PIC microcontroller in the design resulted in a cost-effective robot which had the capability of moving small to medium payloads (1 kg) in a manufacturing setting.

The robotic arm was designed to be part of an overall system that included a drilling machine (for production) and an automated guided vehicle (AGV, for transport).

18.3.1 Overview

The pick-and-place robot was designed around the PIC16F877 microcontroller. An important functionality of the robot was that it was able to determine when an AGV arrived at the arm's workstation so that it would activate to pick up or drop off parts. The robot was also capable of determining the action to be taken (pick-up or drop off). The robot was equipped with a sensor (Figure 18.6) to detect the arrival of an AGV. On detection of an AGV, the sensor sends a signal to the PIC which in turn energizes the motors that control the robotic arm. The arm then moves to the object on the AGV and picks it up and places it at the destination position, which in this case is the workstation. The robot's movements and coordination of joints are preprogrammed in the PIC microcontroller.

18.3.2 The robot body design

The design specification required that the arm successfully moved to a correct position to pick and place an object, and that it reached a distance of 565 mm from the home position. The rotation around the base depended on the location of the pick and place stations. The stationary robot comprised a base, a shoulder,

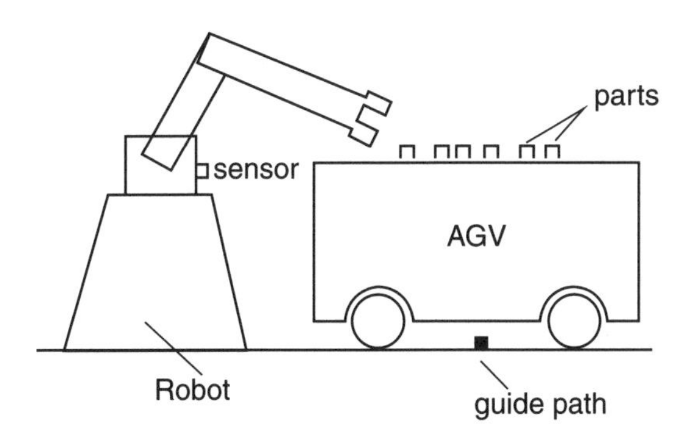

Figure 18.6 Sensor position.

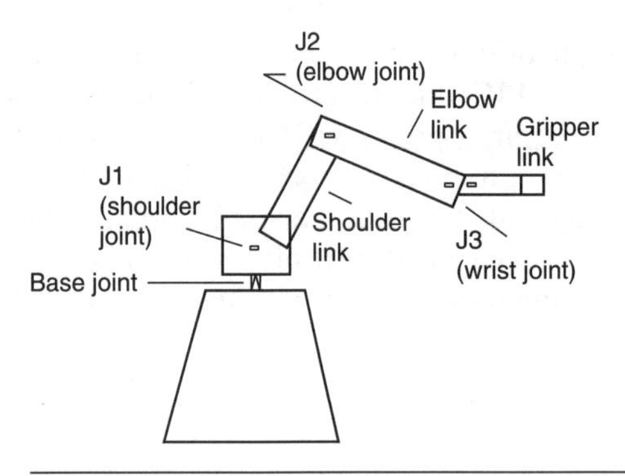

Figure 18.7 Robot configuration.

a shoulder link, an elbow link, and a gripper, as shown in Figure 18.7. Stepper motors were used to attain the precision required for pick and place positioning.

A tapered bottom base was constructed in steel since a stable platform was required and weight was not an issue here. The base motor, giving the arm its horizontal movement, was fixed to this structure. The other joint stepper motors were housed above the base joint. The complete base configuration is shown in Figure 18.8.

The arm (consisting of the shoulder, elbow, and wrist) was made from Perspex as it needed to be as light as possible to reduce torque, as well as being reasonably long to achieve its functionality. Since the length between the motor and the wrist changed as the system operated, the sprocket could not go directly from the motor to the gripper. Instead, it went through the shoulder and the elbow, and at each joint the sprocket drove another sprocket. Since the stepper motors were the heaviest component of the arm, all the motors were placed at the base to reduce the weight and hence reduce the torque required by each motor to rotate the arms. These motors used sprockets and timing belts to drive the shafts that are fixed to the arms at the shoulder, elbow and the wrist.

Placing the motors at the base reduced the weight of the shoulder, elbow and the wrist. Therefore, less torque was required by the motor to move these links. The arrangement also balanced the weight of the arm with the weight of the base and hence reduced the turning moment at the top and bottom base joints. Figure 18.9 shows the design of the arm links.

18.3.2.1 *Loading*

The torque, T_i, acting on each joint, is

$$T_i = (M_{ip/2} + M_p)gL_{ip}, \tag{18.1}$$

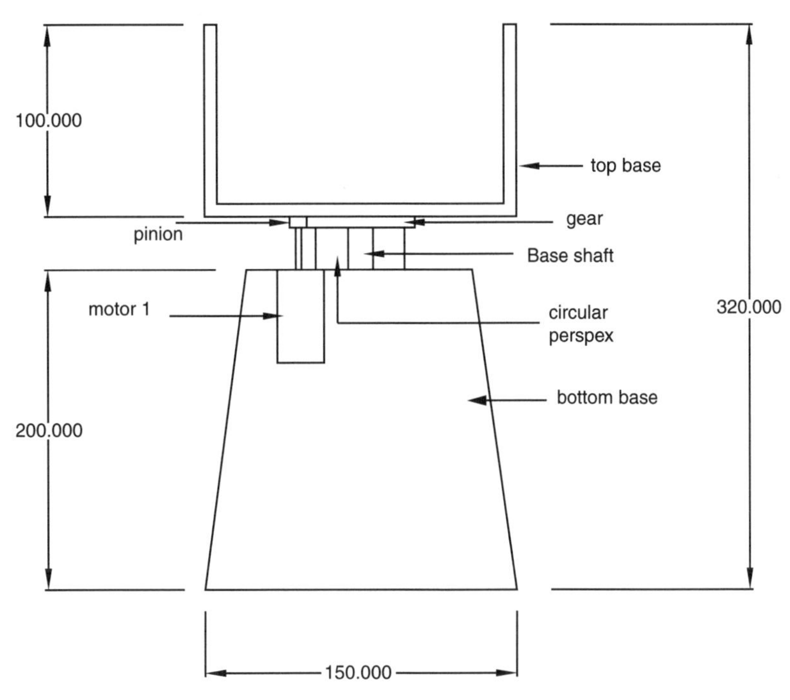

Figure 18.8 Base configuration.

where M_{ip} is the mass of the links between joint i and the payload position, given as

$$M_{1p} = M_{\text{shoulder}} + M_{\text{elbow}} + M_{\text{wrist}} \tag{18.2}$$

$$M_{2p} = M_{\text{elbow}} + M_{\text{wrist}} \tag{18.3}$$

$$M_{3p} = M_{\text{wrist}} \tag{18.4}$$

The length of the links between joint i and payload position L_{ip} are similarly calculated.

For the robot in this study, $M_{\text{base}} = 3$ kg, $M_{\text{shoulder}} = 0.365$ kg, $M_{\text{elbow}} = 0.365$ kg, $M_{\text{wrist}} = 0.111$ kg and $M_{\text{payload}} = 1$ kg. Consequently,

$$M_{1p} = 0.365 + 0.365 + 0.111 = 0.841\,\text{kg}$$

$$M_{2p} = 0.365 + 0.111 = 0.476\,\text{kg}$$

$$M_{3p} = 0.111\,\text{kg}$$

$$M_p = 1\,\text{kg}$$

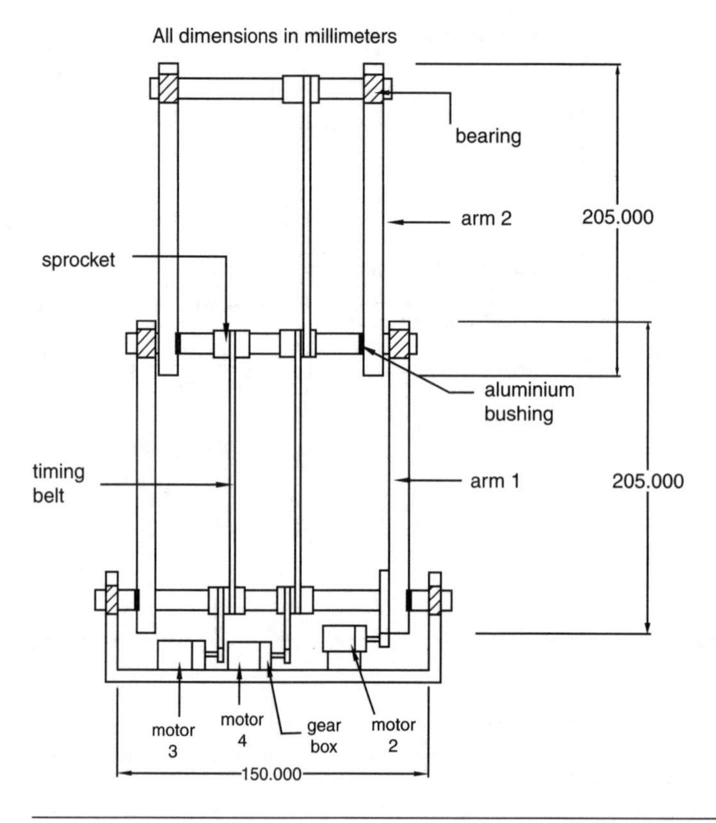

Figure 18.9 Arm links.

Further component parameters are shown in Table 18.2.

$$L_{1p} = |J_1J_2| + |J_2J_3| + |J_3J_p| = 250 + 205 + 120 = 575\,\text{mm}$$

$$L_{2p} = L_{1p}/|J_1J_2| = 205 + 120 = 325\,\text{mm}$$

$$L_{3p} = L_{2p}/|J_2J_3| = 120\,\text{mm}$$

$$R_b = 75\,\text{mm}$$

So from Equation 18.1, the torque acting on each joint, T_i, is

$$T_1 = [(M_{1p/2} + M_p)gL_{1p}]n_g = 1.6\,\text{Nm}$$

$$T_2 = [(M_{2p/2} + M_p)gL_{2p}]n_g = 0.8\,\text{Nm}$$

$$T_3 = [(M_{3p/2} + M_p)gL_{3p}]n_g = 0.25\,\text{Nm}$$

$$T_{\text{base}} = M_{\text{base}}g/n_g + T_1 = 2.04\,\text{Nm}$$

Table 18.2 Robot parameters

Component	*Specification*
Base shaft	30 mm
Shoulder shaft	180 mm
Elbow shaft	150 mm
Wrist shaft	125 mm
Belt 1	220 mm
Belt 2	300 mm
Belt 3	576.5 mm
Belt 4	576.5 mm
Belt 5	486.5 mm
Length of shoulder	250 mm
Length of elbow	205 mm
Length of wrist	120 mm

18.3.2.2 ***Design simulation***

Inverse kinematics was used to calculate the values of possible joint angles for given (x, y) coordinates of the centroid of the gripper from the base. Since all the joints are revolute, the locations were calculated using the cylindrical coordinate system. R, z and α represent the coordinates of the end effector with respect to the origin as shown in Figure 18.10. A second angle α is defined as the angle of the third link to the horizontal.

The following definitions are used for the inverse transformation:

$$\mathrm{CB} = x_3;\ \mathrm{OC} = z_3;\ \text{and}\ \mathrm{OB} = \sqrt{x_3{}^2 + z_3{}^2} \tag{18.1}$$

$$x_3 = x - L_3 \cos\alpha \tag{18.2}$$

$$z_3 = z - L_3 \sin\alpha \tag{18.3}$$

$$\text{From } \Delta\mathrm{OAB},\ \mathrm{OB}^2 = \mathrm{AB}^2 + \mathrm{OA}^2 - 2\mathrm{AB} \times \mathrm{OA}\ \cos(180^\circ - \theta_2) \tag{18.4}$$

leading to

$$x_3{}^2 + z_3{}^2 = L_1{}^2 + L_2{}^2 + 2L_1L_2c_2 \tag{18.5}$$

$$c_2 = \frac{x_3{}^2 + z_3{}^2 - \left(L_1{}^2 + L_2{}^2\right)}{2L_1L_2} \tag{18.6}$$

$$s_2 = \pm(1 - c_2) \tag{18.7}$$

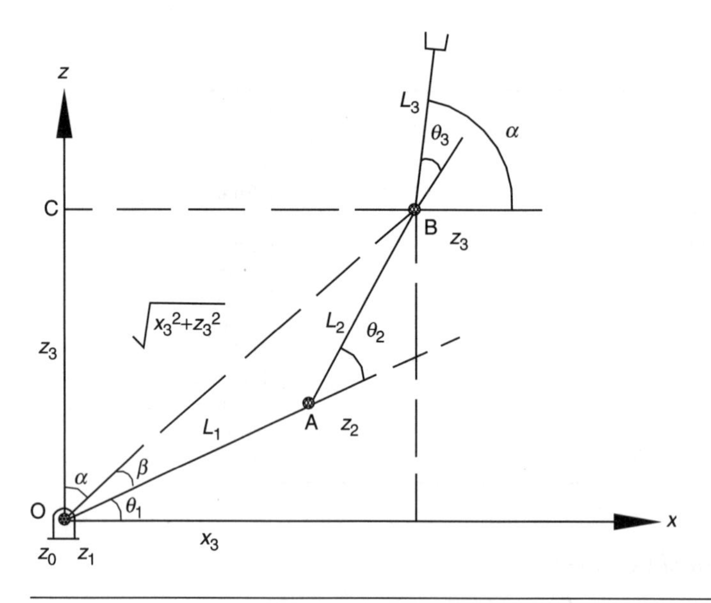

Figure 18.10 Backward analysis of robotic arm.

$$s_1 = \frac{(L_1 + L_2)c_2 z - L_2 s_2 x}{z^2 + x^2} \tag{18.8}$$

$$c_1 = \frac{(L_1 + L_2)c_2 x + L_2 s_2 z}{z^2 + x^2} \tag{18.9}$$

$$\theta_1 = \text{A tan } 2(s_1, c_1) \tag{18.10}$$

$$\theta_2 = \text{A tan } 2(s_2, c_2) \tag{18.11}$$

$$\theta_3 = \alpha - \theta_1 - \theta_2. \tag{18.12}$$

The equations were solved using MATLAB® and the arm movements are shown in Figure 18.11. Figure 18.12 shows the final plot of the simulated arm positions with a specified and unspecified angle α.

Problems

Q18.1 Re-design the PC-based drilling machine described in this chapter so that it is controlled using a microcontroller.

Q18.2 Design a PC-based CNC milling machine.

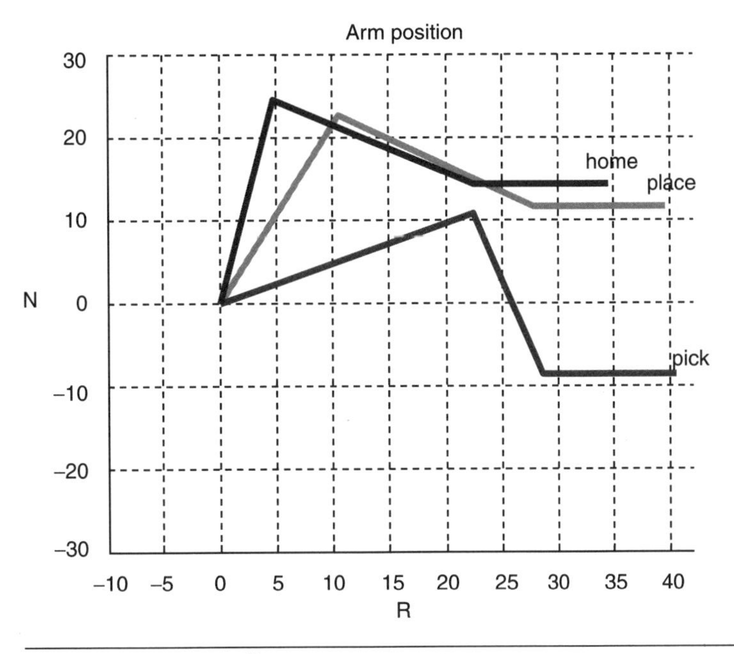

Figure 18.11 MATLAB® plot of arm simulation.

Q18.3 Design a microcontroller-based CNC milling machine.

Q18.4 Model and analyze the cutting forces in the *x*-, *y*-, *z*-directions for the drilling machine described in this chapter.

Q18.5 Model and analyze the cutting forces in the *x*-, *y*-, *z*-directions for a milling machine.

Q18.6 Design a switching circuit for a d.c. motor using a pulse width modulated signal input from a microcontroller and a closed-loop system based on an encoder integrated with a rechargeable power supply.

Q18.7 What specific adjustments need to be made to the d.c. motor speed controller in Q18.6 in order to design a motor position controller?

Q18.8 Describe the modifications that need to be made to convert a PC-based CNC drilling machine to one that is microcontroller based.

Q18.9 How accurate is the time prediction model discussed in section 18.2?

Q18.10 Design a speed control unit for a robotic application.

Q18.11 Design a more efficient gripper for the robotic manipulator.

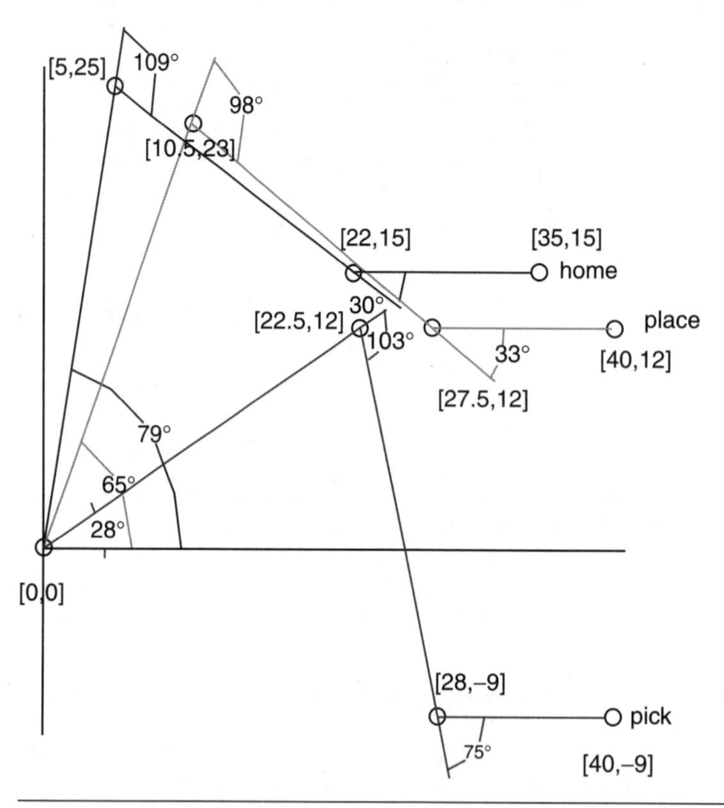

Figure 18.12 Specified angles of the robot arm for home, pick, and place positions.

Further reading

Automation

[1] Chang, C.-W. and Melkanoff, M.A. (1989) *NC Machine Programming and Software Design*, Prentice-Hall International Editions.

[2] DeGarmo, E.P., Black, J.T. and Kohser, R.A. (1990) *Materials and Processes in Manufacturing*, New York: Macmillan.

[3] Noaker, P.M. (1995) The PC's CNC transformation, *Manufacturing Engineering*, August, 49–53.

[4] Kohonen, T. (1984) *Self-organization and Associative Memory*, Berlin: Springer Verlag.

[5] Pham, D.T. and Wang, X. (2000) Prediction and reduction of build times for the selective laser sintering process, *Proceedings of the Inst. Mech. Eng.*, Part B: *J. Engineering Manufacture*, **214**, 425–30.

[6] Onwubolu, G.C. *et al.* (2002) Development of a PC-based computer numerically controlled drilling machine, *Proceedings of the Inst. Mech. Eng.*, Part B: *J. Engineering Manufacture*, **216**, 1–7.
[7] Onwubolu, G.C. (2002) *Emerging Optimization Techniques in Production Planning and Control*, London: Imperial College Press.

Robotics

[8] Craig, J.J. (1989) *Introduction to Robotics: Mechanics and Control* (2nd. ed.), Addison Wesley.
[9] Groover, M.P. (1996) *Fundamentals of Modern Manufacturing*, Prentice Hall.
[10] Groover, M.P. (2001) *Automation, Production Systems, and Computer-integrated Manufacturing*, Prentice Hall.
[11] Hall, A.S., Holowenko, A.R. and Lauhlin, H.G. (1980) *Machine Design*, New York: McGraw-Hill.
[12] Kumar, S. (2003) *Development of a mobile robot with obstacle avoidance system*, M.Sc. Thesis, University of the South Pacific.
[13] Narayan, S. and Sharan, R.V. (2003) *Development of a microcomputer-based pick and place robot*, B.Sc. Dissertation, University of the South Pacific.
[14] Onwubolu, G.C., Narayan, S. and Sharan, R.V. (2004) *Development of a microcontroller-based pick and place robot for FMS application*, (awaiting publication).

Internet resources

- http://www.national.com/ds/LM/LM124.pdf
- http://www.es.geocities.com/astrohyperion/HybridstepperMotors.pdf
- http://www.acroname.com/robotics/parts/R64-P5587.html
- http://www.ece.ubc.ca/~elec474/sections/summer02/team5.pdf

APPENDIX 1

The engineering design process

Chapter objectives

When you have finished this chapter you should be able to:

- understand the engineering design process;
- apply the design process to mechantronics system design.

The engineering design process applies to all engineering disciplines. The process stages are:

- establishment of need and goal recognition;
- specification;
- system conception;
- detailed design;
- prototype;
- testing;
- review and documentation.

A1.1 Establishment of need and goal recognition

Before any product development takes place, it is imperative that the need for the product is established. This requirement is usually driven by a customer or by an enterprise trying to anticipate a market. If the need is not clear then, although the product may be technically sound, it may fail commercially.

A1.2 Specification

The product needs to be fully specified before detailed design starts. Part of the specifications will be the clear statement of the need, because without this the project is liable to drift away from the original intention. Specifications should include a design specification, testing specifications, acceptance specifications and development specifications, although others may well be required to fully document what is required and how the end product will be verified to meet those requirements. The specifications must also include details of how the end product will be maintained to ensure that it continues to fully meet the customer's requirement throughout its lifetime. For instance, the design specification topics should include:

- introduction;
- performance;
- life;
- useful life (period);
- life between overhauls;
- environment;
- power;
- reliability;
- cost;
- maintentance;
- manufacturing process;
- physical size and weight.

The specification stage is arguably the most important stage of the design process because if this is not clear from the outset the end product may not satisfy the customer's needs, or time and money could also be wasted on modifications during the project lifetime.

A1.3 System conception

This is the stage in which the specification is realized in conceptual designs. Normally, several design solutions are obtained during this stage, and using different decision-making processes, the best design is chosen. Some detailed design may be necessary to prove 'ground-breaking' concepts, but the overall cost/benefit will need to be analyzed before embarking on this route. All stakeholders might be involved in choosing the final design. Decision-making techniques include:

- comparison charts;
- category weighting;
- decision networks;
- morphological analysis of decision making (MADM).

A1.4 Detailed design

Once the conceptual design has been agreed, the detailed design to meet the specification within the conceptual framework can begin. Engineering drawings showing all the necessary views and dimensions are prepared. The material for each component is specified and assembly drawings are made. Everything that will be required by the production teams and maintenance teams must appear in the detailed design. An important part of the design process is the inclusion of a mechanism that allows the design to be reviewed either continually or at key stages to ensure that it does not deviate from the original intention. This process also caters for problems that might arise potentially affecting the end product's ability of meeting the specified requirements. These processes will be part of the designer's quality assurance system.

A1.5 Prototyping

The designer will work with the production department to produce models (prototypes) of a design. Prototypes are built to verify that key or novel aspects of the design work as expected or physically fit into existing plant or equipment. A close relationship with the production department also ensures that the design can actually be manufactured using the techniques available to the department. If new production equipment and training are needed, it should be identified early in the project.

A1.6 Testing

Complete advance prototypes, key components, and pre-production models of a design are thoroughly tested to verify that the product meets the specification requirements. The tests form part of the quality assurance procedures of the project and usually have to be signed off as being complete and satisfactory before the product goes into production.

A1.7 Review and documentation

The design is not complete until it has been thoroughly documented and reviewed.

The level of documentation required will depend on the nature of the project but may include the following:

- Rationale behind the project (need, market, cost/benefit, etc.);
- specification;
- detailed design to meet specification (explaining how the design meets the specifications and how it works);
- user guides (how to use the product);
- maintenance guides (how to keep the product working, e.g. spares, servicing schedules);
- et cetera.

Designers are not always the best people to write the documentation but they should work closely with qualified technical writers who can help them complete their design.

The project review should take place to close the design process. The aim is to see how good the design process is. Any lessons learnt should be fed back into future projects. The process may be a one-off review or a regular process that allows production problems or design changes to be addressed quickly.

APPENDIX 2

Mechanical actuator systems design and analysis

Chapter objectives

When you have finished this chapter you should be able to:

- understand a more detailed analysis of mechanical actuators;
- understand fatigue failure mode for mechanical actuators subjected to dynamic loading.

A2.1 Introduction

Chapters 9 and 10 presented an overview of electrical and mechanical actuators and systems. This appendix introduces more mechanical actuators giving details of the analysis involved in these systems. Specifically, the following are covered: springs, gears, roller bearings selection, design and analysis for fatigue strength, shafts, flexible mechanical elements, and screws.

A2.2 Helical springs

Springs are mechanical elements used in machines for exerting force, storing and absorbing energy, and providing flexibility. Springs are classified as wire, flat, or special-shaped. In the following W is applied load, T is torque, and α is the helix angle (usually less than 7.5° for a closed helical spring) (see Figure A2.1).

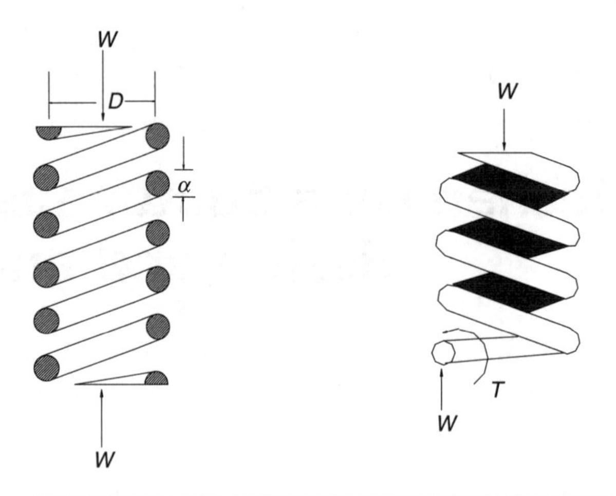

Figure A2.1 The helical spring.

A2.2.1 Shear stress

The analysis of a spring element shows that two types of loads are involved: direct shear force due to the applied load, W, and a torque, T.

Treating the spring element as a curved bar, and not a straight bar, the maximum stress is

$$\tau_{\max} = \frac{Tr}{J} + \frac{W}{A}. \tag{A2.1}$$

Since $T = WD/2$, $r = d/2$, $J = \pi d^4/32$ and $A = \pi d^2/4$, the stress is then expressed as

$$\tau_{\max} = \frac{8WD}{\pi d^3} + \frac{4W}{\pi d^2}. \tag{A2.2}$$

Defining the spring index as $C = D/d$, leads to the following expression for the maximum stress:

$$\tau_{\max} = \frac{8WD}{\pi d^3}\left(1 + \frac{0.5}{C}\right). \tag{A2.3}$$

Defining the shear stress concentration factor, K_s, as

$$K_s = \left(1 + \frac{0.5}{C}\right), \tag{A2.4}$$

the maximum shear stress in the spring element is then given as

$$\tau_{\max} = K_s \frac{8WD}{\pi d^3}. \quad \text{(A2.5)}$$

A2.2.2 Deflection and stiffness

The deflection and stiffness of springs are obtained from the assumption that the strain energy due to the applied load and the strain energy due to the applied torque are the same. The effective coil number is $L = \pi Dn$, so the deflection of the spring is obtained by equating the strain energy due to the direct load equal to the strain energy due to the induced torque:

$$\begin{aligned} \frac{1}{2} W\delta &= \frac{1}{2} T\theta \\ &= \frac{1}{2} \frac{WD}{2} \frac{TL}{GJ} \\ \delta &= \frac{1}{2} \frac{WD}{2} \frac{WD}{2} \frac{(\pi Dn) \times 32}{G\pi d^4}. \end{aligned} \quad \text{(A2.5A)}$$

The spring deflection is therefore given as follows:

$$\delta = \frac{8WD^3 n}{Gd^4}. \quad \text{(A2.6)}$$

$$k = \frac{W}{\delta} = \frac{Gd^4}{8D^3 n}. \quad \text{(A2.7)}$$

A2.2.3 Materials

The commonly used materials for springs are music wire, oil-tempered wire, hard-drawn wire, chrome-vanadium and chrome silicone. Chrome-vanadium (SAE 6150) is the best material for springs. These materials, the exponent and the constant for estimating the tensile strength of the materials are given in Table A2.1.

The equation for estimating the tensile strength of spring materials is given as follows:

$$S_{ut} = \frac{\beta}{d^m}. \quad \text{(A2.8)}$$

Table A2.1 Spring materials and constants for estimating tensile strength

Material	*Size range (mm)*	*Exponent, m*	*Constant, β (MPa)*
Music wire	0.10–6.5	0.146	2170
Oil-tempered wire	0.50–12	0.186	1880
Hard-drawn wire	0.70–12	0.192	1750
Chrome-vanadium	0.80–12	0.167	2000
Chrome silicone	1.60–10	0.112	2000

The yield strength of the material is normally approximated as

$$S_y = 0.75S_{ut}. \tag{A2.9}$$

The torsional yield strength of the material is given as

$$S_y = 0.577S_y. \tag{A2.10}$$

A2.3 Spur gears

Spur gears are cylindrical toothed wheels used to transmit rotary motion as well as power between parallel shafts. Their teeth are straight and parallel to the axis of the shaft. A typical spur gear is shown in Figure A2.2

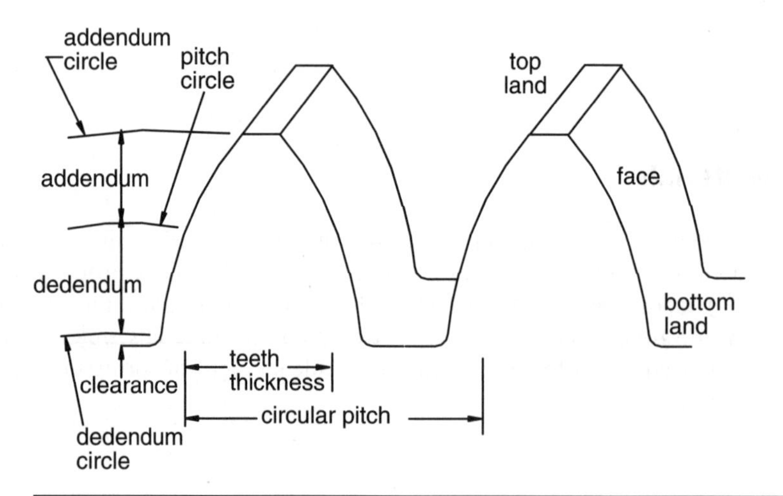

Figure A2.2 Spur gear configuration.

The following basic definitions are essential:

- **Pitch circle:** meshed gears are always in contact along the pitch circle;
- **Circular pitch, *p*:** the distance between corresponding points on adjacent points on adjacent teeth. It is given by $p = \pi d/T$, where d is the circular pitch diameter and T is the number of the teeth on the gear;
- **Diametral pitch, *P*:** the number of teeth per millimeter (inch) of pitch circular diameter and is given by $P = T/d$. The circular pitch and diametral pitch are related by the following definition: $P \times p = \pi$.
- **Module, *m*:** the number of millimeter (inch) of pitch circle diameter per teeth and is given by $m = d/T = 1/P$.
- **Pressure angle, ψ:** the line along which all points of contact of two meshing teeth lie and the normal load acting on each tooth lies along it. A value of $\psi = 20°$ is assumed for most gears.
- **Addendum, *a*:** the radial distance between the tip of the gear and the pitch circle. It is equal in value to module.
- **Dedendum, *b*:** it is the radial distance between the bottom land and the pitch circle. A value of $1.25m$ (or $1.25/P$) is assumed for most gears.
- **Whole depth, h_h:** the sum of the addendum and dedendum, $(a + b) = 2.25/P$.
- **Working depth, h_k:** is twice the module, $2m = 2/P$.
- **Circular teeth thickness, *t*:** is half the circular pitch, $p/2 = \pi d/2T = \pi/2P = \pi m/2$.
- **Clearance, *c*:** is the difference between the addendum and dedendum, and is given by $c = (b - a) = 0.25/P$.
- **Fillet radius, r_f :** the radius of the arc between the flank and the bottom land, given as $0.3m = 0.3/P$.

A2.3.1 Design considerations

It is assumed that gears can fail in any of the three different modes:

- static failure due to bending strength;
- fatigue failure due to bending strength (in motion); and
- fatigue failure due to wear.

The following methods are used to determine the bending strength of a gear:

- the Lewis equation;

- the American Gear Manufacturers Association (AGMA) method; and
- British Standards (BS) method.

The Lewis method is very simplistic and the AGMA method is very cumbersome. The BS method is based on a series of charts and is straightforward. Here we use the Lewis equation to show the basic principles of gear design.

A2.3.1.1 *The Lewis method for bending stress*

The gear tooth shown in Figure A2.3 is modeled as a cantilever having a length, l, equal to the depth of the gear (i.e. $l = a + b$), a uniform tooth thickness, t, width equal to the face width, F, and a load F_t acting tangentially at the point of mesh.

A major assumption in this method is that the profile has a constant thickness, which is not true. The bending stress is given as $\sigma = My/I$, where $I = Ft^3/12$, $y = t/2$ and $M = F_t \times l$.

Hence, assuming that maximum stress occurs at the root, the bending stress can be expressed as

$$\sigma = \frac{F_t l \times t/2}{Ft^3/12} = \frac{6F_t l}{Ft^2}. \tag{A2.11}$$

By similar triangles, $\tan\alpha = x/(t/2) = (t/2)/l$, so the expression for x is given as $x = t^2/4l$. Substituting this expression for x into Equation A2.11 gives

$$\sigma = \frac{6F_t l}{F \times 4l} = \frac{F_t \times p}{F \times 2/3p}. \tag{A2.12}$$

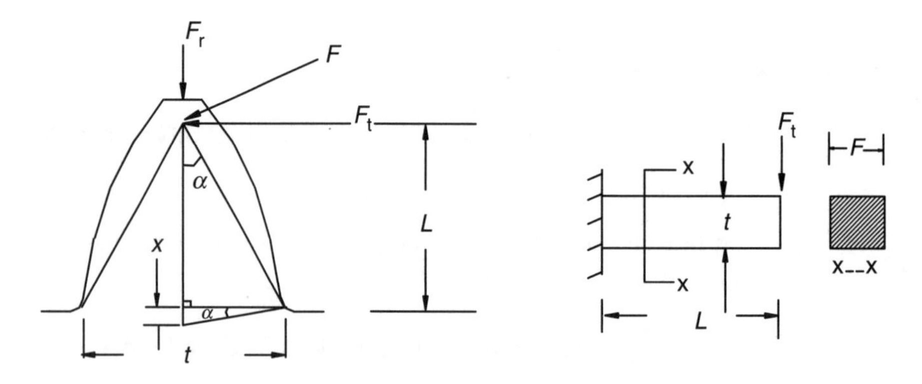

Figure A2.3 Gear geometry with applied load.

Defining $y = 2x/3p$, we get the stress acting on a spur gear, $\sigma = F_t/Fpy = (F_t \times P)/F\pi y$.

Defining $Y = \pi y = 2\pi x/3p$ as the Lewis form factor, the stress acting on a spur gear is given as $(F_t \times P)/FY$.

A2.3.1.2 *The modified Lewis equation*

In addition to the dynamic effects due to the elastic nature of the teeth and shaft during operation, the velocity factor, k_v, is included in the Lewis equation. The modified form of the Lewis equation becomes $k_v F_t P/FY$, where $k_v = 1 + (V/200)$, and V is in meters per minute. This equation is usually used to obtain a rough value of the face width, F.

The AGMA velocity factor $k_v = 1 + (\sqrt{V}/33)$, and the geometry factor, J, are introduced in the Lewis equation. $J = Y/K_f$, where $K_f = 1 + q(K_t - 1)$ is the fatigue stress concentration factor, q is the notch sensitivity, and K_t is the stress concentration factor. Consequently, the stress value can be expressed as $\sigma = k_v F_t Y/FJ$.

A2.3.1.3 *Some design notes*

- The form factor, Y, is selected for the pinion if both gears are of the same material. If the pinion is made from a stronger material than the gear, then Y must be found for the pinion and the gear.
- A safety factor, n, based on the yield stress of the material, is between 3 and 5, that is $\sigma = S_y/n$, for $3 \leq n \leq 5$. Greater values ($n > 5$) are used when shock or vibration is involved.
- The face width, F, is between 3 to 5 times the circular pitch, $3p \leq F \leq 5p$.
- The minimum number of pinion teeth is 18, and the maximum number of teeth per pair is 36.
- A value of $K_f = 1.5$ is acceptable for an approximate design. If Y for the gear is known, J is obtained as being equal to Y/K_f. For a more rigorous design, the AGMA graph for J should be used.

A2.4 Rolling contact bearings

Rolling contact bearings transfer the main load through either ball or roller elements in rolling contact unlike in journal bearings in which the load is transferred by sliding contact. The running friction in rolling contact bearings is about half that of the starting friction. Hence they are particularly useful in engineering applications that are frequently started and stopped.

Basically, there are two classes of rolling contact bearing: the ball bearing, and the roller bearing.

Designers do not usually design bearings, rather they select them from available catalogs. This section outlines a methodology for bearing selection, but it should be noted that selection methods vary from manufacturer to manufacturer.

A2.4.1 Types of ball bearing

A non-exhaustive list of different types of ball bearings follows.

- **Deep groove:** This takes a relatively high thrust load (Figure A2.4).
- **Angular contact:** This takes a higher thrust load than the deep groove ball bearing (Figure A2.5).
- **Double row:** This is similar to the single row deep groove ball bearing except that the outer and inner rings have double grooves. This type of bearing takes heavier radial and thrust loads than the single row type.
- **Self-aligning:** This is used mainly where radial loads are predominant and moderate thrust in either direction is possible (Figure A2.6). It compensates for angular misalignments due to erroneous shaft or similar mountings.

A2.4.2 Types of roller bearing

Roller bearings normally support greater loads than ball bearings of a similar size due to the fact that they have greater area of contact. However, if there are slight

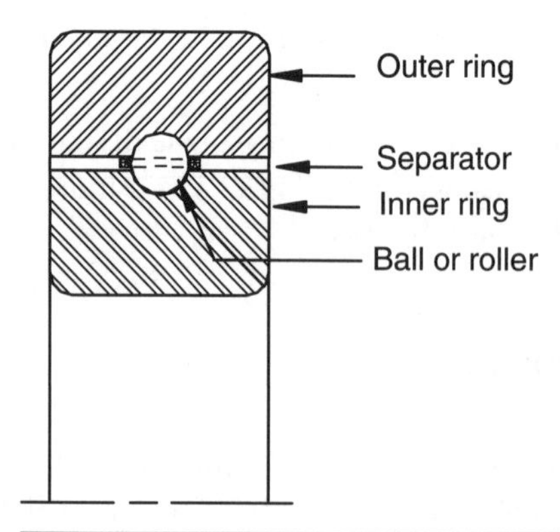

Figure A2.4 A typical rolling bearing.

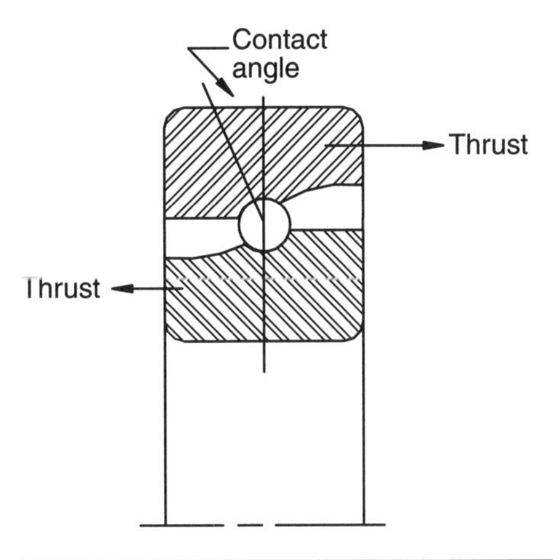

Figure A2.5 Angular contact ball bearing.

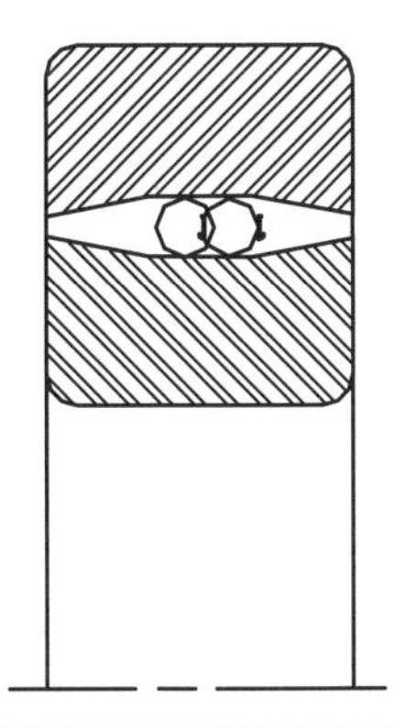

Figure A2.6 Self-aligning ball bearing.

geometrical differences between the ball and raceways, misalignment results. The balls are usually cylindrical, and are straight, spherical, or tapered.

- **Straight cylindrical:** The area of a roller carrying the load is greater than that of the ball, hence straight cylindrical roller bearings sustain a larger radial capacity (Figure A2.7).
- **Spherical:** The outer race is spherical (Figure A2.8). The area of the roller in contact with the load is greater than with the cylindrical roller. Hence the spherical roller bearing supports greater loads.
- **Tapered:** The tapered roller is essentially the frustrum of a cone (Figure A2.9). They support large axial and thrust loads.

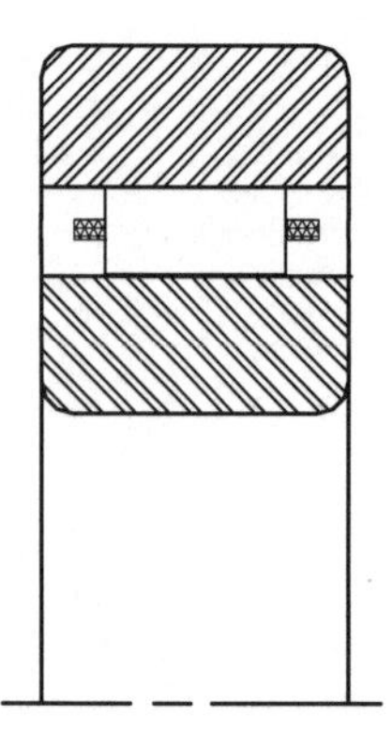

Figure A2.7 Straight cylindrical roller bearing.

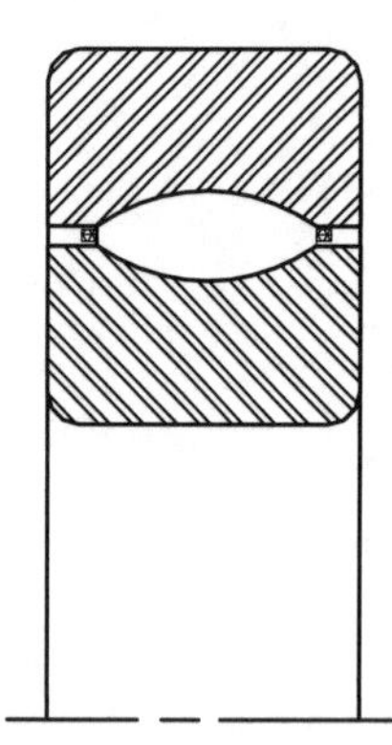

Figure A2.8 Spherical roller bearing.

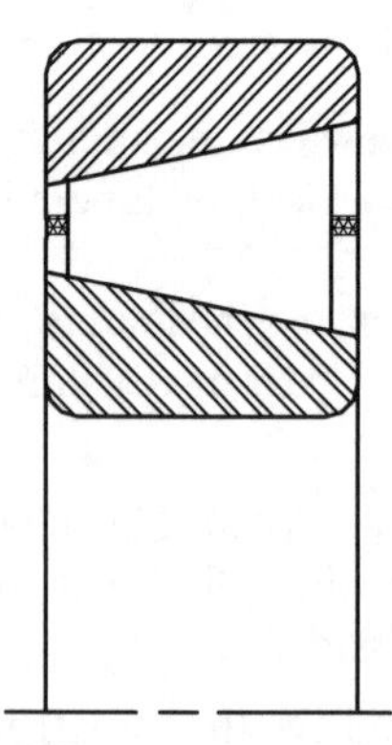

Figure A2.9 Tapered roller bearing.

A2.4.3 The life of a bearing

High stresses result when a bearing is in operation under load. Failure due to fatigue is predominant and since this occurs after many millions of stress applications, we refer to bearing life as the total number of revolutions or hours of bearing operations, at a constant speed, before failure occurs.

Generally, the life of a bearing, L, varies inversely as the power of the load, F, that it sustains. Hence the following definition:

$$\frac{F_1}{F_2} = \left(\frac{L_2}{L_1}\right)^k, \tag{A2.13}$$

where k is 10/3 for a roller bearing and 1/3 for a ball bearing.

A2.4.3.1 *The rating life of a bearing*

The rating life, otherwise know as its '90 percent life' is defined as the number of revolutions or hours at a given speed in which 10 percent will fail before the failure criterion developed. It is also referred to as L_{10}, which is given as

$$L_{10} = \frac{60LN}{10^6}, \tag{A2.14}$$

where L is the life before failure and N is the speed of the bearing in revolutions per minute.

A2.4.4 The reliability of a bearing

The Weibull probability distribution is used for estimating reliabilities other than 90 percent. The reliability of bearing, R, is expressed as follows:

$$R = \mathrm{e}^{-(L/aL_{10})^b}, \tag{A2.15}$$

where a and b are constants and

$$R = \begin{cases} 90\% & \text{implies} \quad L = L_{10} \\ 50\% & \text{implies} \quad L = 5L_{10}. \end{cases} \tag{A2.15A}$$

We can easily obtain the values of a as 6.84 and b *as* 1.17 using these two conditions and Equation A2.15.

A2.4.5 Static load capacity

Static load capacity, F_s, is load that if exceeded can cause permanent deformation when the bearing rotates. It is proportional to the number of balls or rollers, N_b, and the square of the ball diameter, D_b, so

$$F_s = C_s N_b D_b^2, \tag{A2.16}$$

where C_s depends on the type and material of bearing under consideration.

A2.4.6 Dynamic load capacity

A rotating bearing fails due to fatigue and the dynamic load capacity, F_d, of such bearing is given as

$$F_d = C_d N_b^{2/3} D_b^{1.8} (N_r \mathrm{Cos}\ \alpha)^{0.7}, \tag{A2.17}$$

where C_d depends on the type and material of the bearing, N_r is the speed of rotation, and α is the angle between face of bearing and line of action of resultant force as in the angular contact bearing.

A2.4.7 Equivalent dynamic load

Catalog ratings are based on radial load or thrust load, however, apart from pure thrust bearings, roller bearings support both types of load. Therefore, it is necessary to combine the loads to give an equivalent load, $F_e = (0.56C_rF_r + C_tF_t)S_f$, where C_r is 1 for a rotating inner race and 1.2 for a rotating outer race, $(F_t/C_rF_r) > Q$, and F_r is the applied radial load, and F_t, is the applied thrust load. The service factor, S_f, is 1.1–1.5 for rotating parts, 1.3–1.9 for reciprocating parts, 1.6–4.0 for high impact parts. The thrust factor, C_t, is obtained from manufacturers' tables.

A2.5 Fatigue failure

In real life, mechanical elements are not only loaded statically, but they are also loaded in such a manner that the stresses in the members vary from a maximum value to a minimum value during an infinite number of cycles. A shock absorber in a car is a typical example where the springs are loaded cyclically as the car is driven on a rough road. The springs are repeatedly loaded by forces that can vary from a maximum value to a minimum value. The same can be said of a rotating shaft that experiences bending moments resulting in cyclical compressive and tensile stresses that may be repeated several times a minute. Stresses of this nature are referred to as fluctuating stresses and they result in mechanical members failing under *fatigue failure* mode.

In fatigue failure, ten million cycles are referred to as an infinite life. What this means is that if a shaft rotates ten million times, then it is assumed that it has attained its design life. Fatigue failure is very dangerous to mechanical parts because the stress required to cause it is normally below the ultimate strength and the yield strength of the material. The designer needs to be familiar with fatigue failure and take steps to make sure that a machine part is resistant to this failure mode.

The stress concentration factor is linked with fatigue failure. A small crack in a turbine blade can cause a major failure because it can propagate very easily under fluctuating stresses.

The concept of fluctuating stresses is shown in Figure A2.10. For fluctuating stresses, the stress at a point, is given by a range stress, σ_r, and a mean stress, σ_m. These stresses are functions of the maximum and minimum stresses, respectively.

The maximum stress is given as:

$$\sigma_m = \left(\frac{\sigma_{max} - \sigma_{min}}{2}\right) + \sigma_{min} = \frac{\sigma_{max} + \sigma_{min}}{2}. \tag{A2.18}$$

The stress range is given as

$$\sigma_r = \frac{\sigma_{max} - \sigma_{min}}{2}. \tag{A2.19}$$

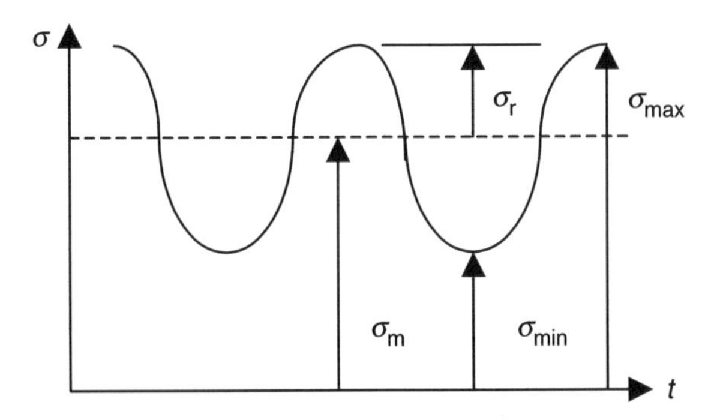

Figure A2.10 Range and mean stresses.

A2.5.1 The endurance limit

Engineering parts, which fail under fatigue loading, experience extreme stresses, σ_{max} and σ_{min}. Such parts are more likely to fail than those parts that experience only maximum stress, σ_{max}. The stress amplitude, at which the machine member will fail after a given number of stress cycles is known as the *fatigue strength*, designated S_n. For $n \to \infty$, the fatigue strength approaches the *endurance limit*, designated as S_n', as shown in Figure A2.11.

However, the imperfection of surfaces due to manufacturing processes and environment affects the endurance limit, and results are lower than those published by the manufacturers. Therefore a modified endurance limit is given as $S_e = k_{sf}k_r k_s k_t k_m \times (S_u/2)$, where k_{sf} is the surface finish factor, k_r is the reliability factor, k_s is the size factor, k_t is the temperature factor, k_m is the stress concentration modifying factor, and S_u is the ultimate strength of the material.

The surface finish factor, k_{sf}, depends on the quality of the finish and the tensile strength of the material. Graphs of these are normally available.

The reliability factor, k_r, can be obtained from the Table A2.2.

The size factor, k_s, is obtained from:

$$k_s = \begin{cases} 1 & \text{if } d \le 8\,\text{mm} \\ 0.85 & \text{if } 8\,\text{mm} \le d \le 50\,\text{mm} \\ 0.75 & \text{if } d > 50\,\text{mm}. \end{cases} \tag{A2.20}$$

The temperature factor, k_t, is obtained from

$$k_t = \begin{cases} \dfrac{620}{460 + T} & \text{if } T > 160^\circ\text{F} \\ 1 & \text{if } T < 160^\circ\text{F} \end{cases} \tag{A2.21}$$

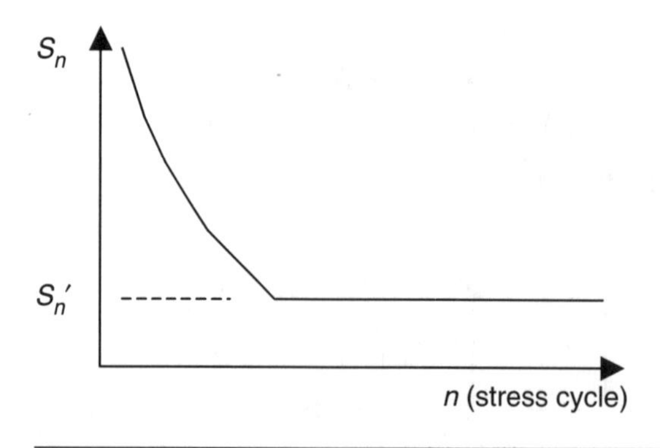

Figure A2.11 Endurance limit.

Table A2.2 Reliability factors

Reliability	*Reliability factor, k_r*
50%	1
90%	0.897
95%	0.868
99%	0.814

The stress concentration modifying factor, k_m, is $1/k_f$, where k_f is the fatigue stress concentration factor:

$$k_f = 1 + q(k_{st} - 1), \tag{A2.22}$$

and q is the notch sensitivity usually given graphically and is a function of the material's ultimate strength and notch radius of the part. The stress concentration factor, k_{st}, is also a function of the part's geometry.

For a no notch round shaft, $k_t = 1$, which implies $k_f = 1$, and hence $k_m = 1$.

A2.5.2 Fatigue strength

When the mean stress and the stress range are varied, the fatigue resistance of parts subjected to these fluctuating stresses can be studied using a Goodman diagram (Figure A2.12). The diagram is drawn by plotting the yield strength of a material on both the x- and the y-axis. The ultimate strength of the material is marked out in the x-axis, and this is usually greater than the yield strength. The endurance limit

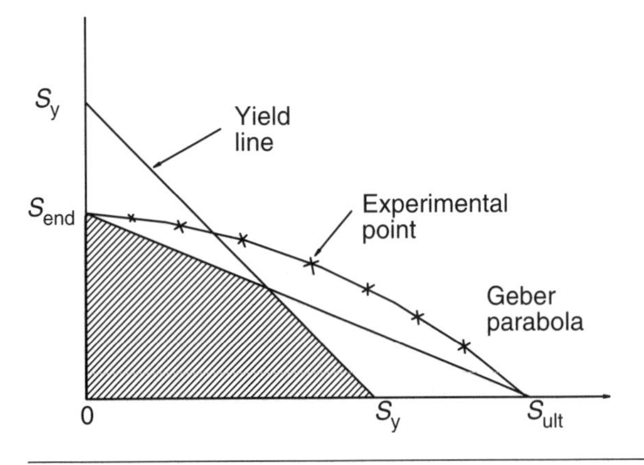

Figure A2.12 The Goodman diagram.

Table A2.3 Material properties

Material	Denisty $Kg\,m^{-3}$	Unit cost Kg^{-1}	Temp. coefficient $m/m°C\,T \times 10^{3}$	Yield strength $N\,m^{-2} \times 10^{6}$		Ultimate strength $N\,m^{-2} \times 10^{6}$		Elasticity modulus $N\,m^{-2} \times 10^{9}$		Remarks
				Tension $(St)_s$	Shear $(Ss)_s$	Tension $(St)_{alt}$	Shear $(Ss)_{alt}$	E	G	
AISI 1020 HRS Stl.	7800	0.33	11.70	241	145	414	310	200	83	
AISI 1020 CR Stl.	7800	0.40	11.70	414	248	552	248	200	83	Oil quenched 455°C
AISI 1095 Stl.	7800	0.70	11.34	669	379	745	–	200	83	Oil quenched 205°C
AISI 2340 Stl.	7800	0.90	11.52	1200	662	1944	–	200	83	Water quenched 205°C
AISI 4130 Stl.	7800	0.95	11.52	1358	–	1600	–	200	83	Oil quenched 590°C
Nitralloy 135 Stl.	7800	–	11.52	1138	–	1248	–	200	83	
304 Stainless Stl.	8000	1.90	17.28	228	124	517	–	200	83	
416 Stainless Stl.	7700	150	9.90	276	–	517	–	200	83	
446 Stainless Stl.	7550	1.70	10.44	310	–	517	–	200	83	
Titanium Alloy	4500	26.5	9.54	896	–	1034	–	110	41	$Se = 586 \times 10^{6}\,N/m^{2(*)}$
24 S-T Alum. Plate	2800	1.20	23.22	317	–	469	283	73	27.6	$Se = 124 \times 10^{6}\,N/m^{2(*)}$
75 S-T Alum. Plate	2800	1.30	23.58	496	–	565	324	71.7	26.9	$Se = 145 \times 10^{6}\,N/m^{2(*)}$
A M-C 585 Mag Alloy	1825	2.50	26.10	221	–	317	148	44.8	16.6	$Se = 121 \times 10^{6}\,N/m^{2(*)}$
A M-C 655 Mag Alloy	1850	2.40	26.10	193	–	276	110	44.8	16.6	$Se = 75 \times 10^{6}\,N/m^{2(*)}$
Phosphor Bronze Strip	8850	3.00	17.28	552	–	627	–	114	45.4	$Se = 190 \times 10^{6}\,N/m^{2(*)}$
Beryllium Copper	8300	3.00	16.56	965	–	138	–	126	51	$Se = 310 \times 10^{6}\,N/m^{2(*)}$
Spring Steel Strip	7800	3.00	11.70	1034	–	1241	–	207	80	
Hevimet	16900	–	5.50	517	359	655	483	345	138	Tungst-Nickel Cu. alloy
Mallory 1000	16300	–	5.40	517	–	655	745	276	132	Tungst-Nicket Cu. alloy
Stl. Oilite	7000	–	12.60	–	–	241	–	–	–	Relatively new material
Cast Phenolic	1330	1.00	90.00	52	–	69	65	4.89	1.93	
Polystyrene	1050	2.50	70.20	45	–	62	57	2.76	1.44	
Nylon Fm 100001	1130	5.20	99.00	–	–	76	–	2.76	1.24	
Nylatron GS	1130	–	41.40	–	–	83	–	4.14	1.86	
Lexan	–	–	–	–	–	55	–	1.10	–	Relatively new material

is marked out in the y-axis. The points corresponding to the yield points on the x-axis and y-axis are connected by a straight line, as are the points corresponding to the ultimate strength along the x-axis and the endurance limit along the y-axis. A region is defined by the intersection of the endurance limit–ultimate strength line and the yield strength–yield strength lines, within which the design is feasible (shaded area). The stress range is associated with the y-axis and the mean stress is associated with the x-axis. By plotting a line defined by the gradient obtained from the loads related to the stress range and mean stress, respectively, it is possible to locate the stress range and mean stress values.

As the mean stress increases, the semi-range reduces. Within the region bounded by the yield strength–yield strength line and ultimate strength–endurance limit line, the sum of the mean stress and the stress range is equal to the yield strength. The yield strength, ultimate strength, and other properties for various materials are given in Table A2.3.

The stress range and the mean stress in a machine member can be estimated by superimposing a line whose gradient is given by the ratio of the stress range to mean load as shown in Figure A2.13.

The maximum stress is given as

$$\hat{S} = S_{\text{mean}} + S_{\text{sr}} = \frac{W_{\text{mean}}}{A} + \frac{KW_{\text{sr}}}{A}. \tag{A2.23}$$

The mean stress is given as

$$S_{\text{mean}} = \frac{W_{\text{mean}}}{A}. \tag{A2.24}$$

The stress range is given as

$$S_{\text{sr}} = \frac{KW_{\text{sr}}}{A}. \tag{A2.25}$$

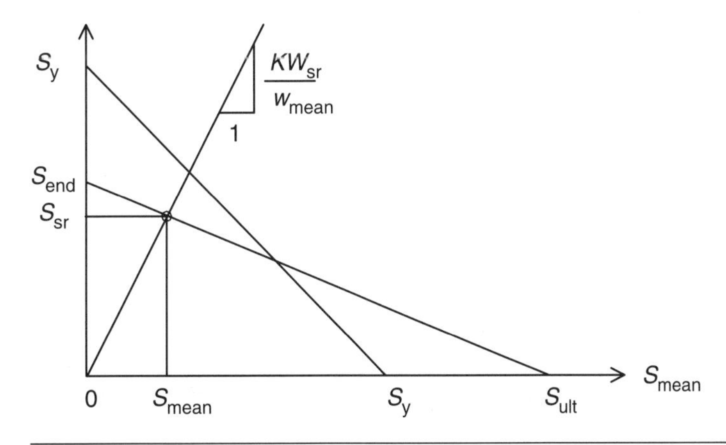

Figure A2.13 Deriving the mean stress and stress range from a Goodman diagram.

The ratio of the stress range to the mean stress is given as follows:

$$\frac{S_{sr}}{S_{mean}} = \frac{KW_{sr}}{W_{mean}}. \tag{A2.26}$$

The ratio of the stress range to the mean stress given in Equation A2.26 is a much quicker way of obtaining the stress range and mean stress than using the Goodman diagram, if the associated loads are known.

A2.6 Shafts

A shaft is a rotating member used for the purpose of transmitting power. It could be subjected to constant bending moment or torsional stress or a combination of these due to fluctuating loads.

A2.6.1 Shaft design based on static load

From maximum shear stress theory,

$$\tau_{max} = \sqrt{\left(\frac{\sigma_x}{2}\right)^2 + \tau_{xy}{}^2}, \tag{A2.27}$$

where $\sigma_x = 32M/\pi d^3$ and $\tau_{xy} = 16T/\pi d^3$.

Substituting these equations into Equation A2.27 and taking $\tau_{xy} = S_y/2N_y$, where S_y is the yield strength of the material and N_y is the safety factor, gives the following:

$$\frac{S_y}{2N_y} = \frac{16}{\pi d^3}\sqrt{M^2 + T^2}. \tag{A2.28}$$

The maximum bending moment and torque must be found to facilitate the evaluation of the section diameter of the shaft in Equation A2.28.

A2.6.2 Shaft design based on fluctuating load

This is more involved than the static load situation since the effect of stress concentration and endurance limit are significant factors that should be taken into account to ensure the shaft possesses adequate strength.

Assume a tensile stress at a point is given by a range stress, σ_r, and a mean stress, σ_m, as shown in Figure A2.10, then, as in section A2.4, $\sigma_m = (\sigma_{max} + \sigma_{min})/2$.

The Soderberg failure line (Figure A2.14) is drawn between the yield point and the endurance limit. The yield point is considered to be associated with σ_m and the endurance limit is considered to be associated with σ_r. Most failures, which are usually associated with a combination of σ_r and σ_r can be below or above the Soderberg failure line. When a factor of safety, N, is applied to the yield strength and endurance limit, a line known as the safe stress line is obtained, which is parallel to the Soderberg failure line.

When stress concentration is considered for a point on the surface of a part, the stress range is multiplied by the stress concentration factor. Assume such a point is on the safe stress line as shown in Figure A2.14, then from similar triangles,

$$\frac{(S_y/N) - \sigma_m}{k_f \sigma_r} = \frac{S_y}{S_e}, \tag{A2.29}$$

where S_e is the modified endurance limit already discussed.

A2.7 Power screws

The power screw is a device used in industrial machinery for converting rotational motion into linear motion, usually for transmitting power. They are key elements in presses, vises, jacks, and numerically controlled (NC) machines.

A2.7.1 The mechanics of power screws

Figure A2.15 shows a single thread power screw with square threads and having a mean diameter d_m, a pitch p, a helix angle ψ, and a lead angle λ, which is loaded axially. Let this axial compressive force be F.

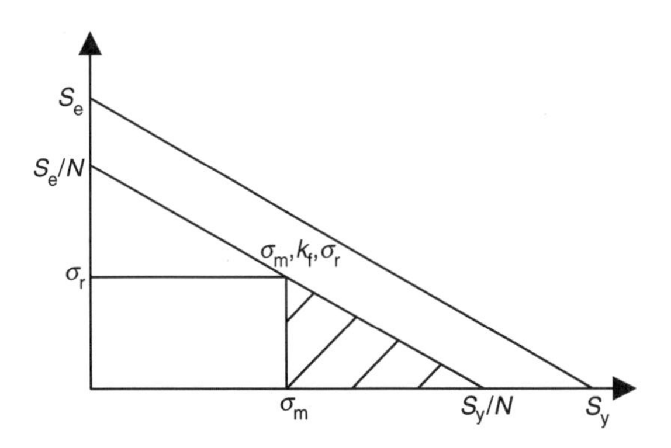

Figure A2.14 The Soderberg failure line.

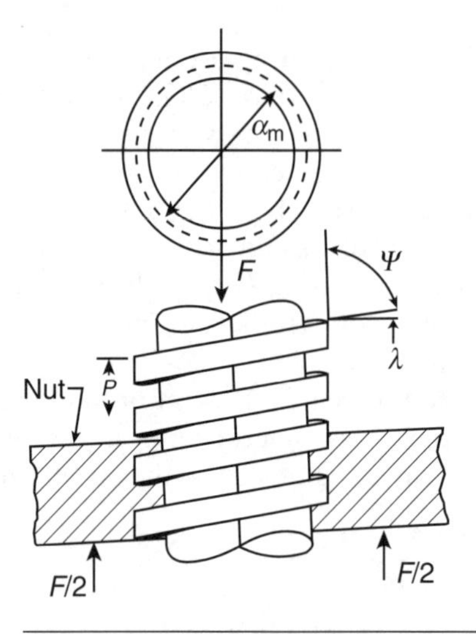

Figure A2.15 The power screw.

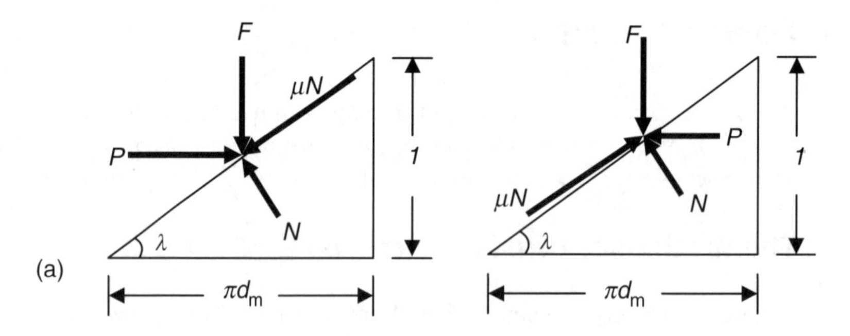

Figure A2.16 Raising and lowering axial force, *F*.

In one revolution of the power screw, the screw moves by one thread. A triangle (Figure A2.16) shows that one edge of the thread forms the hypotenuse of a right-angled triangle having the base as the circumference of the mean-thread-diameter circle and the height as the lead. Let the force P be applied to raise or lower the axial force, F.

In equilibrium, the sum of the forces is considered for raising and lowering the axial force, F.

Resolving horizontally (H) and vertically (V), the raising load is given as

$$\sum F_H = P - N \sin \lambda - \mu N \cos \lambda = 0 \tag{A2.30}$$

$$\sum F_V = F + \mu N \sin \lambda - N \cos \lambda = 0, \tag{A2.31}$$

and the lowering load is given as

$$\sum F_H = -P - N \sin \lambda + \mu N \cos \lambda = 0 \quad \text{(A2.32)}$$

$$\sum F_V = F - \mu N \sin \lambda - N \cos \lambda = 0. \quad \text{(A2.33)}$$

A2.7.1.1 *Raising load*

Multiplying Equation A2.30 by μ gives:

$$\mu P - \mu N \sin \lambda - \mu^2 N \cos \lambda = 0. \quad \text{(A2.34)}$$

Add Equations A2.31 and A2.34:

$$F + \mu P = N(\cos \lambda + \mu^2 \cos \lambda), \quad \text{(A2.35)}$$

giving

$$N = \frac{F + \mu P}{(\cos \lambda + \mu^2 \cos \lambda)}. \quad \text{(A2.36)}$$

From Equations A2.31 and A2.36:

$$F = N \cos \lambda - \mu N \sin \lambda = \frac{(F + \mu P) \cos \lambda}{(\cos \lambda + \mu^2 \cos \lambda)} - \frac{(F + \mu P)\mu \sin \lambda}{(\cos \lambda + \mu^2 \cos \lambda)}. \quad \text{(A2.37)}$$

Simplifying:

$$F(\cos \lambda + \mu^2 \cos \lambda) = F \cos \lambda + \mu P \cos \lambda - F\mu \sin \lambda - P\mu^2 \sin \lambda \quad \text{(A2.38)}$$

$$F(\mu^2 \cos \lambda + \mu \sin \lambda) = P(\mu \cos \lambda - \mu^2 \sin \lambda) \quad \text{(A2.39)}$$

$$P = \frac{F(\mu \cos \lambda + \sin \lambda)}{(\cos \lambda - \mu \sin \lambda)} = \frac{F(\mu + \tan \lambda)}{(1 - \mu \tan \lambda)}. \quad \text{(A2.40)}$$

Therefore, the force required for raising the load is:

$$P = \frac{F[\mu + (l/\pi d_m)]}{[1 - \mu(l/\pi d_m)]}. \quad \text{(A2.41)}$$

The applied torque is given as:

$$T = P\frac{d_m}{2} = \frac{Fd_m}{2}\frac{(\mu\pi d_m + l)}{(\pi d_m - \mu l)}. \tag{A2.42}$$

A2.7.1.2 *Lowering load*

The load is given as:

$$P = \frac{F[\mu - (l/\pi d_m)]}{[1 + \mu(l/\pi d_m)]}. \tag{A2.43}$$

The applied torque is given as:

$$T = P\frac{d_m}{2} = \frac{Fd_m}{2}\frac{(\mu\pi d_m - l)}{(\pi d_m + \mu l)}. \tag{A2.44}$$

From Equation A2.44, the condition for self-locking is $\mu\pi d_m > l$.
Since $l/\pi d_m = \tan\lambda$, then $\mu > \tan\lambda$.
For $\mu = 0$,

$$T_0 = \frac{Fl}{2\pi}. \tag{A2.45}$$

The efficiency of the power screw is then given as:

$$\eta = \frac{T_0}{T} = \frac{Fl}{2\pi T}. \tag{A2.46}$$

A2.8 Flexible mechanical elements

Flexible mechanical elements, such as belts and chains, are used to usually replace a group of gears, bearings, and shafts, or similar power transmission devices. These flexible mechanical transmission devices are employed for power transmission when comparatively long distances are involved. They have the following functions:

- increase torque by reducing speed;
- reduce torque by increasing speed;
- change axis of rotation;

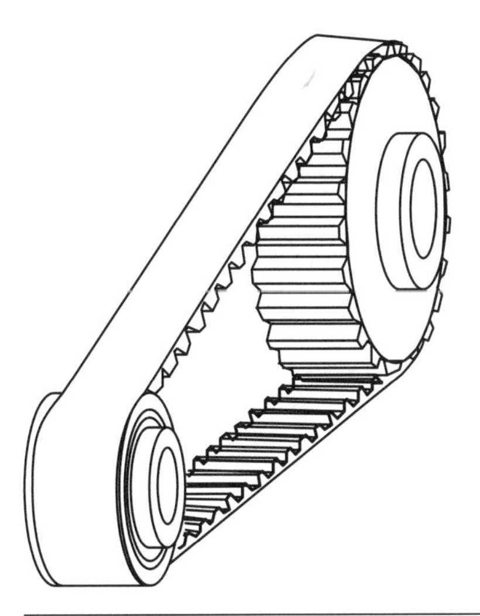

Figure A2.17 Transmission belt.

- convert linear motion into rotary motion; and
- convert rotary motion into linear motion.

Figure A2.17 shows two pulleys and a belt. The driving pulley is called the input pulley, and the driven pulley is called the output pulley. Pulley sizes are usually different and the belt and the pulley have matching teeth to prevent any slippage. Chain is usually used in place of belt when heavy loads and torques are involved.

A2.8.1 Analysis of flat belts

The input and output torques are related to the input and output rotation speeds as follows:

$$\frac{N_{in}}{N_{out}} = \frac{T_{out}}{T_{in}} = \frac{D_{out}}{D_{in}} = \text{pulley ratio}, \qquad \text{(A2.47)}$$

where N is the rotational speed, T is the torque, and D is the diameter of the pulley. From this equation we can deduce that if the output pulley is twice as large as the input pulley (a pulley ratio of 2), the output will rotate at half the speed of the input but with twice the torque. Hence, a pulley ratio that is greater than 1 will provide a reduction in speed and increase in torque, enabling a small capacity but fast motor to drive a heavy actuator (output). Such an arrangement is referred to as a *reduction transmission*.

A2.8.2 Length of belt

The open-belt and crossed-belt systems are shown in Figure A2.18

The length of open belt is:

$$\theta_s = \pi - 2\sin^{-1}\left(\frac{D-d}{2C}\right) \quad \text{(A2.48)}$$

$$\theta_L = \pi + 2\sin^{-1}\left(\frac{D-d}{2C}\right) \quad \text{(A2.49)}$$

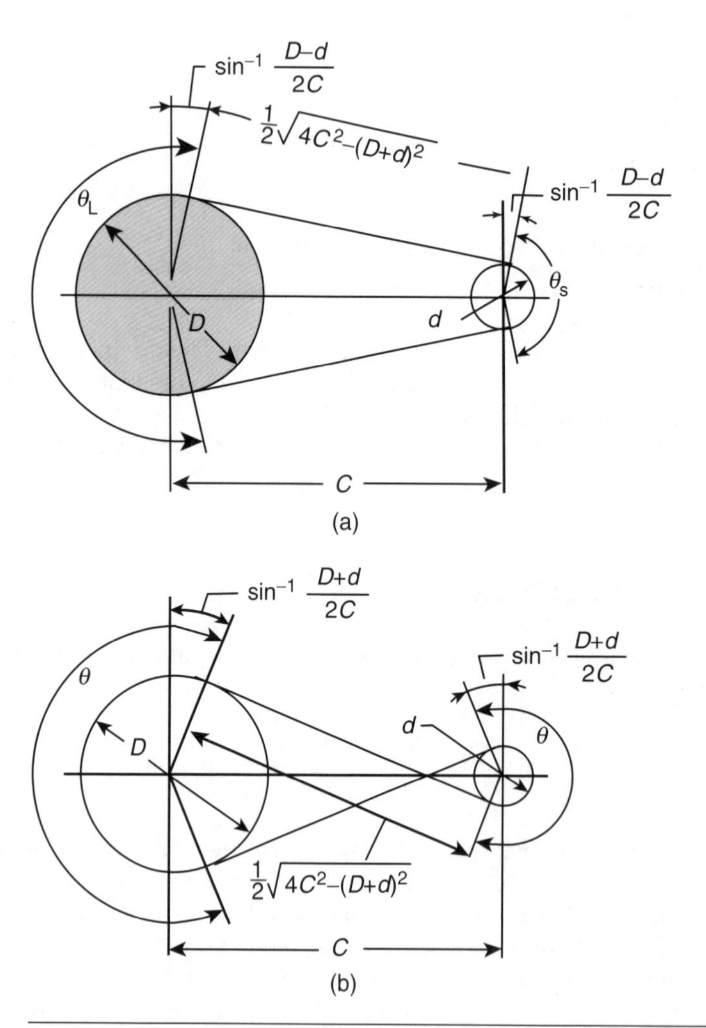

Figure A2.18 Belt length: (a) open; (b) crossed.

$$L = \sqrt{4C^2 - (D - d)^2} + \frac{1}{2}(D\theta_L + d\theta_s), \tag{A2.50}$$

where L is the length, θ is the angle of contact, d and D are the small and large diameters, respectively, of the pulleys.

The length of crossed belt is:

$$\theta = \pi + 2\sin^{-1}\left(\frac{D + d}{2C}\right) \tag{A2.51}$$

$$L = \sqrt{4C^2 - (D + d)^2} + \frac{\theta}{2}(D + d). \tag{A2.52}$$

Problems

QA2.1 (a) Briefly discuss some industrial applications of springs.

(b) An engine valve made from SAE 6150 (chrome vanadium) is to exert a minimum load of 150 N and a maximum load of 300 N. The valve lift is 7.5 mm. The modulus of elasticity of the spring material is 175 GPa. Figure A2.19 shows the allowable torsional stress for SAE 6150 (chrome vanadium). Design the spring for infinite life.

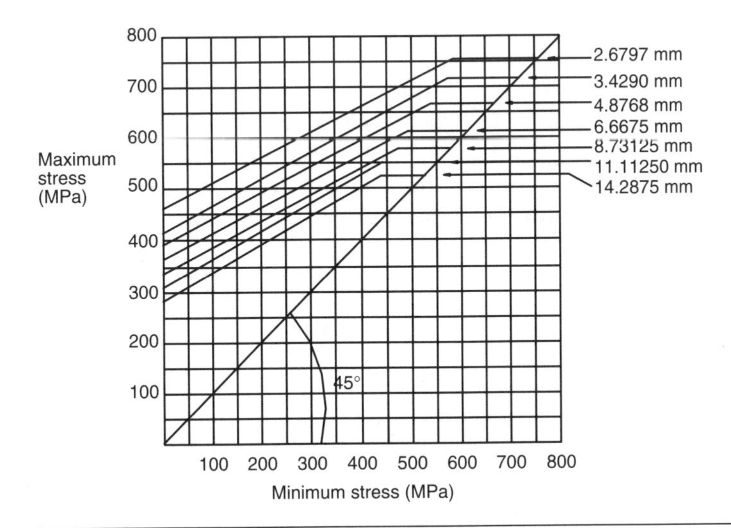

Figure A2.19 Allowable torsional stress range for chrome vanadium (QA2.1).

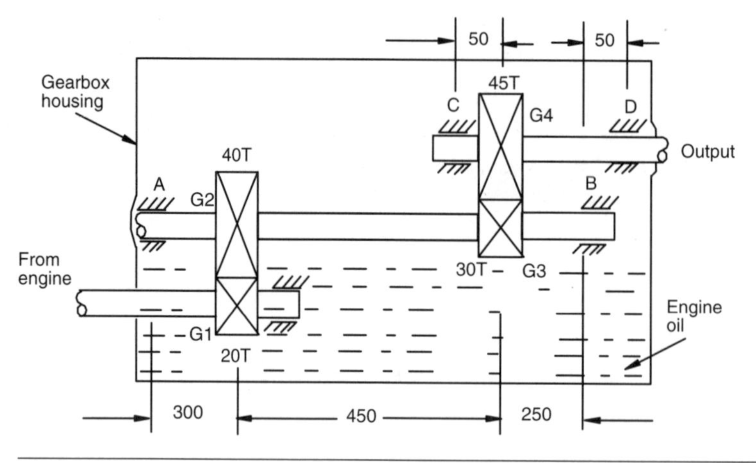

Figure A2.20 Gearbox for QA2.2 and QA2.3.

QA2.2 Figure A2.20 shows an automobile gearbox. The power input from the engine is 74.6 kW, which drives the pinion G1 at 2000 revolutions per minute in a clockwise direction when viewed from the engine side. The module for gears G1 and G2 is 9 mm and the module for gears G3 and G4 is 8 mm. The pressure angle for all gears is 20°. Select bearings A and B for a life of 18,000 hours, with the inner bearing races rotating.

QA2.3 (a) Why is the knowledge of fatigue strength important in design of shaft?

(b) Design the output shaft, CD, of the automobile gearbox in Figure A2.20 using the Soderberg criterion and distortion energy theory. The shaft is to be made from AISI 2340 steel. Assume a stress concentration factor of 1.5, a modified endurance limit of 40% of the endurance limit, and a safety factor of 2.

QA2.4 (a) Distinguish between spur, bevel, and rack and pinion gears. Sketch some practical applications.

(b) Figure A2.21 shows a shaft, AB, driven by an input shaft rotating at 2000 revolutions per minute. The power input from an electric motor is 74.6 kW. The pressure angle for the gears is 20°. Design the pair of spur gears, G1 and G2, using the modified Lewis equation. The gear material is AISI 1020 CR steel. Assume the safety factor, N_y, is 4.

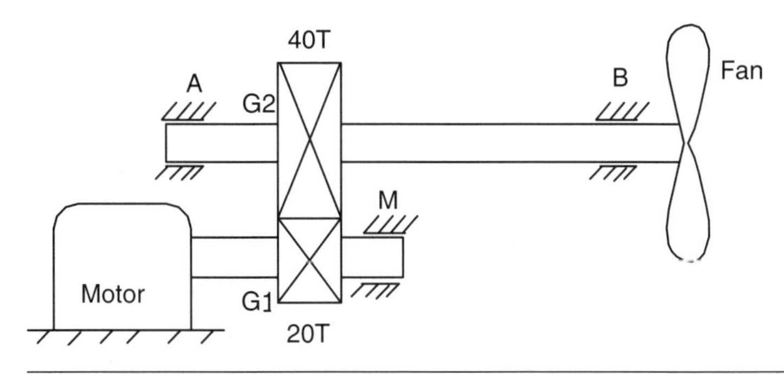

Figure A2.21 Gearing system for QA2.4.

QA2.5 (a) Briefly discuss the functions of flexible mechanical elements.

(b) A flat belt is 150 mm wide, 7 mm thick, and transmits 11 kW of power at a belt rotational speed of 1750 revolutions per minute. The driving pulley is 150 mm in diameter and the driven pulley is 450 mm in diameter. The pulley axes are parallel in a horizontal plane and 2400 mm apart. The open belt has a mass of 1017 kg m^{-3} of the belt volume, and the coefficient of friction is 0.3. Determine: (i) the tension in the tight and slack sides of the belt; (ii) the length of the belt.

Further reading

[1] Bolton, W. (1993) *Mechanical Science*, Blackwell Scientific Publications.
[2] Bolton, W. (1995) *Mechatronics: Electronic Control Systems in Mechanical Engineering*, Essex: Longman.
[3] Bralla, J.G. (1999) *Design for Manufacturability Handbook* (2nd. ed.), McGraw-Hill.
[4] Dieter, G.E. (2000) *Engineering Design: A materials and Processing Approach* (3rd. ed.), McGraw-Hill.
[5] Norton, R.L. (1992) *Design of Machinery*, McGraw-Hill.

APPENDIX 3

CircuitMaker 2000 tutorial

The appendix details the fundamental elements of the electronic design software package *CircuitMaker 2000* (from Altium Limited).

A3.1 Drawing and editing tools

A3.1.1 Wiring the circuit

CircuitMaker will only simulate circuits and generate PCB layouts accurately if the components in your circuit are correctly wired together. The software's *Auto routing*, *Manual routing* and *Quick connect* methods are integrated automatic so that you do not have to choose or switch between wiring modes. The term *valid connect point* is any device pin or wire.

- **Auto routing:** Click with the *Wire tool* from any valid connection point and drag to another connection point.
- **Manual routing:** Click with the *Wire tool* to start a new wire, click to change directions, and click on a connection point (or double click anywhere) to end a wire.
- **Quick connect:** When enabled, place or move a device with the *Arrow tool* so that unconnected pins touch a wire or other device pins.

A3.1.1.1 *Wire tool*

Use *Wire tool* to place wires in the work area. Draw bus wires by holding down the *Shift* key when starting to draw the wire. (Refer to sections [wiring the circuit] and [working with bus wires] for more information.) Draw a dashed line by holding down the *Alt* key while drawing. Dashed lines are similar to full lines but if they are not connected to anything they will not be included in the netlist. *Wire tool* can be activated by right clicking in the schematic background and choosing *Wire* from the shortcut menu.

A3.1.1.2 *Auto wire routing*

To quickly and easily auto route wires:

1. Select *Wire tool* from the *Toolbar*.
2. Move the tool over a valid connection point (i.e. any device pin or wire).
3. Click and hold the left mouse button.
4. Drag the cursor to another connection point and release the mouse button. The wire is automatically routed between the two selected points.

Note that auto routing required two valid connection points (i.e. a wire that does not connect to something at both ends cannot be auto routed). Also, bus wires cannot be drawn using auto routing.

A3.1.1.3 *Manual wire routing*

Manual routing allows wires to be specifically placed rather than automatically placed. *Free wires*, that is wires that are not connected to anything, can be drawn using manual routing. Bus wires are drawn using this method.

To manually route wires:

1. Select *Wire tool* from the *Toolbar*.
2. Move the tool to the position where you want to start the wire.
3. Click and release the left mouse button. The *Wire tool* cursor is replaced with the *Extended wiring* cursor which simplifies the task of precisely aligning wires with other objects.
4. Click the left button of the mouse to turn 90° or double click it to end the wire. (Single click to terminate the wire at a valid connection point if it has been enabled in the *Schematic options* dialog box.)
5. Cancel a wire at any time by pressing any keyboard key or right clicking the mouse.

A3.1.1.4 *Text tool*

The circuit can be annotated using the *Text tool.*

1. Select the *Text tool* from the *Toolbar* or by right clicking in the *Work area* and selecting *Text* from the shortcut menu.
2. Click in the work area and type the text.
3. Choose *Options* > *Schematic*. Use *Text font* and *Colors* to format the text.

4. Resize the text rectangle by clicking it.
5. Resize the text rectangle by clicking it.

A3.1.1.5 *Grid option*

The *Grid* option (*Options* > *Schematic*) allows an alignment grid to be displayed. The grid can help objects to be precisely aligned. *Snap to* allows new devices (i.e. devices not already in the circuit) to be placed on the specified grid. It also allows old devices (i.e. devices already in the circuit) to be accurately moved. Note that when you place a device exactly on the grid, it always remains on the grid regardless of the scroll position. It does not guarantee that component pins will be aligned.

A3.1.1.6 *Extending, joining and cutting wires*

To extend a wire:

1. Select *Wire tool* from the *Toolbar*.
2. Position the *Wire* tool at the end of the wire you wish to extend and start a new wire from this position.

To join two wires:

1. Select *Wire tool* from the *Toolbar*.
2. Draw a wire from the end of the first wire to the end of the second wire.

To cut a single wire:

1. Select the *Delete tool* from the *Toolbar*.
2. Place the tool over the point(s) where the cut is to be made.
3. Press the *Shift* key.
4. Click the left mouse button to cut the wire.

A3.1.2 Placing a device

After you have searched and found a device, you can place the device or reselect it.
To place a device:

1. Select the device.
2. Press the r key or right click the device to rotate it.

3. Press the m key to 'mirror' the device.
4. Left click the *Work area* at the desired position to place the device OR press any key (except r or m) to cancel the placement. Note that to repeatedly place identical items, select the *Auto repeat* check box under *Options* > *Schematic*.

A3.1.3 Highlighting an entire circuit node

Highlighting an entire circuit node is useful when looking for wiring errors. Even wires which are not physically connected on the schematic (for example the ground node) can be highlighted if they are in the same circuit node.

To highlight an entire node:

1. Select the *Arrow* tool from the *Toolbar*.
2. At the same time, press *Alt* and click one of the wires of the node.

A3.2 Simulation modes

CircuitMaker allows both analog and digital simulations.

A3.2.1 Analog mode

This is an accurate, 'real world', simulation that can be used for analog, digital and mixed signal circuits with results akin to those from a breadboard circuit (for example digital ICs have accurate propagation delays, loading effects on device outputs are modeled, etc.).

A3.2.2 Digital mode

This mode is purely for logic simulation. Here, unit delays are modeled instead of actual propagation delays. A power supply is not required, and the device output levels are constant. CircuitMaker's digital logic simulator allows you to flip switches and alter a circuit while a simulation is running, and see the response.

INDEX